Nanobiotechnology

Inorganic Nanoparticles vs Organic Nanoparticles

FRONTIERS OF NANOSCIENCE

Series Editor: Richard E. Palmer
The Nanoscale Physics Research Laboratory,
The School of Physics and Astronomy,
The University of Birmingham, UK

Vol. 1 **Nanostructured Materials** edited by Gerhard Wilde

Vol. 2 **Atomic and Molecular Manipulation** edited by Andrew J. Mayne and Gérald Dujardin

Vol. 3 **Metal Nanoparticles and Nanoalloys** edited by Roy L. Johnston and J.P. Wilcoxon

Vol. 4 **Nanobiotechnology** edited by Jesus M. de la Fuente and V. Grazu

Nanobiotechnology

Inorganic Nanoparticles vs Organic Nanoparticles

Edited by

Jesus M. de la Fuente
Nanotherapy and Nanodiagnostics Group (GN^2), Instituto de Nanociencia de Aragón-ARAID, Universidad de Zaragoza, Edificio I+D, Mariano Esquillor s/n, Campus Rio Ebro, Zaragoza, Spain

V. Grazu
Biofunctionalization of nanoparticles and surfaces (BioNanoSurf), Instituto de Nanociencia de Aragón, Universidad de Zaragoza, Mariano Esquillor, Zaragoza, Spain

ELSEVIER
Amsterdam • Boston • Heidelberg • London • New York • Oxford
Paris • San Diego • San Francisco • Singapore • Sydney • Tokyo

Elsevier
The Boulevard, Langford Lane, Kidlington, Oxford OX5 1GB, UK
Radarweg 29, PO Box 211, 1000 AE Amsterdam, The Netherlands

British Library Cataloguing in Publication Data
A catalogue record for this book is available from the British Library

Library of Congress Cataloging-in-Publication Data
A catalog record for this book is available from the Library of Congress

ISSN: 1876-2778
ISBN: 978-0-12-415769-9

Contents

Part IV
Nanoparticles for Therapy 307

Colour version of figures in this book can be found at http://www.elsevierdirect.com/product.jsp?isbn=9780124157699

Contributors

Numbers in Parentheses indicate the pages on which the author's contributions begin.

David Alcantara (269), Center for Nuclear Medicine and Molecular Imaging, Massachusetts General Hospital-Harvard Medical School, Charlestown, Boston, Massachusetts, USA

Pilar Batalla (247), Department of Analytical Chemistry and Chemical Engineering, University of Alcalá, Alcalá de Henares, Madrid, Spain

Catherine C. Berry (159), Centre for Cell Engineering, Glasgow University, Glasgow, United Kingdom

Susana Carregal-Romero (181), Fachbereich Physik, Philipps Universität Marburg, Marburg, Germany

Amparo Casal (443), SOD Medicina del Lavoro, Azienda Ospedaliera Universitaria Carreggi, Florence, Italy

Rocío Costo (35), Instituto de Ciencias de Materiales de Madrid, CSIC, C/Sor Juana Ines de la Cruz, Cantoblanco, 28049 Madrid, Spain

Pablo del Pino (3, 309), Nanotherapy and Nanodiagnostics Group (GN^2), Instituto de Nanociencia de Aragón, Universidad de Zaragoza, Edificio I+D, Mariano Esquillor s/n, Campus Rio Ebro, Zaragoza, Spain

Pedro M. Enriquez-Navas (233), Bionand-Centro Andaluz de Nanomedicina y Biotecnología, Parque Tecnológico de Andalucía, Malaga, Spain

Helene Feracci (197), Centre de Recherche Paul Pascal, CNRS UPR 8641, Université Bordeaux 1, Pessac, France

Maria L. Garcia-Martin (233), Bionand-Centro Andaluz de Nanomedicina y Biotecnología, Parque Tecnológico de Andalucía, Malaga, Spain

Isabel Segura Gil (197), Instituto de Nanociencia de Aragón, Universidad de Zaragoza, Edificio I+D, Mariano Esquillor s/n, Campus Rio Ebro, Zaragoza, Spain

Pilar Rivera_Gil (181), Fachbereich Physik, Philipps Universität Marburg, Marburg, Germany

Ignasi Gispert (487), Researcher at the Institute for Law and Technology (Universitat Autònoma de Barcelona) and Advisor of the Legal Observatory of the Center for NanoBioSafety and Sustainability (CNBSS) at the Catalan Institute of Nanotechnology (ICN). Barcelona, Spain

África González-Fernández (443), Immunology, Biomedical Research Centre (CINBIO), University of Vigo, Campus Lagoas Marcosende, Vigo, Spain

Valeria Grazú (337), Nanotherapy and Nanodiagnostics Group (GN^2), Instituto de Nanociencia de Aragón, Universidad de Zaragoza, Edificio I+D, Mariano Esquillor s/n, Campus Rio Ebro, Zaragoza, Spain

Berta Saez Gutierrez (197), Instituto de Nanociencia de Aragón, Universidad de Zaragoza, Edificio I+D, Mariano Esquillor s/n, Campus Rio Ebro, Zaragoza, Spain and Instituto Aragones de Ciencias de la Salud, Avenida de Alcalde Gómez Laguna, 25, Zaragoza, Spain

Raimo Hartmann (181), Fachbereich Physik, Philipps Universität Marburg, Marburg, Germany

William Hempel (197), Biosystems International, 1 rue Pierre Fontaine, Evry, France

Niko Hildebrandt (291), Université Paris-Sud, Institut d'Electronique Fondamentale, Orsay, France

Zongwen Jin (291), Université Paris-Sud, Institut d'Electronique Fondamentale, Orsay, France

Lee Josephson (269), Center for Nuclear Medicine and Molecular Imaging, Massachusetts General Hospital-Harvard Medical School, Charlestown, Boston, Massachusetts, USA

Tamara Lozano (443), Immunology, Biomedical Research Centre (CINBIO), University of Vigo, Campus Lagoas Marcosende, Vigo, Spain

María del Puerto Morales (35), Instituto de Ciencias de Materiales de Madrid, CSIC, C/Sor Juana Ines de la Cruz, Cantoblanco, 28049 Madrid, Spain

María Moros (337), Nanotherapy and Nanodiagnostics Group (GN^2), Instituto de Nanociencia de Aragón, Universidad de Zaragoza, Edificio I+D, Mariano Esquillor s/n, Campus Rio Ebro, Zaragoza, Spain

Sergio E. Moya (115), CIC biomaGUNE, Paseo Miramón, San Sebastian, Gipuzkoa, Spain

Wolfgang J. Parak (181), Fachbereich Physik, Philipps Universität Marburg, Marburg, Germany

Beatriz Pelaz (3, 309), Nanotherapy and Nanodiagnostics Group (GN^2), Instituto de Nanociencia de Aragón, Universidad de Zaragoza, Edificio I+D, Mariano Esquillor s/n, Campus Rio Ebro, Zaragoza, Spain

Mercedes Peleteiro (443), Immunology, Biomedical Research Centre (CINBIO), University of Vigo, Campus Lagoas Marcosende, Vigo, Spain

Cécile Philippot (81), CEA Grenoble, INAC/SPrAM (UMR 5819 CEA-CNRS-UJF)/ LEMOH, Grenoble Cedex, France

Ester Polo (247), Nanotherapy and Nanodiagnostics Group (GN^2), Instituto de Nanociencia de Aragón, Universidad de Zaragoza, Edificio I+D, Mariano Esquillor s/n, Campus Rio Ebro, Zaragoza, Spain

Sara Puertas (247), Nanotherapy and Nanodiagnostics Group (GN^2), Instituto de Nanociencia de Aragón, Universidad de Zaragoza, Edificio I+D, Mariano Esquillor s/n, Campus Rio Ebro, Zaragoza, Spain

Peter Reiss (81), CEA Grenoble, INAC/SPrAM (UMR 5819 CEA-CNRS-UJF)/LEMOH, Grenoble Cedex, France

Javier Rojo (143), Glycosystems Laboratory, Instituto de Investigaciones Químicas, CSIC — Universidad de Sevilla, Seville, Spain

Gabriela Romero (115), CIC biomaGUNE, Paseo Miramón, San Sebastian, Gipuzkoa, Spain

Christian Sánchez-Espinel (337), NanoImmunoTech SL, Pza. Fernando Conde, Montero Ríos 9,36201 Vigo, Spain

Macarena Sánchez-Navarro (143), Glycosystems Laboratory, Instituto de Investigaciones Químicas, CSIC — Universidad de Sevilla, Seville, Spain

Gorka Salas (35), Instituto de Ciencias de Materiales de Madrid, CSIC, C/Sor Juana Ines de la Cruz, Cantoblanco, 28049 Madrid, Spain and IMDEA Nanociencia, Facultad de Ciencias, Av. Tomas y Valiente 7, Cantoblanco, 28049 Madrid, Spain

Rosana Simón-Vázquez (443), Immunology, Biomedical Research Centre (CINBIO), University of Vigo, Campus Lagoas Marcosende, Vigo, Spain

Preface

During the past 20 years, the number of applications of nano- and microdevices in the fields of biotechnology and biomedicine has been increased drastically. Actually it is very common to find products in our laboratories based on nanotechnology for applications such as cell separation, catalysis, supports, labeling, etc. In fact, during the past years, few nanosystems have been approved by FDA to their use as contrast agents or drug delivery devices. From all of these nanodevices developed, the most popular and most used are probably the nanoparticles. These nanoparticles could be based on organic or inorganic materials or hybrid structures. But, what is the main reason why the research focused in these nanoparticles has been developed so fast? Probably, the main reason is the fast development of characterization techniques during the past years. New electronic microscopes, more precise, with better resolution, and easier to manipulate, have been developed to improve and simplify the characterization of novel nanoparticles. DLS and Z-potential measurements have been more and more popular during the past years with the availability of economical and excellent machines. All of these make easier the production and characterization of the new nanostructures, getting a better understanding of the mechanisms that control the growing processes driving their formation. Once the synthesis of different sizes and shapes of organic and inorganic nanoparticles has been understood and under control, the number of applications grow very fast, mainly in the field of biotechnology.

Which are the reasons why these nanoparticles are very interesting for biotechnological applications? There are a few aspects that we should take into consideration. The first one is that these nanomaterials are in the same size range scale as biomolecules. For this reason, they can interact quite well with them and we can control and intervene in biological processes. We should also take into account the high specific surface that these nanoparticles have. This specific surface allows the accommodation of a high concentration of drugs or required molecules in them. In the case of inorganic nanoparticles, due to their nanometric size, nanoparticles have interesting physical properties such as different fluorescent or magnetic properties. The use of these novel properties has allowed the development of most sensitive and selective biosensors, new therapy approaches, and also merge diagnosis and therapy in the same device. Organic and inorganic nanoparticles can also be used as a multifunctional scaffold allowing the incorporation of different biomolecules to their surface or inside hollow nanoparticles to improve the crossing of biological membranes, for example. The increase of solubility or stability of biomolecules or drugs is also other of nanoparticle features most used to develop new drug formulations. For these reasons, we thought it could be interesting to publish a book focused in the last biotechnological applications of nanoparticles. Taking into account the huge differences between organic and inorganic nanoparticles, this book is oriented to describe the differences, advantages, and disadvantages between both types of nanoparticles for their use in biotechnological applications. New drugs and diagnosis tools are being discovered using nanotechnology as base. Every year hundreds of new ideas using nanomaterials are used for the development of biosensors. More and more new enterprises are also searching for market opportunities using these technologies. Nanomaterials world for biotechnological applications is a very complex field. Thousands of different nanoparticles could be used for these purposes. Some of them are very different; their synthesis, characterization, and potentiality are very diverse. With this book, we have pretended to establish a route guide for nonerudite researchers in the field showing the advantages and disadvantages of the different kind of nanomaterials.

This book is divided into 5 parts including 16 chapters encompassing from the synthesis and characterization of organic and inorganic nanoparticles to all different kinds of their applications in biotechnology and medicine. It begins with a part devoted on the synthesis of nanoparticles. Chapter 1 is a revision of the most used methods to prepare gold nanoparticles controlling their size and their shape. Chapter 2 is mainly devoted in the synthesis and characterization of magnetic nanoparticles. This chapter presents different methodologies to prepare different size and shape magnetic nanoparticles and also a final section oriented to the conjugation of organic molecules to improve the colloidal stability of these nanoparticles. To finish with the synthesis of inorganic nanoparticles, Chapter 3 is mainly focused on the synthesis of semiconductor nanocrystals, the so-called quantum dots. Chapters 4 and 5 are up-to-date reviews of literature on advances strategies for the synthesis of liposomes, micelles, vesicles, polymers, and dendrimers searching for biotechnological applications.

The next part of the book is oriented to present specific examples of biotechnological applications of inorganic nanoparticles (Chapter 6) and organic nanoparticles (Chapter 8). Both chapters present examples of the use of nanoparticles for DNA transfection, organelle labeling, molecular interactions, cellular separations, etc. All these applications are mainly oriented to cellular or molecular biology laboratories. Chapter 7 presents a practical approach to get a better understanding of the mechanisms and also to quantify the nanoparticle internalization in cells.

The next part is divided into four chapters, and it is mainly oriented to review the different applications of nanoparticles for diagnosis and biosensing, including *in vitro*, *ex vivo*, and *in vivo* experimentation. Two chapters are oriented to present applications of magnetic nanoparticles by means of magnetic resonance technology. Chapter 9 is mainly focused on the application of inorganic nanoparticles, mainly magnetic nanoparticles for magnetic resonance imaging techniques. A good explanation of the advantages of superparamagnetic nanoparticles instead Gd-based complexes for this application is presented. Chapter 11, under the scope of diagnostics-biosensing, presents a different approach based on the monitoring of aggregation of magnetic nanoparticles due to molecular recognition events to be used as biosensor. This approach is called magnetic resonance switches. Two more chapters oriented to describe different examples of nanoparticles as biosensors for *in vitro* experimentation are also included. Chapter 10 is mainly focused on metallic and metal oxide nanoparticles and Chapter 12 is mainly oriented to the use of semiconductor nanocrystals.

The last part oriented to present applications of nanoparticles is focused on therapeutical approaches mainly oriented to cancer treatment. Both chapters included here are from the University of Zaragoza, and they are very complete and updated reviews on the use of metallic, metal oxide, and organic nanoparticles' use of nanoparticles for hyperthermia treatment and drug delivery. Chapter 13 explains the physical properties of metal oxide and metallic nanoparticles for their use in magnetic hyperthermia and photothermal ablation. It is a very interesting revision of the recent approaches using this kind of novel therapy strategy to destroy specifically tumoral cells. Chapter 14 constitutes a very complete revision of the different approaches to use inorganic and organic nanoparticles as drug delivery vehicles. The first part of this chapter makes a description of the different features that a nanoparticle should have to be used for biomedical applications. The revision includes several options to trigger the drugs and what it is very interesting, several examples of devices already probed and already in the market. There are several tables with all the most relevant nanosystems in the market for drug delivery.

We did not want to end the book with a part oriented to toxicological and regulatory aspects of this novel technologies based on nanoparticles. Chapters 15 and 16 are mainly focused on these two important aspects.

A book like this is impossible without the support and effort of thc authors who have taken time to submit their manuscripts even with their busy schedules. We gratefully acknowledge the offer of Prof. R. Palmer to participate in this book series and the support from Elsevier, in particular, Derek Coleman and Susan Dennis, who have been working with us to make this book a reality.

Jesus M. de la Fuente
Instituto de Nanociencia de Aragon-ARAID, University of Zaragoza,
Zaragoza, Spain

Part I

Synthesis and Characterization of Nanoparticles

Chapter 1

Synthesis Applications of Gold Nanoparticles

Beatriz Pelaz and Pablo del Pino
Nanotherapy and Nanodiagnostics Group (GN[2]), Instituto de Nanociencia de Aragón, Universidad de Zaragoza, Edificio I+D, Mariano Esquillor s/n, Campus Rio Ebro, Zaragoza, Spain

1. INTRODUCTION

As bulk materials, the noble metal gold (Au) has been for centuries a very popular subject of study in science. However, the technological development of nanoscience (nanotechnology) has revolutionized the impact of Au in the scientific community and, more recently, in our society in the form of very beneficial technological applications in fields such as materials science, energy, and medicine. Currently, the development of new synthetic routes to obtain Au colloids, complex multicomponent materials embedding Au colloids, and the following novel applications based on these materials are the subjects of work for many scientists; consequently, reports related to gold colloids, also referred to as Au nanoparticles or nanocrystals (GNPs), have become very prolific in the literature. Furthermore, the popularity of synthetic methods yielding GNPs with on-demand size, shape, structure (solid or hollow), coating, etc., continues growing exponentially; as the degree of control to fabricate such structures evolves, new applications based on these materials can be explored. Photothermal therapy (PTT), photoacoustic (PA) imaging, and surface-enhanced resonance spectroscopy (SERS) are examples of applications where NPs such as nanorods, hollow nanospheres, and nanostars, among others, can act as transducers.

Among material scientists, the interesting optical features of GNPs are their key to success; in general, photons can couple to the plasma resonance of the conduction electrons at the surface of metallic NPs to produce their collective oscillation, in a phenomenon usually referred as localized surface plasmon resonance (LSPR). Indeed, the optical properties of metallic NPs are mainly dictated by surface plasmons. For NPs made of Au (noble metals in general), the LSPR band lies in the visible-NIR part of the spectrum; the position of the LSPR band is determined by a combination of NP attributes such as composition, size, and shape, and also by the dielectric environment. The interaction of light with

Frontiers of Nanoscience, Vol. 4. DOI: 10.1016/B978-0-12-415769-9.00001-7

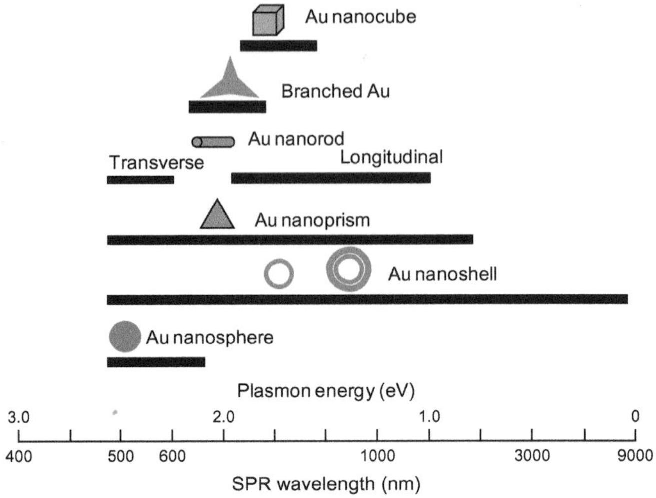

FIGURE 1.1 Range of surface plasmon resonances of GNPs as a function of their shape (reproduced from Treguer-Delapierre).[4]

surface plasmons at very confined volumes like that of Au colloids is currently being explored in fields such as optics and energy.[1–3]

The ability for tuning the color of colloidal solutions of GNPs by changing their size has been known for long time. However, in the past decade, a consistent body of work has highlighted the fact that shape plays also a crucial role in determining the optical properties of nanomaterials. Ultimately, the energy confinement of electrons, holes, excitons, phonons, and plasmons with respect to the NPs shape is the reason for this observation. Colors can be attributed to changes in surface plasmon resonance, energy at which light couples to the surface plasmons of NPs (Figure 1.1).

In the context of biology and medicine, in addition to the widely explored optical properties, GNPs exhibit two extra value properties which make them the "stars" among a wide variety of NPs for bioapplications, that is, biocompatibility and ease of surface modification by a wide range of thiolated molecules. In summary, GNPs represent an ideal platform for many applications in very different areas. The interest on Au is reflected by the increasing number of publications related to GNPs in the past years (Figure 1.2A); the most recent interest on anisotropic GNPs such as nanorods, nanoplates, branched nanostructures, nanoprisms, or nanoshells can also be seen in Figure 1.2B.

2. ANCIENT USES OF GNPS

Throughout human history, gold has been valued for its beauty and for its resistivity against corrosion; also, it was hammered into articles of jewelry. Gold extraction started in the fifth millennium BC However, in contrast to the long

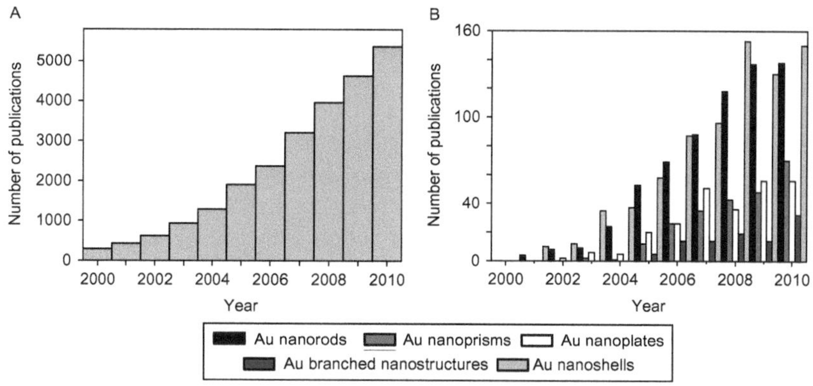

FIGURE 1.2 Publication during the past decade related to (A) GNPs and (B) different types of anisotropic GNPs (Source Isi Web of Knowledge, June 2011).

history of gold, the development of its chemistry was delayed because of its noble character. Gold can be only dissolved in oxidizing media such as aqua regia or by complexing species such as cyanides under air.[5] Long before the development of modern gold chemistry, the unique "golden" appearance of gold was highly appreciated. Probably, "soluble" gold appeared around the fifth or fourth century BC in Egypt and China; colloidal gold was used then to make ruby glass and for coloring ceramics, applications which continue nowadays. In the Middle Ages, gold was greatly valued due to the belief of its curative powers for various diseases. All these aspects were compiled in which is considered the first book on colloidal gold wrote by F. Antonii in 1618.[6] Communications regarding the curative properties and optical features of "soluble gold" continued steadily during the seventeenth and eighteenth centuries. In 1857, M. Faraday reported on the synthesis of gold colloids by the reduction of a gold (III) salt, in a vastly cited work which may be well considered the pioneering work on the synthesis of GNPs by wet-chemistry methods.[7] However, a definitive breakthrough in the synthesis of GNPs (among other NPs) had to wait until the 1990s with a series of works by Schiffrin and coworkers.[8–11] It is worth mentioning that, as for GNPs, during the 1990s, definitive and substantial advances were achieved in the field of nanoscience.[12–14]

3. SYNTHESIS OF GNPS

In the literature, one can find a great diversity of synthetic methods, both bottom-up and top-down approaches, to obtain GNPs with different size and shape; however, most of these synthetic routes have been reported for the synthesis of "spherical" GNPs (polyhedral). Although there are physical methods to obtain GNPs,[15–17] wet-chemical methods have proven so far more efficient to yield NPs with controlled size, shape, and coating; these are based on solution-phase colloidal chemistry. For GNPs, these methods rely on the chemical reduction of a gold salt into its metallic form which can grow then to form nanocrystals. To control the growth of the initially formed seeds and to prevent nanoparticle aggregation, donor ligands, polymers, and surfactants, to name a few, are used as stabilizers. Once anchored onto NPs, these molecules can provide NPs with colloidal stability by electrostatic repulsion (e.g., citrate) or steric stabilization (e.g., polymers).

A variety of redox reactions, by using reducing agents such as sodium borohydride, hydrogen, and alcohols, have been used to synthesize metal nanoparticles. As previously stated, as early as in 1857, Faraday reported the preparation of a colloidal gold sol (NPs) from the reduction of $HAuCl_4$ with phosphorous ligands as stabilizers.[7] In 1951, Turkevich and coworkers synthesized a stable deep-red dispersion of uniform 20-nm gold nanoparticles by using sodium citrate as both a reductant and a stabilizer.[18] Typically, molecules containing phosphanes and thiols are used as stabilizers by means of Au—P and Au—S bonds. Water-soluble phosphanes have been successfully used to shield GNPs[16]; however, thiolated chains appear to have been established as the prevalent stabilizer for GNPs of any size and shape. The stabilization of GNPs with alkanethiols was first reported by Mulvaney and Giersig[19]; shortly after, this group described the use of a silica coating for stabilization of GNPs.[20]

In 1994, Schiffrin and coworkers reported on a novel method (since then, normally referred to as the Brust–Schiffrin method) for the synthesis and stabilization of GNPs[10]; this method has had a tremendous impact because it allows the facile synthesis of thermally and air-stable monodispersed GNPs with controlled size. This technique is based on the Faraday's two-phase system and uses thiolated chains that strongly bind to GNPs. Shortly after, Brust and coworkers also described a single-phase system by which the synthesis and stabilization of GNPs can be extended to a variety of thiolated chains.[9,10,21] Alternative methods make use of the "alcohol reduction process" developed by Hirai and Toshima[22]; this approach is widely used for preparation of NPs stabilized by organic polymers such as poly(vinylpyrrolidone) (PVP), poly(vinyl alcohol), and poly(methyl vinyl ether).[23]

Regarding the synthesis of anisotropic GNPs, a number of synthetic approaches have been described including photochemical and electrochemical methods,[24–26] and others based on the chemical reduction of the gold precursor.[27] The most relevant synthetic methods for the synthesis of anisotropic GNPs in high yield are briefly discussed in the following:

i. The seed-mediated method is nowadays the most extended aqueous synthetic approach to produce a variety of Au anisotropic NPs such as nanorods and nanoprisms. This method was developed in basis of the Zsigmondy's "nuclear" method, which involves a two-step process for making NPs.[28] First of all, seed NPs are formed by a reduction process in which the metal salt is reduced in the presence of stabilizing agents. Typically, this step involves the use of a strong reductant agent like sodium borohydride. After this, seeds are grown using a growth solution which contains the gold precursor, surfactants, and a mild reducing agent. In this step, the newly formed metal atoms bind onto seed surfaces. The surfactant molecules can form suitable templates that facilitate the growth process to yield nanoparticles of desired morphology. The size of the nanoparticles can be typically tuned by adjusting the amount of growth solution. External agents such as molecules or ions can alter the growth direction of GNPs to yield NPs with very different shapes.[29–31] For instance, with the addition of a small amount of iodide ion into the growth solution to obtain nanorods, the shape of the resultant NPs can be changed to be triangular nanoprisms instead.[32]

ii. Polyol synthesis or thermal decomposition is a simple and versatile route, developed in 1989 by Fievet and coworkers.[33] Probably, the most successful reaction for producing noble metal colloids with controlled shape has been the polyol process by using PVP as a surface-capping agent. The metal precursor is dissolved in a polyol solution that is brought to near-reflux temperatures. Typical choices of polyols include ethylene glycol, 1,2-propylene glycol, and 1,5-pentanediol. Because viscosity (h) is largely determined by the length of the hydrocarbon chain of the polyol (for the solvents listed, $h = 16.1$, 40.4, and 140 mPa s, respectively), the solvent of choice can greatly impact the diffusion and growth processes associated with metal nanocrystal formation. In addition, the polyol acts as the reducing agent for metal reduction and, at the relatively high temperatures for the reactions described herein, is likely oxidized to various aldehyde and ketone species.

iii. Template-mediated synthesis represents a straightforward route to nonspherical GNPs. In this approach, the template serves as a scaffold within (or around) which gold is generated *in situ* and shaped into a morphology complementary to that of the template. A wealth of templates have been successfully developed and utilized, including channels within porous materials, mesoscale structures self-assembled from organic surfactants or block copolymers, biological macromolecules such as DNA strains or viruses, and existing nanostructures synthesized using other approaches. When the template is only physically involved, it is often necessary to remove it selectively using a postsynthesis treatment in order to harvest the resulting gold nanoparticles. It is widely accepted that template-directed synthesis is a simple and cost–effective procedure that also allows a complex topology present on the surface of a template to be duplicated. As a major drawback, the nanoparticles synthesized by this route are often polycrystalline and the quantity that can be produced in each run is relatively limited.

iv. Galvanic replacement reaction ("electroless deposition") is a method introduced by Brenner and Riddell to describe the spontaneous reduction of metal ions to metallic particles and films in the absence of an external electric field.[34] Recently, this method has attracted attention due to its simplicity of operation and cost–effectiveness. Xia and coworkers have prepared a number of anisotropic GNPs (e.g., nanorings, prism-shaped nanoboxes, nanotubes, and multiple-walled nanoshells or nanotubes) using the galvanic replacement reaction between solutions containing Au precursor salts and Ag nanostructures prepared through polyol reduction. The electrochemical potential difference between the two species drives the reduction reaction; concurrently, metallic gold can deposit epitaxially on the surface of the AgNPs which act as anisotropic templates. In parallel to the gold deposition, the inner part made of Ag is oxidized and ultimately removed to produce hollow and, eventually, porous structures. Moreover, using Pt and Pd precursor salts instead of gold salt allows for the preparation of Pt- and Pd-containing hollow nanostructures, respectively.[35]

4. COMMON ANISOTROPIC GNPS FOR BIOAPPLICATIONS

Among a great variety of anisotropic GNPs, those which can be prepared in a controlled way, high yield, reproducibility, etc., and are currently in use for bioapplications are discussed in the following sections; they are presented in order of structural complexity, that is, (4.1) nanorods, (4.2) nanoprisms, (4.3) nanoshells, (4.4) nanocages and hollow nanostructures, and (4.5) nanostars (also called branched nanostructures, nanoflowers, etc.).

4.1. Nanorods

The first reported synthesis of gold nanorods (GNRs) was performed by using the hard-template method (electrochemical deposition in a porous membrane).[36,37] In a pioneering work, Wang and coworkers introduced a combination of surfactant additives and an electrochemical oxidation/reduction procedure to prepare colloidal GNRs with tunable aspect ratio.[38] However, the most extended methodology to generate GNRs was described independently by the groups of Murphy and El-Sayed in 2001 and 2003, respectively.[29,39] This methodology provides an excellent degree of control over the size, shape, and monodispersity of the resulting NPs. The method is a two-step seed-mediated protocol which requires the use of the surfactant cetyltrimethylammonium bromide (CTAB) (Figure 1.3).

In the first step, "seeds" (small GNPs of ca. 4 nm) are formed by adding aqueous sodium borohydride to a solution of CTAB and the gold (III) salt. In the second step, GNRs are grown by adding the gold seeds to a growth solution containing CTAB, silver nitrate, gold salt, and ascorbic acid. Usually, GNRs are allowed to growth overnight without stirring. In this methodology, silver nitrate is found to be necessary due to its effect on the final shape and crystallinity of the resulting GNRs.[39,40] Besides the seeding methodology, single-crystal GNRs (also nanowires) can be synthesized by electrochemical and photochemical methods.[38,41] Interestingly, in all cases, bromide is required. This fact highlights

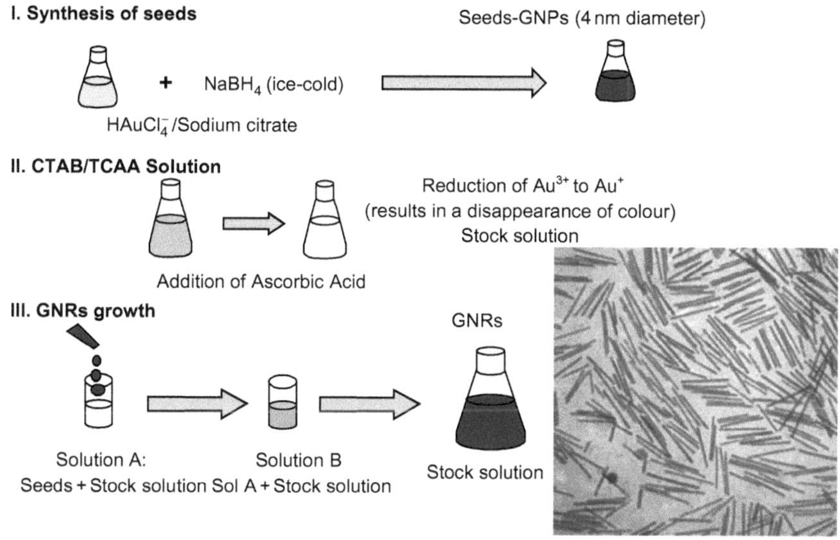

FIGURE 1.3 Scheme of GNRs synthesis (adapted from Murphy).[30]

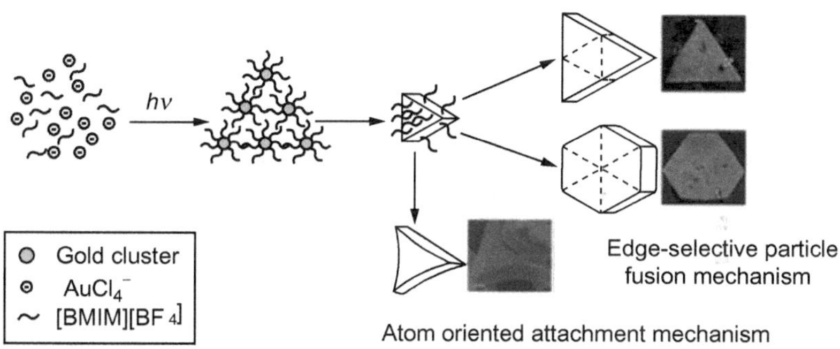

FIGURE 1.4 Formation of nanoplates using the photoinduced method (adapted from Zhu).[26]

the importance of the presence of bromide to which it might be given a more critical role in the formation of GNRs than that of CTAB.

4.2. Nanoplates

In the past decade, there have been described several methodologies to produce gold nano- and microplates. One strategy involves the use of light (photoinduced methods); although this kind of light-triggered reactions is usually used for AgNPs, it is possible to find some examples of Au nanoplates.[26,42] A representative example is shown in Figure 1.4 where one can see a scheme of the synthetic route developed by Zhang and coworkers to produce giant Au microplates.

In aqueous solution, the formation of a specific shape is usually restricted to a specific choice of the reaction parameters. For instance, Sau and coworkers showed that either hexahedra (cubes) or octahedral shapes can be obtained through variations of the silver nitrate mediated synthesis of GNRs.[40,43,44] The seed-mediated growth of GNRs is frequently modified to obtain gold nanoplates. Some parameters such as an increase in the pH or surfactant concentration and the addition of halide ions have been studied to promote the formation planar nanostructures. Unfortunately, all current procedures still result in rather low yields (40–65%) compared to those for GNRs (ca. 95–99%), and a purification step is required in order to eliminate isotropic structures.

To date, the most relevant approach to synthesize gold nanoprisms appears to be that developed in Mirkin's lab (Figure 1.5); trigonal nanoprisms were synthesized by the group of Mirkin using a procedure adapted from the seed-mediated method to produce GNRs where the use of silver nitrate is prevented.[45,47] Citrate-capped AuNPs (ca. 5 nm) were used as templates to grow into nanoprisms in a three-step protocol (Figure 1.5); the growing solution consisted here in a saturated CTAB solution containing the gold precursor, ascorbic acid, and NaOH. Nanoprisms of average edge length and thickness of 144 and 8 nm, respectively, were obtained with yields up to 65%. Larger prisms, with edge lengths up to 220 nm, were also grown using smaller prisms as seeds.[46]

Other useful variations to the seed-mediated method include the addition of salts or polymers to the growth solution. Ha and coworkers showed that the addition of KI at pH 4 to the CTAB-based growth solution increases the yield of gold nanoprisms up to 45% (Figure 1.6).[32] Based on a series of control experiments, the authors proposed a possible growth mechanism for nanoprisms involving growth inhibition of the {111} facets by strongly bonded halide ions. In principle, the addition of traces of iodine is enough to change the morphology of the particles to nanoprisms.[48] Nanoprism formation in the presence of iodide ions has also been demonstrated to be dependent on the pH of the growth solution and the temperature. Nevertheless, a relatively residual population of polyhedral NPs is always observed. In 2008, Mirkin's lab also reported on the iodine effect; they found that using no iodine at all, the reaction products were exclusively polyhedral NPs, whereas as the iodine concentration was increased, the shape passed through GNRs (2.5–5 μM I^-, 45% yield), mixture of shapes (10 μM I^-), nanoprisms (50 μM I^-,65% yield), and plate-like structures for higher iodine concentrations (Figure 1.6C).[48]

Fan and coworkers highlighted the importance of the crystalline structure of the seeds to produce stacking faults which, ultimately, are responsible for nanoprisms generation.[49] Other work studied the effect of temperature and the presence of traces of silver ions in the shape-generation process (Figure 1.7)[50]; their results showed that temperature is critical in order to generate kinetic-controlled

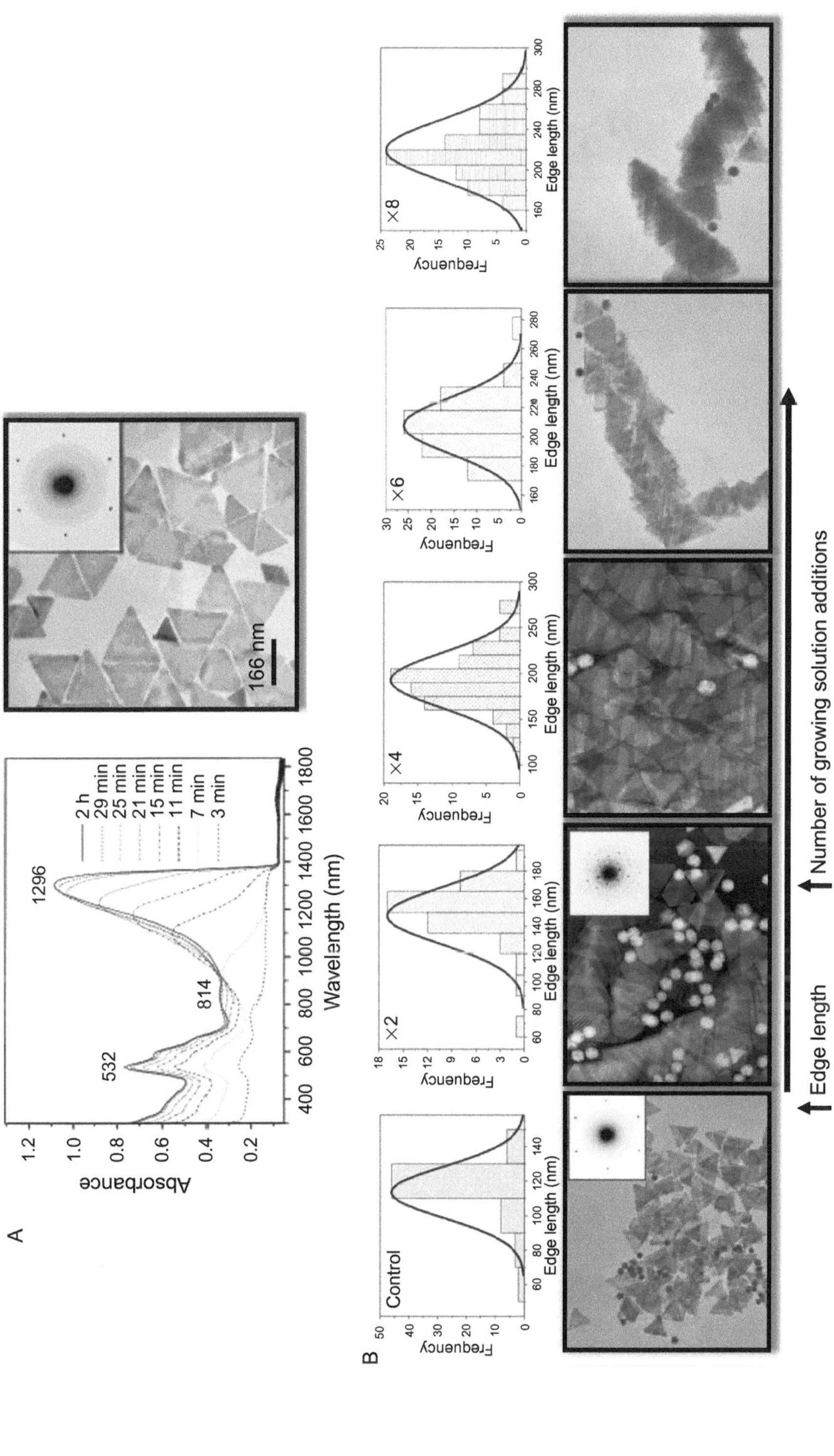

FIGURE 1.5 (A) Growth process of gold nanoplates followed by Vis–NIR (left), and TEM images from the nanoplates obtained (right) (adapted from Millstone)[45]; (B) Effect on the edge length of nanoprisms of the addition of different amounts of the growing solution.[46]

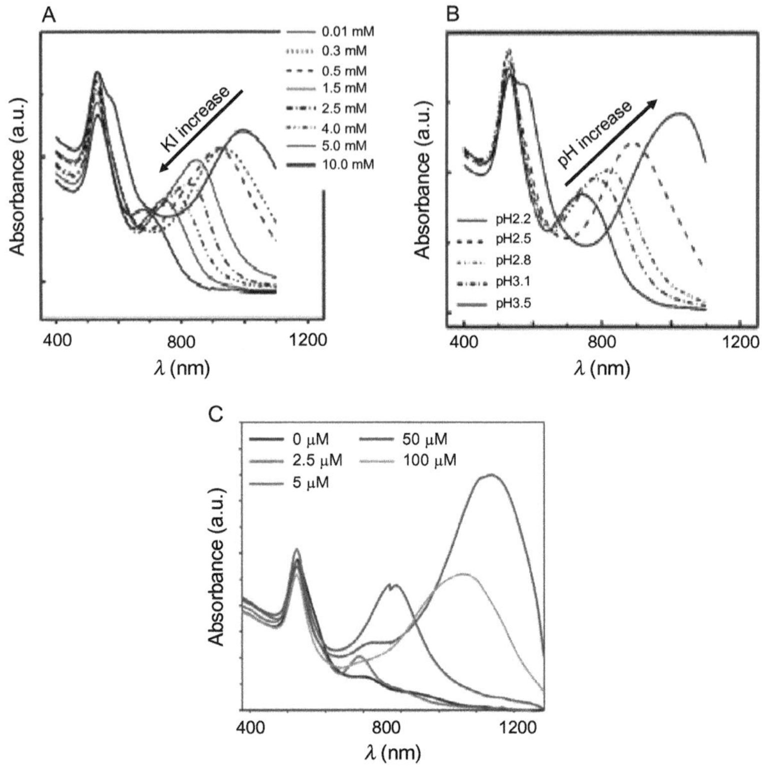

FIGURE 1.6 (A) Nanoprisms formation dependence with iodine concentration, and B shows pH contribution to nanoprisms formation (both adapted from Ha).[32] (C) Different products spectra obtained changing iodine concentration (adapted from Millstone).[48]

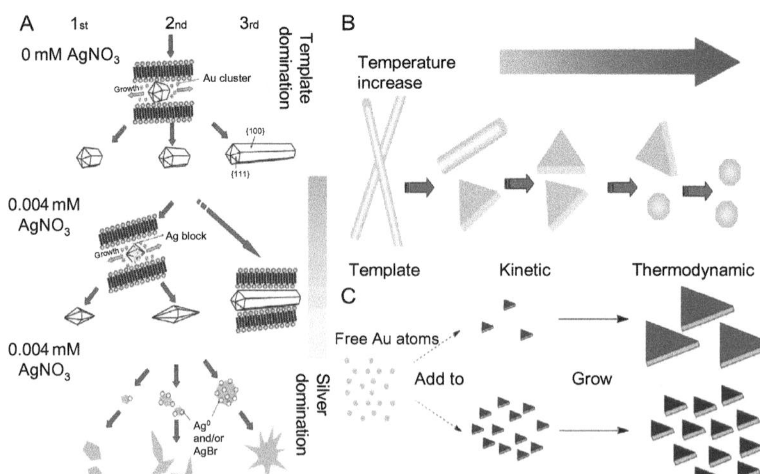

FIGURE 1.7 (A) Influence of silver ions on seeds and the final shape and (B) influence of temperature (adapted from Chen *et al.*)[50]; (C) scheme showing the influence of the concentration of seeds on the size of nanoplates (adapted from Fan *et al.*).[49]

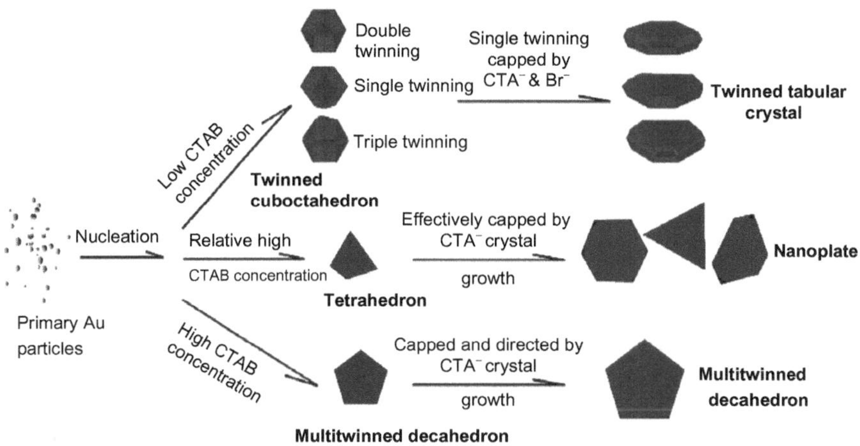

FIGURE 1.8 Scheme of the CTAB directed synthesis of gold particles (adapted from Zhang).[51] Chu and coworkers published a work where in a single step, avoiding preformed seeds, and using the thermal decomposition method in the presence of CTAB, they were able to produce three different sizes of nanoplates.[52]

products, by changing the yield of different shapes. They also proposed an effect of the silver concentration on the shape of the seeds which ultimately determine the shape of the final product.

The influence of CTAB in the nucleation step to form NPs was studied by Zhang and coworkers using the seed-mediated approach.[51] They were able to modulate the shape of the seeds by varying the CTAB concentration; these "controlled seeds" evolved into different products (Figure 1.8).

The use of polymers as surfactants to help plate formation during the metal salt reduction has also been studied. PVP is the most popular polymer in the chemical synthesis of a wide variety of metal nanocrystals. Typically, PVP acts as capping agent, although it has been demonstrated that it may also act as a mild reducing agent.[53] Also, PVP has often been claimed to play a key role in determining the final particle morphology, through stronger adsorption onto certain crystalline facets. In the gold case, the use of PVP is linked to the use of ethylene glycol as solvent, using temperatures above 100 °C. Size-controlled gold nanoplates were synthesized with high purity by the reduction of the gold precursor by sodium citrate (in an unusual low concentration) in the presence of PVP (Figure 1.9).[54]

With the insufficient addition of the reductant, the molar ratio of sodium citrate and PVP relative to the gold salt played an important role in determining the geometric shape and size of the product. These nanoplates were single crystals edge-length in the range of 80–500 nm and thickness of 10–40 nm; these exhibited remarkable plasmon bands in the range of the EM spectrum of 700–2000 nm. Surprisingly, this methodology has not been further explored in the context of bioapplications, maybe due to the presence of PVP or irreproducibility in terms of high yield of nanoplates.

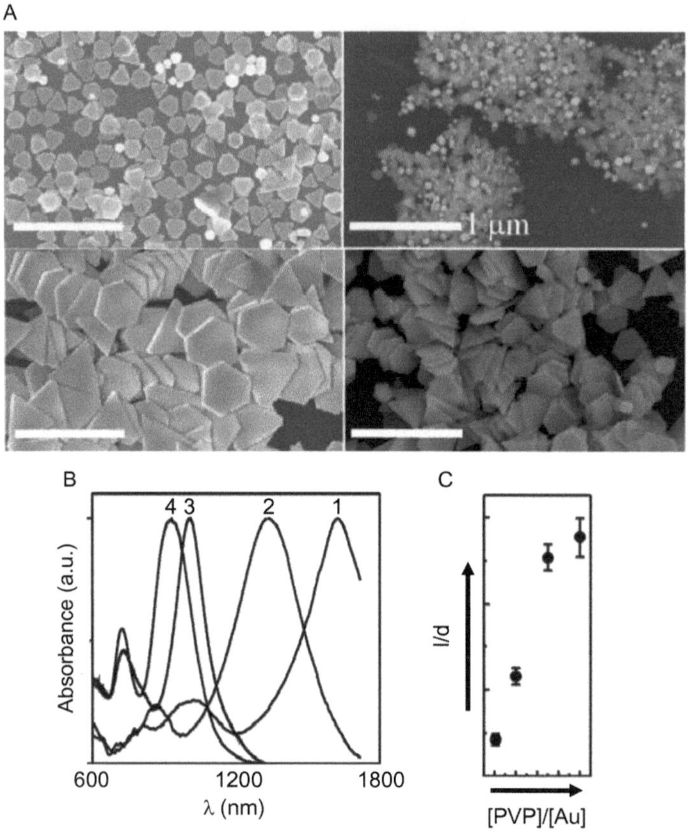

FIGURE 1.9 (A) SEM images, (B) absorption spectra from the resultant plates with different ratios of PVP/Au, and (C) aspect ratio for these products which increase as the ratio PVP/Au increases (adapted from Ah).[54]

Shortly after, in 2006, Xia and coworkers proved this method to be applicable for different metals such as Ag, Pd, and Pt, in addition to Au. They also studied the effect of the molecular weight of PVP, the ratio of the reagents, a negative role of the light, and more importantly, they assessed to PVP the role of mild reductant in addition to its role as surfactant. Also surprisingly, the yield of the Au nanoplates was only 30%,[23] far from the previously obtained by Ah and coworkers.[54] Micrometer triangular plates (Au nanosheets) have also been synthesized using this PVP-based methodology. Li and coworkers studied the effect of changing parameters such as temperature, PVP, or gold salt concentration[55]; recently, the effect of ions in the final shape of the sheets has been reported.[56] Micrometer nanoplates have also been produced using linear polyethylenimine instead of PVP with an relatively good yield.[57]

In 2008, Lavarta and coworkers used amphiphilic polyethylene oxide–polystyrene oxide block copolymers as both reductant and stabilizing agents in water solution to form NPs; the production of spherical or pseudospherical nanoparticles was shown to depend on parameters such as the copolymer block length and copolymer and gold salt concentrations. By moderately increasing the temperature up to 65 °C, they observed a progressive formation of single-crystalline gold nanoplates in good yields (up to 70%) without further addition of any reactants or growing solutions. These nanoplates are characterized by edge-length values in the range of 0.1–1.2 μm depending on copolymer concentration and reaction temperature.[58]

In the context of green chemistry, another kind of chemical reduction of gold can be achieved by using natural extracts from different organisms in approaches normally referred to as biological synthesis. For the production of Au nanoplates, the yields obtained with the reported methodologies are so far relatively low; this fact can be inferred from the optical features in the NIR range from different biological synthesis (Figure 1.10). The pioneer work in this area was published in 2004 by Sastry and coworkers; they used the extract of lemongrass plant (*Cymbopogon flexuosus*) to produce gold triangular nanoplates.[59] In other work, these authors also studied the combined effect of a halide ions and the extract of lemongrass; surprisingly, it was founded that chloride ions promote nanoplate formation whereas iodine promoted pseudospherical NPs.[60] A variety of other extracts, such as those from brown seaweed (*Sargassum* sp.),[61] aloe vera,[62] a green alga (*Chlorella vulgaris*),[63] and, more recently, an extract of pear fruit, have been used to produce gold nanoplates.[65]

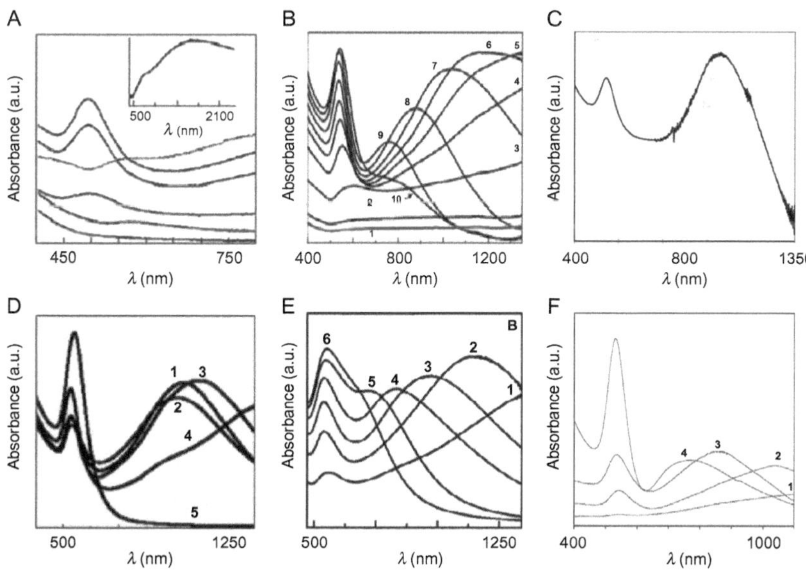

FIGURE 1.10 Spectra of biological synthesized gold nanoplates using extracts from (A, B, and D) lemongrass plant, (C) brown seaweed, (E) aloe vera plant, and (F) a green alga. Reproduced from different works.[59–64]

Also in the context of methods based on green chemistry, there have been further successful attempts to produce nanoplates; in one-pot synthesis and based on the biomineralization ability of BSA in aqueous solutions, Lee and coworkers reported the synthesis of different plate-like (triangular and hexagonal shapes) Au microstructures and the influence of temperature, pH, and Ag^{+} in the shape and size of the resulting NPs.[66] Malikova and coworkers found that a gold (III) salt can be reduced by salicylic acid to yield a mixture of triangular, hexagonal, and polyhedral NPs.[67] In other work, by using green chemistry as well, the gold salt was reduced at room temperature by L-aspartate to produce plates, whereas using L-tryptophan instead resulted in spherical NPs.[68]

4.3. Gold Nanoshells

These are spherical nanoparticles consisting of a dielectric core (e.g., silica, polystyrene, or sodium sulfide) covered by a thin layer of gold (Figure 1.11). Typically, the optical features of these structures characterized by a tunable plasmon band that gives rise to intense optical absorption and scattering.[73] The modulation of the plasmon band can be reached controlling the thickness of the Au shell relative to the diameter of the core; thus, the plasmon band of nanoshells can be tuned in the range of 700–1300 nm.[74]

Gold nanoshells (GNSs) based on a silica core were originally developed by Halas and coworkers.[75] Typically, the scaffolds (silica cores) are prepared by the Stöber method, and later on, small GNPs are grown onto the silica to form a relative uniform thin shell.[69,74–76] Due to their biocompatibility and on demand plasmon band, these structures have been widely applied in PTT by NIR light.[77–83] Moreover, the commercial product AuroLase® Therapy uses AuroShell® particles (nanoshells) with a near-infrared laser source to thermally destroy cancer tissue without significant damage to surrounding healthy tissue; these particles are already in advanced stages of FDA clinical trials.

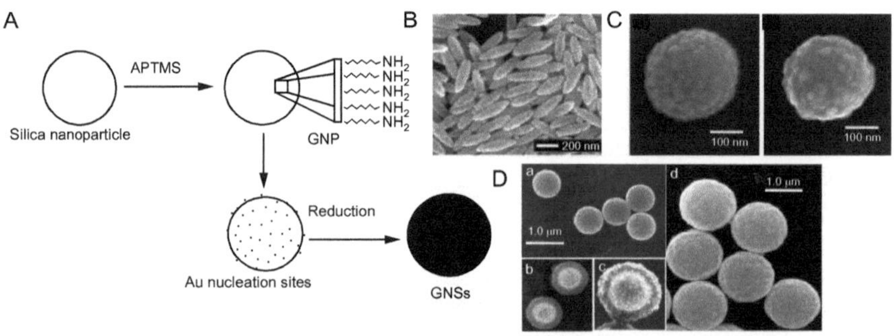

FIGURE 1.11 (A) Scheme of GNSs synthesis,[69] (B) image of nanorice,[70] (C) images of GNSs with different roughness,[71] and (D) images of the sequence of synthesis of concentric GNSs.[72]

Using similar techniques, Halas and coworkers have developed different anisotropic structures[84]; for example, nanorices contain elongated hematite cores as scaffolds and the thin gold shell in nanocomposites.[70] In other works, nanoshells were based on a gold sulfide core[85,86]; this nanostructure has been widely studied and applied although its core–shell nature is still very controversial.[87–91] Recently, Halas and coworkers have discussed a variety of gold-coated semiconductor cores.[92]

4.4. Nanocages and Hollow Nanostructures

These nanostructures have attracted a great deal of attention due to their high surface to volume ratio and large pore volume. The most successful synthetic route to obtain these is based on the galvanic replacement method. The group of Xia has used this method extensively to obtain hollow nanostructures made of different metals, that is, nanocages (hollow nanocubes), nanorings, prism-shaped nanoboxes, nanotubes, and multiple-walled nanoshells or nanotubes.[93]

Galvanic replacement is based in the different electrochemical potential between two metals which drives the redox reaction. Regarding gold nanocages, Ag nanocubes serve as "reducing agents" and templates where Au^0 deposits; the reduction potential of $AuCl_4^-/Au$ (0.99 V) is more positive than AgCl/Ag (0.22 V). Redox reaction is the following (Equation 1.1):

$$3Ag(s) + HAuCl_4 \leftrightarrow Au(s) + 3AgCl(s) + HCl(aq) \qquad (1.1)$$

Xia and coworkers have been able to control the reaction to obtain complete hollow gold cages or noncomplete hollow cages without corners or with holes in their faces (Figure 1.12). To obtain holes in the faces, they used wet etching of silver with $Fe(NO_3)_3$ and/or NH_4OH.[35]

4.5. Branched Nanostructures, Nanostars, or Nanoflowers

These structurally complex structures represent an important subcategory of anisotropic structures which show remarkable optical features; sharp edges and tips exhibit large enhancements of the electric field (hot spots) which are very sensitive to local changes in the dielectric environment; these structures can be used as SERS platforms of enhanced sensitivity.[94,95] These gold nanostructures represent excellent "light confiners."[96]

The synthesis of branched nanostructures represents a complex scenario where the control of nucleation and growth processes is crucial. A highly branched structure with the same volume of a polyhedral nanocrystal has a larger surface area and thus higher surface energy. Due this high-energy structure, their synthesis must be done under kinetics conditions.[95] A great diversity of parameters have been shown to play an important role in the formation of this NPs, that is, the reactivity of the metal precursor,[97] the effect of the surface stabilization through a capping agent, the solvent of choice, the

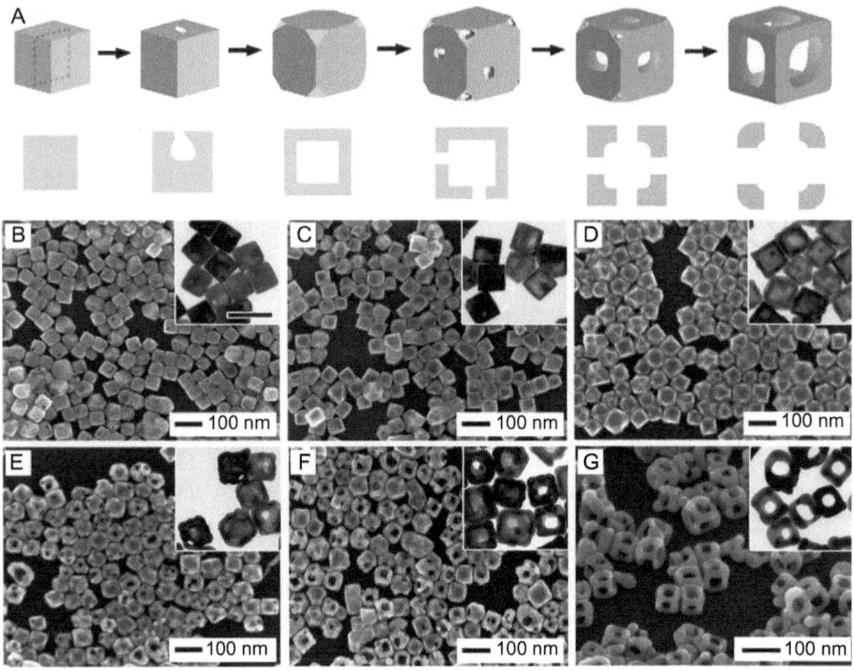

FIGURE 1.12 Scheme of galvanic replacement gold nanocages formation (A) and SEM micrographs showing different structures (B–G). Adapted from Cobley.[93]

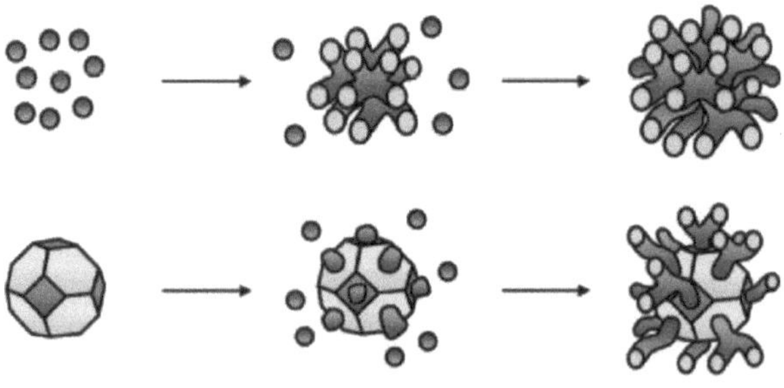

FIGURE 1.13 Scheme of highly branched structures (adapted from Lim).[99]

nature (strong, mild, or soft) of the reducing agent, the reaction temperature, or the effect of the additives.[95,98]

In the literature, one can find two main wet-chemical approaches to produce Au-branched nanostructures, that is, one-pot methods and those based on the two-step seed-growth method (Figure 1.13).

The seed-growth method used for branched nanostructures is a variant of that used to produce nanorods.[39] In the simplest approach, preformed seeds work as scaffolds to support the growth of a number of "fingers" or tips.[100,101]

Variables such as the concentration of bromide and/or silver ions,[44,102] addition steps, order of reactant mixing, amount and/or type of capping agent have been shown to greatly influence in the successful generation of multi-tips.[41,99]

As previously stated, branched AuNPs can be made in one-pot synthesis where the gold precursor is reduced in the presence of an appropriate capping agent which eventually can also be act as reductant.[103–105]

5. DERIVATIZATION OF GNPS

The colloidal stability of NPs is an important issue in the context of any application in which the organic coating is the key to success. For bioapplications, avoiding aggregation and providing NPs with a load of molecules with biological relevance is of the most importance; material scientists have to ask themselves how hybrid nanomaterials made of nanocrystals (plasmonics, superparamagnetic, semiconductors, etc.) and their constituents (surfactants, precursors, macromolecules or organic polymers, etc.) behave in a physiological environment that normally is far from the environment where the synthesis of the nanomaterials took place, that is, polar or nonpolar solvents with very particular attributes such as ionic strength, pH, or presence of multivalent ions, salts, or polymers, to name a few.

The most straightforward approach to derivatize NPs is attaching molecules onto the surface. Many synthetic routes of NPs use surfactants or molecules that can act as ligands. These molecules can avoid the aggregation by ionic stabilization (e.g., citrate) or steric hindrance (e.g., CTAB, PVP, PEG, etc.). Typically, ligand molecules must be bound onto the NPs surface by some attractive interaction such as chemisorption, electrostatic attraction, or hydrophobicity.[5,106] Different groups (thiol, phosfine, amine, carboxy, etc.) present in a molecule of interest can be used to promote the derivatization of NPs.[107] As previously discussed, in the case of GNPs, one of the features that make GNPs so attractive for bioapplications is the easy of functionalization with thiolated chains. Thiol groups are considered to show the highest affinity to novel metals, principally with gold.[108]

To stabilize gold NP in organic phase solutions, thiolated hydrocarbon chains are the most popular choice. On the other hand, for GNPs in aqueous solution, the common approach is the use of bifunctional chains bearing a thiol group in one end and a hydrophilic group in the outermost end, facing the aqueous solvent. In order to create organic monolayers onto the NPs surface, amphiphilic chains are commonly used to promote the correct packaging; once stabilized, the organic layer can be exchanged by other made of molecules with higher affinity for Au^0. The group of Murray is pioneering on ligand exchange strategies.[109]

Techniques of ligand exchange have also been used for water-to-organic phase transfer of GNPs originally coated with CTAB[80]; CTAB chains strongly attached to Au nanorods in the aqueous phase can be exchanged by using dodecanethiol to promote its transfer to organic phase, whereas thiolated PEG chains can be used to organic phase-to-water retransfer.

Other protocols reported on the use of amphiphilic molecules to provide GNPs with colloidal stability in polar and nonpolar conditions. Small molecules are able to change their orientation depending on the solvent, or polymers that can be ionic or nonionic depending on surrounding conditions (e.g., PEI, poly(acrylic acid)).[110,111]

In another approach developed by Parak and coworkers, amphiphilic polymers are intercalated in the aliphatic shell of the NPs in the organic phase; this method is usually referred to as the polymer-coating method.[112,113] This strategy solves the drawbacks of some ligand exchange protocols where it is required a huge excess of the new ligand, and furthermore, it provides with a general route to water-soluble nanocrystals which can be then further derivatized for biological applications.[113]

Growing a glass shell around the GNPs, also referred to as the silanization method developed by Liz-Marzan and coworkers, is another approach widely used to stabilized NPs.[114]

Once achieved stabilization in physiological environments by one of the aforementioned methods, GNPs can be engineered into complex functional materials for a particular bioapplication. Nowadays, there is a wide variety of chemical methods to anchor molecules of biological relevance onto GNPs, that is, fluorescence tags, PEG chains, proteins, carbohydrates, DNA, peptides, siRNA, enzymes, antibodies, cyclodextrins, biotin, etc.[13,115–119]

In all these cases, heterogeneity in surface coverage can result in patches on the surface, or in ligand conformational changes depending on the surrounding conditions which will determine their behavior in physiological media.[120,121] This difference in the surface coverage will drive protein absorption to form the known protein corona, or even the nanoparticles fate in biological systems.[122,123]

Purity of the final derivatized NPs is a really crucial aspect for many bioapplications; the physical properties of nanocrystals as well as the biofunctional coating should be as controlled as possible; in many cases, both the physical properties (plasmon bands, superparamagnetism, fluorescence, etc.) and the biological activity depend on the homogeneity of the sample. Therefore, a variety of methods are currently used for separation of functional NPs with different size, charge, shape, number of biomolecules, etc.[124]

6. BIOAPPLICATIONS OF GNPS

The most outstanding property of GNPs is the presence of surface plasmons; these determine the optical properties of GNPs (in general, of metallic NPs) and open a wide range of potential applications.[125] Relative to the use of the surface plasmons of GNPs, the different applications can be categorized in different areas, that is, energy, environment, information technologies, and bioapplications, to name a few.

Recently, the application of GNPs in energy has become very popular due to the energy problem associated with the ending of fossil fuels; GNPs are currently investigated to rise the efficiency in photovoltaic devices.[1] The use of Ag and AuNPs has increased the efficiency of photovoltaic cells by an order of 10–15%.[126]

The catalytic nature of some metallic NPs,[127,128] including GNPs, has made them very suitable candidates to be applied in the industry associated with the elimination of hazardous side products; the products derived from industrial processes represent a problem for humans and the environment. The catalytic properties of some metallic NPs can be fostered by the excitation on their SPR; thus, GNPs can be used in photocatalysis in processes that can use solar light and mimics other biological processes. Moreover, GNPs are involved in the oxidation of CO into CO_2[129] and can be used in homogeneous catalysis showing different catalytic behavior related with crystal structure differences.[130]

In the area of information technology, the use of metallic NPs has allowed for new methods for light manipulation, reading, and writing information in sizes significantly smaller than light wavelengths, while the huge absorption and scattering cross-section allows detection of very small signals.[131] Nowadays, these materials can be used to design antennas, lenses, and resonators.[132,133]

Although the application of GNPs has extended to very different areas, GNPs have found a remarkable variety of applications, especially, in biomedicine; this has been motivated by their optical properties, biocompatibility, ease of production of multitask nanosystems, and the ability for tuning the plasmon band in the NIR biological window. Applications in diagnosis and therapy have been widely developed. Some bioapplications for *in vivo* systems of GNPs are summarized in Table 1.1.

Optical properties of GNPs can be complemented by its modification with molecules bearing a variety of biofunctions such as peptides, genetic material, antibodies, proteins, enzymes, fluorophores, carbohydrates, etc. In general, a correct derivatization is the key of success of these applications. In Table 1.2, imaging techniques related to GNPs are summarized.

The application of GNPs in imaging includes dark-field imaging which uses light scattering of GNPs; this technique has been widely used for imaging of GNPs in tumoral cells.[82] For instance, this technique has been used to image GNRs while they were being used to treat and diagnose head and neck cancer.[135]

Optical coherence tomography (OTC) can be used to produce 3D images with micrometer resolution by analysis of a cross-section of tissues under a short-coherence light source. The potential of nanoshells as OTC agents to detect tumors has been demonstrated by Halas and coworkers.[78] Also, other gold nanostructures such as rods, spheres, or nanocages have been successfully applied in this direction.[136]

Two-photon luminescence (TPL) has been identified as a serial process involving sequential absorption of photons and emission from the recombination of electrons in the sp-band and holes in the d-band.[137] This effect is enhanced in

TABLE 1.1 Imaging and Therapeutic Capabilities of NPs

	Modality	Nanoparticle/Agent
Imaging	Optical scattering/OCT	Au nanoshells, GNRs, nanocages, NPs
	PET, SPECT	Radioisotope 198Au
	CT	GNPs[134]
Therapeutic actuation	Photothermal	Au nanoshells, GNRs, nanocages
	Photoacoustic	NIR-absorbing GNPs
	Chemotherapy	GNPs loaded with anticancer drugs (doxorubicin, paclitaxel, etc.)
	Gene therapy	GNPs loaded with siRNA, DNA

Adapted from Bardhan.[77]
OCT, optical coherence tomography, MRI, magnetic imaging resonance; PET, positron electron tomography; SPECT, single photon emission computed tomography; CT, computed tomography.

TABLE 1.2 Imaging Principle, Imaging Technique, and Description of Application of GNPs

Imaging Principle	Imaging Technique	Description
Scattering	Coherent optical tomography (OTC)	GNPs embedded in the medium change the backscattering properties
	Reflectance spectroscopy	Similar to above
Emission	Fluorescence	Emission of light with wavelength longer than excitation laser
	Two-photon luminiscence	Nonlinear fluorescence
Absorption	Photothermal imaging	The absorption of laser energy by GNPs induces local increment of temperature that results in changes in the optical path length
	Photothermal OTC	The above effect measured by phase sensitive OCT
	Photoacoustic tomography	The absorption of laser energy by GNPs generate acoustic waves

Adapted from Bardhan.[77]

GNPs, especially in NIR-absorbing GNPs. TPL presents some advantages with respect to confocal microscopy such as higher spatial resolution and reduced background signal. GNRs have been used in TPL imaging *in vitro*[137] and *in vivo*.[138]

GNRs exhibit twofold higher contrast than clinical iodine contrast agents for X-ray computed tomography[134]; this is motivated by the high atomic weight of gold. The use of GNPs has lead this imaging technique in a diagnosis tool since GNPs can be derivatized for specific purposes.

Photothermal (PT) and PA imaging techniques make use of the interaction of laser-induced heating of materials, measuring directly the heat generated in the former and detecting acoustic waves in the latter.[117] Due to the strong light absorption of GNPs, showing extinction coefficients around $10^9\ M^{-1} \times cm^{-1}$, GNPs are very suitable agents for both techniques. NIR-absorbing GNPs are ideal for these applications because NIR light can be used in deeper tissue-imaging applications; the penetration of NIR light is optimal for this. For instance, NIR irradiation of NIR-absorbing NPs provokes heat generation in the illuminated place. The heat dissipation to the adjacent tissue undergoes with a thermoplastic expansion creating an acoustic wave that propagates through the tissue and can be mechanically detected.[139] GNRs have been applied in tumor imaging[140] and PT flow cytometry.[117,141] In other study, 40 nm gold nanocages have been applied to imaging blood vessels by PT,[142] and they have been used also as contrast agents for photoacoustic tomography.[143,144] Gold nanostars also demonstrated their potential as PA agents.[145]

PT imaging can be combined with PTT which uses the generated heat to produce localized hyperthermia to destroy cancer cells, bacteria, or virus. GNPS exposed to laser irradiation can act as therapeutic agents as themselves. NIR-absorbing particles are the most appropriate agents for this therapy. In the literature, it is possible to find examples of PTT at molecular level, cellular lever, tumor level, and *in vivo*. Most employed particles are GNRs, GNSs, and gold nanocages. Currently, Aurolase, a product based on GNSs, is being commercialized by Nanospectra® Bioscience and it is currently in the process of FDA-sanctioned human pilot studies.[81] A similar PTT in clinical trials is also being designed using gold nanocages (nanocages LLC).

The ease to attach tumoral markers or drugs to GNPs has also been used to add specificity to multitask GNPs-based nanosystems. El-Sayed and coworkers have carried out extensive work in this direction; for instance, they proved the ability of GNPs functionalized with a monoclonal antibody against epidermal growth factor receptor (anti-EGRF) to target only malignant epithelial cells.[146] Different labeling molecules such as platinum prodrugs, tamoxifen, folic acid, and transcription necrosis factor (α-TNF) have been used to functionalize GNPs to target different cancer cells such as lung and colon, breast,[141] KB, and MC-38 colon carcinoma, respectively.[147]

Since the original work of Mirkin,[13] GNPs functionalized with genetic material, DNA or siRNA, have been widely used in sensing and genetic therapy[148–155]; in a very general definition, gene therapy is the medical process by which inserting a small DNA or RNA sequence allows to modify the expression of certain specific proteins associated with a disease. GNPs functionalized with the genetic material of interest can be used to shuttle the genetic payload into cells to be repaired. Moreover, GNSs and Au nanorods loaded with DNA have been recently proven as efficient drug delivery carriers. In two studies, Halas and coworkers have shown the ability of GNPs to generate heat upon laser illumination which can be used to release genetic materials into cells.[80,156]

Other important applications of GNPs in drug delivery involve the use of laser irradiation, pH change, ionic concentration change, etc., to trigger a drug release. For instance, Xia and coworkers have achieved controlled drug release of gold nanocages upon laser irradiation.[157] Also, Parak and coworkers have used the heat capacity of GNPs upon laser illumination to release cargo from polyelectrolyte capsules.[91]

The plasmon band of GNPs is very sensitive to changes in the environment; a variety of sensors based on this sensitivity have been developed. The most popular are these based on GNPs aggregation which provoke a visible color change. Different kinds of interactions capable of producing aggregation of GNPs have been used, for example, monoclonal antibodies to detect various antigenic sites or lectin–carbohydrate interactions. GNPs aggregation was also used to detect chorionic gonadotropin in the urine of pregnancy women, or to detect immunoglobulins in the blood serum of HIV patients.[158]

In 2003, Mirkin develops a single-analyte bio-barcode assay.[159] It is carried out in a disposable chip through basic steps. In this sensor, when GNPs are illuminated with light matching their plasmon band, target DNA sequences can be detected.

Other sensors to visualize specific interaction are based on the dot assay; this is a very useful assay due to the small volume required. Recently, the sensitivity of this assay was improved using Ag and GNPs.[160] Most important kind of assays is the enzyme-linked immunosorbent assay due to its simple methodology, specificity, and sensibility. In this kind of assays, the recognition is amplified by the use of an enzyme-marked reagent which will recognize the analyte and will reveal its presence by a simple colorimetric enzymatic reaction. Result reading is performed in a plate reader, using changes in absorbance. Previously described strategies have been adapted to be used in this assays which amplified the method sensibility.[158]

SERS is one of the most powerful analytical techniques for identification of molecular species, providing complete vibrational information of the molecular system under study. This spectroscopic technique is based on the enhancement of Raman-active vibrations associated with proximity to surface plasmons; the enhancement of the Raman signal has been attributed to electromagnetic enhancement derived from the proximity to surface plasmons in theory; others have attributed the Raman enhancement to chemical contributions.[125]Gold nanostars are excellent SERS agents; sharp edges and tips exhibit large enhancements of the electric field (hot spots) which are very sensitive to local changes in the dielectric environment; these structures are used as SERS platforms of enhanced sensitivity.[94,95] In the past decade, SERS based on GNPs has been widely used to detect biomarkers in cancer[161]and other diseases,[162] bacteria[163] or viral agents.[164] Recently, SERS has been revealed as a promising technique to detect cancer cell in circulation.[161] The advantage of this tool compared to fluorescence detection is based on the reduction of biological background interactions.[117]

REFERENCES

1. Atwater HA, Polman A. Plasmonics for improved photovoltaic devices. *Nat Mater* 2010;**9**:205–13.
2. Lal S, Link S, Halas NJ. Nano-optics from sensing to waveguiding. *Nat Photonics* 2007;**1**:641–8.
3. Ozbay E. Plasmonics: merging photonics and electronics at nanoscale dimensions. *Science* 2006;**311**:189–93.
4. Treguer-Delapierre M, Majimel J, Mornet S, Duguet E, Ravaine S. Synthesis of non-spherical gold nanoparticles. *Gold Bull* 2008;**41**:195–207.
5. Schmid G, Corain B. Nanoparticulated gold: syntheses, structures, electronics, and reactivities. *Eur J Inorg Chem* 2003;**2003**:3081–98.
6. Antonii F. *Panacea Aurea-Auro potabile*. Hamburg: Bibliopolio Frobeniano; 1618.
7. Faraday M. The Bakerian lecture: experimental relations of gold (and other metals) to light. *Philos Trans R Soc Lond* 1857;**147**:145–81.
8. Brust M, Bethell D, Schiffrin DJ, Kiely CJ. Novel gold-dithiol nano-networks with nonmetallic electronic-properties. *Adv Mater* 1995;**7**:795.
9. Brust M, Fink J, Bethell D, Schiffrin DJ, Kiely C. Synthesis and reactions of functionalised gold nanoparticles. *J Chem Soc Chem Commun* 1995;1655–6.
10. Brust M, Walker M, Bethell D, Schiffrin DJ, Whyman R. Synthesis of thiol-derivatised gold nanoparticles in a two-phase liquid-liquid system. *J Chem Soc Chem Commun* 1994;801–2.
11. Kiely CJ, Fink J, Brust M, Bethell D, Schiffrin DJ. Spontaneous ordering of bimodal ensembles of nanoscopic gold clusters. *Nature* 1998;**396**:444–6.
12. Bruchez M, Moronne M, Gin P, Weiss S, Alivisatos AP. Semiconductor nanocrystals as fluorescent biological labels. *Science* 1998;**281**:2013–6.
13. Mirkin CA, Letsinger RL, Mucic RC, Storhoff JJ. A DNA-based method for rationally assembling nanoparticles into macroscopic materials. *Nature* 1996;**382**:607–9.
14. Caruso F, Caruso RA, Mohwald H. Nanoengineering of inorganic and hybrid hollow spheres by colloidal templating. *Science* 1998;**282**:1111–4.
15. Stoeva S, Klabunde KJ, Sorensen CM, Dragieva I. Gram-scale synthesis of monodisperse gold colloids by the solvated metal atom dispersion method and digestive ripening and their organization into two- and three-dimensional structures. *J Am Chem Soc* 2002;**124**:2305–11.
16. Bonnemann H, Richards RM. Nanoscopic metal particles—synthetic methods and potential applications. *Eur J Inorg Chem* 2001;2455–80.
17. Uccello-Barretta G, Evangelisti C, Balzano F, Vanni L, Aiello F, Jicsinszky L. Water soluble heptakis(6-deoxy-6-thio)cyclomaltoheptaose capped gold nanoparticles via metal vapour

synthesis: NMR structural characterization and complexation properties. *Carbohydr Res* 2011;**346**:753–8.
18. Turkevich J, Stevenson PC, Hillier J. A study of the nucleation and growth processes in the synthesis of colloidal gold. *Discuss Faraday Soc* 1951;**11**:55–75.
19. Giersig M, Mulvaney P. Preparation of ordered colloid monolayers by electrophoretic deposition. *Langmuir* 1993;**9**:3408–13.
20. Giersig M, Liz-Marzan LM, Ung T, Su DS, Mulvaney P. Chemistry of nanosized silica-coated metal particles EM-study. *Phys Chem Chem Phys* 1997;**101**:1617–20.
21. Chen SW. 4-Hydroxythiophenol-protected gold nanoclusters in aqueous media. *Langmuir* 1999;**15**:7551–7.
22. Hirai H, Nakao Y, Toshima N. Preparation of colloidal transition metals in polymers by reduction with alcohols or ethers. *J Macromol Sci A Chem* 1979;**13**:727–50.
23. Xiong Y, Washio I, Chen J, Cai H, Li Z-Y, Xia Y. Poly(vinyl pyrrolidone): a dual functional reductant and stabilizer for the facile synthesis of noble metal nanoplates in aqueous solutions. *Langmuir* 2006;**22**:8563–70.
24. Pastoriza-Santos I, Alvarez-Puebla RA, Liz-Marzán LM. Synthetic routes and plasmonic properties of noble metal nanoplates. *Eur J Inorg Chem* 2010;**2010**:4288–97.
25. Furuya K, Hirowatari Y, Ishioka T, Harata A. Protective agent-free preparation of gold nanoplates and nanorods in aqueous HAuCl4 solutions using gas–liquid interface discharge. *Chem Lett* 2007;**36**:1088–9.
26. Zhu J, Shen Y, Xie A, Qiu L, Zhang Q, Zhang S. Photoinduced synthesis of anisotropic gold nanoparticles in room-temperature ionic liquid. *J Phys Chem C* 2007;**111**:7629–33.
27. Park J, Joo J, Kwon SG, Jang Y, Hyeon T. Synthesis of monodisperse spherical nanocrystals. *Angew Chem Int Ed* 2007;**46**:4630–60.
28. Zsigmondy R, Spear EB. *The chemistry of colloids.* New York: John Wiley; 1917.
29. Jana NR, Gearheart L, Murphy CJ. Wet chemical synthesis of high aspect ratio cylindrical gold nanorods. *J Phys Chem B* 2001;**105**:4065–7.
30. Murphy CJ, Sau TK, Gole AM, Orendorff CJ, Gao J, Gou L, et al. Anisotropic metal nanoparticles: synthesis, assembly, and optical applications. *J Phys Chem B* 2005;**109**:13857–70.
31. Orendorff CJ, Murphy CJ. Quantitation of metal content in the silver-assisted growth of gold nanorods. *J Phys Chem B* 2006;**110**:3990–4.
32. Ha TH, Koo HJ, Chung BH. Shape-controlled syntheses of gold nanoprisms and nanorods influenced by specific adsorption of halide ions. *J Phys Chem C* 2007;**111**:1123–30.
33. Fievet F. Preparing monodisperse metal powders in micrometer and submicrometer sizes by the polyol process. *MRS Bull* 1989;**14**:29–34.
34. Brenner A, Riddell GE. Nickel plating on steel by chemical reduction. *J Res Natl Bur Stand* 1946;**37**:31–4.
35. Skrabalak SE, Chen J, Sun Y, Lu X, Au L, Cobley CM, et al. Gold nanocages: synthesis, properties, and applications. *Acc Chem Res* 2008;**41**:1587–95.
36. Martin CR. Nanomaterials: a membrane-based synthetic approach. *Science* 1994;**266**:1961–6.
37. Khlebtsov N, Dykman L. Biodistribution and toxicity of engineered gold nanoparticles: a review of in vitro and in vivo studies. *Chem Soc Rev* 2011;**40**:1647.
38. Yu-Ying Yu, Chang S-S, Lee C-L, Wang CRC. *Gold nanorods: electrochemical synthesis and optical properties. J Phys Chem B* 1997;**101**:6661–4.
39. Nikoobakht B, El-Sayed MA. Preparation and growth mechanism of gold nanorods (NRs) using seed-mediated growth method. *Chem Mater* 2003;**15**:1957–62.
40. Grzelczak M, Pérez-Juste J, Mulvaney P, Liz-Marzán LM. Shape control in gold nanoparticle synthesis. *Chem Soc Rev* 2008;**37**:1783.

41. Tao AR, Habas S, Yang P. Shape control of colloidal metal nanocrystals. *Small* 2008;**4**:310–25.
42. Ibano D, Yokota Y, Tominaga T. Preparation of gold nanoplates protected by an anionic phospholipid. *Chem Lett* 2003;**32**:574–5.
43. Jain PK, Huang XH, El-Sayed IH, El-Sayed MA. Noble metals on the nanoscale: optical and photothermal properties and some applications in imaging, sensing, biology, and medicine. *Acc Chem Res* 2008;**41**:1578–86.
44. Sau TK, Murphy CJ. Room temperature, high-yield synthesis of multiple shapes of gold nanoparticles in aqueous solution. *J Am Chem Soc* 2004;**126**:8648–9.
45. Millstone JE, Park S, Shuford KL, Qin L, Schatz GC, Mirkin CA. Observation of a quadrupole plasmon mode for a colloidal solution of gold nanoprisms. *J Am Chem Soc* 2005;**127**:5312–3.
46. Millstone JE, Métraux GS, Mirkin CA. Controlling the edge length of gold nanoprisms via a seed-mediated approach. *Adv Funct Mater* 2006;**16**:1209–14.
47. Busbee BD, Obare SO, Murphy CJ. An improved synthesis of high-aspect-ratio gold nanorods. *Adv Mater* 2003;**15**:414–6.
48. Millstone JE, Wei W, Jones MR, Yoo H, Mirkin CA. Iodide ions control seed-mediated growth of anisotropic gold nanoparticles. *Nano Lett* 2008;**8**:2526–9.
49. Fan X, Guo ZR, Hong JM, Zhang Y, Zhang JN, Gu N. Size-controlled growth of colloidal gold nanoplates and their high-purity acquisition. *Nanotechnology* 2010;**21**:105602.
50. Chen HM, Liu R-S, Tsai DP. A versatile route to the controlled synthesis of gold nanostructures. *Crystal Growth Des* 2009;**9**:2079–87.
51. Zhang L, Huang CZ, Li YF, Li Q. Morphology control and structural characterization of Au crystals: from twinned tabular crystals and single-crystalline nanoplates to multitwinned decahedra. *Crystal Growth Des* 2009;**9**:3211–7.
52. Chu H-C, Kuo C-H, Huang MH. Thermal aqueous solution approach for the synthesis of triangular and hexagonal gold nanoplates with three different size ranges. *Inorg Chem* 2005;**45**:808–13.
53. Washio I, Xiong Y, Yin Y, Xia Y. Reduction by the end groups of poly(vinyl pyrrolidone): a new and versatile route to the kinetically controlled synthesis of Ag triangular nanoplates. *Adv Mater* 2006;**18**:1745–9.
54. Ah CS, Yun YJ, Park HJ, Kim WJ, Ha DH, Yun WS. Size-controlled synthesis of machinable single crystalline gold nanoplates. *Chem Mater* 2005;**17**:5558–61.
55. Li CC, Cai WP, Cao BQ, Sun FQ, Li Y, Kan CX, et al. Mass synthesis of large, single-crystal Au nanosheets based on a polyol process. *Adv Funct Mater* 2006;**16**:83–90.
56. Jiu J, Suganuma K, Nogi M. Effect of additives on the morphology of single-crystal Au nanosheet synthesized using the polyol process. *J Mater Sci* 2011;**46**:4964–70.
57. Sun X, Dong S, Wang E. High-yield synthesis of large single-crystalline gold nanoplates through a polyamine process. *Langmuir* 2005;**21**:4710–2.
58. Goy-López S, Castro E, Taboada P, Mosquera Vc. Block copolymer-mediated synthesis of size-tunable gold nanospheres and nanoplates. *Langmuir* 2008;**24**:13186–96.
59. Shankar SS, Rai A, Ankamwar B, Singh A, Ahmad A, Sastry M. Biological synthesis of triangular gold nanoprisms. *Nat Mater* 2004;**3**:482–8.
60. Rai A, Singh A, Ahmad A, Sastry M. Role of halide ions and temperature on the morphology of biologically synthesized gold nanotriangles. *Langmuir* 2005;**22**:736–41.
61. Liu B, Xie J, Lee JY, Ting YP, Chen JP. Optimization of high-yield biological synthesis of single-crystalline gold nanoplates. *J Phys Chem B* 2005;**109**:15256–63.

62. Chandran SP, Chaudhary M, Pasricha R, Ahmad A, Sastry M. Synthesis of gold nanotriangles and silver nanoparticles using Aloe vera plant extract. *Biotechnol Prog* 2006;**22**:577–83.
63. Xie J, Lee JY, Wang DIC, Ting YP. Identification of active biomolecules in the high-yield synthesis of single-crystalline gold nanoplates in algal solutions. *Small* 2007;**3**:672–82.
64. Shankar SS, Rai A, Ahmad A, Sastry M. Controlling the optical properties of lemongrass extract synthesized gold nanotriangles and potential application in infrared-absorbing optical coatings. *Chem Mater* 2005;**17**:566–72.
65. Ghodake GS, Deshpande NG, Lee YP, Jin ES. Pear fruit extract-assisted room-temperature biosynthesis of gold nanoplates. *Colloids Surf B Biointerfaces* 2010;**75**:584–9.
66. Xie J, Lee JY, Wang DIC. Synthesis of single-crystalline gold nanoplates in aqueous solutions through biomineralization by serum albumin protein. *J Phys Chem C* 2007;**111**:10226–32.
67. Malikova N, Pastoriza-Santos I, Schierhorn M, Kotov NA, Liz-Marzán LM. Layer-by-layer assembled mixed spherical and planar gold nanoparticles: control of interparticle interactions. *Langmuir* 2002;**18**:3694–7.
68. Jin y Y, Dong S. Synthesis of gold nanoplates by aspartate reduction of gold chlorideElectronic supplementary information (ESI) available: Fig. S1. UV/Visible-NIR extinction spectra of an aqueous dispersion of gold nanoparticles synthesized by tyrosine (a), phenylalanine (b), lysine (c), aspartate (d) and tryptophan (e). See http://www.rsc.org/suppdata/cc/b3/b315732f. *Chem Commun* 2004;1104.
69. Pham T, Jackson JB, Halas NJ, Lee TR. Preparation and characterization of gold nanoshells coated with self-assembled monolayers. *Langmuir* 2002;**18**:4915–20.
70. Wang H, Brandl DW, Le F, Nordlander P, Halas NJ. Nanorice: a hybrid plasmonic nanostructure. *Nano Lett* 2006;**6**:827–32.
71. Wang H, Goodrich GP, Tam F, Oubre C, Nordlander P, Halas NJ. Controlled texturing modifies the surface topography and plasmonic properties of Au nanoshells. *J Phys Chem B* 2005;**109**:11083–7.
72. Prodan E. A hybridization model for the plasmon response of complex nanostructures. *Science* 2003;**302**:419–22.
73. Nehl CL, Grady NK, Goodrich GP, Tam F, Halas NJ, Hafner JH. Scattering spectra of single gold nanoshells. *Nano Lett* 2004;**4**:2355–9.
74. Radloff C, Halas NJ. Plasmonic properties of concentric nanoshells. *Nano Lett* 2004;**4**:1323–7.
75. Oldenburg SJ, Jackson JB, Westcott SL, Halas NJ. Infrared extinction properties of gold nanoshells. *Appl Phys Lett* 1999;**75**:2897–9.
76. Averitt RD, Westcott SL, Halas NJ. Linear optical properties of gold nanoshells. *J Opt Soc Am B Opt Phys* 1999;**16**:1824–32.
77. Bardhan R, Lal S, Joshi A, Halas NJ. Theranostic nanoshells: from probe design to imaging and treatment of cancer. *Acc Chem Res* 2011;**44**:936–46.
78. Gobin AM, Lee MH, Halas NJ, James WD, Drezek RA, West JL. Near-infrared resonant nanoshells for combined optical imaging and photothermal cancer therapy. *Nano Lett* 2007;**7**:1929–34.
79. Hirsch LR, Stafford RJ, Bankson JA, Sershen SR, Rivera B, Price RE, et al. Nanoshell-mediated near-infrared thermal therapy of tumors under magnetic resonance guidance. *Proc Natl Acad Sci USA* 2003;**100**:13549–54.

80. Huschka R, Zuloaga J, Knight MW, Brown LV, Nordlander P, Halas NJ. Light-induced release of DNA from gold nanoparticles: nanoshells and nanorods. *J Am Chem Soc* 2011;**133**:12247–55.
81. Lal S, Clare SE, Halas NJ. Nanoshell-enabled photothermal cancer therapy: impending clinical impact. *Acc Chem Res* 2008;**41**:1842–51.
82. Loo C, Lowery A, Halas NJ, West J, Drezek R. Immunotargeted nanoshells for integrated cancer imaging and therapy. *Nano Lett* 2005;**5**:709–11.
83. O'Neal DP, Hirsch LR, Halas NJ, Payne JD, West JL. Photo-thermal tumor ablation in mice using near infrared-absorbing nanoparticles. *Cancer Lett* 2004;**209**:171–6.
84. Knight MW, Halas NJ. Nanoshells to nanoeggs to nanocups: optical properties of reduced symmetry core-shell nanoparticles beyond the quasistatic limit. *New J Phys* 2008;**10**:105006.
85. Zhou HS, Honma I, Komiyama H, Haus JW. Controlled synthesis and quantum-size effect in gold-coated nanoparticles. *Phys Rev B* 1994;**50**:12052–6.
86. Averitt RD, Sarkar D, Halas NJ. Plasmon resonance shifts of Au-coated Au2S nanoshells: insight into multicomponent nanoparticle growth. *Phys Rev Lett* 1997;**78**:4217–20.
87. Raschke G, Brogl S, Susha AS, Rogach AL, Klar TA, Feldmann J, et al. Gold nanoshells improve single nanoparticle molecular sensors. *Nano Lett* 2004;**4**:1853–7.
88. Schwartzberg AM, Grant CD, van Buuren T, Zhang JZ. Reduction of HAuCl(4) by Na(2)S revisited: the case for Au nanoparticle aggregates and against Au(2)S/Au Core/Shell particles. *J Phys Chem C* 2007;**111**:8892–901.
89. Mikhlin Y, Likhatski M, Karacharov A, Zaikovski V, Krylov A. Formation of gold and gold sulfide nanoparticles and mesoscale intermediate structures in the reactions of aqueous HAuCl4 with sulfide and citrate ions. *Phys Chem Chem Phys* 2009;**11**:5445–54.
90. Skirtach AG, Dejugnat C, Braun D, Susha AS, Rogach AL, Parak WJ, et al. The role of metal nanoparticles in remote release of encapsulated materials. *Nano Lett* 2005;**5**:1371–7.
91. Munoz Javier A, del Pino P, Bedard MF, Ho D, Skirtach AG, Sukhorukov GB, et al. Photoactivated release of cargo from the cavity of polyelectrolyte capsules to the cytosol of cells. *Langmuir* 2008;**24**:12517–20.
92. Bardhan R, Grady NK, Ali T, Halas NJ. Metallic nanoshells with semiconductor cores: optical characteristics modified by core medium properties. *ACS Nano* 2010;**4**:6169–79.
93. Cobley CM, Xia Y. Engineering the properties of metal nanostructures via galvanic replacement reactions. *Mater Sci Eng R Rep* 2010;**70**:44–62.
94. Kumar PS, Pastoriza-Santos I, Rodriguez-Gonzalez B, Garcia de Abajo FJ, Liz-Marzan LM. High-yield synthesis and optical response of gold nanostars. *Nanotechnology* 2008;**19**:015606.
95. Guerrero-Martínez A, Barbosa S, Pastoriza-Santos I, Liz-Marzán LM. Nanostars shine bright for you: colloidal synthesis, properties and applications of branched metallic nanoparticles. *Curr Opin Colloid Interface Sci* 2011;**16**:118–27.
96. Alvarez-Puebla RN, Liz-Marzán LM, García de Abajo FJ. Light concentration at the nanometer scale. *J Phys Chem Lett* 2010;**1**:2428–34.
97. Henglein A. Small-particle research: physicochemical properties of extremely small colloidal metal and semiconductor particles. *Chem Rev* 1989;**89**:1861–73.
98. Gasser U, Weeks ER, Schofield A, Pusey PN, Weitz DA. Real-space imaging of nucleation and growth in colloidal crystallization. *Science* 2001;**292**:258–62.
99. Lim B, Xia Y. Metal nanocrystals with highly branched morphologies. *Angew Chem Int Ed* 2011;**50**:76–85.
100. Nehl CL, Liao H, Hafner JH. Optical properties of star-shaped gold nanoparticles. *Nano Lett* 2006;**6**:683–8.

101. Wu H-L, Chen C-H, Huang MH. Seed-mediated synthesis of branched gold nanocrystals derived from the side growth of pentagonal bipyramids and the formation of gold nanostars. *Chem Mater* 2008;**21**:110–4.
102. Wu HL, Chen CH, Huang MH. Seed-mediated synthesis of branched gold nanocrystals derived from the side growth of pentagonal bipyramids and the formation of gold nanostars. *Chem Mater* 2009;**21**:110–4.
103. Xie J, Lee JY, Wang DIC. Seedless, surfactantless, high-yield synthesis of branched gold nanocrystals in HEPES buffer solution. *Chem Mater* 2007;**19**:2823–30.
104. Liao H-G, Jiang Y-X, Zhou Z-Y, Chen S-P, Sun S-G. Shape-controlled synthesis of gold nanoparticles in deep eutectic solvents for studies of structure–functionality relationships in electrocatalysis. *Angew Chem Int Ed* 2008;**47**:9100–3.
105. Burt JL, Elechiguerra JL, Reyes-Gasga J, Martin Montejano-Carrizales J, Jose-Yacaman M. Beyond Archimedean solids: star polyhedral gold nanocrystals. *J Crystal Growth* 2005;**285**:681–91.
106. Sperling RA, Parak WJ. Surface modification, functionalization and bioconjugation of colloidal inorganic nanoparticles. *Philos Trans R Soc Math Phys Eng Sci* 2010;**368**:1333–83.
107. Khlebtsov NG, Dykman LA. Optical properties and biomedical applications of plasmonic nanoparticles. *J Quant Spectrosc Radiative Trans* 2010;**111**:1–35.
108. Love JC, Estroff LA, Kriebel JK, Nuzzo RG, Whitesides GM. Self-assembled monolayers of thiolates on metals as a form of nanotechnology. *Chem Rev* 2005;**105**:1103–69.
109. Ackerson CJ, Jadzinsky PD, Kornberg RD. Thiolate ligands for synthesis of water-soluble gold clusters. *J Am Chem Soc* 2005;**127**:6550–1.
110. Song HT, Choi JS, Huh YM, Kim S, Jun YW, Suh JS, et al. Surface modulation of magnetic nanocrystals in the development of highly efficient magnetic resonance probes for intracellular labeling. *J Am Chem Soc* 2005;**127**:9992–3.
111. Jans H, Jans K, Lagae L, Borghs G, Maes G, Huo Q. Poly(acrylic acid)-stabilized colloidal gold nanoparticles: synthesis and properties. *Nanotechnology* 2010;**21**:455702.
112. Pellegrino T, Manna L, Kudera S, Liedl T, Koktysh D, Rogach AL, et al. Hydrophobic nanocrystals coated with an amphiphilic polymer shell: a general route to water soluble nanocrystals. *Nano Lett* 2004;**4**:703–7.
113. Moros M, Pelaz B, Lopez-Larrubia P, Garcia-Martin ML, Grazu V, de la Fuente JM. Engineering biofunctional magnetic nanoparticles for biotechnological applications. *Nanoscale* 2010;**2**:1746–55.
114. Liz-Marzan LM, Giersig M, Mulvaney P. Synthesis of nanosized gold-silica core-shell particles. *Langmuir* 1996;**12**:4329–35.
115. de la Fuente JM, Berry CC, Riehle MO, Curtis ASG. Nanoparticle targeting at cells. *Langmuir* 2006;**22**:3286–93.
116. Rojo J, Diaz V, de la Fuente JM, Segura I, Barrientos AG, Riese HH, et al. Gold glyconanoparticles as new tools in antiadhesive therapy. *Chembiochem* 2004;**5**:291–7.
117. Dreaden EC, Alkilany AM, Huang X, Murphy CJ, El-Sayed MA. The golden age: gold nanoparticles for biomedicine. *Chem Soc Rev* 2012;**41**:2740–79.
118. Sokolov K, Follen M, Aaron J, Pavlova I, Malpica A, Lotan R, et al. Real-time vital optical imaging of precancer using anti-epidermal growth factor receptor antibodies conjugated to gold nanoparticles. *Cancer Res* 2003;**63**:1999–2004.
119. El-Sayed IH, Huang XH, El-Sayed MA. Surface plasmon resonance scattering and absorption of anti-EGFR antibody conjugated gold nanoparticles in cancer diagnostics: applications in oral cancer. *Nano Lett* 2005;**5**:829–34.

120. Verma A, Uzun O, Hu Y, Hu Y, Han H-S, Watson N, et al. Surface-structure-regulated cell-membrane penetration by monolayer-protected nanoparticles. *Nat Mater* 2008;**7**:588–95.
121. Nel AE, Madler L, Velegol D, Xia T, Hoek EMV, Somasundaran P, et al. Understanding biophysicochemical interactions at the nano-bio interface. *Nat Mater* 2009;**8**:543–57.
122. Gref R, Lück M, Quellec P, Marchand M, Dellacherie E, Harnisch S, et al. 'Stealth' corona-core nanoparticles surface modified by polyethylene glycol (PEG): influences of the corona (PEG chain length and surface density) and of the core composition on phagocytic uptake and plasma protein adsorption. *Colloids Surf B Biointerfaces* 2000;**18**:301–13.
123. Monopoli MP, Walczyk D, Campbell A, Elia G, Lynch I, Baldelli Bombelli F, et al. Physical–chemical aspects of protein corona: relevance to in vitro and in vivo biological impacts of nanoparticles. *J Am Chem Soc* 2011;**133**:2525–34.
124. Kowalczyk B, Lagzi I, Grzybowski BA. Nanoseparations: strategies for size and/or shape-selective purification of nanoparticles. *Curr Opin Colloid Interface Sci* 2011;**16**:135–48.
125. Garcia MA. Surface plasmons in metallic nanoparticles: fundamentals and applications. *J Phys D Appl Phys* 2011;**44**:283001.
126. Sundararajan SP, Grady NK, Mirin N, Halas NJ. Nanoparticle-induced enhancement and suppression of photocurrent in a silicon photodiode. *Nano Lett* 2008;**8**:624–30.
127. Zhou S, Varughese B, Eichhorn B, Jackson G, McIlwrath K. Pt–Cu Core–Shell and alloy nanoparticles for heterogeneous NOx reduction: anomalous stability and reactivity of a Core–Shell nanostructure. *Angew Chem Int Ed* 2005;**44**:4539–43.
128. Wang DS, Li YD. Bimetallic nanocrystals: liquid-phase synthesis and catalytic applications. *Adv Mater* 2011;**23**:1044–60.
129. Hashmi ASK, Rudolph M. Gold catalysis in total synthesis. *Chem Soc Rev* 2008;**37**:1766–75.
130. Burda C, Chen XB, Narayanan R, El-Sayed MA. Chemistry and properties of nanocrystals of different shapes. *Chem Rev* 2005;**105**:1025–102.
131. Sau TK, Rogach AL. Nonspherical noble metal nanoparticles: colloid-chemical synthesis and morphology control. *Adv Mater* 2010;**22**:1781–804.
132. Martin OJF, Hecht B, Pohl DW. Resonant optical antennas. *Science* 2005;**308**:1607–9.
133. Barnes WL, Dereux A, Ebbesen TW. Surface plasmon subwavelength optics. *Nature* 2003;**424**:824–30.
134. von Maltzahn G, Park JH, Agrawal A, Bandaru NK, Das SK, Sailor MJ, et al. Computationally guided photothermal tumor therapy using long-circulating gold nanorod antennas. *Cancer Res* 2009;**69**:3892–900.
135. Huang X, El-Sayed IH, Qian W, El-Sayed MA. Cancer cell imaging and photothermal therapy in the near-infrared region by using gold nanorods. *J Am Chem Soc* 2006;**128**:2115–20.
136. Chen J, Saeki F, Wiley BJ, Cang H, Cobb MJ, Li Z-Y, et al. Gold nanocages: bioconjugation and their potential use as optical imaging contrast agents. *Nano Lett* 2005;**5**:473–7.
137. Durr NJ, Larson T, Smith DK, Korgel BA, Sokolov K, Ben-Yakar A. Two-photon luminescence imaging of cancer cells using molecularly targeted gold nanorods. *Nano Lett* 2007;**7**:941–5.
138. Tong L, Zhao Y, Huff TB, Hansen MN, Wei A, Cheng JX. Gold nanorods mediate tumor cell death by compromising membrane integrity. *Adv Mater* 2007;**19**:3136–41.

139. Eghtedari M, Oraevsky A, Copland JA, Kotov NA, Conjusteau A, Motamedi M. High sensitivity of in vivo detection of gold nanorods using a laser optoacoustic imaging system. *Nano Lett* 2007;**7**:1914–8.
140. Agarwal A, Huang SW, O'Donnell M, Day KC, Day M, Kotov N, et al. Targeted gold nanorod contrast agent for prostate cancer detection by photoacoustic imaging. *J Appl Phys* 2007;**102**:064701.
141. Dreaden EC, Mackey MA, Huang X, Kang B, El-Sayed MA. Beating cancer in multiple ways using nanogold. *Chem Soc Rev* 2011;**40**:3391.
142. Lu W, Huang Q, Geng KB, Wen XX, Zhou M, Guzatov D, et al. Photoacoustic imaging of living mouse brain vasculature using hollow gold nanospheres. *Biomaterials* 2010;**31**:2617–26.
143. Wang Y, Xie X, Wang X, Ku G, Gill KL, O'Neal DP, et al. Photoacoustic tomography of a nanoshell contrast agent in the in vivo rat brain. *Nano Lett* 2004;**4**:1689–92.
144. Yang X, Skrabalak SE, Li Z-Y, Xia Y, Wang LV. Photoacoustic tomography of a rat cerebral cortex in vivo with Au nanocages as an optical contrast agent. *Nano Lett* 2007;**7**:3798–802.
145. Kim C, Song HM, Cai X, Yao JJ, Wei A, Wang LHV. In vivo photoacoustic mapping of lymphatic systems with plasmon-resonant nanostars. *J Mater Chem* 2011;**21**:2841–4.
146. El-Sayed IH, Huang XH, El-Sayed MA. Selective laser photo-thermal therapy of epithelial carcinoma using anti-EGFR antibody conjugated gold nanoparticles. *Cancer Lett* 2006;**239**:129–35.
147. De M, Ghosh PS, Rotello VM. Applications of nanoparticles in biology. *Adv Mater* 2008;**20**:4225–41.
148. Cao YWC, Jin RC, Mirkin CA. Nanoparticles with Raman spectroscopic fingerprints for DNA and RNA detection. *Science* 2002;**297**:1536–40.
149. Han MS, Lytton-Jean AKR, Oh BK, Heo J, Mirkin CA. Colorimetric screening of DNA-binding molecules with gold nanoparticle probes. *Angew Chem Int Ed* 2006;**45**:1807–10.
150. Lee JS, Han MS, Mirkin CA. Colorimetric detection of mercuric ion (Hg2+) in aqueous media using DNA-functionalized gold nanoparticles. *Angew Chem Int Ed* 2007;**46**:4093–6.
151. Park SJ, Taton TA, Mirkin CA. Array-based electrical detection of DNA with nanoparticle probes. *Science* 2002;**295**:1503–6.
152. Rosi NL, Giljohann DA, Thaxton CS, Lytton-Jean AKR, Han MS, Mirkin CA. Oligonucleotide-modified gold nanoparticles for intracellular gene regulation. *Science* 2006;**312**:1027–30.
153. Rosi NL, Mirkin CA. Nanostructures in biodiagnostics. *Chem Rev* 2005;**105**:1547–62.
154. Stoeva SI, Lee JS, Thaxton CS, Mirkin CA. Multiplexed DNA detection with biobarcoded nanoparticle probes. *Angew Chem Int Ed* 2006;**45**:3303–6.
155. Storhoff JJ, Elghanian R, Mucic RC, Mirkin CA, Letsinger RL. One-pot colorimetric differentiation of polynucleotides with single base imperfections using gold nanoparticle probes. *J Am Chem Soc* 1998;**120**:1959–64.
156. Huschka R, Neumann O, Barhoumi A, Halas NJ. Visualizing light-triggered release of molecules inside living cells. *Nano Lett* 2010;**10**:4117–22.
157. Yavuz MS, Cheng Y, Chen J, Cobley CM, Zhang Q, Rycenga M, Xie J, Kim C, Song KH, Schwartz AG, Wang LV, Xia Y. Gold nanocages covered by smart polymers for controlled release with near-infrared light. *Nat Mater* 2009;**8**:935–9.
158. Cao X, Ye Y, Liu S. Gold nanoparticle-based signal amplification for biosensing. *Anal Biochem* 2011;**417**:1–16.
159. Nam J-M, Thaxton CS, Mirkin CA. Nanoparticle-based bio-bar codes for the ultrasensitive detection of proteins. *Science* 2003;**301**:1884–6.
160. Gupta S, Huda S, Kilpatrick PK, Velev OD. Characterization and optimization of gold nanoparticle-based silver-enhanced immunoassays. *Anal Chem* 2007;**79**:3810–20.

161. Qian X, Peng X-H, Ansari DO, Yin-Goen Q, Chen GZ, Shin DM, et al. In vivo tumor targeting and spectroscopic detection with surface-enhanced Raman nanoparticle tags. *Nat Biotechnol* 2008;**26**:83–90.
162. Dufek EJ, Ehlert B, Granger MC, Sandrock TM, Legge SL, Herrmann MG, et al. Competitive surface-enhanced Raman scattering assay for the 1,25-dihydroxy metabolite of vitamin D3. *Analyst* 2010;**135**:2811–7.
163. Stasko NA, Johnson CB, Schoenfisch MH, Johnson TA, Holmuhamedov EL. Cytotoxicity of polypropylenimine dendrimer conjugates on cultured endothelial cells. *Biomacromolecules* 2007;**8**:3853–9.
164. Zhang H, Harpster MH, Park HJ, Johnson PA, Wilson WC. Surface-enhanced Raman scattering detection of DNA derived from the west nile virus genome using magnetic capture of Raman-active gold nanoparticles. *Anal Chem* 2010;**83**:254–60.

Chapter 2

Synthesis of Inorganic Nanoparticles

Gorka Salas[*,†], Rocío Costo[*] and María del Puerto Morales[*]
[*]Instituto de Ciencias de Materiales de Madrid, CSIC, C/Sor Juana Ines de la Cruz, Cantoblanco, 28049 Madrid, Spain
[†]IMDEA Nanociencia, Facultad de Ciencias, Av. Tomas y Valiente 7, Cantoblanco, 28049 Madrid, Spain

The development of highly uniform and biocompatible inorganic nanoparticles with optimized functional properties is critical to bring new biotechnological innovations. In general, inorganic nanoparticles are more stable than organic nanostructures, which still face unsolved problems, such as their limited chemical and mechanic stability and swelling. In the case of magnetic nanoparticles, their high performance could play an important role not only in cancer diagnosis and treatment but also in the real-time visualization of other biological events such as cell trafficking, cancer metastasis, cellular signalling and interactions at the molecular and cellular level.

There are two approaches for manufacturing inorganic nanomaterials. The first one is the 'top-down' approach, which involves breaking down large pieces of material, generally by attrition or milling, to generate nanoparticles. Etching and lithography are probably the most typical examples of this approach. The biggest problem of the 'top-down' approach is the heterogeneity of the resulting product. Particles produced in this way usually present broad size distribution and varied particle shape. Moreover, they might contain a significant amount of impurities from the milling medium. In addition, the material surface presents important defects and crystallographic damage. Such materials are frequently used in the preparation of ceramic nanograined bulk materials, where defects may be annealed during sintering and particle size and shape distributions are relatively not essential for its applications.

The second approach is known as 'bottom-up' and implies assembling single atoms and molecules into larger nanostructures. In this way, we have a better chance to obtain particles with less defects, more homogeneous chemical composition and narrower size and shape distribution. Nanoparticulated colloidal dispersions or quantum dots (QDs) are good examples of this synthesis approach, which is routinely used in the nanoparticle fabrication

Frontiers of Nanoscience, Vol. 4. DOI: 10.1016/B978-0-12-415769-9.00002-9

and processing on a large scale. In this chapter, we focus on the 'bottom-up' approach, which is basically the one and only approach used to produce nanoparticles for biotechnological applications and offers a better chance to obtain uniform nanoparticulated systems.

The ability to control the nanoparticle size and size distribution requires the knowledge of the mechanism of particle formation and growth, which could be very different depending on the synthesis method.[1] Strategies for the synthesis of uniform inorganic nanocrystals will be the subject addressed in the first part of this section. Then, we describe in detail different methods for obtaining inorganic nanoparticles, in particular, magnetic nanoparticles, hybrids and multicomponent systems.

1. GENERAL STRATEGIES FOR THE SYNTHESIS OF UNIFORM INORGANIC NANOPARTICLES

In this context, the word uniform (also known as monodisperse, homodisperse or isodisperse) refers to particles with the same or nearly the same characteristics in terms of composition, size and shape. Such uniformity is of paramount importance because in many cases the properties of the nanoparticulated system depend directly on its homogeneity. For instance, the sizes of silver halide colloids for photographic emulsions need to be monodisperse to present photographic properties.[2]

Strictly speaking, a system can be considered monodisperse when its standard deviation of the size distribution (σ) is lower than 5%,[3] or 20% in a more general way. For many practical situations, as in paints or ceramic processing, a slight polydispersity is not so critical. However, for the study of theories or basic principles, highly monodisperse systems are needed. In the case of the biomedical applications, uniformity is essential. For instance, for the use of intravenously administrated contrast agents for MRI, the volume injected into the patient must contain a specific amount of particles which must have the same size and composition to produce the same response and signal. Thus, uniformity is one of the main goals to achieve when synthesizing nanoparticles for biotechnological applications.

1.1. Mechanism of Nanoparticle Formation

The basic model proposed to explain the precipitation from solution is based on the classical nucleation theory published by LaMer and Dinegar in 1950.[4] The so-called LaMer diagram illustrates the variation of the solute concentration as a function of time (Figure 2.1). LaMer and Dinegar theory is based on nucleation–growth mechanisms and considers nucleation as the limiting step in the process. For practical reasons, the formation process and the resulting LaMer diagram are divided into three stages (Figure 2.1).

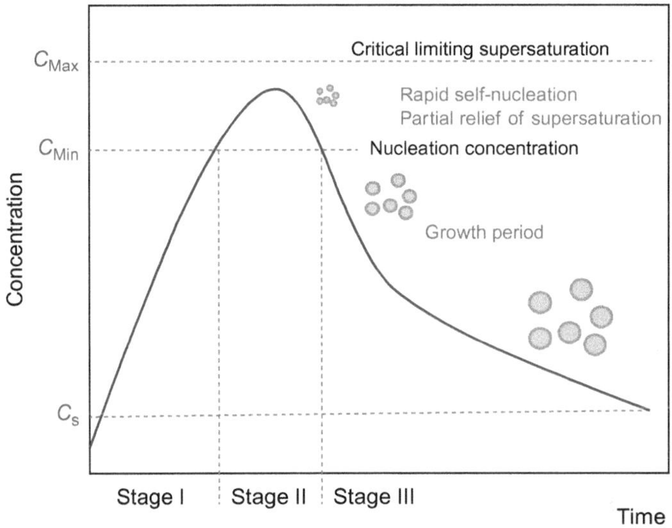

FIGURE 2.1 Schematic representation of the concentration of molecular dissolved sulphur before and after nucleation as a function of time.[4] (For colour version of this figure, the reader is referred to the online version of this chapter.)

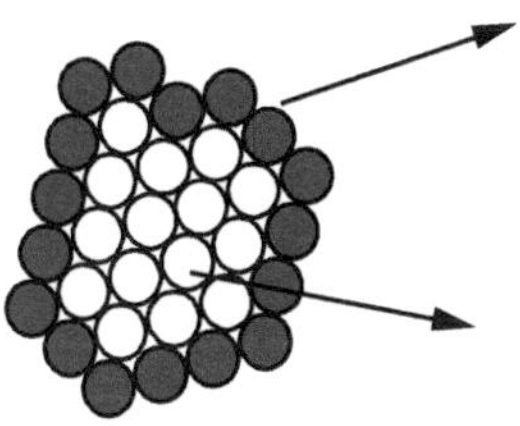

$\Delta G_{surface} = \gamma \beta N^{2/3}$ (2.2)

where γ is the interfacial tension between the nucleus and the solution, β is the geometric factor (a constant depending on the shape of the nucleus) and N is the number of molecules.

$\Delta G_{bulk} = -Nk_B T \ln S$ (2.3)

where N is the number of molecules, k_B is the Boltzmann constant, T is the temperature and S is the supersaturation ratio.

FIGURE 2.2 Stability of a critical nucleus depends on the height of the free-energy barrier (ΔG). The overall free energy is the sum of two terms due to the bulk and the surface of the nucleus.[5] (For colour version of this figure, the reader is referred to the online version of this chapter.)

During stage I, the concentration of salts increases until the nucleation concentration (C_{MIN}, Figure 2.1) is reached. On reaching this concentration, the system becomes heterogeneous, and the solute molecules combine to produce embryos or nuclei. These embryos are the result of collisions of ions or molecules in the solution in a process known as self-nucleation. Nuclei are continuously being formed and dissolved. The stability of the critical nuclei depends on the height of the free-energy barrier (ΔG): if it is high compared with the thermal energy (kT), the chance of forming such nuclei is negligible and no particles will be formed. However, if the free-energy barrier is low enough, the rate of nucleation tends to be infinite. The overall free energy is the sum of two terms due to the bulk and the surface of the nucleus (Figure 2.2).[5]

$$\Delta G = \Delta G_{\text{surface}} + \Delta G_{\text{bulk}} \tag{2.1}$$

If the critical concentration is not achieved, a spongy phase with a characteristic wavelength is formed instead of particles.[6] This is known as spinodal decomposition.[7]

In stage II, the appearance of nuclei partially reduces the supersaturation, at least rapidly enough that the rate of nucleation falls almost immediately to zero. Nucleation will continue as the concentration falls from the critical limiting supersaturation (C_{MAX}, Figure 2.1) to C_{MIN}, where particle growth replaces particle nucleation. Consequently, by controlling the initial concentrations of precursors, the period in which nucleation can occur is made so short that a monodisperse colloid results from the subsequent uniform growth of the existing nuclei. More nuclei formed results in smaller particles because a given amount of mass is distributed among a larger number of centres.

If the concentration decreases but the system is still supersaturated (stage III), the growth of stable nuclei to form discrete particles proceeds by diffusion of species in solution to the nuclei. Only when the nucleus exceeds a certain size, ΔG_{bulk} (Equations (2.1 and 2.3, Figure 2.2) predominates and the nucleus grows by a decrease of the free energy. The final size of the particles and the grow rate will depend upon three variables: the number of nuclei that are grown, the total amount of precursor and its diffusion coefficient in the medium, which depends inversely on the viscosity and directly on the temperature.

This process is, however, not as simple as this brief description implies, and uniform colloidal dispersions are obtained only if the conditions are properly controlled. In fact, LaMer's mechanism is strictly true only for the system for which it was developed, sulphur sols. However, regardless of the rigorous validity of LaMer's prediction, it provides an overall idea of the chemical mechanism behind the process. Moreover, the LaMer–Dinegar idea of separating the nucleation and growth stages in time has been widely used to produce monodisperse particles.

To better understand the real mechanisms of nanoparticle formation, we consider now the different types of nucleation and growth processes, both of them critical to produce uniform nanoparticles useful for biotechnological applications.

1.1.1. Nucleation Mechanisms

As described before, the nucleation rate plays an important role in controlling the uniformity of the resulting nanoparticles. However, this step is the most poorly understood due to the limitations in detecting the embryos or nuclei in an accurate and effective way, because most of the time nucleation is so fast that nuclei are just detected after some growth. Basically, the nucleation process may be divided into three different categories: primary homogeneous nucleation, primary heterogeneous nucleation and secondary nucleation.

The primary homogeneous nucleation is the ideal scenario described by LaMer and Dinegar where molecular solutes are completely dissolved in a liquid in the absence of any solid interface. The primary heterogeneous nucleation is induced by solid surfaces (i.e. seeds) present in the system. These foreign surfaces may have a lower surface energy than that of a new particle, in such a way that the nucleation takes place at a lower critical concentration. In this case, the interfacial tension of the solid/liquid interface (γ) from Equation 2.2 is replaced by the interfacial tension of the precipitated solid/foreign interface. Since primary homogeneous and primary heterogeneous nucleation processes are not likely to happen at the same time (because C_{MIN} of heterogeneous nucleation is larger than C_{MIN} of homogeneous nucleation), the total nucleation rate will be the combination of these two nucleation rates.

The secondary nucleation results from the presence of solute particles in the solution. This nucleation mechanism only takes place when the level of supersaturation is higher than the critical level for the solute particles present in solution. Secondary nucleation is not negligible; in most of the cases, a growing particle may contact the walls of the container, the stirrer or other particles and leave behind residual solute particles. Secondary nucleation may lead to polydisperse systems and should be avoided when uniform nanoparticulated systems are aimed.[8]

1.1.2. Growth Mechanisms

A deeper analysis of the crystal growth by diffusion of the solute molecules to the nuclei surface allows the separation and identification of the different growing stages. Basically, the solute molecules must be transported from the bulk solution to the particle surface. Then, these molecules are adsorbed on the particle surface and diffuse and attach to the surface until they are integrated into the crystal. In this step, the coordination shell of solvent molecules or counter ions diffuse far from the particle surface. The final step is the liberation of heat of crystallization and its transport away from the particle.

In this context, the growth is mainly limited by the diffusion of the solute molecules to the particle surface which in short depends on the solute concentration. So basically, the particles simply stop growing when supersaturation has dropped enough. This is known as 'Diffusion Limited Growth' or Lifshitz–Slyozov–Wagner theory. On the other hand, in the 'Reaction-Limited Growth', the growth of the particles is considered to be limited by the reaction between the diffusing solute and the particle surface. Both theories are widely accepted. Thus, we will apply either one theory or the other, or even a mixture of both, depending on the specific system we are studying.[9]

A fortunate consequence of particle growth is that, in many cases, the size distribution is self-sharpening. Once the initial particles are formed, the size distribution may further change in time due to cluster–cluster aggregation and Ostwald ripening. In the cluster–cluster aggregation, a certain number of

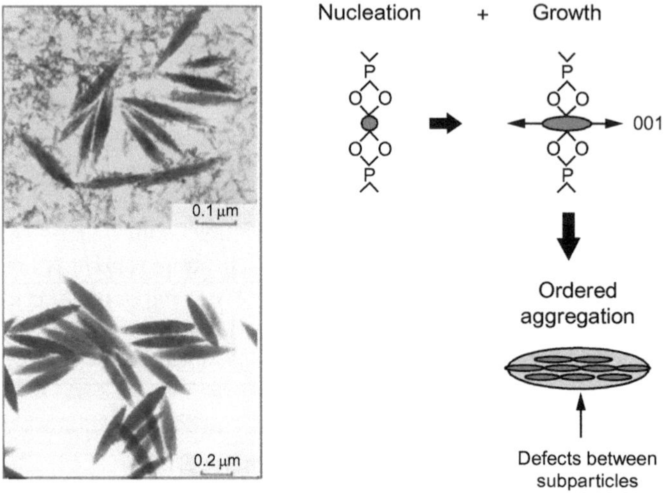

FIGURE 2.3 Schematic representation of the formation mechanism of uniform ellipsoidal nanoparticles by directional aggregation of smaller ellipsoids (right).[10,11] (For colour version of this figure, the reader is referred to the online version of this chapter.)

tiny nanoparticles aggregate and form larger and uniform particles. This aggregation may be at random (giving rise to spherical particles of amorphous or polycrystalline character) or directional (producing elongated, ellipsoidal, etc., particles, usually due to the adsorption of a capping agent),[10] as shown in Figure 2.3.

During the Ostwald ripening process, the smallest nanoparticles are dissolved and recrystallized onto the surface of larger particles. The smallest nanoparticles present higher surface to volume ratio. Surface atoms are less stable than bulk atoms because they have less number of neighbours. This implies that the smallest nanoparticles are more unstable than the larger ones. To minimize the overall energy, the molecules at the smallest nanoparticle surface detach from the particle and diffuse into the solution. These free molecules in solution tend to condense on the surface of the largest particles. After a certain time, all the small nanoparticles will be finally dissolved, so the particle population becomes monodisperse.

1.2. Strategies for Size and Shape Control

Despite the existence of long-time self-sharpening phenomena as Ostwald ripening, when no precautions are taken, precipitation from a supersaturated solution inevitably produces polydisperse colloids. Here, we outline some general strategies (as schematized in Figure 2.4) to tune the size and shape of inorganic nanoparticles by controlling several parameters concerning nucleation and growth steps.

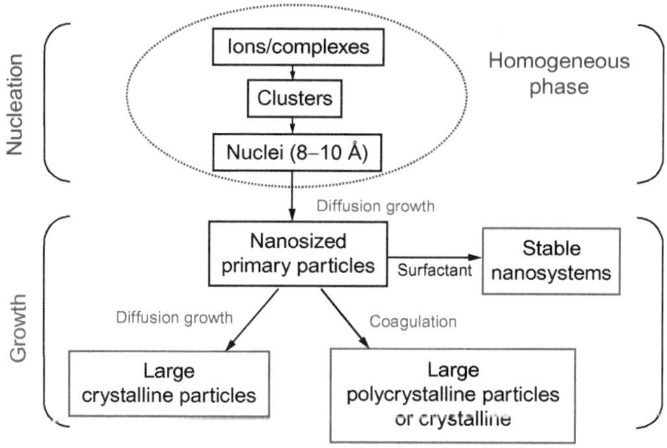

FIGURE 2.4 Different types of growth processes to produce uniform nanoparticles. (For colour version of this figure, the reader is referred to the online version of this chapter.)

1.2.1. Control at the Nucleation Step

As advanced by LaMer, to obtain monodisperse particulated systems, it is important to ensure that a very large number of nuclei are produced in a time as short as possible. These reactions are usually governed by parameters affecting thermodynamics and kinetics of the process, such as temperature, reaction time and the presence of surfactants. Some of these parameters may be tuned to control the particle size.

Temperature affects the diffusion coefficient of molecules from solution to the embryos or nuclei (it concerns the solubility). It also influences the overall free energy, so lower temperatures give rise to less and smaller nucleus. When it is necessary to mix two different precursors, small particle sizes may be obtained by short reaction times. In some cases, this is reached by a fast injection of one of the precursors in a diluted solution of the other. Sometimes longer reaction times favour particle growth. Finally, surfactants induce a reduction in the interfacial tension (γ) between the particle and the solution. The low interfacial tension cannot compete with the spontaneous bulk formation driving the precipitation, unless clusters are very small.[12]

1.2.2. Control at the Growth Step

Once we have produced homogeneous nuclei, the key condition to adjust the size and shape of the final particles is the control of the growth conditions. Monodisperse colloidal nanoparticles of a wide range of materials can be easily synthesized by means of several approaches, all of them based on the control of the growth conditions.

Capping agents have been used to limit the particle growth by adsorbing them on the surface of the particle. For example, the coprecipitation of ferrous and ferric salts in basic aqueous medium in presence of polyvinyl alcohol (PVA) limits the size of the final iron oxide nanoparticle to a few nanometres.[13,14] If the capping agent is large enough, as in the case of surfactants, it may provide steric hindrance and prevent aggregation of the particles.[15] Facet selective binding agents and surfactants may selectively adsorb to specific crystal planes, stabilizing that facet. In this way, the growth is inhibited in the crystal plane where the interaction is stronger and favoured in the rest producing anisotropic particles as wires, rods or needles.[16]

In a heterogeneous nucleation process where seeds are used, the nucleation and growth steps are definitely separated. The seeds are added to a solution with a subcritical concentration, preventing new nucleation phenomena. Thus, it enables a wider range of growth conditions (i.e. different temperatures or solvents) to control the final size and shape of the particles. The quality of the initial seed is also critical because seeds with crystal defects usually result in polycrystalline and polydisperse nanoparticles.[17] In some cases, it is possible to produce anisotropic inorganic particles by taking advantage of some of the material properties. For instance, performing the synthesis of magnetic materials in the presence of a magnetic field or producing some metal nanowires under an electric field.[18,19]

1.2.3. Nanoreactors

The term 'nanoreactor' is generally applied to nanometric physical spaces where chemical reactions take place. In this case, the reaction is the nucleation and growth of inorganic nanoparticles. The small space inside the reactor limits the particle growth, so it enables to control the particle size and shape. The concept of 'nanoreactor' may be applied to inorganic templates, for instance, to the pores of materials such as activated carbon or porous silica.[20] Once the particles are formed, the templates are removed by calcination, dissolution or any other chemical process. Micelles have been commonly used as 'nanoreactors'. The control of the particle size is possible by controlling the micelle diameter which, in short, depends on the solvents and the type of surfactant. Moreover, with the proper surfactant, it is possible to have a certain degree of shape control.[21]

1.3. Synthesis Methods

In the past decades, inorganic nanoparticles have been successfully prepared by many different synthetic routes. All of them are able to produce nanoparticles, but there are important differences in terms of uniformity, size, yield or crystallinity. In this section, we briefly describe some of these methods of synthesis as well as their strengths and weaknesses.

1.3.1. Precipitation of Salts in Aqueous Medium

This is the most widely used method to obtain magnetic nanoparticles in aqueous media, and it is routinely used to produce iron oxide nanoparticles for biomedical applications. Two main strategies based on this synthetic approach have been reported. In the first case, a ferrous hydroxide suspension is oxidized by an oxidizing agent, that is, H_2O_2 or KNO_3.[22–24] The second approach is based on the coprecipitation of stoichiometric mixtures of ferrous and ferric salts in basic medium.[25–27] Particles from 2 to 20 nm can be prepared by this synthesis route where the control of the experimental conditions is critical to tune the particle size and polydispersity. Through the precipitation of salts in aqueous medium, it is possible to synthesize large amounts of magnetic nanoparticles in a simple and non-expensive way, although other synthesis routes produce particles with better particle crystallinity and size control.

1.3.2. Hydrothermal Synthesis

Under hydrothermal conditions, a broad range of nanoparticulated materials can be formed. Hydrothermal process uses water as solvent and takes advantage from pressure and temperature to increase the solubility of the precursors and speed up the reaction time. The hydrothermal synthesis can be carried out above or below the supercritical point of water. A suitable reactor or sealed reaction vessel is needed to control the temperature and the pressure during the reaction. This synthesis route is able to produce monodisperse particles with good crystallinity; however, in some cases, it may be hardly scalable.

1.3.3. Microemulsions

A microemulsion is a thermodynamically stable isotropic dispersion of two immiscible liquids, where the microdomains of either or both liquids are stabilized by an interfacial film of surfactant molecules.[28] In water-in-oil (W/O) microemulsions, the aqueous phase is dispersed as microdroplets (typically 1–50 nm in diameter) surrounded by a monolayer of surfactant molecules in the continuous hydrophobic phase. The size of the droplet is determined by the water/surfactant ratio. When we mix two W/O microemulsions containing the desired reactants, the microdroplets continuously collide, coalesce and break again until a precipitate is formed in the micelles. The droplets are then nanoreactors for the formation of particles. Well-crystallized nanoparticles can be obtained by this method, but the versatility in terms of size, shape and nature of the nanoparticles is relatively limited. Further, the yield of the process is low compared to other methods, and although the scaling up is possible, large amounts of solvent would be needed to synthesize appreciable amounts of material, making this process expensive and non-environmentally friendly.

1.3.4. Polyol Process

Magnetic nanoparticles can be obtained by the high temperature reduction of dissolved metallic salts in a polyol, which works as solvent, surfactant and reducing agent (by the weakly reducing ability of hydroxyl groups). The average size can be tuned by controlling the kinetics of the precipitation or by adding seeds to promote heterogeneous nucleation. This process was firstly developed to produce noble metal particles[29] and was later extended to produce magnetic nanoparticles.

1.3.5. Decomposition in Organic Media

Highly monodisperse magnetic nanoparticles with good crystallinity can be synthesized through the thermal (high temperature) decomposition of metallic compounds in organic solvents with high boiling points and containing stabilizing surfactants. The mean particle size can be shifted by controlling factors as the reaction temperature, concentration and nature of the precursors, or the addition of seeds. The metal precursor can be added to a preheated mixture of the other reagents in an approach known as 'hot injection'. If the metal precursor is stable below the reaction temperature, then it can be heated together with the other reagents (heating up). The main advantage of this method is the extremely good particle crystallinity and the control of the size distribution. However, the major drawback for biomedical applications is that nanoparticles are usually hydrophobic, so it is necessary to stabilize the particles in aqueous medium. Some of the strategies to disperse these nanoparticles in aqueous solution are to use of more hydrophilic molecules, as 2-pyrrolidone, and modification or substitution of the surfactant coating.[30–32]

1.3.6. Aerosol Pyrolysis

Aerosol pyrolysis methods, as the spray or laser pyrolysis, are a powerful technique to synthesize pure and non-expensive magnetic nanoparticles at commercial scales of kg/h. In the spray pyrolysis method, a solution of salts is sprayed into a series of reactors where the aerosol droplets undergo evaporation of the solvent and solute concentration within the droplet. Particles that precipitate in this way are further dried and transformed at high temperature. The small nanoparticles synthesized by spray pyrolysis (between 2 and 10 nm in size) are agglomerated and give rise to microporous particles (up to 300 nm) that can be sintered to form dense particles.

In the laser pyrolysis method, a laser heats a dilute mixture of vaporized precursors to decompose them and initiate particle nucleation in a reaction zone. The final size of the particles depends on the aggregation of the initial nuclei (Figure 2.5).

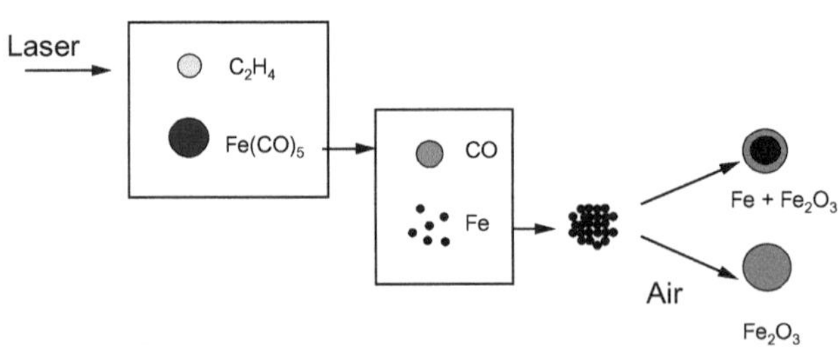

FIGURE 2.5 Mechanism of nanoparticle formation by laser pyrolysis. (For colour version of this figure, the reader is referred to the online version of this chapter.)

These particles are further transported to a filter by an inert gas, and depending on the amount of air, iron oxide or core/shell metal/iron oxide nanoparticles are obtained. As the nucleation is very fast and the residence time in the reaction chamber is very short, particles prepared by this technique are usually smaller (under 10 nm) and less crystalline than those prepared by other synthetic routes.

Advantages and disadvantages of the different synthesis methods are briefly summarized in Table 2.1. In terms of price and simplicity of the synthesis, coprecipitation is the preferred route because there is no need of special devices, and cheap precursors and solvents are used (Figure 2.6A).[26,33] In terms of size and morphology control of the nanoparticles, thermal decomposition seems to be the best method developed to date (Figure 2.6B).[34,35] Besides, the gas phase routes, as the spray and laser pyrolysis, allow scalability for mass production but they require sophisticated and expensive equipment.[14,36,37] Inorganic nanoparticles have been obtained also by hydrothermal treatments[38] or electrochemically.[39,40] Thus, the selection of the method of synthesis will be determined as a function of the special features required for the ulterior use.

2. MAGNETIC NANOPARTICLES

Here, we describe methods of synthesis and characterization of metal and oxide magnetic nanoparticles for biotechnology applications with tunable shapes and controlled sizes. It should be noted that there are a lot of reviews dealing with this topic.[3,41–43] Many of these works are focused on iron-based nanoparticles, because of its low toxicity and the existence of natural routes for its biodegradation. Further ongoing work has profited other magnetic nanoparticles with better magnetic properties to improve detection and performance. Metallic nanoparticles have higher magnetization values than metal oxide nanoparticles, although the formers have problems of toxicity due to the high reactivity of zero-valent metals. In general, the development of large-scale methods for the production of uniform magnetic nanoparticles remains a challenge.

Another area of considerable interest is the coating process, particularly in biotechnological applications such as targeting and drug delivery. Non-toxic coatings are able to transform magnetic nanoparticles into smart materials.[44,45] This important step will be addressed for each synthesis method. In general, the bonding between particle and coating molecules can be done through electrostatic interaction with surface charges or through coordination to the surface if the molecules contain an appropriate group (Figure 2.7).

Finally, we describe the synthesis of inorganic hybrid nanostructures, which integrate multiple inorganic nanocrystals of different materials in a nanoobject.[46–48] The challenge in this direction is the control of the size of the final object at the nanoscale, in order to preserve a high surface to volume ratio for specific functionalization and the attractive properties of the individual components, for example, magnetism and fluorescence.

TABLE 2.1 Summary Comparison of the Synthetic Methods.

Synthesis Method	Reaction Time	Solvent	Surface-Capping Agent	Sizes	Size Distribution	Shape Control	Yield
Coprecipitation	Minutes	Water	No	2–15	Broad	Not good	Medium
Thermal decomposition	Hours–days	Organic compound	Yes	4–30	Very narrow	Very good	Medium
Polyol process	Hours	Polyglycol	Yes	5–150	Narrow–broad	Good	Medium
Microemulsion	Hours	Organic compound	Yes	5–50	Narrow	Good	Low
Spray pyrolysis	Seconds	Water and volatile solvents	No	2–10	Broad	Not good	High
Laser pyrolysis	Milliseconds	Gases	No	2–10	Very narrow	Good	High

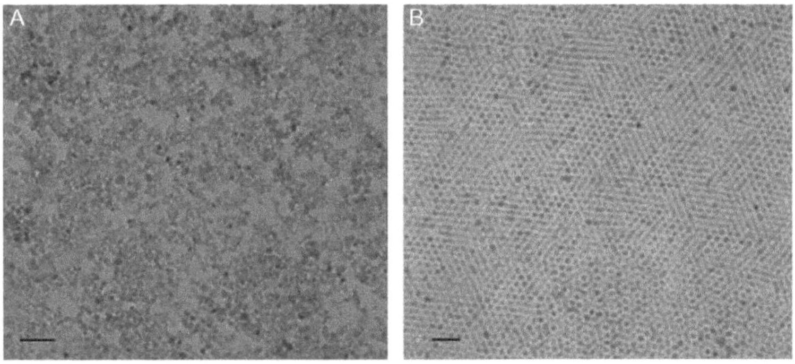

FIGURE 2.6 Magnetic nanoparticles prepared by (A) coprecipitation in the presence of amino dextran (scale bar is 150 nm) and (B) thermal decomposition of $Fe(acac)_3$ in octadecene in the presence of oleic acid (scale bar is 50 nm).

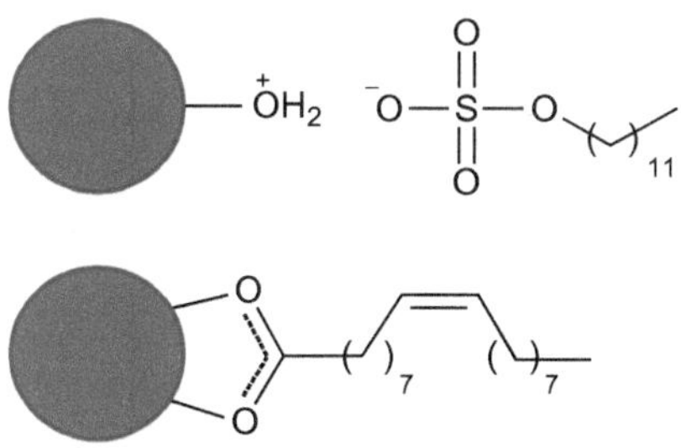

FIGURE 2.7 Coating magnetic nanoparticles via electrostatic interaction (sodium dodecyl sulphate) or via superficial complexation (oleic acid). (For colour version of this figure, the reader is referred to the online version of this chapter.)

2.1. Precipitation of Salts in Aqueous Medium

Iron oxide particles have been synthesized since 1852 by mixing an aqueous Fe^{2+}/Fe^{3+} salt solutions with a base.[49,50] Since then, the synthesis process has been widely studied and improved,[25,51–53] and nowadays, it is used to prepare most of the commercially available iron oxide nanoparticles. Commercial Fe_2O_3 particles mainly coated by dextran (Endorem®, Sinerem®) or carboxydextran (Resovisr®, Supravist®) are currently and routinely used as contrast agents for MRI. Silicon is also used to produce contrast agents for MRI (Lumirem®) and many other biocompatible coatings are subject of research (citrate,[54] polyethylene glycol (PEG),[55,56] starch, etc.).

The advantages of this method have been overwhelmingly demonstrated. For iron oxide nanoparticles, the precipitation or coprecipitation of cations in aqueous solution is an easy and cheap route of synthesis. Particles between 5 and 15 nm can be prepared at stoichiometry $Fe^{2+}/Fe^{3+} = 0.5$.[26] Larger particles (above 20 nm) can be prepared by the controlled oxidation of ferrous hydroxide suspensions.[22,24,53] Experimental parameters as changes in the pH, stirring rate, temperature or ionic strength have been used to control the particle size (Figure 2.8),[27,57] obtaining relatively narrow particle size distributions. The crystallinity of the samples can be also improved by dissolution–recrystallization mechanisms.[58] Other

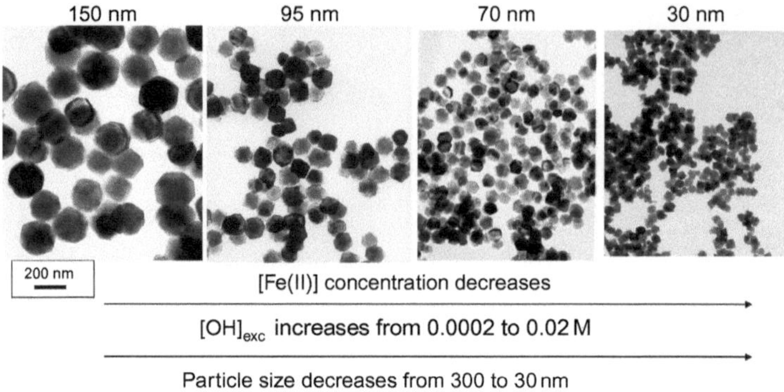

FIGURE 2.8 Nanoparticles larger than 20 nm prepared by controlled oxidation of ferrous hydroxide suspensions.[24]

strategies are based on the precipitation of iron salts in the presence of a polymer.[59] In this case, the polymer not only controls the particle growth and reduces the size but also provides colloidal stability. However, it is necessary to take into consideration the nature of such polymers because they often induce surface effects that strongly affect the magnetic behaviour of the particles or lead to changes in the toxicological profile of the samples.

Other nanoparticulated magnetic materials apart from iron oxide may be prepared by this synthesis route as cobalt ferrite,[60–62] manganese ferrite,[63,64] magnesium ferrite,[65] nickel ferrite,[66,67] cobalt oxide,[68] manganese oxide,[69] etc.

Coating of the magnetic nanoparticles can be achieved during or after the synthesis and, in general, improves the dispersion and increases further functionality, which opens up the possibility of creating multimodal and multifunctional nanostructures. Polysaccharides such as agarose, alginate, chitosan, dextran, heparin and starch are some of the most commonly used biopolymers for coating magnetic nanoparticles.[15,70] Dimercaptosuccinic acid (DMSA) and silica have also been used to stabilize particles prepared by coprecipitation at pH 7.[71,72]

2.2. Hydrothermal

Hydrothermal synthesis method has been recently proposed as a promising route for the synthesis of either hydrophilic[73] or hydrophobic magnetic nanoparticles[74] with controlled shape and size.

Hydrophilic magnetite nanoparticles have been synthesized by the so-called citric acid-assisted hydrothermal route. It uses ammonium iron citrate as a precursor in the presence of $N_2H_4{\cdot}H_2O$ and leads to isolated particles of ~4 nm.[75] Larger particles of around 20 nm, aggregated in spheres of around 250 nm, were obtained in the presence of urea. Urea decomposes into NH_3 and CO_2 and provides an alkaline environment to the system.[76] Citric acid is responsible for reducing Fe^{3+} to Fe^{2+}, leading to the formation of magnetite. The extent of the reduction depends on the citric acid amount. Therefore, low amounts of citric acid lead to the formation of hematite, while excess of citric inhibits the formation of magnetite. Larger hydrophilic magnetite particles with high crystallinity and narrow size distribution have been prepared by hydrothermal treatment of nanoparticles previously prepared by coprecipitation.[77]

Hydrothermal treatment of organic precursors in the presence of fatty acids and, using a mixture of water and ethanol as solvents, has been shown to lead to high-quality hydrophobic particles of various types (noble metals, dielectric, semiconductor and luminescent nanoparticles),[74,78–80] including magnetic ferrite particles.[79,81–83] In the case of ferrite nanoparticles, a mixture of water and ethanol is used to dissolve a mixture of sodium oleate and NaOH. Then, a solution of metal nitrates which serves as precursor is added and kept at temperatures between 180 and 200 °C in an autoclave, for a time that was varied between 10 and 15 h. Metal oleates are hydrolyzed, resulting in hydrophobic nanoparticles capped with oleic acid. The size of the particles is always smaller than 10 nm. These particles cannot be immediately used in biological applications because they are only soluble in organic solvents.

In general, this method allows the synthesis of nanostructures requiring relatively high temperatures for its formation, and therefore, the synthesis of small nanoparticles is almost impossible, that is, the case of barium hexaferrite powders prepared at temperatures around 900 °C. Nanoparticles less than 10 nm wide have been obtained at 150 °C by using the hydrothermal method.[84,85]

2.3. Microemulsions

The quest for narrow size distributions led to the use of microemulsions as nanoreactors. The synthesis of nanoparticles in W/O microemulsions, which can be seen as reverse micelles, typically consists of microdroplets of metal salt solution in water, stabilized by a surfactant and dispersed in a non-polar solvent. The cavities formed into these systems have diameters in the sub-micrometre scale, so they appear to be a suitable method to prepare nanoparticles. The most common surfactants are sodium bis(2-ethylhexyl)sulphosuccinate (AOT), sodium dodecyl sulphate (SDS) and cetyltrimethylammonium bromide (CTAB).

The first example of magnetic iron oxide nanoparticles prepared in micelles consisted in the air oxidation of an iron(II) sulphate in an organic solvent, with AOT as surfactant.[86] Magnetite nanoparticles were always very small in size (3–5 nm), independent of the addition of other surfactants such as hexa(oxyethylene) nonylphenyl ether or the use other Fe salts such as $FeCl_2$ and $FeCl_3$ and NH_4OH as the base.[87,88] Larger particles of γ-Fe_2O_3 and Fe_3O_4 (4–12 nm) with very narrow size distributions were lately obtained by using SDS,[89] also cobalt ferrite nanoparticles from 2 to 15 nm at room temperature.[90–95] More recently, it has been demonstrated that the large-scale production of small MFe_2O_4 ferrite nanoparticles (sizes: 2–10 nm; M = Fe, Mn, Co, Ni, Zn), with very narrow size distributions ($\sigma \leq 10\%$), is possible in microemulsion. Solutions of the metal in ethanol were added to a microemulsion of sodium dodecylbenzene sulphate as surfactant and xylene as oil phase.[96] Direct synthesis of polymer-coated Fe_3O_4 nanoparticles (80 nm) has been achieved by simultaneous polymerization of methacrylate.[97]

Metallic nanoparticles can also be prepared by the microemulsion technique. Thus, Co and CoPt alloy nanoparticles have been prepared in reverse micelles of CTAB as the surfactant, 1-butanol as co-surfactant and octane as the organic (oil) phase.[98] A similar approach has been used to prepare Fe nanoparticles passivated with a thin layer of gold, with CTAB as the surfactant and octane as the organic phase.[99,100] The Au layer protects metallic iron against oxidation.

2.4. Polyol Process

The polyol process is the synthesis of metal-containing compounds in polyalcohols by reduction of a metal precursor. The polyalcohols used in this methodology are glycols (1,2-diols) or polyglycols. The term 'polyol' is used in a broad sense covering a wide range of compounds: ethylene glycol, propylene glycol, 1,2,3-propanetriol and PEG derivatives (which, in fact, are poly-ethers with different functional groups at the end of the polymer chain). They usually have high dielectric constants and boiling points, so metal precursors can be total or partially dissolved and can undergo reaction in a wide range of temperatures (from room temperature to boiling point, usually 100–200 °C). The polyol plays the role of solvent and reducing agent at the same time. It can also be the stabilizer, preventing particle aggregation, or contribute to the stabilization together with other compounds (e.g. polyvinylpyrrolidone, PVP). The reaction is usually carried out in the presence of an alkali, such as NaOH, that assists the reduction. This method allows the synthesis of non-agglomerated nanoparticles with well-controlled sizes and shapes. In general, size distribution of nanoparticles obtained by this method is good and in some cases was possible to achieve $\sigma < 10\%$. It should be emphasized that the term 'polyol process' is sometimes used in a wrong way, that is, when a long-chain diol (e.g. 1,2-hexadecanediol or 1,2-octadecanediol) is added in the synthesis of nanoparticles by thermal decomposition in other organic solvents.

The polyol process was first reported by Fiévet *et al.* for the synthesis of metallic cobalt, copper and nickel particles, in the range of micro- and submicrometres, by reduction of the corresponding oxides, hydroxides or acetates in refluxing ethylene glycol.[29] Smaller particles were obtained at higher temperatures or by heterogeneous nucleation with Pd or Ag seeds. The procedure was extended to obtain metallic and metal alloy nanoparticles with other compositions such as Fe, Ni, FePt, FeCo, CoNi.[101–104] If PVP is added to the reaction mixture, the so-obtained particles are more stable against oxidation without greatly affecting their magnetic properties.[105,106] The synthesis of Fe(0) nanoparticles by the polyol process has been described with very good control of the size (in the range of 10–150 nm) and shape (spherical or cubic) using H_2PtCl_6 as nucleating agent.[107] Because of the observation of both metallic and iron(III) oxide in Fe nanoparticle synthesis, it was initially thought that metallic nanoparticles were originated by disproportionation of the precursor.[101] However, it was later demonstrated that the role of the polyol is to reduce the iron precursor[108] and that the observation of iron oxide was due to partial oxidation of the particles.

Iron oxide nanoparticles which are directly dispersible in water have been synthesized in ethylene glycol, diethylene glycol, triethylene glycol and tetraethylene glycol.[109] However, only in the case of triethylene glycol, the nanoparticles were not agglomerated. Interestingly, long-chain polyols like PEG derivatives lead to larger particles.[110] This method allows direct functionalization with PVP or amines by simply adding the corresponding PVP or amine to the reaction medium.[111,112]

Great control of the morphology is also possible and Fe_3O_4 nanoparticles with different shapes (cube, truncated octahedron, octahedron, sphere, truncated cube and equilateral octahedron) have been obtained changing the concentration of the alkali.[113] Other ferrites can be synthesized through the polyol process, as it is the case of $CoFe_2O_4$ nanoparticles (5.5 nm) with very narrow size distribution ($\sigma < 10\%$) obtained in relatively mild conditions (160 °C)[114] and non-magnetic iron oxides such as hematite with sizes between 80 and 100 nm in diethylene glycol with a small volume of water.[115]

2.5. Decomposition in Organic Media

Thermal decomposition of metal precursors in organic solvents gives nanoparticles with the narrowest size distributions and high degree of crystallinity. However, recent works show the lack of homogeneity in chemical composition under certain condition or the polycrystalline character of larger particles obtaining by a seeding process, which affect the magnetic response.[116–118]

In general, this method involves the use of a metal precursor (usually, a coordination complex or an inorganic salt), one or more stabilizers (typically, long-chain carboxylic acids or carboxylates, amines, phosphines or phosphine oxides) and a reducing or oxidizing agent (if necessary). Decomposition of the precursors needs elevated temperatures, so the reaction is carried out in solvents with high boiling point (e.g. phenyl ether, benzyl ether, 1-octadecene, trioctylamine, etc.). Several parameters must be adjusted to control the size and morphology of the particles: stirring, temperature ramp, boiling point and nature of the solvent, concentration and nature of the reagents (metal precursor, reducing agent and stabilizers) and ratio between metal precursors and stabilizers. For the synthesis of Fe_3O_4 and metallic nanoparticles, it is necessary to perform the reaction under an inert atmosphere such as nitrogen or argon to avoid oxidation of the product.

2.5.1. Metallic Nanoparticles

Zero-valent metallic precursors, typically carbonyl complexes, are usually preferred for the synthesis of metallic nanoparticles. These are toxic and flammable compounds, and the alternative is the reduction of precursors in higher oxidation states.[119,120] Once the metal particles are formed, they are easily oxidized, so special attention must be paid to keep an inert atmosphere during the synthesis and to the subsequent surface protection.

Co and Fe nanoparticles were first prepared by thermal decomposition of dicobalt octacarbonyl and iron pentacarbonyl, respectively, in the presence of functional polymers.[121,122] The size of the particles was varied between 5 and 100 nm for Co,[121] and below 20 nm for Fe,[122,123] depending on the nature of the polymers and the metal:polymer ratio. Other stabilizers successfully employed to obtain Co and Fe nanoparticles with sizes below 20 nm and different shapes from the corresponding carbonyl complexes are fatty acids, long-chain amines, phosphines, phosphine oxides, etc.[124–128] The size of the particles can also be tuned by varying the size of the stabilizer, that is, larger particles can be obtained with larger alkyl chains.[119,129]

Much smaller particles (1.5–27 nm) were prepared by decomposition of an Fe(II) precursor $Fe[N(SiMe_3)_2]_2$ in relatively mild conditions (150 °C), under a dihydrogen reductive atmosphere and in the presence of long-chain surfactants.[120,130–132]

Alloy nanoparticles can be prepared by thermal decomposition of the corresponding precursors. FePt nanoparticles (3–10 nm) with very narrow size distribution ($\sigma < 5\%$) have been obtained by decomposition and reduction of $Pt(acac)_2$ with $Fe(CO)_5$, in the presence of oleic acid and oleylamine as stabilizers.[133] This procedure, developed by Sun *et al.*,[133] has been widely employed and adapted to obtain nanoparticle alloys with other compositions such as FePd, CoPt or $CoPt_3$.[134] Inspired in the polyol process, the synthesis employs 1,2-hexadecanediol as reducing agent. Some authors have confused this genuine thermal decomposition procedure with the polyol process, in which the polyol is the solvent and the reducing and the stabilization agent.

2.5.2. Metal Oxide Nanoparticles

It is possible to prepare metal oxide nanoparticles by thermal decomposition of zero-valent metallic compounds or with precursors in higher oxidation states. The first approach requires oxidation of the metal (in air or adding an oxidant), while the latter may require a reducing agent depending on the desired product.

Maghemite (γ-Fe_2O_3) and magnetite (Fe_3O_4) nanoparticles (<20 nm) have been obtained by decomposition of $Fe(CO)_5$, in the presence of fatty acids, phosphines and phosphine oxides, adding oxidants such as trimethylamine oxide or pyridine-*N*-oxide.[30,135] Strategies to modify the size of the nanoparticles included variations of the Fe:surfactant molar ratio or the nature of the surfactant.[30] Also a seed-growth procedure has been described to allow tuning of the size of the final nanoparticles between 6 and 15 nm, with very high precision.[136] This 'oleate approach' has been proven to be highly versatile allowing the preparation of other nanomaterials such as Cr, Mn, Co, Ni and Zn oxide nanoparticles,[137,138] or Co, Mn and Ni ferrites.[138,139] Modification of the procedure adding $Mn_2(CO)_{10}$ leads to $MnFe_2O_4$ ferrite nanocrystals (5–13 nm).[140]

The stabilizer can influence not only the size but also the shape of the particles, as reported for the decomposition of $Fe(CO)_5$ with dodecylamine (DDA) or trioctylphosphine oxide (TOPO) as capping ligands.[16] With DDA, a mixture of diamond-, triangle- and spherical-shaped 12 nm maghemite nanoparticles were obtained. Increasing the molar ratio of DDA, mainly hexagonal nanoparticles with mean diameters of 50 nm were observed. With TOPO instead of DDA, smaller nanoparticles (6 nm) were obtained. The difference was attributed to the higher affinity of TOPO for the nanocrystal surface, which prevents further growth or ripening processes.

Thermal decomposition of Fe(III) is more common than Fe(0) precursors for the synthesis of iron oxide nanoparticles. Several iron(III) salts such as $Fe(cup)_3$ (cup = cupferron, $NH_4[C_6H_5N(O)NO]$),[31] $Fe(acac)_3$ (acac = acetylacetonate),[32] FeO(OH)[141] or iron carboxylates[137,138,142–144] have been decomposed in solvents with high boiling point (e.g. phenyl ether, benzyl ether, 1-octadecene, trioctylamine, etc.) and in the presence of different stabilizers as fatty acids, long-chain amines and phosphines, phosphine oxides or polymers. Sizes of the obtained nanoparticles are usually below 20 nm, with some exceptions (Figure 2.9). The synthesis of magnetite (Fe_3O_4) nanoparticles (4 nm) was first reported by Sun and Zeng.[32] This procedure has been widely applied by several research groups and consists of the decomposition of $Fe(acac)_3$ (acac = acetylacetonate) in diphenylether (bp 265 °C), in the presence of oleic acid, oleylamine and 1,2-hexadecanediol. The diol facilitates the partial reduction of the Fe(III) complex, needed to obtain Fe_3O_4. The reaction can be carried out also without diol, with the oleylamine acting as reducing agent and stabilizer,[145] or even with the mixture of oleic acid and oleylamine itself acting as solvent.[146] This versatile procedure has been also adapted to synthesize other nanomaterials such as cobalt and manganese ferrites (MFe_2O_4; M = Co, Mn) with sizes ranging from 3 to 20 nm,[35,147] or ferrimagnetic magnetite nanocubes (30–100 nm).[46]

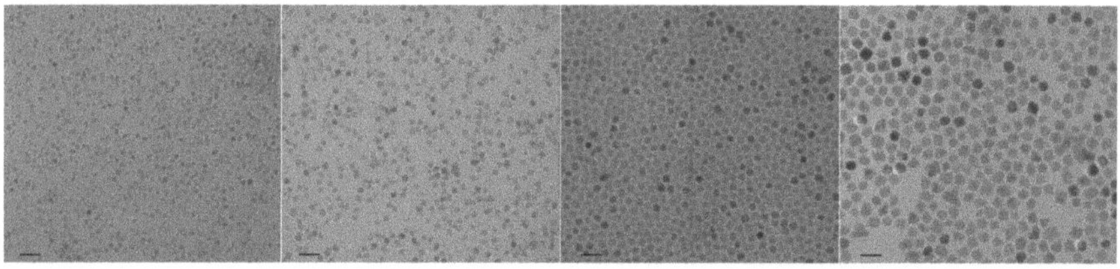

FIGURE 2.9 From left to right: Fe_3O_4 nanoparticles of 9, 12, 14 and 21 nm in size, synthesized by thermal decomposition of an Fe(III) precursor in 1-octadecene. Scale bars are 40 nm.

Besides seed-growth methods,[32] strategies to prepare larger nanoparticles include longer reaction times, changing surfactant:iron molar ratios or increasing the temperature of the reaction by using solvents with higher boiling points,[117,137,138,141,148,149] although it is worth mentioning that the solvent may not be innocent.

Based on UV–vis and IR spectroscopy data, some authors proposed that an iron oleate complex is generated prior to nanoparticle formation.[30,141] Researchers reasoned that a presynthesized and isolated iron oleate complex would be a good precursor for the controlled synthesis of magnetic nanoparticles. Hence, decomposition of Fe(oleate)$_3$ and Fe(oleate)$_2$, with excess of oleic acid, has been reported to yield magnetite nanoparticles with sizes in the range of 3–50 nm and very narrow size distributions.[137] In a similar way, the Fe(oleate)$_3$ alone has been shown to be an appropriate precursor for the synthesis of iron oxide nanoparticles in large scale.[138] Modifications of the experimental conditions, including using other iron carboxylate complexes, like iron palmitate and iron myristate, make it possible to obtain nanoparticles with different shapes: spheres, cubes, bipyramids, stars and whiskers.[142–144]

The very narrow size distributions observed are related to a good separation of the nucleation and growth processes.[34,117,144] Although the mechanism of the reaction is not clear yet, results suggest that it follows the following steps: (i) decomposition of iron oleate forming intermediate species (probably iron-oleate-oxide compounds), (ii) expeditious nucleation at high temperature and (iii) fast growth process in which the size distribution is rapidly narrowed. Formation of small amounts of CO, H_2 and carbon during decomposition has been suggested to be responsible for the partial reduction of Fe(III) to Fe(II).[34]

It is worth mentioning that magnetite nanoparticles prepared by this method may actually consist of a mixture of γ-Fe_2O_3 and Fe_3O_4, as evidenced by X-ray absorption spectroscopy and X-ray magnetic circular dichroism techniques,[136,138] with higher proportions of Fe_3O_4 in larger particles. Wüstite (FeO) has also been found as a minor component of the particles,[117] which may lower the saturation magnetization.

In addition to the reagents described above, other compounds have been used for the stabilization of iron oxide nanoparticles obtained by decomposition of Fe(III) precursors. For example, benzyl alcohol or 2-pyrrolidone was used as both the solvent and the stabilizer in the synthesis of magnetite nanoparticles.[148,150,151] Amine-coated Mn, Fe, Co and Ni oxide nanoparticles (<20 nm) can be obtained by decomposition of their acetylacetonate complexes in hexadecylamine.[152] Smaller particles (<5 nm) coated with PVP have been synthesized by thermal decomposition of Fe(acac)$_3$ in octylether with PVP and 1,2-hexadecanediol.[153] Finally, water-dispersible and biocompatible nanoparticles (<10 nm), directly coated with monocarboxyl-terminated poly (ethylene glycol) (PEG–COOH), are obtained if the reaction is carried on in the presence of PEG–COOH.[154]

2.5.3. Surface Modification

Although nanoparticles obtained by thermal decomposition in organic media have very good magnetic properties and size distribution, due to their hydrophobic nature, an efficient phase transfer to water process is required before its use in biotechnology. Moreover, zero-valent magnetic nanoparticles have a great reactivity, so they need to be protected. The surface modification will affect the hydrodynamic radius of the particles and their net surface charge (Figure 2.10). It can be used for further functionalization if the objective is to target the nanoparticles to specific sites in the human body or drug delivery.

The nature of the union between the nanoparticle core and the coating molecules can be either chemical or physical. Most of the developed methods follow three strategies: ligand exchange, ligand oxidation or encapsulation into a polymer. Ligand exchange methods using small molecules such as DMSA,[155,156] or amino modified PEG,[157] (also referred to as polyethylene oxide) are between the most promising ones.

Small molecules provide smaller hydrodynamic radii if compared with polymers. In addition to DMSA, other small molecules such as carboxylates, phosphates, phosphonates, sulphonates and silicon compounds are used for surface modification of magnetic nanoparticles for biomedical applications (Figure 2.10). Citrate is the most common carboxylate for stabilizing nanoparticles leading to negatively charged surfaces.[158–161] Silica is an inorganic material frequently used for biocompatible coatings. It has low toxicity and can be easily functionalized. Silica layer is often formed on hydrophobic particles after prefunctionalization with an amphiphilic molecule.[162,163] The oleic acid has also been removed from magnetic metal particles by TMAOH leading to water-stable suspensions.[164]

Some of the polymers used for modification of hydrophobic magnetic nanoparticles are described below. Dextran is a branched polysaccharide that has been extensively used for nanoparticle coating because of its long plasma

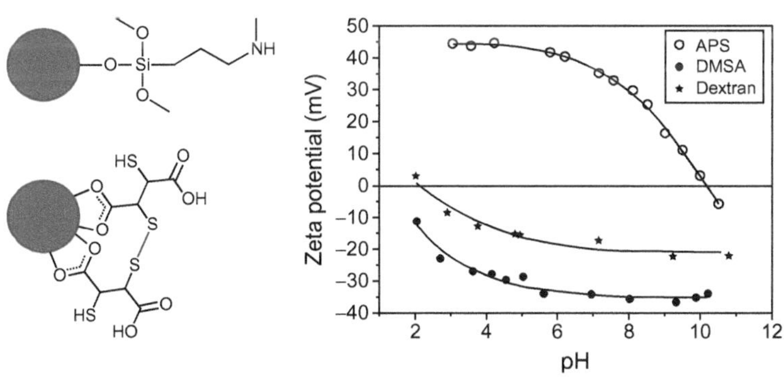

FIGURE 2.10 APS-coated magnetite, DMSA-coated magnetite and surface charge variation as a function of pH for those magnetite nanoparticles and a commercial sample coated with dextran. (For colour version of this figure, the reader is referred to the online version of this chapter.)

half-life. Dextran-coated iron oxide nanoparticles are the only iron oxide nanoparticles approved for clinical use as contrast agent for magnetic resonance.[165] PEG derivatives are very promising polymers for surface modification because of their biocompatibility, long plasma half-life and easy functionalization.[165] Polyethylene imine has attracted a lot of attention in the past years because of its ability to bind DNA, which makes it an efficient agent for transfection *in vitro*.[166,167] Its polycationic nature provides a net cationic surface charge to the particle, enhancing the cellular uptake of magnetic nanoparticles.[168] However, it has a short plasma half-life and research is undergoing to overcome this serious limitation.[169] Polypeptide dendrimers with protein-like structures, multivalence and excellent biocompatibility, degradability and low immunogenicity such as L-lysine- and L-glutamic acid-based have been used to transform hydrophobic magnetic nanoparticles to hydrophilic by ligand exchange, leading to aggregate sizes smaller than 30 nm.[170] In general, multidentate polymers having numerous functional groups (such as nitrogen, oxygen and sulphur atoms) capable of binding nanoparticles are preferred. Some other examples are PVAs, poly(methyl methacrylate)s, poly(vinyl pyrrolidone) and poly(lactic-*co*-glycolic) acid, which are approved by the FDA.

A suitable approach to obtain water-stable nanoparticles from the oleic coated ones is to oxidize the double bond of the ligand with permanganate (see Figure 2.11), which leads to nanoparticles with a dicarboxylic functionalization that provides water stability and very low hydrodynamic sizes (<50 nm).[171]

Finally, amphiphilic surfactants have been used to encapsulate oleic-coated nanoparticles. The hydrophobic chain interacts with the hydrophobic surface of the nanoparticle, while the hydrophilic groups of the surfactant are pointing out (to the water medium) and forming a bilayer. Surfactants employed can be long-chain carboxylic acids,[172] CTAB,[163] amphiphilic polymers,[173] etc.

2.6. Aerosol Pyrolysis

Magnetic nanoparticles can be synthesized by pyrolysis of an aerosol, generated by spraying a solution or by ultrasound. The ultrafine droplets forming the aerosol are carried by a gas (nitrogen, ethylene or ammonium) up to the reaction zone where the pyrolysis takes place. The residence time of the particles in the reaction zone and the concentration are controlled by the gas

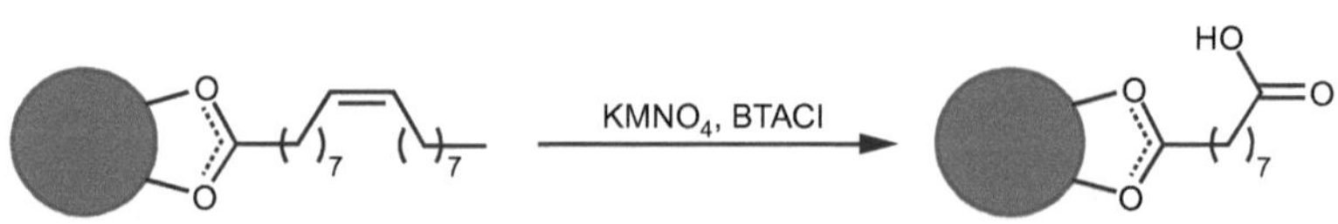

FIGURE 2.11 Oxidation of the oleate's double bond at the surface of the nanoparticle yields water-stable nanoparticles.[171] (For colour version of this figure, the reader is referred to the online version of this chapter.)

flow. The nature of the gas and the flow determine the nature and the size of the resulting particles. Two kinds of devices have been developed based on thermal pyrolysis, which are called spray pyrolysis and laser pyrolysis.

Spray pyrolysis is a powerful technique to synthesize highly pure and non-expensive iron oxide nanopowder. Iron salts are sprayed into a series of reactors where the aerosol droplets undergo evaporation of the solvent and solute concentration within the droplet. The particles precipitated in this way are further dried and transformed at a higher temperature. The small nanoparticles synthesized by spray pyrolysis are agglomerated and give rise to microporous particles that can be sintered to form dense particles. The diameter of the final particle will be usually similar to that of the initial droplet, so the main particle size is easily tunable. Initial nanoparticles with sizes between 2 and 10 nm can be isolated, depending on the nature and concentration of the precursor. However, in most of the cases, these small crystallites are aggregated forming large dense particles, microporous particles or hollow spheres with up to 300 nm in diameter (Figure 2.12).

Laser pyrolysis was pioneered and developed by Cannon and coworkers in 1982.[174] Since then, a hard engineering work has been done to turn this process into a viable manufacturing platform. Three characteristics of this method must be emphasized: the small particle size, the narrow particle size distribution and the near absence of aggregation. It has been successfully employed in synthesizing various nanoscale materials including iron oxide nanoparticles.[36,175,176] The nucleation is very fast and the carrier gases immediately push the particles out from the reaction chamber in such a way that the particles have no time to grow or properly crystallize. As a result of the short residence time in the reaction chamber, magnetic nanoparticles prepared by laser pyrolysis are usually smaller and less crystalline than those prepared by other synthetic routes. Laser pyrolysis is chosen as the most convenient synthesis process to produce very small (<5 nm) and homogeneous particles in a continuous way with low cost and with no use of surfactants or hazardous additives.

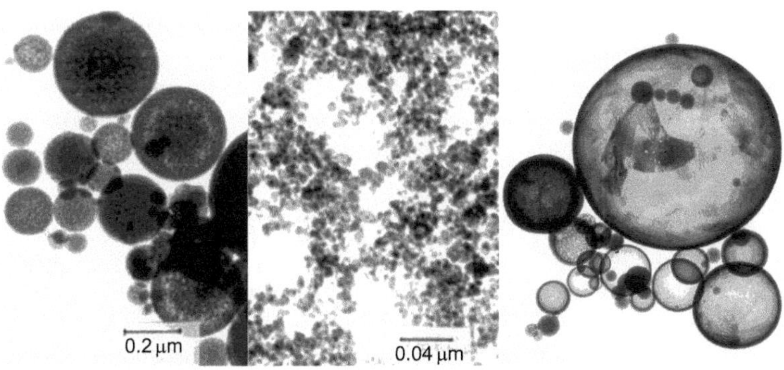

FIGURE 2.12 Magnetite nanoparticles prepared by spray pyrolysis from Fe(III) nitrate and acetylacetonate. (For colour version of this figure, the reader is referred to the online version of this chapter.)

Laser pyrolysis of a mixture of iron carbonyl and ethylene vapours, followed by controlled oxidation, has also been shown to be an excellent technique for the direct and continuous production of iron nanoparticles.[36,177,178] The composition of the core is found to change with the particle size from the α-Fe structure to a highly disordered Fe phase (probably containing C atoms in its structure).[179]

2.7. Others: Anisometric Particles

A great deal of progress has been made in the synthesis of nanoparticles with controllable morphology and properties, in particular, for one-dimensional structures such as nanorods and nanowires for semiconductors, metals and metal oxides.[47,180,181]

In the case of magnetic nanoparticles, diameter and length control can be achieved in organic solution by modifying the molar ratio of surfactants to metal precursors, the addition sequence of the surfactants and metal precursors, heating rate and reflux temperature. Thus, FePt nanoparticles with average diameters of 3 nm and lengths tunable from 15 to 150 nm can be obtained.[182] On the other hand, the appropriate choice of ligands and alkyl chain lengths enables the synthesis of ferromagnetic nanorods of monodisperse diameter and length from a non-carbonyl metal complex as precursor. This is the case of cobalt and nickel nanorods and nanowires and the control of their aspect ratio through the decomposition of metal complexes in the presence of a mixture of long-chain amines and oleic acid.[183,184] The exact mechanism controlling both the formation of the rods and their dimensions remain unclear.

Magnetite nanostructures such as nanowires,[185] nanorods,[186] nanotubes,[187] nanosheets[188] and nanocubes[189] and hollow spheres[190] have also been prepared. Superparamagnetic iron oxide nanorods have also been obtained by the controlled thermal reduction of goethite nanorods.[191] The resulting particles consist of clusters of maghemite (γ-Fe_2O_3) nanoparticles embedded in porous colloidal nanorods of slightly canted antiferromagnetic hematite (α-Fe_2O_3). Briefly, goethite (α-FeOOH) nanorods (80×20 nm) are produced by a carbonate route and subsequently coated an alumina external layer.[192] Alumina avoids interparticle sintering during the thermal reduction treatment needed to develop the maghemite phase. Finally, alumina is dissolved in basic media to leave the porous matrix accessible for aqueous infiltration.

Hollow structures of inorganic materials have aroused an extensive research interest owing to their promising biotechnological applications.[193] There are three general approaches for the synthesis of these structures: (a) using hard templates such as silica, polymer spheres and metal oxides[194] and (b) using soft templates such as microemulsion droplets, micelles visible or gas bubbles[195] and free template methods relying on the Ostwald ripening mechanism, Kirkendall effect and self-attachment.[196] Starting from nanoparticles having a metal core (either Au or AuPt) and an iron oxide shell, and upon treatment with iodine to leach

Au out of these structures, asymmetric core–shells evolve into 'nanocontainers', that is, iron oxide nanoparticles enclosing a cavity accessible through nanometre-sized pores.[197]

3. HYBRIDS

Multicomponent nanostructures or nanostructures composed of two or more different materials have recently exhibited great interest for biomedical applications.[46–48] The integration of multiple materials in one structure offers opportunities for enhanced physical and chemical properties and for targeting many functions within a single nanostructure.[198] However, stabilization of multiple materials within the nanostructure is not a simple issue. One way is to build brick-by-brick nanocomposite materials in a bottom-up approach. The first step is the stabilization of the colloids in the reaction medium to coat them by a controlled ultrathin layer of another material (silica, carbon or noble metals). Encapsulation is another type of multicomponent structure designed for inorganic materials to be incorporated in polymers, lipids and peptide or proteins. These vesicles can be coated with ligands for specific receptors, functional molecules such as antibodies or small molecules such as folic acid. The final structure of the resulting material could be core/shell type, with a gradual composition or no, depending on the synthesis method and the particle mechanism of growth and its properties (colloidal, magnetic, luminescent, etc.) will be strongly dependent on this nanostructure and the spatial distribution of the components.

3.1. Inorganic–Inorganic

3.1.1. SiO_2

Functionalization of triethoxysilane (TEOS) or aminopropylsilane (APS) onto the magnetite nanoparticles surface occurs via a hydrolysis and condensation mechanism in which the first step is the hydrolysis of the ethoxy groups in the Si precursor which leads to the formation of silanol (Si—OH) groups. Those groups then condense around the nanoparticle surface through siloxane (Si—O) bonds.[199] The hydrolysis of the silane molecules is very fast while their condensation reaction is slower, taking several hours. Modification of iron oxide particles prepared by precipitation can be carried out with TEOS in ethanol/water,[71] (Figure 2.13A) or with APS under mechanically stirring at room temperature for 3 days in DMSO.[200] Magnetic particles modified by these methods have been widely used for biotechnological applications.[201]

Hydrophobic particles have been first coated with an APS monolayer and then the subsequent condensation of TEOS on the nanoparticle surface in organic media (Figure 2.13B).[202] SiO_2 layer thickness can be controlled simply by varying the ratio of nanocrystals to the silica precursor. SiO_2 coating can also be performed in W/O microemulsions and by spray pyrolysis.[203–207]

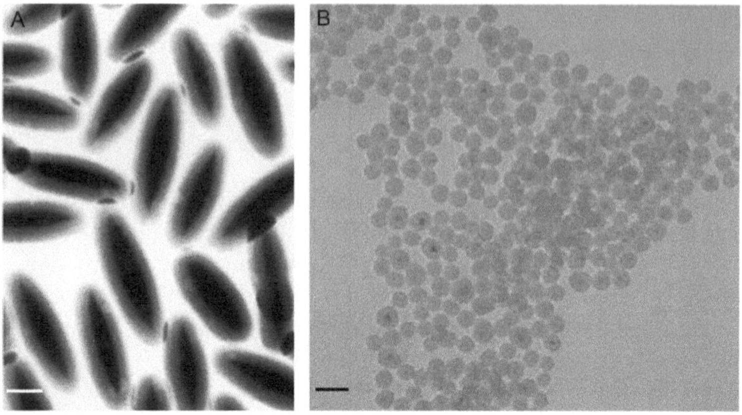

FIGURE 2.13 Silica coating by (A) hydrolysis and condensation of TEOS in ethanol/water of acicular iron oxide particles prepared by precipitation (scale bar = 200 nm) and (B) by water-in-oil microemulsions and ligand exchange reactions of hydrophobic magnetite nanoparticles (scale bar is 50 nm).

More complicated structure can be generated by combining silica-coated particles modified by an aminosilane coupling agent with a positively charged surface leading to nanoraspberries formed by electrostatic assembly. In addition, aggregates of $MnFe_2O_4$ nanocrystals, wrapped first in polymer and then coated with silica and gold, with interesting nuclear magnetic resonance (NMR) relaxivity properties have been prepared.[208] However, these systems present very low magnetization values (2 emu/g), which, for example, precludes its use in guiding. Besides, it requires an extensive number of complex steps significantly reducing the production yield of these composites.

Other methods to obtain these hybrid inorganic materials are spray pyrolysis, which leads to spherical particles heterogeneous in size with diameters between 0.3 and 3 μm,[209] and the supercritical fluid-assisted method.[210] Hybrid nanocomposites based on a core of silica nanospheres (160, 330 and 660 nm in diameter) covered by an outer shell of superparamagnetic nanoparticles, either iron oxide or heterodimeric FePt–iron oxide nanocrystals, have also been prepared. The nanocomposites show potential as contrast agents for dual-mode imaging purposes, combining magnetic resonance imaging and ultrasonography.[211]

3.1.2. Carbon

Commercial activated carbon has been used as matrix for the incorporation of magnetic nanoparticles. The method consists of the formation of iron oxide nanoparticles of uniform and tunable particle size (i.e. 4–8 nm range), which are highly dispersed along the porous structure of the activated carbon.[198] The preparation procedure consists firstly in filling the porosity of the activated carbon with an appropriate amount of $Fe(NO_3)_3 \cdot 9H_2O$ dissolved in ethanol. The filling is followed by the impregnation of the dried sample with an organic reducing agent (ethylene glycol). Finally, the impregnated sample is heat-treated under nitrogen at a moderate temperature to implement the superparamagnetic functionality. A variation of this method allows the preparation of superparamagnetic α-Fe homogeneously dispersed in the porous structure of the activated carbon. Stabilization of the α-Fe nanoparticles against corrosion is achieved by coating these nanoparticles with a carbon layer (this layer is produced by chemical vapour deposition with benzene as precursor). This methodology has been applied to the insertion of different ferrites such as iron oxide spinel nanoparticles on the shell of carbon capsules of diameter around 400 nm.[212]

Finally, carbon nanotubes (MWNTs) have been recently proposed as 'needle-like' nanovehicles capable of carrying drug molecules via exo- and endohedral functionalization and steerable by an external magnetic field due to the presence of ferromagnetic nanoparticles in the nanotube core.[213,214] Moreover, carbon nanotubes have been proposed as magnetically controlled drug delivery systems where drug molecules are encapsulated in the interior of a tube capped by the magnetic nanoparticles. Uncapping of the nanotube can be achieved through interaction with a magnetic field, and hence drug release can be triggered in a target site.[215]

3.1.3. Metals

The combination of magnetic and metallic nanoparticles is promising to functionally serve as contrast agent for magnetic resonance imaging and as platform to carry and deliver drugs in the body. Moreover, local temperature around core/shell structures with the noble metal in the shell can be sharply increased by near-IR (NIR) laser illumination because of the surface plasmon resonance effect.[208,216,217]

One method involves depositing gold onto iron oxide nanoparticle in an aqueous solution via hydroxylamine seeding resulting in around 60 nm particles. Using the reverse micelle method, gold-coated and silver-coated iron nanoparticles of mean size 10 nm have also been synthesized.[98,99,218] Au nanoparticles, 2–3 nm in size were also deposited forming a shell onto aminosilane-coated magnetic nanoparticles.[219]

Essentially, a reactive moiety attached to the surface of the magnetic nanoparticles directs the formation of the coating.[220,221] Owing to the difference in surface energies, gold does not wet an iron oxide surface. The strategy developed by the group of Majetich for preparing bifunctional plasmonic magnetic nanoparticles of around 20 nm was to synthesize monodisperse magnetic nanoparticles and then attach discrete Au clusters to the surface with bidentate ligands to finally complete the shell through electrodeless deposition of more gold onto the Au clusters.[222,223] 11-Mercaptoundecanoic acid (MUA) was used as the bidentate ligand that serves as a bridge between the iron oxide and the gold seed clusters. The thiol group of MUA promotes binding of small gold nanoparticles, while the carboxylic acid group attaches to the iron oxide surface resulting in magnetic nanoparticles coated with Au with a size of about 20 nm. Gold nanoparticles have also been deposited on silica-coating magnetic nanoparticles and an important Raman enhancement was observed.[224]

In organic media, gold is deposited onto the oleic acid-coated magnetic nanoparticles by reduction of $Au(ac)_3$ (ac = acetate) in the presence of magnetic seeds, a capping agent and a reducing agent. The process involves a thermally activated desorption of the capping agent, deposition of the Au on the magnetic nanoparticle surface and subsequent re-encapsulation of the Au surface by the capping agent.[225,226] Dumbbell-like bifunctional Au–magnetite nanoparticles were also obtained by epitaxial growth of iron oxide on the Au seeds. Iron carbonyl is decomposed in octadecene in the presence of oleic acid and oleylamine over the surface of Au nanoparticles previously synthesized.[227]

Alternative strategies have been developed to combine fluorescent and magnetic particles together using densely packed emulsion droplet,[228,229] silica encapsulation,[230,231] polymer or polyelectrolyte encapsulation.[232–237] The size of the fluorescent magnetic nanoparticles described in the literature can be rather large, that is, several hundred nanometres.[238] Fluorescent-magnetic-biotargeting multifunctional nanobioprobes have been used in detection and isolation of multiple tumour markers or tumour cells from complex samples sensitively and with high efficiency, which is critical for the early diagnosis of tumours, especially malignant tumours or cancers, which will improve clinical diagnosis outcomes and help to select effective treatment approaches.[239]

3.1.4. Bimagnetic Nanostructures

Ferro- or ferrimagnetic (FM or FiM) and antiferromagnetic (AFM) materials can lead to enhanced superparamagnetic blocking temperatures or tunable coercivities.[191,240] Some of these bimagnetic systems have been obtained from the oxidation of transition-metal nanoparticles, leading to an FM core and the corresponding AFM or FiM shell. The main inconvenience when considering these conventional bimagnetic core/shell nanoparticle systems comes from the difficulty of controlling core and shell dimensions due to the reduced size. AFM shell usually grows highly disordered. In view of that, recently, ‘inverse’ core/shell nanostructures with AFM cores and FiM or spin-glass shells have been synthesized, leading to a number of novel magnetic properties.[241,242] For example, the AFM component was found to increase the ordering temperature of the FiM component.[243]

Bimagnetic core/shell $FePt/Fe_3O_4$ particles have been synthesized by simultaneous thermal decomposition of iron pentacarbonyl and reduction of platinum salt, followed by growing an iron oxide shell on the preformed FePt core based on a heterogeneous seeded growth reaction.[244] It seems that Pt cations play a critical role in inducing and accelerating the reduction of associated metal cations.

3.2. Organic–Inorganic

Considerable efforts have been undertaken by several groups in developing procedures for the controlled synthesis of polymer nanobeads encasing in their interior inorganic nanoparticles[245] and, in particular, magnetic nanoparticles.[246–250]

Magnetic *liposomes* containing an aqueous dispersion of iron oxide nanoparticles have sizes between 200 nm and 1.2 μm.[251–253] PEG incorporated on the lipid surface help stabilizing these vesicles in physiological conditions. A different approach that helps to reduce the final size of the vesicle is the coating of each nanoparticle with a lipid bilayer, which is called lipoparticles. They usually have smaller sizes (30–250 nm) and do not contain water inside (so nor the typical active substances dissolved in water).[254–257]

Magnetic nanoparticles have also been encapsulated in *artificial capsules* that incorporate electrostatic interactions between negatively and positively charged polymers on to surfaces.[258,259] The layer-by-layer approach is a flexible, simple and inexpensive way for encapsulation of structures with variable size and shapes. Thickness and diameter can be controlled and have been demonstrated to be biocompatible.[260] Spatial distribution of the magnetic nanoparticles is important since magnetic properties can be diminished due to the magnetic interparticle interactions.[261,262]

Biodegradable polymers such as poly(D,L-lactic-*co*-glycolic acid) (PLGA) have been used as matrix for loading magnetic nanoparticles and drugs.[263] Those capsules have been synthesized by using an oil-in-water emulsion and a subsequent solvent evaporation method, resulting in polymer nanoparticles with sizes between 100 and 200 nm. The negative surface charge of PLGA allows anchoring positively charged biomolecules to render stability or specificity. Stability assays at high ionic strengths and in cell culture media indicate that only poly(acrylic acid) fully preserved the dispersion stability over the long term (longer than weeks) in comparison to other ligands such as citric acid and phosphonate-terminated poly(ethylene oxide). The improved stability was explained by the multipoint attachments of the chains onto the particle surface and by the ad layer-mediated electrostatic interactions.[264]

In recent years, magnetic nanoparticles have been encapsulated in *thermally sensitive polymers* so that when the core is self-heated in response to an external magnetic stimulus it triggers the release of the drug contained within the micelle. For example, pluronic nanocapsules that exhibit a reversible volume and hydrophilic to hydrophobic transition at a tunable temperature have been prepared by a self-assembly procedure and iron oxide nanoparticles were incorporated via internal precipitation.[265,266] Thermoresponsive polymer microgels and iron oxide nanoparticles have also been combined. Most of the works deals with lower critical solution temperature (LCST)-type microgels such as PNIPAM, prepared by precipitation polymerization, in which, for the encapsulation of nanoparticles, an intermediate step of surface functionalization of the magnetic nanoparticles is needed to enhance the biocompatibility with the polymer.[267] LCST-type systems collapse upon heating. UCST-type (upper critical solution temperature) microgels, such as poly(acrylamide-acrylic acid) that exhibit a positive swelling, are prepared via an inverse emulsion polymerization technique and offer the advantage of avoiding this intermediate step.[268]

3.3. Multicomponents

More sophisticated devices have been designed to improve specificity and drug release control. It involves the use of nanoparticles with different properties such as magnetic and luminescent, and the use of polymers as a 'glue' for the whole.[269–272]

Bioactivated fluorescent magnetic micelles with a hydrodynamic diameter of around 25 nm were obtained by coencapsulation of both hydrophobic CdSe/ZnS QDs and γ-Fe_2O_3 nanocrystals into aliquid crystal phase composed of synthetic amphiphilic PEG gallate derivatives bearing a terminal active chemical group. The present method permits to both functionalize and solubilize magnetic and fluorescent nanocrystals into the same micelle starting from a condensed hybrid hexagonal phase. Reactive groups, such as amine, carboxylic acid and biotin, are introduced on the surface of the nanoparticle with a controlled molar ratio.[273]

Multifunctional magnetic-fluorescent nanobeads made of poly(maleic anhydride-alt-1-octadecene) by embedding both superparamagnetic manganese iron oxide nanoparticles (MNPs) and core–shell CdSe/ZnS nanocrystals (QDs), and bead diameter from 70 to 160 nm. The nanobeads are formed by adding a destabilizing solvent to a starting solution of polymer and nanoparticles, which induces controlled aggregation of the components. A careful choice of this solvent allows tuning both the total bead size and the geometry of the beads in terms of nanoparticle distributions within the polymer, while the fluorescence signal of the nanobeads can be tuned by varying the relative ratio of fluorescent to magnetic nanoparticles added (QD:MNP).[274]

The bioconjugation of aptamers, ligands with high specificity and precision in drug targeting, with MNPs could help the rapid screening and detection of acute leukaemia cells for a pool of whole blood sample.[275] Moreover, the selective removal of target cells can be achieved by using aptamer-conjugated MNP and an external applied three-dimensional rotational magnetic field.[276]

Biodegradable polymer–metal multilayer half-shell nanoparticles have been proposed for drug delivery control. Drug is encapsulated within biocompatible and biodegradable polymer nanoparticles such as PLGA, and metal multilayers are deposited on these nanoparticles. Drug is released through the open half of the shell. Moreover, the release is modulated in response to NIR irradiation.[277]

4. CONCLUDING REMARKS

Inorganic nanoparticles with tailored optical, electronic, chemical, colloidal and magnetic properties can be synthesized by different methods that allow the control of the nanoparticle size and shape. In particular, magnetic iron oxide nanoparticles with appropriate surface modification have been widely used experimentally for *in vivo* applications such as magnetic resonance imaging contrast enhancement, drug delivery and hyperthermia.[278] Current methods for nanoparticle synthesis and surface modification match the requirements for biomedical applications. However, there is still room for improvements in terms of narrower size distributions, improved crystallinity and homogeneity in chemical composition, which will strongly affect the magnetic properties. Also, it remains the challenge of developing cleaner, less contaminating synthetic routes preserving good colloidal and magnetic properties.

To enhance the performance in existing applications and to explore novel applications, understanding and controlling the effects of surface functional groups on those nanoparticles are increasingly important. Implementation of NMR to study organic coatings of magnetic nanoparticles can be foreseen as an excellent tool for the characterization of nanoparticle surfaces.

We envisage two main subjects that hold great promise in the application of magnetic nanoparticles in biotechnology. One is focused on stimuli-responsive magnetic systems for drug delivery based on magnetic nanoparticles and 'smart' polymers, which are able to respond to environmental changes, such as changes in the temperature and pH.[279] The other one is the use of magnetic nanoparticles in gene therapy. Nanoparticles could facilitate targeted delivery of DNA into tissues and cells, binding DNA to nanoparticles using polyethylene imine as intermediate linker.[280]

REFERENCES

1. Pileni MP. Metal particles made in various colloidal self-assemblies: syntheses and properties. In: Sugimoto T, editor. *Fine particles: synthesis, characterization, and mechanism of growth. Surfactant science series*, vol. 92. New York: Marcel Dekker, Inc.; 2000. pp. 497–511.
2. Leubner IH. Particle nucleation and growth models. *Curr Opin Colloid Interface Sci* 2000;**5**:151–9.
3. Schladt TD, Schneider K, Schild H, Tremel W. Synthesis and bio-functionalization of magnetic nanoparticles for medical diagnosis and treatment. *Dalton Trans* 2011;**40**:6315–43.
4. Lamer VK, Dinegar RH. Theory, production and mechanism of formation of monodispersed hydrosols. *J Am Chem Soc* 1950;**72**:4847–54.
5. Everett DH. *Basic principles of colloid science*. London: Royal Society of Chemistry; 1988.
6. Kluijtmans SGJM, Dhont JKG, Philipse AP. A light-scattering contrast-variation study of bicontinuous porous glass media. *Langmuir* 1997;**13**:4976–9.
7. Cahn JW, Hilliard JE. Spinodal decomposition: a reprise. *Acta Metal* 1971;**19**:151–61.
8. Dirksen JA, Ring TA. Fundamentals of crystallization: kinetic effects on particle size distributions and morphology. *Chem Eng Sci* 1991;**46**:2389–427.

9. Viswanatha R, Santra P, Dasgupta C, Sarma D. Growth mechanism of nanocrystals in solution: ZnO, a case study. *Phys Rev Lett* 2007;**98**:98–101.
10. Ocaña M, Rodriguez-Clemente R, Serna CJ. Uniform colloidal particles in solution: formation mechanisms. *Adv Mater* 1995;**7**:212–6.
11. Morales MP, González-Carreño T, Serna CJ. Formation of Fe_2O_3 monodispersed particles in solution. *J Mater Res* 1992;**7**:2538–45.
12. Philipse A. Particulate colloids: aspects of preparation and characterization. In: Lyklema J, editor. *Fundamentals of interface and colloid science*, vol. 4. London: Elsevier Ltd.; 2005. pp. 2.1–2.71 Chapter 2.
13. Lee J, Isobe T, Senna M. Preparation of ultrafine Fe_3O_4 particles by precipitation in the presence of PVA at high pH. *J Colloid Interface Sci* 1996;**177**:490–4.
14. Morales MP, Veintemillas-Verdaguer S, Montero MI, Serna CJ, Roig A, Casas LI, et al. Surface and internal spin canting in Fe_2O_3 nanoparticles. *Chem Mater* 1999;**11**:3058–64.
15. Cozzoli PD, Pellegrino T, Manna L. Synthesis, properties and perspectives of hybrid nanocrystal structures. *Chem Soc Rev* 2006;**35**:1195–208.
16. Cheon J, Kang N-J, Lee S-M, Lee J-H, Yoon J-H, Oh SJ. Shape evolution of single-crystalline iron oxide nanocrystals. *J Am Chem Soc* 2004;**126**:1950–1.
17. Tao AR, Habas S, Yang P. Shape control of colloidal metal nanocrystals. *Small* 2008;**4**:310–25.
18. Cheng C, Gonela RK, Gu Q, Haynie DT. Self-assembly of metallic nanowires from aqueous solution. *Nano Lett* 2005;**5**:175–8.
19. Wang J, Zeng C. Growth of SrFe12O19 nanowires under an induced magnetic field. *J Cryst Growth* 2004;**270**:729–33.
20. Fuertes AB, Sevilla M, Valdes-Solis T, Tartaj P. Templated synthesis of mesoporous superparamagnetic polymers. *Chem Mater* 2007;**19**:5418–23.
21. Pileni MP, Tanori J, Filankembo A. Biomimetic strategies for the control of size, shape and self-organization of nanoparticles. *Colloids Surf A Physicochem Eng Asp* 1997;**123–124**:561–73.
22. Sugimoto T, Matijevic E. Formation of uniform spherical magnetite particles by crystallization from ferrous hydroxide gels. *J Colloid Interface Sci* 1980;**74**:227–43.
23. Tada M, Hatanaka S, Sanbonsugi H, Matsushita N, Abe M. Method for synthesizing ferrite nanoparticles 30 nm in diameter on neutral pH condition for biomedical applications. *J Appl Phys* 2003;**93**:7566–8.
24. Vergés MA, Costo R, Roca AG, Marco JF, Goya GF, Serna CJ, et al. Uniform and water stable magnetite nanoparticles with diameters around the monodomain-multidomain limit. *J Phys D: Appl Phys* 2008;**41**:134003.
25. Jolivet JP, Tronc E, Chanéac C. Synthesis of iron oxide- and metal-based nanomaterials. *Eur Phys J Appl Phys* 2000;**10**:167–72.
26. Massart R. Preparation of aqueous magnetic liquids in alkaline and acidic media. *IEEE Trans Magn* 1981;**17**:1247–8.
27. Vayssières L, Chanéac C, Tronc E, Jolivet J. Size tailoring of magnetite particles formed by aqueous precipitation: an example of thermodynamic stability of nanometric oxide particles. *J Colloid Interface Sci* 1998;**205**:205–12.
28. Langevin D. Micelles and microemulsions. *Annu Rev Phys Chem* 1992;**43**:341–69.
29. Fiévet F, Lagier J, Blin B, Beaudoin B, Figlarz M. Homogeneous and heterogeneous nucleations in the polyol process for the preparation of micron and submicron size metal particles. *Solid State Ion* 1989;**32–33**:198–205.

30. Hyeon T, Lee SS, Park J, Chung Y, Na HB. Synthesis of highly crystalline and monodisperse maghemite nanocrystallites without a size-selection process. *J Am Chem Soc* 2001;**123**:12798–801.
31. Rockenberger J, Scher EC, Alivisatos AP. A new nonhydrolytic single-precursor approach to surfactant-capped nanocrystals of transition metal oxides. *J Am Chem Soc* 1999;**121**:11595–6.
32. Sun S, Zeng H. Size-controlled synthesis of magnetite nanoparticles. *J Am Chem Soc* 2002;**124**:8204–5.
33. Bomatí-Miguel O, Morales MP, Tartaj P, Ruiz-Cabello J, Bonville P, Santos M, et al. Fe-based nanoparticulate metallic alloys as contrast agents for magnetic resonance imaging. *Biomaterials* 2005;**26**:5695–703.
34. Kwon SG, Piao Y, Park J, Angappane S, Jo Y, Hwang N-M, et al. Molecular orbital theory study on surface complex structures of phosphates to iron hydroxides: calculation of vibrational frequencies and adsorption energies. *J Am Chem Soc* 2007;**129**:12571–84.
35. Sun S, Zeng H, Robinson DB, Raoux S, Rice PM, Wang SX, et al. Monodisperse MFe2O4 (M = Fe, Co, Mn) nanoparticles. *J Am Chem Soc* 2004;**126**:273–9.
36. Martelli S, Mancini A, Giorgi R, Alexandrescu R, Cojocaru S, Crunteanu A, et al. Production of iron-oxide nanoparticles by laser-induced pyrolysis of gaseous precursors. *Appl Surf Sci* 2000;**154**:353–9.
37. Morjan I, Alexandrescu R, Soare I, Dumitrache F, Sandu I, Voicu I, et al. Nanoscale powders of different iron oxide phases prepared by continuous laser irradiation of iron pentacarbonyl-containing gas precursors. *Mater Sci Eng C* 2003;**23**:211–6.
38. Teja AS, Koh P-Y. Synthesis, properties, and applications of magnetic iron oxide nanoparticles. *Prog Cryst Growth Charact Mater* 2009;**55**:22–45.
39. Marques R, Garcia C, Lecante P, Ribeiro S, Noe L, Silva N, et al. Electro-precipitation of Fe_3O_4 nanoparticles in ethanol. *J Magn Magn Mater* 2008;**320**:2311–5.
40. Cabrera L, Gutierrez S, Menendez N, Morales M, Herrasti P. Electrochemical synthesis and characterization. *Electrochim Acta* 2007;**53**:3436–41.
41. Figuerola A, Corato RD, Manna L, Pellegrino T. From iron oxide nanoparticles towards advanced iron-based inorganic materials designed for biomedical applications. *Pharmacol Res* 2010;**62**:126–43.
42. Sun C, Lee JSH, Zhang M. Magnetic nanoparticles in MR imaging and drug delivery. *Adv Drug Deliv Rev* 2008;**60**:1252–65.
43. Lu A-H, Salabas EL, Schüth F. Magnetic nanoparticles: synthesis, protection, functionalization, and application. *Angew Chem Int Ed Engl* 2007;**46**:1222–44.
44. Namiki Y, Namiki T, Yoshida H, Ishii Y, Tsubota A, Koido S, et al. A novel magnetic crystal-lipid nanostructure for magnetically guided in vivo gene delivery. *Nat Nanotechnol* 2009;**4**:598–606.
45. Dias AMGC, Hussain A, Marcos AS, Roque ACA. A biotechnological perspective on the application of iron oxide magnetic colloids modified with polysaccharides. *Biotechnol Adv* 2011;**29**:142–55.
46. Kim J, Piao Y, Hyeon T. Multifunctional nanostructured materials for multimodal imaging, and simultaneous imaging and therapy. *Chem Soc Rev* 2009;**38**:372–90.
47. Jun Y-W, Lee J-H, Cheon J. Chemical design of nanoparticle probes for high-performance magnetic resonance imaging. *Angew Chem Int Ed Engl* 2008;**47**:5122–35.
48. Cheon J, Lee J-H. Synergistically integrated nanoparticles as multimodal probes for nanobiotechnology. *Acc Chem Res* 2008;**41**:1630–40.

49. Lefort J. Mémoire sur les oxydes ferroso-ferriques et leur combinaisons. *C R Hebd Seances Acad Sci* 1852;**T34**:488–91.
50. Lefort J. Mémoire sur l'oxyde de fer magnétique et ses combinaisons salines. *C R Hebd Seances Acad Sci* 1869;**T69**:179–80.
51. Jolivet J-P, Froidefond C, Pottier A, Chanéac C, Cassaignon S, Tronc E, et al. Size tailoring of oxide nanoparticles by precipitation in aqueous medium. A semi-quantitative modelling. *J Mater Chem* 2004;**14**:3281–8.
52. Massart R, Dubois E, Cabuil V, Hasmonay E. Preparation and properties of monodisperse magnetic fluids. *J Magn Magn Mater* 1995;**149**:1–5.
53. Welo LA, Baudisch O. Active Iron. II. Relationships among the oxide hydrates and oxides of iron and some of their properties. *Chem Rev* 1934;**15**:45–97.
54. Lesieur S, Grabielle-Madelmont C, Ménager C, Cabuil V, Dadhi D, Pierrot P, et al. Evidence of surfactant-induced formation of transient pores in lipid bilayers by using magnetic-fluid-loaded liposomes. *J Am Chem Soc* 2003;**125**:5266–7.
55. Kim DK, Mikhaylova M, Zhang Y, Muhammed M. Protective coating of superparamagnetic iron oxide nanoparticles. *Chem Mater* 2003;**15**:1617–27.
56. Liu X, Guan Y, Ma Z, Liu H. Surface modification and characterization of magnetic polymer nanospheres prepared by miniemulsion polymerization. *Langmuir* 2004;**20**:10278–82.
57. Jolivet JP, Vayssieres L, Chaneac C, Tronc E. Precipitation of spinel iron oxide: nanoparticle size control. *MRS Proc* 1996;**432**:145–50.
58. Costo R, Bello V, Robic C, Port M, Marco JF, Morales MP, et al. Ultrasmall iron oxide nanoparticles for biomedical applications: improving the colloidal and magnetic properties. *Langmuir* 2012;**28**:178–85.
59. Molday RS, Mackenzie D. Immunospecific ferromagnetic iron-dextran reagents for the labeling and magnetic separation of cells. *J Immunol Methods* 1982;**52**:353–67.
60. Vandenberghe RE, Vanleerberghe R, Grave ED, Robbrecht G. Preparation and magnetic properties of ultra-fine cobalt ferrites. *J Magn Magn Mater* 1980;**15–18**:1117–8.
61. Tamura H, Matijevic E. Precipitation of cobalt ferrites. *J Colloid Interface Sci* 1982;**90**:100–9.
62. Veverka M, Veverka P, Kaman O, Lančok A, Závěta K, Pollert E, et al. Magnetic heating by cobalt ferrite nanoparticles. *Nanotechnology* 2007;**18**:345704.
63. Sato T, Iijima T, Seki M, Inagaki N. Magnetic properties of ultrafine ferrite particles. *J Magn Magn Mater* 1987;**65**:252–6.
64. Tourinho FA, Franck R, Massart R. Aqueous ferrofluids based on manganese and cobalt ferrites. *J Mater Sci* 1990;**25**:3249–54.
65. Chen Q, Rondinone AJ, Chakoumakos BC, Zhang ZJ. Synthesis of superparamagnetic $MgFe_2O_4$ nanoparticles by coprecipitation. *J Magn Magn Mater* 1999;**194**:1–7.
66. Regazzoni AE, Matijević E. Formation of spherical colloidal nickel ferrite particles as model corrosion products. *Corrosion* 1982;**38**:212–8.
67. Schuele WJ, Deetscreek VD. Preparation, growth, and study of ultrafine ferrite particles. *J Appl Phys* 1961;**32**:S235–S236.
68. Portehault D, Cassaignon S, Baudrin E, Jolivet J-P. Structural and morphological control of manganese oxide nanoparticles upon soft aqueous precipitation through MnO4−/Mn2+ reaction. *J Mater Chem* 2009;**19**:2407–16.
69. Sinkó K, Szabó G, Zrínyi M. Liquid-phase synthesis of cobalt oxide nanoparticles. *J Nanosci Nanotechnol* 2011;**11**:4127–35.
70. de la Fuente JM, Alcantara D, Penades S. Cell response to magnetic glyconanoparticles: does the carbohydrate matter? *IEEE Trans Nanobioscience* 2007;**6**:275–81.

71. Philipse AP, van Bruggen MPB, Pathmamanoharan C. Magnetic silica dispersions: preparation and stability of surface-modified silica particles with a magnetic core. *Langmuir* 1994;**10**:92–9.
72. Fauconnier N, Pons J, Roger J, Bee A. Thiolation of maghemite nanoparticles by dimercaptosuccinic acid. *J Colloid Interface Sci* 1997;**194**:427–33.
73. Yan J, Mo S, Nie J, Chen W, Shen X, Hu J, et al. Hydrothermal synthesis of monodisperse Fe_3O_4 nanoparticles based on modulation of tartaric acid. *Colloids Surf A Physicochem Eng Asp* 2009;**340**:109–14.
74. Wang X, Zhuang J, Peng Q, Li Y. A general strategy for nanocrystal synthesis. *Nature* 2005;**437**:121–4.
75. Park S. Preparation of iron oxides using ammonium iron citrate precursor: thin films and nanoparticles. *J Solid State Chem* 2009;**182**:2456–60.
76. Cheng W, Tang K, Qi Y, Sheng J, Liu Z. One-step synthesis of superparamagnetic monodisperse porous Fe_3O_4 hollow and core-shell spheres. *J Mater Chem* 2010;**20**:1799–805.
77. Grene JM, Bernhardt P, Leuvrey C, Rogez G, Pasteur L. Hydrothermal synthesis of monodisperse magnetite nanoparticles. *Chem Mater* 2006;**18**:4399–404.
78. Ge J-P, Xu S, Zhuang J, Wang X, Peng Q, Li Y-D. Synthesis of CdSe, ZnSe, and ZnxCd1-xSe nanocrystals and their silica sheathed core/shell structures. *Inorg Chem* 2006;**45**:4922–7.
79. Liang X, Wang X, Zhuang J, Chen Y, Wang D, Li Y. Synthesis of nearly monodisperse iron oxide and oxyhydroxide nanocrystals. *Adv Funct Mater* 2006;**16**:1805–13.
80. Zhang F, Li J, Shan J, Xu L, Zhao D. Shape, size, and phase-controlled rare-earth fluoride nanocrystals with optical up-conversion properties. *Chem Eur J* 2009;**15**:11010–9.
81. Taniguchi T, Nakagawa K, Watanabe T, Matsushita N, Yoshimura M. Hydrothermal growth of fatty acid stabilized iron oxide nanocrystals. *J Phys Chem C* 2009;**113**:839–43.
82. Li X-H, Xu C-L, Han X-H, Qiao L, Wang T, Li F-S. Synthesis and magnetic properties of nearly monodisperse $CoFe_2O_4$ nanoparticles through a simple hydrothermal condition. *Nanoscale Res Lett* 2010;**5**:1039–44.
83. Repko A, Nižňanský D, Poltierová-Vejpravová J. A study of oleic acid-based hydrothermal preparation of $CoFe_2O_4$ nanoparticles. *J Nanopart Res* 2011;**13**:5021–31.
84. Drofenik M, Kristl M, Žnidaršič A, Hanžel D, Lisjak D. Hydrothermal synthesis of Ba-hexaferrite nanoparticles. *J Am Ceram Soc* 2007;**90**:2057–61.
85. Primc D, Makovec D, Lisjak D, Drofenik M. Hydrothermal synthesis of ultrafine barium hexaferrite nanoparticles and the preparation of their stable suspensions. *Nanotechnology* 2009;**20**:315605.
86. Inouye K, Endo R, Otsuka Y, Miyashiro K, Kaneko K, Ishikawa T. Oxygenation of ferrous ions in reversed micelle and reversed microemulsion. *J Phys Chem* 1982;**86**:1465–9.
87. Gobe M, Kon-No K, Kandori K, Kitahara A. Preparation and characterization of monodisperse magnetite sols in W/O microemulsion. *J Colloid Interface Sci* 1983;**93**:293–5.
88. López-Quintela M, Rivas J. Chemical reactions in microemulsions: a powerful method to obtain ultrafine particles. *J Colloid Interface Sci* 1993;**158**:446–51.
89. Feltin N, Pileni MP. New technique for synthesizing iron ferrite magnetic nanosized particles. *Langmuir* 1997;**13**:3927–33.
90. Moumen N, Veillet P, Pileni M. Controlled preparation of nanosize cobalt ferrite magnetic particles. *J Magn Magn Mater* 1995;**149**:67–71.
91. Moumen N, Lisiecki I, Pileni M, Briois V. Micellar factors which play a role in the control of the nanosize particles of cobalt ferrite. *Supramolecular Sci* 1995;**2**:161–8.
92. Moumen N, Pileni MP. Control of the size of cobalt ferrite magnetic fluid. *J Phys Chem* 1996;**100**:1867–73.

93. Moumen N, Pileni MP. New syntheses of cobalt ferrite particles in the range 2–5 nm: comparison of the magnetic properties of the nanosized particles in dispersed fluid or in powder form. *Chem Mater* 1996;**8**:1128–34.
94. Liu C, Rondinone AJ, Zhang ZJ. Synthesis of magnetic spinel ferrite $CoFe_2O_4$ nanoparticles from ferric salt and characterization of the size-dependent superparamagnetic properties. *Pure Appl Chem* 2000;**72**:37–45.
95. Liu C, Zou B, Rondinone AJ, Zhang ZJ. Reverse micelle synthesis and characterization of superparamagnetic $MnFe_2O_4$ spinel ferrite nanocrystallites. *J Phys Chem B* 2000;**104**:1141–5.
96. Lee Y, Lee J, Bae CJ, Park J-G, Noh H-J, Park J-H, et al. Large-scale synthesis of uniform and crystalline magnetite nanoparticles using reverse micelles as nanoreactors under reflux conditions. *Adv Funct Mater* 2005;**15**:503–9.
97. Dresco PA, Zaitsev VS, Gambino RJ, Chu B. Preparation and properties of magnetite and polymer magnetite nanoparticles. *Langmuir* 1999;**15**:1945–51.
98. Carpenter EE, Seip CT, O'Connor CJ. Magnetism of nanophase metal and metal alloy particles formed in ordered phases. *J Appl Phys* 1999;**85**:5184–6.
99. Carpenter EE, Sangregorio C, O'Connor CJ. Effects of shell thickness on blocking temperature of nanocomposites of metal particles with gold shells. *IEEE Trans Magn* 1999;**35**:3496–8.
100. Carpenter EE. Iron nanoparticles as potential magnetic carriers. *J Magn Magn Mater* 2001;**225**:17–20.
101. Viau G, Fiévet-Vincent F, Fiévet F. Monodisperse iron-based particles: precipitation in liquid polyols. *J Mater Chem* 1996;**6**:1047–53.
102. Hegde M, Larcher D, Dupont L, Beaudoin B, Tekaia-Elheissen K, Tarascón J. Synthesis and chemical reactivity of polyol prepared monodisperse nickel powders. *Solid State Ion* 1996;**93**:33–50.
103. Elumalai P. Synthesis and characterization of sub-micron size Co–Ni alloys using malonate as precursor. *Mater Res Bull* 2002;**37**:353–63.
104. Jeyadevan B, Hobo A, Urakawa K, Chinnasamy CN, Shinoda K, Tohji K. Towards direct synthesis of fct-FePt nanoparticles by chemical route. *J Appl Phys* 2003;**93**:7574–6.
105. Saravanan P, Jose TA, Thomas PJ, Kulkarni GU. Submicron particles of Co, Ni and Co-Ni alloys. *Bull Mater Sci* 2001;**24**:515–21.
106. Yu W, Wang Y, Liu H, Zheng W. Preparation and characterization of polymer-protected PtCo bimetallic colloids and their catalytic properties in the selective hydrogenation of cinnamaldehyde. *J Mol Catal A Chem* 1996;**112**:105–13.
107. Joseyphus RJ, Shinoda K, Kodama D, Jeyadevan B. Size controlled Fe nanoparticles through polyol process and their magnetic properties. *Mater Chem Phys* 2010;**123**:487–93.
108. Joseyphus R, Kodama D, Matsumoto T, Sato Y, Jeyadevan B, Tohji K. Role of polyol in the synthesis of Fe particles. *J Magn Magn Mater* 2007;**310**:2393–5.
109. Cai W, Wan J. Facile synthesis of superparamagnetic magnetite nanoparticles in liquid polyols. *J Colloid Interface Sci* 2007;**305**:366–70.
110. Gonçalves RH, Cardoso CA, Leite ER. Synthesis of colloidal magnetite nanocrystals using high molecular weight solvent. *J Mater Chem* 2010;**20**:1167–72.
111. Huang H, Xie Q, Kang M, Zhang B, Zhang H, Chen J, et al. Labeling transplanted mice islet with polyvinylpyrrolidone coated superparamagnetic iron oxide nanoparticles for in vivo detection by magnetic resonance imaging. *Nanotechnology* 2009;**20**:365101.
112. Das M, Dhak P, Gupta S, Mishra D, Maiti TK, Basak A, et al. Highly biocompatible and water-dispersible, amine functionalized magnetite nanoparticles, prepared by a low

temperature, air-assisted polyol process: a new platform for bio-separation and diagnostics. *Nanotechnology* 2010;**21**:125103.
113. Zhao L, Zhang H, Xing Y, Song S, Yu S, Shi W, et al. Morphology-controlled synthesis of magnetites with nanoporous structures and excellent magnetic properties. *Chem Mater* 2008;**20**:198–204.
114. Ammar S, Helfen A, Jouini N, Fiévet F, Rosenman I, Villain F, et al. Magnetic properties of ultrafine cobalt ferrite particles synthesized by hydrolysis in a polyol medium. *J Mater Chem* 2001;**11**:186–92.
115. Jungk H, Feldmann C. Nonagglomerated, submicron α-Fe_2O_3 particles: preparation and application. *J Mater Res* 2000;**15**:2244–8.
116. Levy M, Quarta A, Espinosa A, Figuerola A, Wilhelm C, García-Hernández M, et al. Correlating magneto-structural properties to hyperthermia performance of highly monodisperse iron oxide nanoparticles prepared by a seeded-growth route. *Chem Mater* 2011;**23**:4170–80.
117. Luigjes B, Woudenberg SMC, Groot RD, Meeldijk JD, Galvis HMT, Jong KPD, et al. Diverging geometric and magnetic size distributions of iron oxide nanocrystals. *J Phys Chem C* 2011;**115**:14598–605.
118. Bronstein LM, Huang X, Retrum J, Schmucker A, Pink M, Stein BD, et al. Influence of iron oleate complex structure on iron oxide nanoparticle formation. *Chem Mater* 2007;**19**:3624–32.
119. Sun S, Murray CB. Synthesis of monodisperse cobalt nanocrystals and their assembly into magnetic superlattices. *J Appl Phys* 1999;**85**:4325–30.
120. Dumestre F, Chaudret B, Amiens C, Renaud P, Fejes P. Superlattices of iron nanocubes synthesized from $Fe[N(SiMe_3)_2]_2$. *Science* 2004;**303**:821–3.
121. Hess PH, Parker PH. Polymers for stabilization of colloidal cobalt particles. *J Appl Polym Sci* 1966;**10**:1915–27.
122. Griffiths H, Horo MPO, Smith TW. The structure, magnetic characterization, and oxidation of colloidal iron dispersions. *J Appl Phys* 1979;**50**:7108–15.
123. Butter K. Synthesis and properties of iron ferrofluids. *J Magn Magn Mater* 2002;**252**:1–3.
124. Dinega DP, Bawendi MG. A solution-phase chemical approach to a new crystal structure of cobalt. *Angew Chem Int Ed* 1999;**38**:1788–91.
125. Puntes VF, Krishnan KM, Alivisatos P. Synthesis, self-assembly, and magnetic behavior of a two-dimensional superlattice of single-crystal ε-Co nanoparticles. *Appl Phys Lett* 2001;**78**:2187.
126. Farrell D, Majetich SA, Wilcoxon JP. Preparation and characterization of monodisperse Fe nanoparticles. *J Phys Chem B* 2003;**107**:11022–30.
127. Samia ACS, Hyzer K, Schlueter JA, Qin C-J, Jiang JS, Bader SD, et al. Ligand effect on the growth and the digestion of Co nanocrystals. *J Am Chem Soc* 2005;**127**:4126–7.
128. Lacroix L-M, Huls NF, Ho D, Sun X, Cheng K, Sun S. Stable single-crystalline body centered cubic Fe nanoparticles. *Nano Lett* 2011;**11**:1641–5.
129. Bonnemann H. A size-selective synthesis of air stable colloidal magnetic cobalt nanoparticles. *Inorg Chim Acta* 2003;**350**:617–24.
130. Margeat O, Dumestre F, Amiens C, Chaudret B, Lecante P, Respaud M. Synthesis of iron nanoparticles: size effects, shape control and organisation. *Prog Solid State Chem* 2005;**33**:71–9.
131. Lacroix L-M, Lachaize S, Falqui A, Blon T, Carrey J, Respaud M, et al. Ultrasmall iron nanoparticles: effect of size reduction on anisotropy and magnetization. *J Appl Phys* 2008;**103**:07D521.

132. Lacroix L-M, Lachaize S, Falqui A, Respaud M, Chaudret B. Iron nanoparticle growth in organic superstructures. *J Am Chem Soc* 2009;**131**:549–57.
133. Sun S. Monodisperse FePt nanoparticles and ferromagnetic FePt nanocrystal superlattices. *Science* 2000;**287**:1989–92.
134. (a) Chen M, Nikles DE. Synthesis of spherical FePd and CoPt nanoparticles. *J Appl Phys* 2002;**91**:8477–9. (b) Shevchenko EV, Talapin DV, Rogach AL, Kornowski A, Haase M, Weller H. Colloidal synthesis and self-assembly of CoPt3 nanocrystals. *J Am Chem Soc* 2002;**124**:11480–5.
135. Redl FX, Black CT, Papaefthymiou GC, Sandstrom RL, Yin M, Zeng H, et al. Magnetic, electronic, and structural characterization of nonstoichiometric iron oxides at the nanoscale. *J Am Chem Soc* 2004;**126**:14583–99.
136. Park J, Lee E, Hwang N-M, Kang M, Kim SC, Hwang Y, et al. One-nanometer-scale size-controlled synthesis of monodisperse magnetic iron oxide nanoparticles. *Angew Chem Int Ed Engl* 2005;**44**:2872–7.
137. Palchoudhury S, An W, Xu Y, Qin Y, Zhang Z, Chopra N, et al. Synthesis and growth mechanism of iron oxide nanowhiskers. *Nano Lett* 2011;**11**:1141–6.
138. Jana NR, Chen Y, Peng X. Size- and shape-controlled magnetic (Cr, Mn, Fe, Co, Ni) oxide nanocrystals via a simple and general approach. *Chem Mater* 2004;**16**:3931–5.
139. Bao N, Shen L, Wang Y, Padhan P, Gupta A. A facile thermolysis route to monodisperse ferrite nanocrystals. *J Am Chem Soc* 2007;**129**:12374–5.
140. Kang E, Park J, Hwang Y, Kang M, Park J-G, Hyeon T. Direct synthesis of highly crystalline and monodisperse manganese ferrite nanocrystals. *J Phys Chem B* 2004;**108**:13932–5.
141. Yu WW, Falkner JC, Yavuz CT, Colvin VL. Synthesis of monodisperse iron oxide nanocrystals by thermal decomposition of iron carboxylate salts. *Chem Commun* 2004;2306–7.
142. Park J, An K, Hwang Y, Park J-G, Noh H-J, Kim J-Y, et al. Ultra-large-scale syntheses of monodisperse nanocrystals. *Nat Mater* 2004;**3**:891–5.
143. Kovalenko MV, Bodnarchuk MI, Lechner RT, Hesser G, Schäffler F, Heiss W. Fatty acid salts as stabilizers in size- and shape-controlled nanocrystal synthesis: the case of inverse spinel iron oxide. *J Am Chem Soc* 2007;**129**:6352–3.
144. Bronstein LM, Atkinson JE, Malyutin AG, Kidwai F, Stein BD, Morgan DG, et al. Nanoparticles by decomposition of long chain iron carboxylates: from spheres to stars and cubes. *Langmuir* 2011;**27**:3044–50.
145. Xu Z, Shen C, Hou Y, Gao H, Sun S. Oleylamine as both reducing agent and stabilizer in a facile synthesis of magnetite nanoparticles. *Chem Mater* 2009;**21**:1778–80.
146. Kikuchi T, Kasuya R, Endo S, Nakamura A, Takai T, Metzler-Nolte N, et al. Preparation of magnetite aqueous dispersion for magnetic fluid hyperthermia. *J Magn Magn Mater* 2011;**323**:1216–22.
147. Song Q, Zhang ZJ. Shape control and associated magnetic properties of spinel cobalt ferrite nanocrystals. *J Am Chem Soc* 2004;**126**:6164–8.
148. Li Z, Chen H, Bao H, Gao M. One-pot reaction to synthesize water-soluble magnetite nanocrystals. *Chem Mater* 2004;**16**:1391–3.
149. Vargas JM, Zysler RD. Tailoring the size in colloidal iron oxide magnetic nanoparticles. *Nanotechnology* 2005;**16**:1474–6.
150. Pinna N, Grancharov S, Beato P, Bonville P, Antonietti M, Niederberger M. Magnetite nanocrystals: nonaqueous synthesis, characterization, and solubility. *Chem Mater* 2005;**17**:3044–9.

151. Li Z, Sun Q, Gao M. Preparation of water-soluble magnetite nanocrystals from hydrated ferric salts in 2-pyrrolidone: mechanism leading to Fe_3O_4. *Angew Chem Int Ed Engl* 2004;**44**:123–6.
152. Li Y, Afzaal M, O'Brien P. The synthesis of amine-capped magnetic (Fe, Mn, Co, Ni) oxide nanocrystals and their surface modification for aqueous dispersibility. *J Mater Chem* 2006;**16**:2175–80.
153. Liu H-L, Ko SP, Wu J-H, Jung M-H, Min JH, Lee JH, et al. One-pot polyol synthesis of monosize PVP-coated sub-5nm Fe_3O_4 nanoparticles for biomedical applications. *J Magn Magn Mater* 2007;**310**:e815–e817.
154. Li Z, Wei L, Gao MY, Lei H. One-pot reaction to synthesize biocompatible magnetite nanoparticles. *Adv Mater* 2005;**17**:1001–5.
155. Roca AG, Veintemillas-Verdaguer S, Port M, Robic C, Serna CJ, Morales MP. Effect of nanoparticle and aggregate size on the relaxometric properties of MR contrast agents based on high quality magnetite nanoparticles. *J Phys Chem B* 2009;**113**:7033–9.
156. Miguel-Sancho N, Bomatí-Miguel O, Colom G, Salvador J, Marco M, Santamaría J. Development of stable, water-dispersible, and biofunctionalizable superparamagnetic iron oxide nanoparticles. *Chem Mater* 2011;**23**:2795–802.
157. Nikolic MS, Krack M, Aleksandrovic V, Kornowski A, Förster S, Weller H. Tailor-made ligands for biocompatible nanoparticles. *Angew Chem Int Ed Engl* 2006;**45**:6577–80.
158. Fauconnier N, Roger J, Pons JN. Synthesis of aqueous magnetic liquids complexation of maghemite nanoparticles. *J Mol Liq* 1999;**83**:233–42.
159. Răcuciu M, Creangă DE, Airinei A. Citric-acid-coated magnetite nanoparticles for biological applications. *Eur Phys J E Soft Matter* 2006;**21**:117–21.
160. Srivastava S, Awasthi R, Gajbhiye NS, Agarwal V, Singh A, Yadav A, et al. Innovative synthesis of citrate-coated superparamagnetic Fe_3O_4 nanoparticles and its preliminary applications. *J Colloid Interface Sci* 2011;**359**:104–11.
161. Barreto JA, O'Malley W, Kubeil M, Graham B, Stephan H, Spiccia L. Nanomaterials: applications in cancer imaging and therapy. *Adv Mater* 2011;**23**:H18–H40.
162. Graf C, Vossen DLJ, Imhof A, van Blaaderen A. A general method to coat colloidal particles with silica. *Langmuir* 2003;**19**:6693–700.
163. Kim J, Kim HS, Lee N, Kim T, Kim H, Yu T, et al. Multifunctional uniform nanoparticles composed of a magnetite nanocrystal core and a mesoporous silica shell for magnetic resonance and fluorescence imaging and for drug delivery. *Angew Chem Int Ed Engl* 2008;**47**:8438–41.
164. Salgueiriño-Maceira V, Liz-Marzán LM, Farle M. Water-based ferrofluids from FexPt1-x nanoparticles synthesized in organic media. *Langmuir* 2004;**20**:6946–50.
165. Thorek DLJ, Chen AK, Czupryna J, Tsourkas A. Superparamagnetic iron oxide nanoparticle probes for molecular imaging. *Ann Biomed Eng* 2006;**34**:23–38.
166. McBain SC, Yiu HHP, El Haj A, Dobson J. Polyethyleneimine functionalized iron oxide nanoparticles as agents for DNA delivery and transfection. *J Mater Chem* 2007;**17**:2561–5.
167. Mok H, Veiseh O, Fang C, Kievit FM, Wang FY, Park JO, et al. pH-sensitive siRNA nanovector for targeted gene silencing and cytotoxic effect in cancer cells. *Mol Pharm* 2010;**7**:1930–9.
168. Koch AM, Reynolds F, Kircher MF, Merkle HP, Weissleder R, Josephson L. Uptake and metabolism of a dual fluorochrome Tat-nanoparticle in HeLa cells. *Bioconjug Chem* 2003;**14**:1115–21.

169. Chertok B, David AE, Yang VC. Polyethyleneimine-modified iron oxide nanoparticles for brain tumor drug delivery using magnetic targeting and intra-carotid administration. *Biomaterials* 2010;**31**:6317–24.
170. Zhu R, Jiang W, Pu Y, Luo K, Wu Y, He B, et al. Functionalization of magnetic nanoparticles with peptide dendrimers. *J Mater Chem* 2011;**21**:5464–74.
171. Herranz F, Morales MP, Roca AG, Desco M, Ruiz-Cabello J. A new method for the rapid synthesis of water stable superparamagnetic nanoparticles. *Chem Eur J* 2008;**14**:9126–30.
172. Shen L, Laibinis PE, Hatton TA. Bilayer surfactant stabilized magnetic fluids: synthesis and interactions at interfaces. *Langmuir* 1999;**15**:447–53.
173. Pellegrino T, Manna L, Kudera S, Liedl T, Koktysh D, Rogach AL, et al. Hydrophobic nanocrystals coated with an amphiphilic polymer shell: a general route to water soluble nanocrystals. *Nano Lett* 2004;**4**:703–7.
174. Cannon WR, Danforth SC, Flint JH, Haggerty JS, Marra RA. Sinterable ceramic powders from laser-driven reactions. 1. Process description and modeling. *J Am Ceram Soc* 1982;**65**:324–30.
175. Bomati-Miguel O, Morales MP, Serna CJ, Veintemillas-Verdaguer S. Magnetic nanoparticles prepared by laser pyrolysis. *IEEE Trans Magn* 2002;**38**:2616–8.
176. Swihart MT. Vapor phase synthesis of nanoparticles. *Curr Opin Colloid Interface Sci* 2003;**8**:127–33.
177. Hofmeister H, Huisken F, Kohn B, Alexandrescu R, Cojocaru S, Crunteanu A, et al. Filamentary iron nanostructures from laser-induced pyrolysis of iron pentacarbonyl and ethylene mixtures. *Appl Phys A* 2001;**72**:7–11.
178. Dumitrache F, Morjan I, Alexandrescu R, Ciupina V, Prodan G, Voicu I, et al. Iron–iron oxide core–shell nanoparticles synthesized by laser pyrolysis followed by superficial oxidation. *Appl Surf Sci* 2005;**247**:25–31.
179. Bomatí-Miguel O, Tartaj P, Morales MP, Bonville P, Golla-Schindler U, Zhao XQ, et al. Core-shell iron-iron oxide nanoparticles synthesized by laser-induced pyrolysis. *Small* 2006;**2**:1476–783.
180. Hochbaum AI, Chen R, Delgado RD, Liang W, Garnett EC, Najarian M, et al. Enhanced thermoelectric performance of rough silicon nanowires. *Nature* 2008;**451**:163–7.
181. Contreras-Cáceres R, Pacifico J, Pastoriza-Santos I, Pérez-Juste J, Fernández-Barbero A, Liz-Marzán LM. Au@pNIPAM thermosensitive nanostructures: control over shell cross-linking, overall dimensions, and core growth. *Adv Funct Mater* 2009;**19**:3070–6.
182. Poudyal N, Chaubey GS, Nandwana V, Rong C-B, Yano K, Liu JP. Synthesis of FePt nanorods and nanowires by a facile method. *Nanotechnology* 2008;**19**:355601.
183. Dumestre F, Chaudret B, Amiens C, Fromen M-C, Casanove M-J, Renaud P, et al. Shape control of thermodynamically stable cobalt nanorods through organometallic chemistry. *Angew Chem Int Ed Engl* 2002;**41**:4286–9.
184. Cordente N, Respaud M, Senocq F, Casanove M-J, Amiens C, Chaudret B. Synthesis and magnetic properties of nickel nanorods. *Nano Lett* 2001;**1**:565–8.
185. Mathur S, Barth S, Werner U, Hernandez-Ramirez F, Romano-Rodriguez A. Chemical vapor growth of one-dimensional magnetite nanostructures. *Adv Mater* 2008;**20**:1550–4.
186. Ji R, Lee W, Scholz R, Gösele U, Nielsch K. Monocrystalline spinel nanotube fabrication based on the Kirkendall effect. *Nat Mater* 2006;**5**:627–31.
187. Jia C-J, Sun L-D, Yan Z-G, Pang Y-C, You L-P. Iron oxide tube-in-tube nanostructures. *J Phys Chem C* 2007;**111**:13022–7.
188. Chin KC, Chong GL, Poh CK, Van LH, Sow CH, Lin J, et al. Large-scale synthesis of Fe_3O_4 nanosheets at low temperature. *J Phys Chem C* 2007;**111**:9136–41.
189. Kim D, Lee N, Park M, Kim BH, An K, Hyeon T. Synthesis of uniform ferrimagnetic magnetite nanocubes. *J Am Chem Soc* 2009;**131**:454–5.

190. Peng S, Sun S. Synthesis and characterization of monodisperse hollow Fe_3O_4 nanoparticles. *Angew Chem Int Ed Engl* 2007;**46**:4155–8.
191. Rebolledo AF, Laurent S, Calero M, Villanueva A, Knobel M, Marco JF, et al. Iron oxide nanosized clusters embedded in porous nanorods: a new colloidal design to enhance capabilities of MRI contrast agents. *ACS Nano* 2010;**4**:2095–103.
192. Rebolledo AF, Bomatí-Miguel O, Marco JF, Tartaj P. A facile synthetic route for the preparation of superparamagnetic iron oxide nanorods and nanorices with tunable surface functionality. *Adv Mater* 2008;**20**:1760–5.
193. Caruso F. Nanoengineering of inorganic and hybrid hollow spheres by colloidal templating. *Science* 1998;**282**:1111–4.
194. Kim S-W, Kim M, Lee WY, Hyeon T. Fabrication of hollow palladium spheres and their successful application to the recyclable heterogeneous catalyst for Suzuki coupling reactions. *J Am Chem Soc* 2002;**124**:7642–3.
195. Zhang X, Li D. Metal-compound-induced vesicles as efficient directors for rapid synthesis of hollow alloy spheres. *Angew Chem Int Ed Engl* 2006;**45**:5971–4.
196. Yin Y, Rioux RM, Erdonmez CK, Hughes S, Somorjai GA, Alivisatos AP. Formation of hollow nanocrystals through the nanoscale Kirkendall effect. *Science* 2004;**304**:711–4.
197. George C, Dorfs D, Bertoni G, Falqui A, Genovese A, Pellegrino T, et al. A cast-mold approach to iron oxide and Pt/iron oxide nanocontainers and nanoparticles with a reactive concave surface. *J Am Chem Soc* 2011;**133**:2205–17.
198. Tartaj P. Superparamagnetic composites: magnetism with no memory. *Eur J Inorg Chem* 2009;**2009**:333–43.
199. Bourgeat-Lami E, Lang J. Encapsulation of inorganic particles by dispersion polymerization in polar media. *J Colloid Interface Sci* 1999;**210**:281–9.
200. Herrera AP, Barrera C, Rinaldi C. Synthesis and functionalization of magnetite nanoparticles with aminopropylsilane and carboxymethyldextran. *J Mater Chem* 2008;**18**:3650–4.
201. Arruebo M, Fernández-Pacheco R, Velasco B, Marquina C, Arbiol J, Irusta S, et al. Antibody-functionalized hybrid superparamagnetic nanoparticles. *Adv Funct Mater* 2007;**17**:1473–9.
202. Aslam M, Fu L, Li S, Dravid VP. Silica encapsulation and magnetic properties of FePt nanoparticles. *J Colloid Interface Sci* 2005;**290**:444–9.
203. Tartaj P, Serna CJ. Synthesis of monodisperse superparamagnetic Fe/silica nanospherical composites. *J Am Chem Soc* 2003;**125**:15754–5.
204. Lee DC, Mikulec FV, Pelaez JM, Koo B, Korgel BA. Synthesis and magnetic properties of silica-coated FePt nanocrystals. *J Phys Chem B* 2006;**110**:11160–6.
205. Chang C-L, Fogler HS. Controlled formation of silica particles from tetraethyl orthosilicate in nonionic water-in-oil microemulsions. *Langmuir* 1997;**13**:3295–307.
206. Arriagada F, Osseo-Asare K. Synthesis of nanosize silica in a nonionic water-in-oil microemulsion: effects of the water/surfactant molar ratio and ammonia concentration. *J Colloid Interface Sci* 1999;**211**:210–20.
207. Tartaj P, González-Carreño T, Serna CJ. Synthesis of nanomagnets dispersed in colloidal silica cages with applications in chemical separation. *Langmuir* 2002;**18**:4556–8.
208. Lee J, Yang J, Ko H, Oh S, Kang J, Son J, et al. Multifunctional magnetic gold nanocomposites: human epithelial cancer detection via magnetic resonance imaging and localized synchronous therapy. *Adv Funct Mater* 2008;**18**:258–64.
209. Ruiz-Hernández E, López-Noriega A, Arcos D, Izquierdo-Barba I, Terasaki O, Vallet-Regí M. Aerosol-assisted synthesis of magnetic mesoporous silica spheres for drug targeting. *Chem Mater* 2007;**19**:3455–63.
210. Taboada E, Solanas R, Rodríguez E, Weissleder R, Roig A. Supercritical-fluid-assisted one-pot synthesis of biocompatible core(γ-Fe_2O_3)/shell(SiO_2) nanoparticles as high

relaxivity T_2-contrast agents for magnetic resonance imaging. *Adv Funct Mater* 2009;**19**:2319–24.
211. Malvindi MA, Greco A, Conversano F, Figuerola A, Corti M, Bonora M, et al. Magnetic/silica nanocomposites as dual-mode contrast agents for combined magnetic resonance imaging and ultrasonography. *Adv Funct Mater* 2011;**21**:2548–55.
212. Fuertes AB, Valde T, Sevilla M, Tartaj P. Fabrication of monodisperse mesoporous carbon capsules decorated with ferrite nanoparticles. *J Phys Chem C* 2008;**112**:3648–54.
213. Boncel S, Müller KH, Skepper JN, Walczak KZ, Koziol KKK. Tunable chemistry and morphology of multi-wall carbon nanotubes as a route to non-toxic, theranostic systems. *Biomaterials* 2011;**32**:7677–86.
214. Borowiak-Palen E, Mendoza E, Bachmatiuk A, Rummeli M, Gemming T, Nogues J, et al. Iron filled single-wall carbon nanotubes—a novel ferromagnetic medium. *Chem Phys Lett* 2006;**421**:129–33.
215. Panczyk T, Warzocha TP, Camp PJ. Enhancing the control of a magnetically capped molecular nanocontainer: Monte Carlo studies. *J Phys Chem C* 2011;**115**:7928–38.
216. Hirsch LR, Stafford RJ, Bankson JA, Sershen SR, Rivera B, Price RE, et al. Nanoshell-mediated near-infrared thermal therapy of tumors under magnetic resonance guidance. *Proc Natl Acad Sci USA* 2003;**100**:13549–54.
217. Kim J, Park S, Lee JE, Jin SM, Lee JH, Lee IS, et al. Designed fabrication of multifunctional magnetic gold nanoshells and their application to magnetic resonance imaging and photothermal therapy. *Angew Chem Int Ed Engl* 2006;**45**:7754–8.
218. Cho S-J, Idrobo J-C, Olamit J, Liu K, Browning ND, Kauzlarich SM, et al. Growth mechanisms and oxidation resistance of gold-coated iron nanoparticles. *Chem Mater* 2005;**17**:3181–6.
219. Caruntu D, Cushing BL, Caruntu G, Connor CJO. Attachment of gold nanograins onto colloidal magnetite nanocrystals. *Chem Mater* 2005;**17**:3398–402.
220. Vestal CR, Zhang ZJ. Atom transfer radical polymerization synthesis and magnetic characterization of $MnFe_2O_4$/polystyrene core/shell nanoparticles. *J Am Chem Soc* 2002;**124**:14312–3.
221. Matsuno R, Yamamoto K, Otsuka H, Takahara A. Polystyrene-grafted magnetite nanoparticles prepared through surface-initiated nitroxyl-mediated radical polymerization. *Chem Mater* 2003;**15**:3–5.
222. Lim J, Eggeman A, Lanni F, Tilton RD, Majetich SA. Synthesis and single-particle optical detection of low-polydispersity plasmonic-superparamagnetic nanoparticles. *Adv Mater* 2008;**20**:1721–6.
223. Lim JK, Majetich SA, Tilton RD. Stabilization of superparamagnetic iron oxide core-gold shell nanoparticles in high ionic strength media. *Langmuir* 2009;**25**:13384–93.
224. Spuch-Calvar M, Rodríguez-Lorenzo L, Morales MP, Álvarez-Puebla RA, Liz-Marzán LM. Bifunctional nanocomposites with long-term stability as SERS optical accumulators for ultrasensitive analysis. *J Phys Chem C* 2009;**113**:3373–7.
225. Wang L, Luo J, Fan Q, Suzuki M, Suzuki IS, Engelhard MH, et al. Monodispersed core-shell Fe_3O_4@Au nanoparticles. *J Phys Chem B* 2005;**109**:21593–601.
226. de la Presa P, Rueda T, Morales MP, Hernando A. Ligand exchange in gold-coated FePt nanoparticles. *IEEE Trans Magn* 2008;**44**:2816–9.
227. Yu H, Chen M, Rice PM, Wang SX, White RL, Sun S. Dumbbell-like bifunctional Au-Fe_3O_4 nanoparticles. *Nano Lett* 2005;**5**:379–82.
228. Mandal SK, Lequeux N, Rotenberg B, Tramier M, Fattaccioli J, Bibette J, et al. Encapsulation of magnetic and fluorescent nanoparticles in emulsion droplets. *Langmuir* 2005;**21**:4175–9.

229. Mandal S. Fluorescent magnetic emulsion droplets: potential material for multiplexed optical coding of biomolecules. *J Magn Magn Mater* 2007;**311**:88–91.
230. Yi DK, Selvan ST, Lee SS, Papaefthymiou GC, Kundaliya D, Ying JY. Silica-coated nanocomposites of magnetic nanoparticles and quantum dots. *J Am Chem Soc* 2005;**127**:4990–1.
231. Salgueiriño-Maceira V, Correa-Duarte MA, Spasova M, Liz-Marzán LM, Farle M. Composite silica spheres with magnetic and luminescent functionalities. *Adv Funct Mater* 2006;**16**:509–14.
232. Lee J, Lee Y, Youn JK, Na HB, Yu T, Kim H, et al. Simple synthesis of functionalized superparamagnetic magnetite/silica core/shell nanoparticles and their application as magnetically separable high-performance biocatalysts. *Small* 2008;**4**:143–52.
233. Gaponik N, Radtchenko IL, Sukhorukov GB, Rogach AL. Luminescent polymer microcapsules addressable by a magnetic field. *Langmuir* 2004;**20**:1449–52.
234. Xie H-Y, Zuo C, Liu Y, Zhang Z-L, Pang D-W, Li X-L, et al. Cell-targeting multifunctional nanospheres with both fluorescence and magnetism. *Small* 2005;**1**:506–9.
235. Zebli B, Susha AS, Sukhorukov GB, Rogach AL, Parak WJ. Magnetic targeting and cellular uptake of polymer microcapsules simultaneously functionalized with magnetic and luminescent nanocrystals. *Langmuir* 2005;**21**:4262–5.
236. Hong X, Li J, Wang M, Xu J, Guo W, Li J, et al. Fabrication of magnetic luminescent nanocomposites by a layer-by-layer self-assembly approach. *Chem Mater* 2004;**16**:4022–7.
237. Roullier V, Marchi-Artzner V, Cador O, Dorson F, Aubert T, Cordier S, et al. Synthesis and characterisation of magnetic-luminescent composite colloidal nanostructures. *Int J Nanotechnol* 2010;**7**:46–57.
238. Insin N, Tracy JB, Lee H, Zimmer JP, Westervelt RM, Bawendi MG. Incorporation of iron oxide nanoparticles and quantum dots into silica microspheres. *ACS Nano* 2008;**2**:197–202.
239. Song E-Q, Hu J, Wen C-Y, Tian Z-Q, Yu X, Zhang Z-L, et al. Fluorescent-magnetic-biotargeting multifunctional nanobioprobes for detecting and isolating multiple types of tumor cells. *ACS Nano* 2011;**5**:761–70.
240. Skumryev V, Stoyanov S, Zhang Y, Hadjipanayis G, Givord D, Nogués J. Beating the superparamagnetic limit with exchange bias. *Nature* 2003;**423**:19–22.
241. López-Ortega A, Tobia D, Winkler E, Golosovsky IV, Salazar-Alvarez G, Estradé S, et al. Size-dependent passivation shell and magnetic properties in antiferromagnetic/ferrimagnetic core/shell MnO nanoparticles. *J Am Chem Soc* 2010;**132**:9398–407.
242. Mariño-Fernández R, Masunaga SH, Fontaíña-Troitiño N, Morales MP, Rivas J, Salgueiriño V. Goethite (α-FeOOH) nanorods as suitable antiferromagnetic substrates. *J Phys Chem C* 2011;**115**:13991–9.
243. Golosovsky I, Salazar-Alvarez G, López-Ortega A, González M, Sort J, Estrader M, et al. Magnetic proximity effect features in antiferromagnetic/ferrimagnetic core-shell nanoparticles. *Phys Rev Lett* 2009;**102**:1–4.
244. Zeng H, Li J, Wang ZL, Liu JP, Sun S. Bimagnetic core/shell FePt/Fe_3O_4 nanoparticles. *Nano Lett* 2004;**4**:187–90.
245. Taylor-Pashow KML, Della Rocca J, Huxford RC, Lin W. Hybrid nanomaterials for biomedical applications. *Chem Commun* 2010;**46**:5832–49.
246. Howes P, Green M, Bowers A, Parker D, Varma G, Kallumadil M, et al. Magnetic conjugated polymer nanoparticles as bimodal imaging agents. *J Am Chem Soc* 2010;**132**:9833–42.
247. Ge Y, Zhang Y, He S, Nie F, Teng G, Gu N. Fluorescence modified chitosan-coated magnetic nanoparticles for high-efficient cellular imaging. *Nanoscale Res Lett* 2009;**4**:287–95.
248. Xiao Q, Xiao C. Preparation and characterization of silica-coated magnetic-fluorescent bifunctional microspheres. *Nanoscale Res Lett* 2009;**4**:1078–84.

249. Di Corato R, Piacenza P, Musarò M, Buonsanti R, Cozzoli PD, Zambianchi M, et al. Magnetic-fluorescent colloidal nanobeads: preparation and exploitation in cell separation experiments. *Macromol Biosci* 2009;**9**:952–8.
250. Sukhorukov GB, Rogach AL, Garstka M, Springer S, Parak WJ, Muñoz-Javier A, et al. Multifunctionalized polymer microcapsules: novel tools for biological and pharmacological applications. *Small* 2007;**3**:944–55.
251. Martina M-S, Fortin J-P, Ménager C, Clément O, Barratt G, Grabielle-Madelmont C, et al. Generation of superparamagnetic liposomes revealed as highly efficient MRI contrast agents for in vivo imaging. *J Am Chem Soc* 2005;**127**:10676–85.
252. Soenen SJH, Hodenius M, de Cuyper M. Magnetoliposomes: versatile innovative nanocolloids for use in biotechnology and biomedicine. *Nanomedicine* 2009;**4**:177–91.
253. Meyre M-E, Raffard G, Franconi J-M, Duguet E, Lambert O, Faure C. Production of magnetic multilamellar liposomes as highly T_2-efficient MRI contrast agents. *Nanomed Nanotechnol Biol Med* 2011;**7**:18–21.
254. De Cuyper M, Soenen SJH. Weissig V, editor. *Methods in molecular biology*, vol. 605. Totowa, NJ: Humana Press; 2010. pp. 97–111.
255. Shinkai M, Suzuki M, Iijima S, Kobayashi T. Antibody-conjugated magnetoliposomes for targeting cancer cells and their application in hyperthermia. *Biotechnol Appl Biochem* 1995;**21**:125–37.
256. Ito A, Ino K, Kobayashi T, Honda H. The effect of RGD peptide-conjugated magnetite cationic liposomes on cell growth and cell sheet harvesting. *Biomaterials* 2005;**26**:6185–93.
257. Lesieur S, Gazeau F, Luciani N, Ménager C, Wilhelm C. Multifunctional nanovectors based on magnetic nanoparticles coupled with biological vesicles or synthetic liposomes. *J Mater Chem* 2011;**21**:14387–93.
258. Caruso F. Nanoengineering of particle surfaces. *Adv Mater* 2001;**13**:11–22.
259. De Cock LJ, de Koker S, de Geest BG, Grooten J, Vervaet C, Remon JP, et al. Polymeric multilayer capsules in drug delivery. *Angew Chem Int Ed Engl* 2010;**49**:6954–73.
260. Srivastava S, Kotov NA. Composite Layer-by-Layer (LBL) assembly with inorganic nanoparticles and nanowires. *Acc Chem Res* 2008;**41**:1831–41.
261. Abbasi AZ, Guti L, Mercato LL, Herranz F, Chubykalo-Fesenko O, Veintemillas-Verdaguer S, et al. Magnetic capsules for NMR imaging: effect of magnetic nanoparticles spatial distribution and aggregation. *J Phys Chem C* 2011;**115**:6257–64.
262. Morales MP, Bédard MF, Roca AG, de la Presa P, Hernando A, Zhang F, et al. Relaxation times of colloidal iron platinum in polymer matrixes. *J Mater Chem* 2009;**19**:6381–6.
263. Kim J, Lee JE, Lee SH, Yu JH, Lee JH, Park TG, et al. Designed fabrication of a multifunctional polymer nanomedical platform for simultaneous cancer-targeted imaging and magnetically guided drug delivery. *Adv Mater* 2008;**20**:478–83.
264. Chanteau B, Fresnais J, Berret J-F. Electrosteric enhanced stability of functional sub-10 nm cerium and iron oxide particles in cell culture medium. *Langmuir* 2009;**25**:9064–70.
265. Liu T-Y, Liu K-H, Liu D-M, Chen S-Y, Chen I-W. Temperature-sensitive nanocapsules for controlled drug release caused by magnetically triggered structural disruption. *Adv Funct Mater* 2009;**19**:616–23.
266. Jain TK, Richey J, Strand M, Leslie-Pelecky DL, Flask CA, Labhasetwar V. Magnetic nanoparticles with dual functional properties: drug delivery and magnetic resonance imaging. *Biomaterials* 2008;**29**:4012–21.
267. Rubio-Retama J, Zafeiropoulos NE, Serafinelli C, Rojas-Reyna R, Voit B, Cabarcos EL, et al. Synthesis and characterization of thermosensitive PNIPAM microgels covered with superparamagnetic gamma-Fe_2O_3 nanoparticles. *Langmuir* 2007;**23**:10280–5.

268. Echeverria C, Mijangos C. UCST-like hybrid PAAm-AA/Fe_3O_4 microgels, effect of Fe_3O_4 nanoparticles on morphology, thermosensitivity and elasticity. *Langmuir* 2011;**27**:8027–35.
269. Jaganathan H, Gieseck RL, Ivanisevic A. Transverse relaxivity changes after layer-by-layer encapsulation of multicomponent DNA templated nanostructures. *J Phys Chem C* 2010;**114**:22508–13.
270. Boussif O, Lezoualc'h F, Zanta MA, Mergny MD, Scherman D, Demeneix B, et al. A versatile vector for gene and oligonucleotide transfer into cells in culture and in vivo: polyethylenimine. *Proc Natl Acad Sci USA* 1995;**92**:7297–301.
271. Lee JH, Huh YM, Jun Y, Seo J, Jang J, Song HT, et al. Artificially engineered magnetic nanoparticles for ultra-sensitive molecular imaging. *Nat Med* 2007;**13**:95–9.
272. Fresnais J, Ishow E, Sandre O, Berret J-F. Electrostatic co-assembly of magnetic nanoparticles and fluorescent nanospheres: a versatile approach towards bimodal nanorods. *Small* 2009;**5**:2533–6.
273. Roullier V, Grasset F, Boulmedais F, Artzner F, Cador O. Small bioactivated magnetic quantum dot micelles. *Chem Mater* 2008;**20**:6657–65.
274. Corato RD, Bigall NC, Ragusa A, Dorfs D, Genovese A, Marotta R, et al. Multifunctional nanobeads based on quantum dots and magnetic cell targeting and sorting. *ACS Nano* 2011;**5**:1109–21.
275. Herr JK, Smith JE, Medley CD, Shangguan D, Tan W. Aptamer-conjugated nanoparticles for selective collection and detection of cancer cells. *Anal Chem* 2006;**78**:2918–24.
276. Nair BG, Nagaoka Y, Morimoto H, Yoshida Y, Maekawa T, Kumar DS. Aptamer conjugated magnetic nanoparticles as nanosurgeons. *Nanotechnology* 2010;**21**:455102.
277. Park H, Yang J, Seo S, Kim K, Suh J, Kim D, et al. Multifunctional nanoparticles for photothermally controlled drug delivery and magnetic resonance imaging enhancement. *Small* 2008;**4**:192–6.
278. Chen B, Wu W, Wang X. Magnetic iron oxide nanoparticles for tumor-targeted therapy. *Curr Cancer Drug Targets* 2011;**11**:184–9.
279. Medeiros SF, Santos AM, Fessi H, Elaissari A. Stimuli-responsive magnetic particles for biomedical applications. *Int J Pharm* 2011;**403**:139–61.
280. Kami D, Takeda S, Itakura Y, Gojo S, Watanabe M, Toyoda M. Application of magnetic nanoparticles to gene delivery. *Int J Mol Sci* 2011;**12**:3705–22.

Chapter 3

Synthesis of Inorganic Nanocrystals for Biological Fluorescence Imaging

Cécile Philippot and Peter Reiss
CEA Grenoble, INAC/SPrAM (UMR 5819 CEA-CNRS-UJF)/LEMOH, Grenoble Cedex, France

1. INTRODUCTION

Semiconductor nanocrystals discussed in this chapter are inorganic particles with a size ranging from around 1–10 nm. A surface layer of organic molecules (ligands) stabilizes them in the colloidal state. Ligands consist of two parts: a polar head group with affinity for the nanocrystal surface and a tail assuring solubilization of the particles in an organic solvent or in water. Stabilization against aggregation and precipitation can be achieved either by steric repulsion in case of bulky ligands with apolar tail (e.g. long alkyl chain) or by electrostatic repulsion in case of ligands with a tail containing a charged group (e.g. carboxylic acid function). When reducing the size of semiconductor crystals below the (material dependent) dimensions corresponding to a bound electron–hole pair (exciton), their electronic and optical properties change. In particular, the valence and conduction bands evolve towards discrete energy states and the band gap increases with decreasing particle size. These phenomena are described by the term 'quantum confinement effect' and have been discovered in the early 1980s.[1,2] Semiconductor nanocrystals with a size small enough to be in the quantum confinement regime are also called 'quantum dots' (QDs). One of the most important consequences of quantum confinement is the possibility to adjust the fluorescence emission wavelength of QDs just by changing their size. Figure 3.1 gives an overview of the emission wavelength ranges, which have been reported for different semiconductor QDs as a function of size.

1.1. Semiconductor Nanocrystals as Fluorescent Biological Labels

Although initially supposed to be used in transistors, the first application of semiconductor nanocrystals to be developed was biological labelling. The proof of principle in 1998[3,4] has triggered a large number of studies and nowadays

Frontiers of Nanoscience, Vol. 4. DOI: 10.1016/B978-0-12-415769-9.00003-0

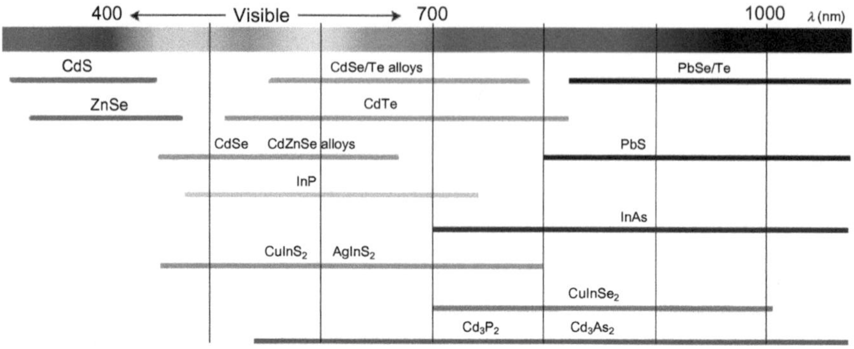

FIGURE 3.1 Reported spectral ranges of emission for different semiconductor nanocrystals. (For colour version of this figure, the reader is referred to the online version of this chapter.)

products from the simple QD without added specific functionalities to the complete labelling kit for specific applications are commercially available. In this field, QDs have to compete with existing technologies using, for example, organic dyes or fluorescent proteins. Just like these solutions, QDs are not the 'ideal' fluorophores for *all* biological applications. But they combine a number of very appealing features, which in some cases open up new horizons in biological imaging. The excellent photostability of QDs constitutes one of their most important advantages over most other fluorophores, as it enables experiments on a much longer timescale. They exhibit high brightness due to the combination of high fluorescence quantum yield (QY) and large absorption coefficient. In contrast to organic dyes, the absorption band of QDs is broad owing to the fact that, as in bulk semiconductors, essentially all photons with an energy exceeding the band gap are absorbed. Therefore, QDs can be excited with a wavelength far from their emission wavelength. This large effective Stokes shift improves the signal-to-noise ratio and avoids the use of cut-off filters for the excitation light in the detection system. Further, QDs emitting at different wavelengths in the visible or near-infrared all can be excited with the same UV or visible light source, which significantly facilitates multiplexing experiments.[5] The possibility of adjusting the fluorescence wavelength precisely with size and the comparably narrow emission line width make QDs excellent candidates for biological assays relying on Förster resonance energy transfer.[6] Finally, depending on the material, the emission colour can be tuned in different spectral regions and, in particular, also in the near-infrared window from 650 to 900 nm (cf. Figure 3.1). The latter is of particular interest for *in vivo* imaging owing to the increased tissue transparency and reduced autofluorescence, resulting in larger imaging depths and improved signal-to-noise ratios.[7] Another important feature of QDs is the possibility to introduce different additional functionalities by surface derivatization. While the primary ligand layer must assure colloidal stability in aqueous medium, additional molecules can be grafted giving specific functions to the QDs. Surface bound antibodies or peptides can, for example, be used to target tumours or to enable cell penetration. Even more, sophisticated scaffolds allow for the tracking and delivery of drugs or for the activation of the probe once the target is reached, for example, by cleavage of a fluorescence quencher. Surface functionalization also allows for the combination of different imaging modes, such as fluorescence and magnetic resonance imaging.[8] The main drawbacks of QDs as biological labels are related to their potential risks for health and environment. Therefore, despite their obvious advantages in many aspects over existing solutions, it is difficult to bring them on the market. While superparamagnetic iron oxide nanoparticles are already used as MRI contrast agents, QDs are far from being introduced as fluorescent labels for *in vivo* imaging of human beings. Their full life cycle, degradation products as well as possible risks related to their small size (passing e.g. the blood–brain barrier) and high surface-to-volume ratio (catalytic effects) need to be assessed very carefully and standards have to be established.

1.2. Basic Properties of Semiconductor Nanocrystals

1.2.1. Optical Properties

Fluorescence is the property of a semiconductor to absorb photons with an energy $h\nu_e$ superior to its band gap and—after charge carrier relaxation via phonons to the lowest excited state—to emit light of a higher wavelength (lower energy $h\nu_f$) after a brief interval, called the fluorescence lifetime. The fluorescence signal of QDs is characterized by a narrow and symmetric peak of Gaussian shape whose line width directly depends on the size dispersion. In the case of Cd chalcogenide nanocrystals, for example, synthesis methods have been developed yielding 'monodisperse' samples, that is, samples showing a deviation from the mean size of less than 5%. By consequence, room temperature fluorescence line widths as narrow as 80–100 meV can be achieved. The intensity of the emission peak is evaluated by measuring the fluorescence QY, which expresses the ratio of photons absorbed to photons emitted by the sample. The QY is always below unity as ensembles generally contain nanocrystals, which do not emit—at least for some time. Besides completely dark dots, nanocrystals showing intermittent emission, the so-called blinking phenomenon, have been discovered in fluorescence studies on individual particles.[9] At the origin of the transition of a QD from a bright 'on' state to a dark 'off' state is most probably the presence of a charge. Several pathways can lead to charging or ionization of a nanocrystal: for example, the generation of two or more excitons in the same nanocrystal followed by Auger recombination. Another possibility is electron or hole tunnelling to trap states outside the nanocrystal core after photogeneration of an exciton. In each case, the resident charge leads to efficient quenching of newly generated excitons via non-radiative Auger processes occurring on a much shorter timescale than radiative recombination. This 'charging model' to explain nanocrystals' fluorescence intermittency has been commonly accepted for a long time but was recently put into question by several research groups.[10–12] Also the mechanism at work to bring back a nanocrystal from the 'off' state to the emissive 'on' state is not clear to date. In any case, the fluorescence QY of individual nanocrystals and of ensembles is closely related to the surface state. While in the core the constituting ions have the same chemical environment as in the bulk crystal, the coordination sphere of surface atoms has to be completed by ligand molecules.

1.2.2. Core/Shell Nanocrystals

An important strategy for increasing the fluorescence QY relies on the growth of an inorganic shell consisting of a second semiconductor on the core nanocrystals. Ref. 13 gives a recent review on this subject. In the case where the valence band edge of the shell material lies energetically lower and the conduction band edge energetically higher than the corresponding band edges of the core material, both charge carriers (electrons and holes) are confined in the core. This situation is called 'type I band alignment' in semiconductor physics. The shell not only assures a better passivation of core nanocrystals' electronic surface states than organic ligands but also improves the photostability, thanks to the physical barrier between the emissive core and the environment (e.g. solvent). For successful shell growth, the crystallographic parameters of the core and shell material should be as close as possible, even though this requirement is not as stringent as in the case of epitaxial 2D growth. One of the most studied systems, CdSe/ZnS core/shell nanocrystals, presents a comparably high lattice mismatch of 10.6%.[14] Therefore, shell growth is limited to around one monolayer; for thicker shells, strain-induced defects decrease the fluorescence QY. On the other hand, materials of intermediate lattice parameter, such as CdS or ZnSe, can be used as buffer layer on the core to enable the growth of thicker ZnS shells.[15,16] The shell can further be used to tune the emission wavelength of the core nanocrystals. In a staggered type II band alignment of the core and shell materials, one of the charge carriers (electron or hole) is located in the core, the other one in the shell. As a consequence, the band gap of the core/shell system is lower than that of its constituents. In CdSe/CdTe nanocrystals, for example, the emission can be tuned from around 700 to more than 1000 nm by varying the CdTe shell thickness and CdSe core diameter.[17]

1.3. Nanocrystal Synthesis

As demonstrated in classical studies by LaMer and Dinegar,[18] the synthesis of monodisperse colloids via homogeneous nucleation requires a temporal separation of nucleation and growth of the seeds. Initially, the concentration of monomers, that is, the minimum subunits of the crystal, constantly increases by addition from exterior or by *in situ* generation within the reaction medium. It should be noted that in this stage no nucleation occurs even in supersaturated solution, due to the very high energy barrier for spontaneous homogeneous nucleation. The latter is overcome for a yet higher degree of supersaturation, where the formation of stable nuclei takes place. As soon as the rate of monomer consumption induced by the nucleation and growth processes exceeds the rate of monomer supply, the monomer concentration and hence the supersaturation decreases below the level at which the nucleation rate becomes zero. In the following stage, particle growth continues under further monomer consumption as long as the system is in the supersaturated regime. Experimentally, the separation of nucleation and growth can be achieved by rapid injection of the reagents into a hot solvent, which raises the precursor concentration in the reaction flask above the nucleation threshold ('hot-injection method').[19] The hot injection leads to an instantaneous nucleation, which is quickly quenched by the fast cooling of the reaction mixture (the solution to be injected is at room temperature) and by the decreased supersaturation after the nucleation burst. Another possibility relies on attaining the degree of supersaturation necessary for homogeneous nucleation via the *in situ* formation of reactive species upon supply of thermal energy ('heating-up method').[20,21] In an ideal case, all nuclei are created at the same time and undergo identical growth. During the growth stage, it is possible to carry out subsequent injections of precursors in order to increase the mean particle size without deterioration of the narrow size distribution as long as the concentration corresponding to the critical supersaturation for nucleation is not exceeded. Crystal growth from solution is in many cases followed by a second distinct growth process, which is called Ostwald ripening.[22] It consists of the dissolution of the smallest particles because of their high surface energy and subsequent redeposition of the dissolved matter onto the bigger ones. Thereby the total number of nanocrystals decreases, whereas their mean size increases.

In the following sections, the synthesis of nanocrystals will be divided into two categories: synthesis in aqueous medium and synthesis in organic solvents. For each case, a (non-exhaustive) overview over the most important synthesis methods for various materials will be given, as summarized in Tables 3.1 and 3.2. Special emphasis will be put on some significant advances of the past few years, which will be discussed in more detail by means of selected examples.

2. AQUEOUS SYNTHESIS OF NANOCRYSTALS

Historically, the synthesis of nanocrystals in water was developed before the synthesis in organic solvents. Initial works by Henglein and co-workers in the early 1980s concerned the synthesis and study of CdS and later ZnS nanocrystals.[1,133] Soon thereafter, the concept of arrested precipitation in the presence of styrene/maleic acid anhydride copolymer, phosphate or polyphosphate stabilizers has been extended to other materials (PbS, Cd_3P_2, Zn_3P_2, Cd_3As_2, CdTe, ZnTe), all showing quantum confinement effects. In parallel to monophase syntheses, a bi-phase technique has been developed, which is based on

TABLE 3.1 Aqueous Synthesis of Selected Types of Nanocrystals

Material	PL (nm)	QY (%)	Ref.
ZnSe/ZnS, Cu-doped ZnSe/ZnS	400–600	66	23,24
ZnSe, Cu-doped ZnSe	400–490	40	25–27
ZnSe(S)	400–600	31	28,29
ZnTe	n.s.	n.s.	30
CuSe	395	n.s.	31
CdTe	520–800	5–65	32–58
CdTe/CdS	500–735	40–65	59–61
CdTe/CdS/ZnS	660	n.s.	62
CdTe/CdSe	510–820	12–40	63,64
CdTe/CdSe/ZnSe	500–700	5	65
CdSe	384–600	3–52	66–68
CdSe/CdS	530–600	30	69–71
CdSeTe	550–690	n.s.	72
CdSeTe/ZnS	600–800	40–70	73
CdSTe	650–780	68	74
CdZnSe	390–460	20–30	75
CdHgTe	640–1600	6–60	76,77
HgTe	800–3700	>40	78–81
PbS	810–1010	n.s.	82

n.s., not specified.

TABLE 3.2 Synthesis of Selected Types of Nanocrystals in Organic Solvents

Materials	PL (nm)	QY (%)	Ref.
In(Zn)P/ZnS	485–586	60–70	83
Cu-doped InP	630–1100	35–40	84
InP/ZnS	480–750	60–70	85–87
InAs/shell	700–1400	90	88
InAs/ZnCdS	700–900	35–50	89,90

TABLE 3.2 Synthesis of Selected Types of Nanocrystals in Organic Solvents

Materials	PL (nm)	QY (%)	Ref.
$CuInS_2/ZnS$	450–800	30–50	91–95
$CuInSe_2$	650–975	25	96
$CuInSe_2/ZnS$	650–1030	25–50	97,98
$AgInS_2$	550–720	8	93
Cd_3P_2	455, 600–1200	7, 38	99–101
Cd_3As_2	530–2000	20–60	102
$Zn_xCd_{1-x}S$	410–450	23	103
ZnSe/ZnS	400–440	15–32	104,105
$Cd_{1-x}Zn_xSe_{1-y}S_y$	500–610	80	106
CdSe/ZnS	480–650	30–50	14,107,108
CdSe/ZnSe	520–630	60–85	109
CdSe/CdS	500–615	50–85	110–116
CdSe/CdS/ZnS	480–650	15–95	16,117–120
CdTe/CdSe	535–1000	40–82	17,63,64,121–124
CdTe/CdSe/ZnS	540–825	94	65,125
CdTe	576–720	65	126
CdSeTe	580–850	53–60	127–129
CdHgTe	800	n.s.	130
PbS	800–1800	20	131
PbS/ZnS	830–1400	26–33	132

n.s., not specified.

the arrested precipitation of nanocrystals within reverse micelles.[2,134] Here, nanometer-sized water droplets (dispersed phase) are stabilized in an organic solvent (continuous phase) by an amphiphilic surfactant. They serve as nanoreactors for the NC growth and prevent at the same time from particle agglomeration. Both methods provide relatively simple experimental approaches using standard reagents as well as room temperature reactions and were of great importance for the development of nanocrystal synthesis. A drawback of the micellar method is its comparably low yield due to the intrinsic requirement to work at rather low concentration. An important step towards the improvement of the emission properties was the growth of a $Cd(OH)_2$ layer on CdS nanocrystals by Spanhel *et al.* in 1987, which is the first example of a core/shell system.[135] A breakthrough in aqueous synthesis was achieved by Nozik's and Weller's groups who developed efficient methods for the growth of CdTe nanocrystals using hydrophilic thiols as stabilizers (cf. below).[32,33] Table 3.1 gives an overview of selected types of QDs synthesized in aqueous phase along with the reported range of emission and fluorescence QY.

2.1. CdTe Nanocrystals

As mentioned before, the most widely studied example of nanocrystals prepared in water are thiol-stabilized CdTe QDs.[34] Under optimized conditions, a PL QY of 40–60% can be obtained for the as-prepared samples and the emission colour can be tuned in the visible and near-infrared range. The synthetic scheme is schematized in Figure 3.2. It comprises the *in situ* generation of H_2Te gas, formation of reactive species by bubbling the gas into the aqueous solution of a mixture of cadmium perchlorate and the thiol ligand, for example, thioglycolic acid (TGA) or mercaptopropionic acid (MPA), followed by nanocrystals' nucleation and growth upon heating to reflux.

In an alternative approach, H_2Te gas is bubbled into a solution containing a stoichiometric amount of NaOH to form NaHTe. The liquid can then be sucked into a syringe and subsequently be injected into the reaction flask containing the soluble Cd salt and thiol ligands. All solutions must be thoroughly degassed as H_2Te and NaHTe are very air-sensitive compounds. A direct influence of the pH value on the PL QY has been observed. Using TGA ligands, an optimum value of 11.2–11.8 has been determined, attained by adding NaOH to the

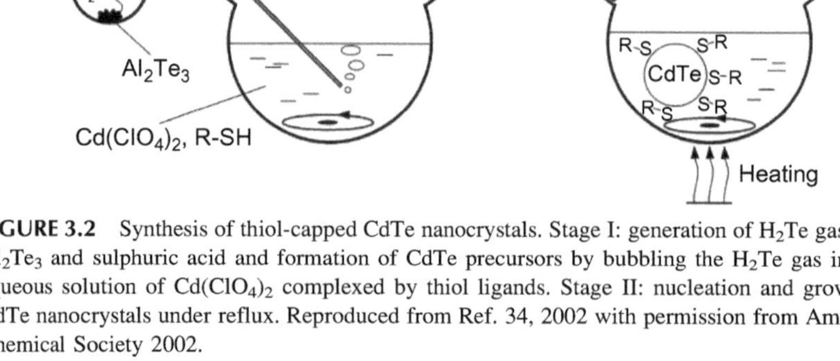

FIGURE 3.2 Synthesis of thiol-capped CdTe nanocrystals. Stage I: generation of H_2Te gas from Al_2Te_3 and sulphuric acid and formation of CdTe precursors by bubbling the H_2Te gas into an aqueous solution of $Cd(ClO_4)_2$ complexed by thiol ligands. Stage II: nucleation and growth of CdTe nanocrystals under reflux. Reproduced from Ref. 34, 2002 with permission from American Chemical Society 2002.

reaction mixture. For Cd:Te chosen as 1:0.5 the best TGA:Cd ratio has been reported to be 1.3:1 leading to the brightest nanocrystals with a QY up to 65%.[37] This value contrasts with earlier reports (TGA:Cd=2.45:1). The enhanced QY was explained by the higher concentration of an uncharged Cd–TGA complex at lower TGA amounts, which is supposed to provide better surface passivation of CdTe nanocrystals.[38] In the light of newer results,[136,137] another plausible explanation would be the reduced amount of disulphides, that is, dithiodiglycolic acid, formed at lower TGA concentration. Disulphide bonds are known to act as hole acceptors and hence as efficient fluorescence quenchers for various types of QDs. Figure 3.3 shows the absorption and emission spectra of a size series of MPA-capped CdTe nanocrystals, along with a sizing curve, reporting the correlation between the excitonic peak position and the diameter.

In 2010, Lesnyak and co-workers described a novel ligand for the aqueous synthesis of CdTe nanocrystals: 5-mercaptomethyltetrazole.[46] Tetrazoles are five-membered cyclic compounds containing four nitrogen atoms of different types (pyrrole and pyridine type). CdTe nanocrystals obtained with the 'standard method' but using mercaptomethyltetrazole instead of TGA as the stabilizer exhibited fluorescence in a range of 510–610 nm depending on the reflux time with a QY reaching up to 60%. Upon addition of a solution of Cd^{2+} ions, they reversibly form hydrogels, that is, highly porous 3D networks. In 2011, He *et al.* reported another modification of the CdTe synthesis yielding nanocrystals emitting in the 700–800 nm range with a QY of 15–20%.[35] In contrast to the 'standard method', NaHTe was generated from tellurium powder by

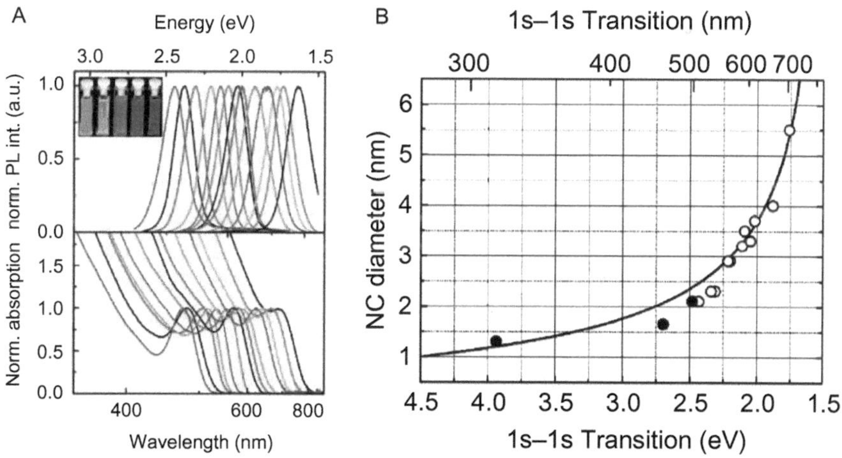

FIGURE 3.3 (A) Absorption (bottom) and photoluminescence (top) spectra of a size series of MPA-capped CdTe nanocrystals. (B) Sizing curve for CdTe nanocrystals (filled circles: determined from XRD data; open circles: determined from TEM images; solid line: calculated dependence of the 1s–1s transition corresponding to the excitonic peak position on CdTe nanocrystals size). Reproduced from Ref. 58 with permission from American Chemical Society 2007. (For colour version of this figure, the reader is referred to the online version of this chapter.)

reduction with sodium borohydride, and the growth reaction was carried out under hydrothermal conditions at 170–180°°C in a microwave reactor. Otherwise similar reaction parameters as reported before were applied: MPA as a stabilizer at pH 8.4, with an MPA:Cd:Te ratio of 2.4:1:0.5. Yet another method for the generation of the tellurium precursor is the electrochemical reduction in acidic medium of the metal. In addition to CdTe, H_2Te generated in this way has, for example, also been used for preparing mid-infrared absorbing HgTe nanocrystals.[78]

2.2. ZnSe Nanocrystals

Due to the comparably large bulk band gap of 2.7 eV, ZnSe nanocrystals exhibit band-edge emission in the UV/blue spectral range.[138] For the aqueous synthesis of ZnSe nanocrystals, very similar approaches have been applied as for CdTe. Several thiol-containing stabilizing molecules have been applied (thioglycerol, TG; TGA; 3-MPA).[139] The reaction was triggered by bubbling H_2Se gas through the aqueous solution of the zinc salt and the stabilizer, followed by refluxing the reaction mixture for several hours to accomplish nanocrystals' growth. Optimal pH values for the formation of stable colloids were found to be 11.5 in the case of TG-stabilizing molecules and 6.5 when TGA or MPA was used. The QY of the obtained 2–3 nm nanocrystals emitting at 390 nm increased from less than 0.1% to 10–30% by the irradiation with white light for several hours. The irradiation process, in addition to improving the QY, produced a bathochromic shift of the nanocrystals' excitonic peak. Both phenomena resulted from the photochemically induced incorporation of sulphur, originating from the TGA stabilizer, into the crystal lattice to give better passivated $ZnSe_{1-x}S_x$ alloyed nanocrystals. In 2009, Fang *et al.* described the synthesis of ZnSe/ZnS core/shell nanocrystals in aqueous media by reacting zinc acetate with NaHSe solution in the presence of L-glutathione stabilizer at pH 11.5.[23] After heating to 90°°C for 1 h, ZnSe QDs with an average size of 2.7 nm and an emission wavelength of 372 nm have been obtained. Shell growth was achieved at pH 10.2 by the addition of a mixture containing zinc acetate, glutathione as well as thiourea acting as the sulphur source. Figure 3.4 shows UV–vis absorption and photoluminescence spectra of aliquots taken during the shell growth as well as the temporal evolution of the emission peak and of the fluorescence QY.

The QY of the obtained water-dispersible ZnSe/ZnS core/shell QDs reached 65% and the stability against photo-oxidation was significantly improved in comparison with plain ZnSe core nanocrystals.

3. SYNTHESIS IN ORGANIC MEDIUM

Table 3.2 gives an overview of various types of fluorescent semiconductor nanocrystals synthesized in organic solvents.

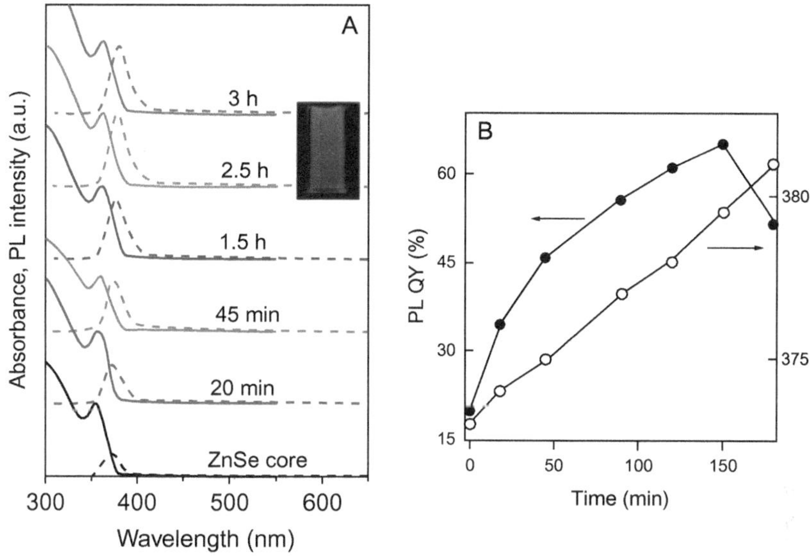

FIGURE 3.4 (A) Evolution of the absorption and photoluminescence spectra with time during the growth of a ZnS shell on ZnSe nanocrystals. (B) Evolution of the emission wavelength and of the fluorescence QY.[23] Reprinted with permission from American Chemical Society 2009. (For colour version of this figure, the reader is referred to the online version of this chapter.)

3.1. Cadmium Chalcogenide Nanocrystals

The introduction of an organometallic hot-injection synthesis method using organic solvents in 1993 constituted an important step towards the fabrication of monodisperse CdS, CdSe and CdTe nanocrystals.[140] For the as-synthesized samples, size dispersion as low as 8–10% has been obtained. In the following, a similar approach has been adapted for the synthesis of zinc and mercury chalcogenide nanocrystals (cf. Table 3.2). Organometallic precursors used in these reactions generally comprised metal alkyls or aryls (dimethylcadmium, diethylzinc, dibenzylmercury), whereas S, Se or Te sources were mostly chosen from trialkylphosphine chalcogenides (R_3PSe, R_3PTe with R = octyl or butyl) or bistrimethylsilylchalcogenides such as $(Me_3Si)_2S$, abbreviated $(TMS)_2S$. The trialkylphosphine chalcogenides can easily be prepared by dissolution of the chalcogenide powder in the phosphine. In the case of sulphur, the silylated compound is preferred over the trialkylphosphine sulphide because the latter exhibits too low reactivity in the temperature range used for the synthesis. The organometallic synthesis uses coordinating solvents, which act as stabilizing ligands for the nanocrystal surface. The choice of these solvents is of crucial importance because it influences the reactivity of the precursors as well as the kinetics of the growth process. A typical example is the mixture of trioctylphosphine (TOP) and trioctylphosphine oxide (TOPO), sometimes with the admixture of hexadecylamine. The described reaction type is not limited to organometallic, pyrophoric reagents and a number of inorganic compounds such as oxides or salts can be applied in the same way. The first example reported was the use of cadmium oxide, complexed with alkylphosphonic acids, as a Cd source in the synthesis of cadmium chalcogenide nanocrystals.[141] The resulting cadmium phosphonates are sufficiently reactive towards Se or Te solutions in TOP and yield nearly monodisperse CdSe and CdTe nanocrystals in the size range of ca. 2.5–5 nm. Larger sized nanocrystals can be prepared by applying cadmium carboxylates as a Cd source, with the crystal growth rate being inversely proportional to the chain length of the carboxylic acid.[142] By proper choice of the cadmium source and solvent, this method allows the synthesis of CdSe nanocrystals with diameters up to 25 nm, while for the organometallic approach, maximum values of ca. 11 nm have been reported.[140] It has also been shown that for the synthesis of group IIB-sulphide nanocrystals (CdS, ZnS)—in contrast to the selenides and tellurides—elemental sulphur can be used as appropriate precursor. Yellow sulphur occurs in discrete S_8 molecules, while its heavier homologues form Se_x and Te_x rings and chains, which are more difficult to solubilize. Therefore, in most syntheses, elemental sulphur, dissolved, for example, in octadecene (ODE) or oleylamine, is used as the S source. An illustrative example is the synthesis of a series of transition metal sulphide nanocrystals comprising CdS and ZnS described by Joo *et al.*[143] At the same time, a number of appropriate monomolecular precursors, containing both the metal and the S source, are commercially available or easy to prepare. In particular, Zn- or Cd-xanthates and -dithiocarbamates have to be mentioned in this context: most of these compounds decompose at temperatures below 200 °C and are therefore suitable precursors for the preparation of transition metal sulphide nanocrystals. Efrima and co-workers reported the use of Zn- and Cd-alkylxanthates for the synthesis of the corresponding sulphide nanocrystals.[144,145]

Core/shell systems are generally fabricated in a two-step procedure, consisting of core nanocrystals' synthesis, followed by a purification step, and the subsequent shell growth reaction, during which a small number of monolayers (typically 1–5) of the shell material are deposited on the cores. The temperature for the core nanocrystal synthesis is generally higher than that used for the shell growth and the shell precursors are slowly added, for example, by means of a syringe pump. The major advantages over a so-called one-pot approach without intermediate purification step is the fact that unreacted precursors or side products can be eliminated before the shell growth. The core nanocrystals are purified by precipitation and redispersion cycles, and finally they are dispersed in the solvent used for the shell growth. In order to calculate the required amount of shell precursors to obtain the desired shell thickness, the knowledge of the concentration of the core nanocrystals is indispensable. It can be obtained by carefully drying the nanocrystal sample, weighing and determination of its composition by elemental analysis using atomic absorption spectroscopy. Correlation of the data with the nanocrystals' size, obtained by transmission electron microscopy (TEM), allows for the calculation of the molar quantity of nanocrystals in the investigated sample. As the nanocrystals' size is directly related to the excitonic peak in the UV–vis absorption spectrum, the size-dependent molar extinction coefficient ε can be determined at the same time. There are several materials for which the correlation between the NC size and ε has been tabulated in the literature, such as CdSe, CdS, CdTe and InP.[13] A variant of the described shell growth method giving a precise control of the shell thickness is the so-called SILAR (successive ion layer adsorption and reaction) method.[111] It is based on the formation of one monolayer at a time by alternating the injections of cationic and anionic precursors and has firstly been applied for the synthesis of CdSe/CdS core/shell nanocrystals. Monodispersity of the samples was maintained for CdS shell thicknesses up to five monolayers on 3.5 nm core CdSe nanocrystals, as reflected by the narrow PL line widths obtained in the range of 23–26 nm (full-width at half-maximum, FWHM). The SILAR approach has later been extended to 'giant' core/shell nanocrystals with a shell thickness of up to around 20 monolayers and a final size of 15–20 nm showing strongly reduced fluorescence intermittency/blinking.[146,147]

Some further recent examples:

$Cd_{1-x}Zn_xSe_{1-y}S_y$ nanocrystals

In 2008, Bae *et al.* described the synthesis of $Cd_{1-x}Zn_xSe_{1-y}S_y$ QDs with a chemical composition gradient in a single step by reacting cadmium oxide and zinc acetate with Se powder and S powder in ODE (cf. Figure 3.5).[106] After heating to 300°°C for 10 min, $Cd_{1-x}Zn_xSe_{1-y}S_y$ QDs with an average size of 6.3 nm and an emission wavelength of 610 nm have been obtained. The highest QY of the $Cd_{1-x}Zn_xSe_{1-y}S_y$ QDs reached 80%, and their photostability was enhanced in comparison with CdSe/ZnS core/shell nanocrystals.

CdTe nanocrystals

Longer emission wavelengths could be obtained by using CdTe nanocrystals. Xing *et al.* reported the synthesis of cubic CdTe nanocrystals with broad emission range from 576 to 720 nm.[126] The CdTe QDs are obtained by mixing cadmium oxide with Te in the presence of oleic acid and paraffin oil and heating to 200°°C. After 45 min, CdTe QDs with an average size of 3.8 nm, an emission wavelength of 633 nm and a QY of 65% have been obtained.

3.2. Doped II–VI Semiconductor QDs

Doping—the introduction of a small amount of 'impurities' into the crystal lattice—is an attractive way to change the nanocrystals' physical properties. An important example is the doping of II–VI semiconductors with paramagnetic Mn^{2+} ions ($S=5/2$), yielding materials denominated dilute magnetic semiconductors, which exhibit interesting magnetic and magnetooptical

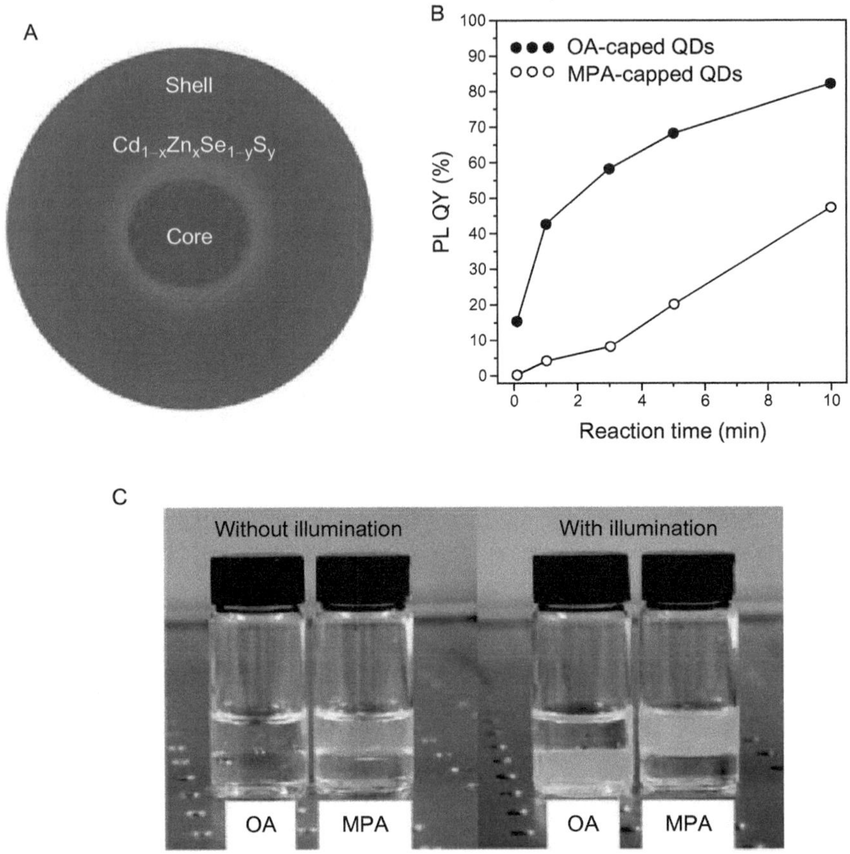

FIGURE 3.5 (A) Scheme of the chemical composition of gradient structure QDs. (B) PL QY of oleic acid (OA)-capped QDs dispersed in chloroform (solid circles) and mercaptopropionic acid (MPA)-capped QDs dispersed in water (empty circles) prepared at different reaction times (5 s, 1 min, 3 min, 5 min, 10 min). (C) Photographs of OA-capped QDs dispersed in chloroform (lower phase) and after phase transfer to water (upper phase) using MPA at pH 8. Reprinted from Ref. 106 with permission from American Chemical Society 2008. (For colour version of this figure, the reader is referred to the online version of this chapter.)

properties.[148] At the same time, the host NC can act as an antenna for the absorption of energy (e.g. light) and excitation of the dopant ions via energy transfer. In this case, mostly UV-absorbing nanocrystals are chosen as the hosts, such as ZnS or ZnSe. Mn-doped ZnSe is an instructive example for the development of doped II–VI semiconductor nanocrystals. Bulk ZnSe:Mn exhibits PL at 582 nm (2.13 eV), commonly assigned to an optically forbidden d–d transition of Mn^{2+} ($^4T_1$ to $^6A_1$).[149] This emission is sensitive to the crystal field splitting being itself dependent on the local chemical environment. A general problem encountered in doping attempts is the fact that the dopant ions are in many cases found to be located on the surface and not as desired in the volume of the nanocrystals. A decisive step towards the understanding of the doping process was achieved by Erwin *et al.* who first succeeded in doping CdSe nanocrystals with Mn^{2+}.[150] They introduced a model of doping based on kinetics and concluded that the doping mechanism is controlled by the initial adsorption of impurities on the surface of growing nanocrystals. Only impurities remaining adsorbed on the surface for a time comparable to the reciprocal growth rate are incorporated into the NC. Three main factors influencing this residence time were determined, namely, the surface morphology, NC shape and surfactants present in the growth solution. It has been shown that (0 0 1) surfaces of zinc blende crystals exhibit much higher impurity binding energies than the other two zinc blende orientations and than any facet of crystals with wurtzite or rock-salt structures. These find ings were fully corroborated by the state of the art, as all nanocrystals successfully doped with Mn ions exhibited the zinc blende crystal structure.

Peng and co-workers explored another approach with the goal to achieve the doping of *all* nanocrystals in a given sample. In the so-called nucleation-doping strategy, MnSe nuclei, formed from manganese stearate and TBPSe in octadecylamine at 280°°C, were overcoated with ZnSe using zinc stearate or zinc undecylenate. No residual ZnSe emission was observed and the doped nanocrystals exhibited thermally stable (up to 300°°C) highly efficient (QY 40–70%) PL in a spectral window of 545–610 nm, depending on the ZnSe shell thickness and on the nature of the surface ligands (charged or neutral).[151] The same approach was extended to the doping of ZnSe with Cu ions yielding emission in the range of 460–525 nm.[152] The range of emission could be further extended to the NIR by using InP as a host material. Xie and Peng reported Cu-doped InP nanocrystals emitting from 630 to 1100 nm, whose QY was 35–40% after growth of a ZnSe shell.[84] Concluding this paragraph, with exception of the comparably broad PL peaks (>50 nm at FWHM), their otherwise very interesting optical properties make transition metal-doped ZnSe nanocrystals promising 'greener' alternatives to the widely studied II–VI semiconductor nanocrystals for a number of applications including biological labelling.[153]

3.3. III–V Semiconductor Nanocrystals

Compared to most of the II–VI and IV–VI nanocrystals, III–V semiconductor nanocrystals are generally referred to as 'greener' compounds because the group III elements such as In or Ga present less risks for the environment and for human health than Cd, Pb or Hg. Nevertheless, the studies and applications of III–V nanocrystals are rather sparse as compared to their II–VI analogues. Their synthesis is more difficult as, due to the stronger covalent bonding of the precursors, generally higher reaction temperatures and longer reaction times are necessary. These conditions favour Ostwald ripening, leading to increased size dispersion. Therefore, highly reactive organometallic precursors or monomolecular precursors, containing both the cation and the anion already chemically bound in the same molecule, are applied in many examples.

3.3.1. InP Nanocrystals

Most of the reports concerning the synthesis of III–V semiconductor nanocrystals deal with indium phosphide. It should be noted however that nanocrystals of the narrower band gap semiconductor InAs can generally be synthesized using similar approaches (see below). InP nanocrystals are an attractive alternative to CdSe or CdTe nanocrystals, due to their size-dependent emission in the visible and near-infrared spectral range combined with the absence of toxic cadmium. In initial synthetic routes,[154–156] the method established for cadmium chalcogenide nanocrystals[140] was adapted to InP, but longer reaction times (3–7 days) were necessary to yield particles of good crystallinity. In 2002, Battaglia and Peng reported a new protocol, also applicable to the synthesis of InAs Nanocrystals, which is based on fatty acids as stabilizers in combination with the non-coordinating solvent 1-ODE instead of TOPO/TOP.[157] The use of this medium provided a fast and controllable reaction, yielding InP nanocrystals of low size dispersion but of limited size range (around 2.5–3.5 nm). Similar results were obtained when organometallic In precursors were used in combination with ester-type solvents.[158] InP/ZnS core/shell nanocrystals have been prepared by thermal cycling during the shell growth[86] and by addition of zinc carboxylate during the core synthesis.[85] In both cases, modifications of the core synthesis led to samples covering the whole visible and NIR range (480–740 nm) and having a maximum QY of 40–60%. One of the simplest methods for the synthesis of high-quality InPZnS nanocrystals consists of mixing both the InP core and ZnS shell precursors (indium myristate, tris(trimethylsilyl)phosphine, abbreviated P $(TMS)_3$, zinc stearate, dodecanethiol (DDT)) at room temperature in ODE and heating this mixture to 250–300°°C. The obtained nanocrystals emit in the range of 480–600 nm with a QY up to 70% (Figure 3.6).[87] Finally, the use of the expensive and pyrophoric phosphorus precursor $P(TMS)_3$ could be avoided by applying *in situ* generation of phosphine gas.[159] In 2011, the example of a solvothermal synthesis was reported by Byun *et al.*[160] They prepared InP/ZnS core/shell QDs in two steps: first the solvothermal synthesis of InP QDs and then the ZnS shell growth under UV irradiation. Indium chloride powder is dissolved in dodecylamine and toluene, then $P(N(CH_3)_2)_3$ is added to the above mixture under inert atmosphere. The reactant solution is transferred to an autoclave and heated to 180°°C for 24 h. The obtained InP QDs exhibited a broad size distribution (2.1–3.1 nm), emission wavelength between 530 and 628 nm and QYs of 5.3–19.3%. Next, phase transfer to water was achieved on size-selected QDs in chloroform by mixing them with

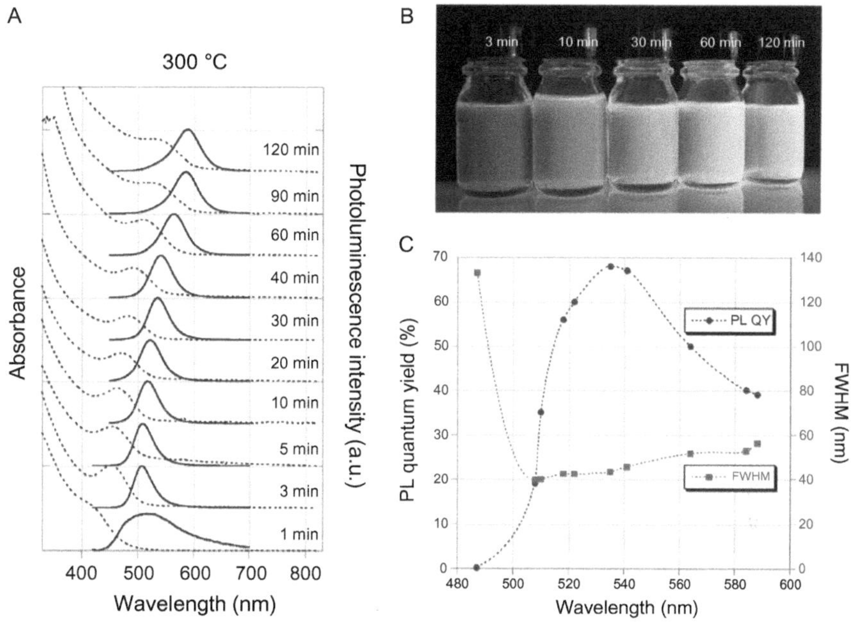

FIGURE 3.6 Single-step one-pot synthesis of InPZnS nanocrystals.[87] (A) Evolution of the PL ($\lambda_{ex}=400$ nm) and absorption spectra with reaction time (vertically shifted for clarity). (B) Photograph of some samples under UV light. (C) Evolution of the fluorescence QY and of the PL line width (increasing reaction time from left to right). Reproduced with permission from American Chemical Society 2008. (For colour version of this figure, the reader is referred to the online version of this chapter.)

an aqueous solution of zinc perchlorate and TGA at pH 11. Room temperature irradiation at $\lambda=365$ nm for 8 h afforded the growth of a ZnS shell via the photochemical decomposition of TGA. As a result, InP/ZnS core/shell QDs with emission wavelengths between 510–625 nm and QYs of 24–39% have been obtained.

3.3.2. InAs Nanocrystals with Different Shell Materials (CdSe, InP, ZnSe)

InAs is an interesting emitting material in the NIR range (700–1400 nm), provided that the nanocrystal size is kept small (1–3 nm). CdSe is an appropriate shell material due to its identical lattice parameter when compared to InAs, but also ZnSe has been investigated (lattice mismatch 6.6%).[161] Aharoni *et al.* reported the first strongly emitting system (QY > 70%) in form of InAs/CdSe/ZnSe core/shell/shell nanocrystals in 2006.[162] Xie and Peng described the synthesis of InAs nanocrystals with different shells, namely, CdSe, InP or ZnSe to limit the sensitivity towards oxidation of the core in 2008 (Figure 3.7).[88] The InAs QDs were synthesized in ODE by reacting

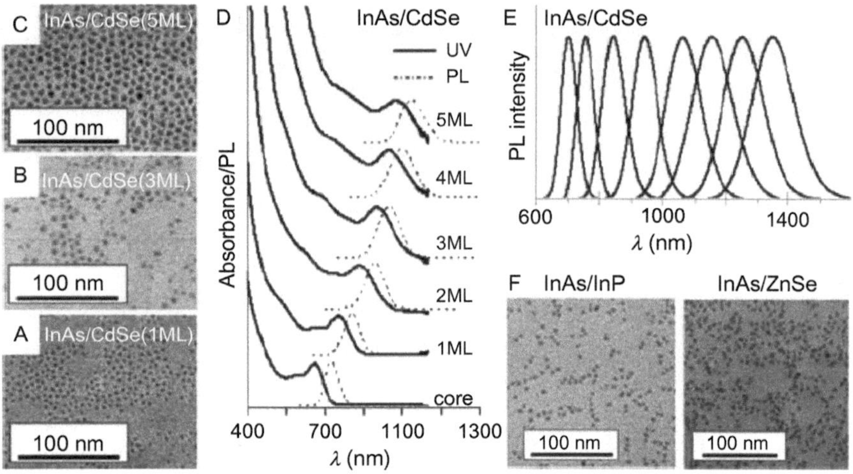

FIGURE 3.7 InAs-based core/shell nanocrystals.[88] (A–C) TEM images of InAs/CdSe nanocrystals with different shell thickness. (D) UV–vis absorption and photoluminescence spectra as a function of the shell thickness. (E) Series of PL spectra showing the spectral range of emission, which can be obtained. (F) TEM images of InAs/InP and of InP/ZnSe core/shell nanocrystals. Reproduced with permission from John Wiley & Sons, Inc., 2008.

indium stearate with $As(TMS)_3$ under inert atmosphere at various temperatures (100–300°°C) depending on the desired size. CdSe (ZnSe) shell growth was achieved on the InAs core QDs at 180°°C by the addition of Se dissolved in TOP and of the Cd (Zn) precursor. The latter was prepared by heating CdO (ZnO) powder with octanoic acid, precipitation with acetone, drying and dissolution in a mixture of ODE and octylamine. After heating to 190°°C for 30 min, the QY increased dramatically from below 1% for the core nanocrystals to as high as 90% for the InAs/CdSe core/shell system. For the synthesis of an InP shell, $P(TMS)_3$ mixed with ODE and octylamine was injected into the crude reaction mixture after formation of the InAs core nanocrystals.

Later, Allen *et al.* proposed a mixed CdZnS shell for capping small (1.4 nm) InAs QDs emitting at 800 nm.[90] After aqueous phase transfer using polymeric imidazole ligands, the retained QY was around 25%.

3.4. Nanocrystals of Ternary Chalcopyrite Semiconductors

Apart from III–V QDs, ternary semiconductor nanocrystals such as I–III–VI_2 type chalcopyrites ($CuInSe_2$–CISe, $CuInS_2$–CIS) have come up in the past few years as further alternative materials to cadmium-based systems. They are direct semiconductors and exhibit a relatively low band gap (1.05 eV for CISe, 1.5 eV for CIS). CIS and CISe nanocrystals were mainly studied because of their high potential for use in photovoltaics.[163,164] To the contrast, their PL properties were rarely investigated before 2008. Castro and co-workers reported a new synthesis method for CISe and CIS via the decomposition of the single source precursor $(PPh_3)_2CuIn(SEt)_4$, yielding luminescent CIS samples with a PL QY of ca. 5%.[165,166] Nakamura *et al.* doped CIS nanocrystals with Zn and were able to vary their PL wavelength from 570 to 800 nm, with QYs in the range of 5%.[167] Increased fluorescence QY upon addition of Zn has also recently been observed in another example of a I–III–VI_2 semiconductor: a solid solution of ZnS and $AgInS_2$ exhibited emission tunable in the visible range and NIR range with a QY of up to 24%.[168] Even higher QYs were obtained by using DDT as both the sulphur source and the surface ligand in ODE, and overcoating the obtained CIS nanocrystals with a ZnS shell.[95] The obtained QDs showed tunable emission in the range of 550–815 nm with a QY up to 60%. QYs up to 67% (80%) have been reported in 2011 when using a very similar approach, but using directly DDT instead of ODE as the solvent and a ZnS (CdS) shell (Figure 3.8).[169]

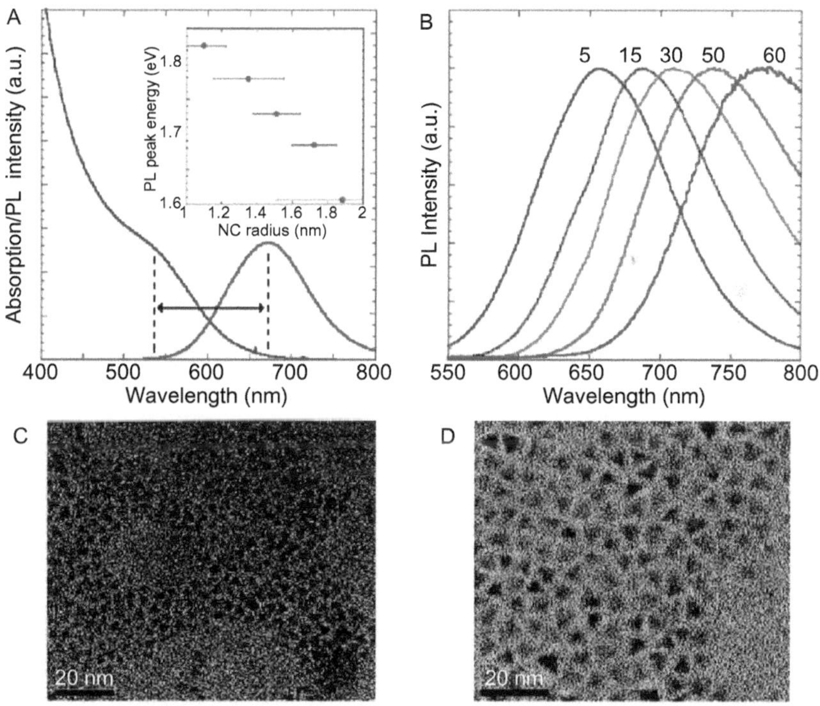

FIGURE 3.8 (A) Absorption and emission spectra of CIS nanocrystals.[169] Inset: PL peak energy as a function of NC size as determined by TEM. (B) Evolution of the emission spectra with reaction time in minutes. TEM images of (C) uncoated CIS nanocrystals and (D) CIS/ZnS core/shell nanocrystals. Reproduced with permission from American Chemical Society 2011. (For colour version of this figure, the reader is referred to the online version of this chapter.)

An early approach for the synthesis of CISe comprised the use of copper(I) and indium chloride in TOPO, and the injection of TOP-Se.[170] In 2007, CISe nanocrystals have been prepared in the non-coordinating solvent ODE.[171] Using a mixture of TOP and oleylamine as the solvent, Allen *et al.* reported in 2008 the synthesis of CISe nanocrystals whose emission ranged from 650 to 975 nm with a maximum QY of 25%.[96] Finally, Cassette *et al.* described the synthesis of $CuInSe_2$/ZnS core/shell nanocrystals in 2010.[97] In a first step, the core nanocrystals were prepared by reacting copper chloride with seleneourea and indium chloride in DDT. The core sizes were controlled by the final temperature of the synthesis or by the speed of the temperature rise. Shell growth was achieved by the slow addition of a mixture containing zinc ethylxanthate and zinc oleate dispersed in ODE, TOP and dioctylamine. The resulting core–shell QDs offer PL emission tunable from 700 to 1000 nm depending on the particle size, with QYs ranging from ~50% for smaller QDs to 10% to larger QDs in organic solvents.

One drawback of all ternary chalcopyrite nanocrystals is their comparably large emission line width, on the order of 100 nm (FWHM) in the best cases. The origin of this behaviour lies in the different emission process as compared to the discussed binary systems. As demonstrated by time-resolved fluorescence measurements, the emission in the ternary nanocrystals is related to recombinations from quantized conduction band states to localized acceptor states and not to excitonic band-edge transition.[95,169]

3.5. Cd_3P_2 and Cd_3As_2 Nanocrystals

II_3–V_2 semiconductor nanocrystals with efficient photoluminescence have been reported only very recently, since 2010. Bulk cadmium phosphide presents a band gap of 0.55 eV and a large exciton Bohr radius of 18 nm. Miao *et al.* reported the synthesis of 3.5–4.5 nm Cd_3P_2 nanocrystals by injection of $P(TMS)_3$ into a solution of cadmium oleate in ODE.[99] The emission could be tuned from 600 to 1200 nm by adding oleylamine and/or TOP and a maximum QY of 38% was obtained. A similar, yet simpler protocol was described by Xie *et al.* who used a mixture of CdO and oleic acid in ODE, in which a mixture of $P(TMS)_3$ and ODE was injected at 250°°C.[101] The size was tuned from 1.6 to 12 nm by varying the amount of oleic acid (Figure 3.9). The obtained QY was around 30% for samples smaller than 8 nm, with the best value being 70% for 2–3 nm nanocrystals.

Bulk cadmium arsenide presents an inverted band structure with a band gap of −0.19 eV; its exciton Bohr radius is estimated to be 47 nm. In 2011, Harris *et al.* reported the synthesis of Cd_3As_2 nanocrystals emitting between 530 and 2000 nm by reacting cadmium myristate with $As(TMS)_3$ at 175°°C (Figure 3.10).[102] The method chosen to inject the As precursor—initial fast injection to form small nuclei followed by the slow, continuous addition to promote growth—played an important role on the final properties of Cd_3As_2 QDs and allowed obtaining emission in the NIR range. The highest QY of 85% was reached for Cd_3As_2 QDs with an average size of 2.5 nm and an emission wavelength of $\lambda \sim 900$ nm, typical

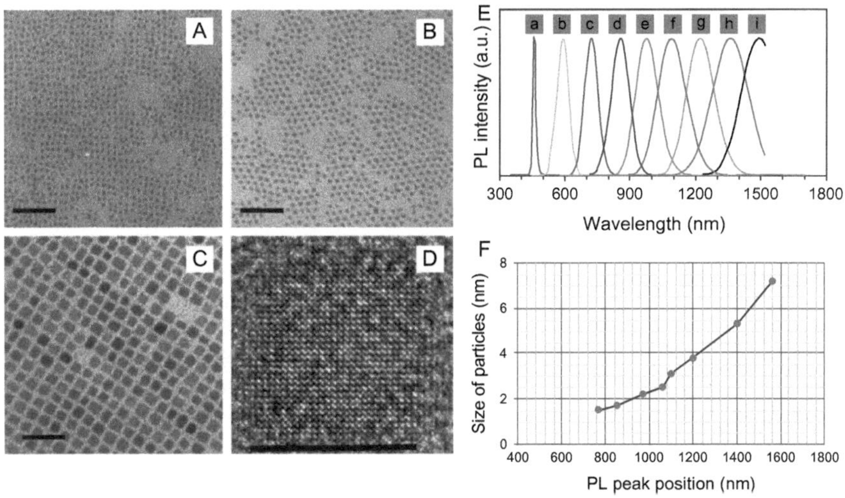

FIGURE 3.9 TEM images of Cd_3P_2 nanocrystals with different sizes: (A) 2.5 nm, (B) 4.0 nm and (C) 12 nm. HR-TEM image (D) of Cd_3P_2 nanocrystals (sample in C). Scale bar is 50 nm for TEM and 10 nm for HR-TEM. (E) Photoluminescence spectra of Cd_3P_2 nanocrystals with various sizes. (a) <1.5 nm, (b) <1.5 nm, (c) 1.5 nm, (d) 1.8 nm, (e) 2.4 nm, (f) 3 nm, (g) 4 nm, (h) 5.5 nm and (i) 7.6 nm. (F) Relation between the size of Cd_3P_2 nanocrystals and their emission peak positions. Reprinted from Ref. 101 with permission from American Chemical Society 2010. (For colour version of this figure, the reader is referred to the online version of this chapter.)

FIGURE 3.10 Cd_3As_2 nanocrystals. (A) Absorption; (B) photoluminescence spectra. Size: 2 nm (emission at 1.65 eV), 5 nm (emission at 0.65 eV). Reprinted from Ref. 102 with permission from American Chemical Society 2011. (For colour version of this figure, the reader is referred to the online version of this chapter.)

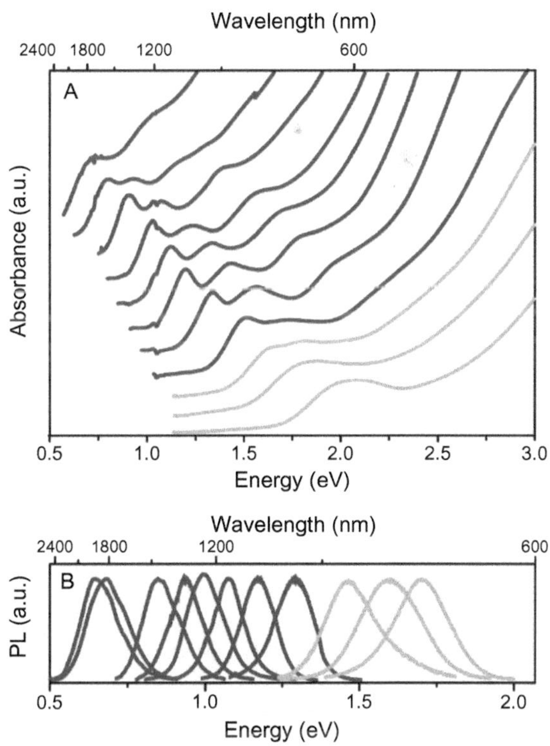

values ranged from 20% to 60%. Air sensitivity was noted, leading to a strong decrease of emission intensity after exposure to ambient conditions.

3.6. Aqueous Phase Transfer of Nanocrystals Prepared in Organics

For biological applications, one drawback of the synthesis in organic solvents is the fact that the obtained nanocrystals are generally hydrophobic. Right from the beginning, different approaches for the aqueous phase transfer have therefore been developed. The most important ones are represented in the scheme in Figure 3.11. They can roughly be divided into two main categories: (i) surface

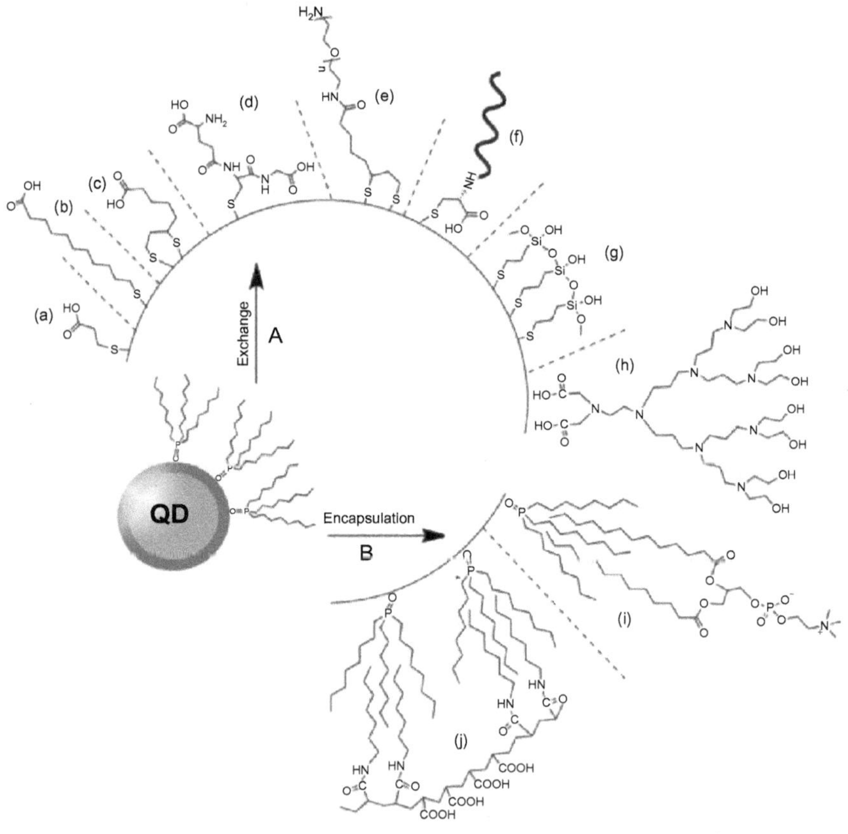

FIGURE 3.11 Strategies for the aqueous phase transfer of nanocrystals prepared in organic solvents. A) Exchange of the organic encapsulating layer with a water-soluble layer; a)-d) thiolated or dithiolated functional monolayers, e) glutathione layer, f) cysteine-terminated peptide, g) thiolated siloxane, h) carboxylic acid-functionalized dendrone. B) Encapsulation of QDs stabilized with an organic encapsulating layer in functional bilayer films composed of i) a phospholipid encapsulating layer, and j) a diblock copolymer. Scheme reproduced from Ref. 172 with permission from John Wiley & Sons, Inc., 2008.

ligand exchange and (ii) encapsulation. In the former case, the original ligands are replaced in an exchange reaction by new hydrophilic ligands. Many bifunctional molecules have been investigated, and in particular mercaptocarboxylic acids have been successfully used.[173] An important step towards improved stability was the development of dithiolated ligands, namely, derivatives of dihydrolipoic acid.[174,175] More recently, increasing stability through precise pH control during the phase transfer[137] as well as the use of metal complexes[176] and tailor-made, more complex ligands with multinary anchoring functions has been proposed.[89] Another elegant approach in the same vein is the use of specific peptides, containing, for example, several cysteine functions capable of grafting to the nanocrystal surface via their thiol groups.[177] This strategy has been extended to control the polyvalency, that is, the number of attached functional groups per nanocrystal.[178] Combined with electrophoretic separation, a precise control of the number of streptavidin molecules per QD could be achieved.[179] When aiming high stability at various pH values and at the same time low tendency for non-specific interactions in biological environment, purely cationic or anionic surface ligands give limited success. Uncharged poly (ethyleneglycol)s of different chain length have been shown to be suitable solubilizing agents yielding high furtivity of the nanocrystals in biological medium. Similarly, zwitterionic coatings achieved, for example, with cysteine yielded promising results. Another approach consists at grafting a first layer of siloxane ligands for the subsequent growth of a silica shell on the nanocrystal surface. Silica has the advantage of being transparent, chemically robust and giving the possibility for a wide variety of surface functionalization thanks to the large number of available functional trimethoxy- and triethoxysilanes. In the second strategy for aqueous phase transfer, the encapsulation method, amphiphilic molecules or macromolecules are wrapped around the nanocrystals with their synthesis ligands. Owing to hydrophobic interactions between the alkyl chains of these surface ligands and of the amphiphilic molecules, the latter organize in a way that their polar headgroups stick to the outside, resulting ultimatively in hydrosolubility of the QD. Stability issues have been overcome by cross-linking this additional ligand shell. Nowadays nanocrystals with an amphiphilic, cross-linked polymer shell containing either PEG groups or functional groups (e.g. amines, carboxylic acids) on the outer surface are commercially available. One difference has to be kept in mind when comparing ligand exchange and encapsulation approaches: in the latter case, the hydrodynamic diameter is significantly increased due to a thick organic layer. Therefore, the overall size of encapsulated nanocrystals can be on the order of 20 nm when starting with an inorganic core of around 7 nm.

For more information on the use in biology of QDs prepared in organic solvents, the interested reader is referred to the review articles in Refs. 173,180–186.

4. CONCLUSIONS

We have summarized the synthesis of semiconductor nanocrystals with a special emphasis on those systems, whose emission properties make them appealing candidates for use in fluorescent biological probes. Both aqueous synthesis and preparation in organic solvents afford a large palette of materials covering the whole visible and NIR range of interest. More progress has been made in recent years in the development of novel, more 'biocompatible' materials without Cd and Hg using synthesis in organic solvents; the overwhelming majority of nanocrystals synthesized in water (still) concern the II–VI semiconductor family and, in particular, cadmium chalcogenides. As a rule of thumb, aqueous synthesis results in most cases in broader size dispersions than syntheses in organics, due to the less efficient separation of nucleation and growth at lower reaction temperatures. Also, the growth of thick inorganic shells has been so far only reported in organics. Finally, it is very challenging to obtain good quality nanocrystals of III–V semiconductors (e.g. InP) using aqueous synthesis. On the other hand, nanocrystals grown in organic solvents require an additional aqueous phase transfer step prior to their use in biology. For many examples, it has been shown that this phase transfer is detrimental for long-term colloidal stability and fluorescence efficiency. Therefore, even though for some materials synthesized in organic medium the reported fluorescence QY approaches unity, after aqueous phase transfer their fluorescence efficacy generally lies below that of nanocrystals directly synthesized in water. Another drawback for some biological applications is related to the strong increase of the hydrodynamic diameter when using encapsulation methods for the aqueous phase transfer of nanocrystals prepared in organics. For all these reasons, QDs directly synthesized in water maintain a considerable interest for biological applications. Main challenges under current research concern the further development of high quality, that is, strongly emitting, photostable and chemically stable nanocrystals of non-toxic materials. Up till now, only a few examples of Cd-, Hg- and Pb-free nanocrystals exist covering the NIR range of 700–950 nm, which is particularly attracting for *in vivo* imaging. Ternary semiconductors, for example, from the chalcopyrite family, novel type II core/shell and doped systems and yet unexplored III–V semiconductors like group III—nitrides and—antimonides will likely play an important role in future developments.

In conclusion, thanks to their unique emission properties, QDs complement the palette of existing fluorophores for biological imaging. They are particularly competitive in studies which require high photostability, NIR emission and/or single-molecule sensitivity. The NIR emission properties of some materials combined with the possibility of anchoring different functionalities for targeting and vectorization on their surface make QDs also very appealing candidates for *in vivo* experiments, provided that toxicity issues related to their chemical nature and to their nanometric dimensions can be resolved.

REFERENCES

1. Henglein A. Photo-degradation and fluorescence of colloidal-cadmium sulfide in aqueous-solution. *Berichte Der Bunsen-Gesellschaft-Physical Chemistry, Chemical Physics* 1982; **86**:301–5.
2. Rossetti R, Nakahara S, Brus LE. Quantum size effects in the redox potentials, resonance Raman-spectra, and electronic-spectra of CdS crystallites in aqueous-solution. *J Chem Phys* 1983;**79**:1086–8.
3. Bruchez M, Moronne M, Gin P, Weiss S, Alivisatos AP. Semiconductor nanocrystals as fluorescent biological labels. *Science* 1998;**281**:2013–6.
4. Chan WCW, Nie SM. Quantum dot bioconjugates for ultrasensitive nonisotopic detection. *Science* 1998;**281**:2016–8.
5. Han M, Gao X, Su JZ, Nie S. Quantum-dot-tagged microbeads for multiplexed optical coding of biomolecules. *Nat Biotechnol* 2001;**19**:631–5.
6. Hildebrandt N. Biofunctional quantum dots: controlled conjugation for multiplexed biosensors. *ACS Nano* 2011;**5**:5286–90.
7. Weissleder R. A clearer vision for in vivo imaging. *Nat Biotechnol* 2001;**19**:316–7.
8. Stasiuk GJ, Tamang S, Imbert D, Poillot C, Giardiello M, Tisseyre C, et al. Cell-permeable Ln(III) chelate-functionalized InP quantum dots as multimodal imaging agents. *ACS Nano* 2011;**5**:8193–201.
9. Nirmal M, Dabbousi BO, Bawendi MG, Macklin JJ, Trautman JK, Harris TD, et al. Fluorescence intermittency in single cadmium selenide nanocrystals. *Nature* 1996;**383**:802–4.
10. Zhao J, Nair G, Fisher BR, Bawendi MG. Challenge to the charging model of semiconductor-nanocrystal fluorescence intermittency from off-state quantum yields and multiexciton blinking. *Phys Rev Lett* 2010;**104**:157403.
11. Rosen S, Schwartz O, Oron D. Transient fluorescence of the off state in blinking CdSe/CdS/ZnS semiconductor nanocrystals is not governed by auger recombination. *Phys Rev Lett* 2010;**104**:157404.
12. Galland C, Ghosh Y, Steinbruck A, Sykora M, Hollingsworth JA, Klimov VI, et al. Two types of luminescence blinking revealed by spectroelectrochemistry of single quantum dots. *Nature* 2011;**479**:203–7.
13. Reiss P, Protière M, Li L. Core/shell semiconductor nanocrystals. *Small* 2009;**5**:154–68.
14. Hines MA, Guyot-Sionnest P. Synthesis and characterization of strongly luminescing ZnS-capped CdSe nanocrystals. *J Phys Chem* 1996;**100**:468–71.
15. Reiss P, Carayon S, Bleuse J, Pron A. Low polydispersity core/shell nanocrystals of CdSe/ZnSe and CdSe/ZnSe/ZnS type: preparation and optical studies. *Synth Met* 2003;**139**:649–52.
16. Talapin DV, Mekis I, Gotzinger S, Kornowski A, Benson O, Weller H. CdSe/CdS/ZnS and CdSe/ZnSe/ZnS core-shell-shell nanocrystals. *J Phys Chem B* 2004;**108**:18826–31.
17. Kim S, Fisher B, Eisler HJ, Bawendi M. Type-II quantum dots: CdTe/CdSe(core/shell) and CdSe/ZnTe(core/shell) heterostructures. *J Am Chem Soc* 2003;**125**:11466–7.
18. LaMer VK, Dinegar RH. Theory, production and mechanism of formation of monodispersed hydrosols. *J Am Chem Soc* 1950;**72**:4847–54.
19. Donega CD, Liljeroth P, Vanmaekelbergh D. Physicochemical evaluation of the hot-injection method, a synthesis route for monodisperse nanocrystals. *Small* 2005;**1**:1152–62.
20. Kwon SG, Hyeon T. Formation mechanisms of uniform nanocrystals via hot-injection and heat-up methods. *Small* 2011;**7**:2685–702.
21. Park J, Joo J, Kwon SG, Jang Y, Hyeon T. Synthesis of monodisperse spherical nanocrystals. *Angew Chem Int Ed* 2007;**46**:4630–60.

22. Voorhees PW. The theory of Ostwald ripening. *J Stat Phys* 1985;**38**:231–52.
23. Fang Z, Li Y, Zhang H, Zhong XH, Zhu LY. Facile synthesis of highly luminescent UV-blue-emitting ZnSe/ZnS core/shell nanocrystals in aqueous media. *J Phys Chem C* 2009;**113**:14145–50.
24. Xu SH, Wang CL, Wang ZY, Zhang HS, Yang J, Xu QY, et al. Aqueous synthesis of internally doped Cu:ZnSe/ZnS core-shell nanocrystals with good stability. *Nanotechnology* 2011;**22**:275605–(7pp).
25. Deng ZT, Lie FL, Shen SY, Ghosh I, Mansuripur M, Muscat AJ. Water-based route to ligand-selective synthesis of ZnSe and Cd-doped ZnSe quantum dots with tunable ultraviolet A to blue photoluminescence. *Langmuir* 2009;**25**:434–42.
26. Han JS, Zhang H, Tang Y, Liu Y, Yao X, Yang B. Role of redox reaction and electrostatics in transition-metal impurity-promoted photoluminescence evolution of water-soluble ZnSe nanocrystals. *J Phys Chem C* 2009;**113**:7503–10.
27. Zhang J, Li J, Zhang JX, Xie RG, Yang WS. Aqueous synthesis of ZnSe nanocrystals by using glutathione as ligand: the pH-mediated coordination of Zn(2+) with glutathione. *J Phys Chem C* 2010;**114**:11087–91.
28. Qian HF, Qiu X, Li L, Ren JC. Microwave-assisted aqueous synthesis: a rapid approach to prepare highly luminescent ZnSe(S) alloyed quantum dots. *J Phys Chem B* 2006;**110**:9034–40.
29. Zan F, Ren JC. Significant improvement in photoluminescence of ZnSe(S) alloyed quantum dots prepared in high pH solution. *Luminescence* 2010;**25**:378–83.
30. Xu SH, Wang CL, Xu QY, Zhang HS, Li RQ, Shao HB, et al. Key roles of solution pH and ligands in the synthesis of aqueous ZnTe nanoparticles. *Chem Mater* 2010;**22**:5838–44.
31. Kumar P, Singh K. Synthesis, characterizations, and optical properties of copper selenide quantum dots. *Struct Chem* 2011;**22**:103–10.
32. Rajh T, Micic OI, Nozik AJ. Synthesis and characterization of surface-modified colloidal cadmium telluride quantum dots. *J Phys Chem* 1993;**97**:11999–2003.
33. Rogach AL, Katsikas L, Kornowski A, Su D, Eychmüller A, Weller H. Synthesis and characterization of thiol-stabilized CdTe nanocrystals. *Bunsenges. Phys. Chem. Bunsengesellschaft für physikalische, Chemie* 1996;**100**:1772–8.
34. Gaponik N, Talapin DV, Rogach AL, Hoppe K, Shevchenko EV, Kornowski A, et al. Thiol-capping of CdTe nanocrystals: an alternative to organometallic synthetic routes. *J Phys Chem B* 2002;**106**:7177–85.
35. He Y, Zhong YL, Su YY, Lu YM, Jiang ZY, Peng F, et al. Water-dispersed near-infrared-emitting quantum dots of ultrasmall sizes for in vitro and in vivo imaging. *Angew Chem Int Ed* 2011;**50**:5694–7.
36. Liu JJ, Shi ZX, Yu YC, Yang RQ, Zuo SL. Water-soluble multicolored fluorescent CdTe quantum dots: synthesis and application for fingerprint developing. *J Colloid Interface Sci* 2010;**342**:278–82.
37. Li C, Murase N. Surfactant-dependent photoluminescence of CdTe nanocrystals in aqueous solution. *Chem Lett* 2005;**34**:92–3.
38. Shavel A, Gaponik N, Eychmüller A. Factors governing the quality of aqueous CdTe nanocrystals: calculations and experiment. *J Phys Chem B* 2006;**110**:19280–4.
39. Dagtepe P, Chikan V, Jasinski J, Leppert VJ. Quantized growth of CdTe quantum dots; observation of magic-sized CdTe quantum dots. *J Phys Chem C* 2007;**111**:14977–83.
40. Dubavik A, Lesnyak V, Thiessen W, Gaponik N, Wolff T, Eychmüller A. Synthesis of amphiphilic CdTe nanocrystals. *J Phys Chem C* 2009;**113**:4748–50.
41. Fang Z, Liu L, Xu LL, Yin XG, Zhong XH. Synthesis of highly stable dyhydrolipoic acid capped water-soluble CdTe nanocrystals. *Nanotechnology* 2008;**19**:235603–(7pp).

42. Ferreira DL, Silva FO, Viol LCD, Licinio P, Valadares M, Cury LA, et al. Growth kinetics of CdTe colloidal nanocrystals. *J Chem Phys* 2009;**131**:084712–(4pp).
43. Han JS, Luo XT, Zhou D, Sun HZ, Zhang H, Yang B. Growth kinetics of aqueous CdTe nanocrystals in the presence of simple amines. *J Phys Chem C* 2010;**114**:6418–25.
44. Hayakawa Y, Nonoguchi Y, Wu HP, Diau EWG, Nakashima T, Kawai T. Rapid preparation of highly luminescent CdTe nanocrystals in an ionic liquid via a microwave-assisted process. *J Mater Chem* 2011;**21**:8849–53.
45. Khalavka Y, Mingler B, Friedbacher G, Okrepka G, Shcherbak L, Panchuk O. Influence of temperature on the synthesis of thiol-stabilized CdTe nanoparticles in aqueous solutions. *Phys Stat Solid A Appl Mater Sci* 2010;**207**:370–4.
46. Lesnyak V, Voitekhovich SV, Gaponik PN, Gaponik N, Eychmüller A. CdTe nanocrystals capped with a tetrazolyl analogue of thioglycolic acid: aqueous synthesis, characterization, and metal-assisted assembly. *ACS Nano* 2010;**4**:4090–6.
47. Li YL, Jing LH, Qiao RR, Gao MY. Aqueous synthesis of CdTe nanocrystals: progresses and perspectives. *Chem Commun* 2011;**47**:9293–311.
48. Liang GD, Shen LP, Zhang XL, Zou GZ. One-pot synthesis of dual-stabilizer-capped CdTe nanocrystals with efficient near-infrared photoluminescence and electrochemiluminescence. *Eur J Inorg Chem* 2011;(25):3726–30.
49. Liang GD, Shen LP, Zou GZ, Zhang XL. Efficient near-infrared electrochemiluminescence from CdTe nanocrystals with low triggering potential and ultrasensitive sensing ability. *Chemistry* 2011;**17**:10213–5.
50. Luo XT, Han JS, Ning Y, Lin Z, Zhang H, Yang B. Polyurethane-based bulk nanocomposites from 1-thioglycerol-stabilized CdTe quantum dots with enhanced luminescence. *J Mater Chem* 2011;**21**:6569–75.
51. Qian HF, Dong CQ, Weng JF, Ren JC. Facile one-pot synthesis of luminescent, water-soluble, and biocompatible glutathione-coated CdTe nanocrystals. *Small* 2006;**2**:747–51.
52. Sheng ZH, Han HY, Hu XF, Chi C. One-step growth of high luminescence CdTe quantum dots with low cytotoxicity in ambient atmospheric conditions. *Dalton Trans* 2010;**39**:7017–20.
53. Smith AM, Nie SM. Nanocrystal synthesis in an amphibious bath: spontaneous generation of hydrophilic and hydrophobic surface coatings. *Angew Chem Int Ed* 2008;**47**:9916–21.
54. Wang CL, Zhang H, Xu SH, Lv N, Liu Y, Li MJ, et al. Sodium-citrate-assisted synthesis of aqueous CdTe nanocrystals: giving new insight into the effect of ligand shell. *J Phys Chem C* 2009;**113**:827–33.
55. Wang CL, Zhang H, Zhang JH, Lv N, Li MJ, Sun HZ, et al. Ligand dynamics of aqueous CdTe nanocrystals at room temperature. *J Phys Chem C* 2008;**112**:6330–6.
56. Yuwen LH, Lu HT, He Y, Chen LQ, Hu M, Bao BQ, et al. A facile low temperature growth of CdTe nanocrystals using novel dithiocarbamate ligands in aqueous solution. *J Mater Chem* 2010;**20**:2788–93.
57. Zhang H, Wang D, Yang B, Moehwald H. Manipulation of aqueous growth of CdTe nanocrystals to fabricate colloidally stable one-dimensional nanostructures. *J Am Chem Soc* 2006;**128**:10171–80.
58. Rogach AL, Franzl T, Klar TA, Feldmann J, Gaponik N, Lesnyak V, et al. Aqueous synthesis of thiol-capped CdTe nanocrystals: state-of-the-art. *J Phys Chem C* 2007;**111**:14628–37.
59. Zeng QH, Kong XG, Sun YJ, Zhang YL, Tu LP, Zhao JL, et al. Synthesis and optical properties of type II CdTe/CdS core/shell quantum dots in aqueous solution via successive ion layer adsorption and reaction. *J Phys Chem C* 2008;**112**:8587–93.

60. Zhao D, He ZK, Chan WH, Choi MMF. Synthesis and characterization of high-quality water-soluble near-infrared-emitting CdTe/CdS quantum dots capped by N-acetyl-L-cysteine via hydrothermal method. *J Phys Chem C* 2009;**113**:1293–300.
61. Liu YF, Xie B, Yin ZG, Fang SM, Zhao JB. Synthesis of highly stable CdTe/CdS quantum dots with biocompatibility. *Eur J Inorg Chem* 2010;**10**:1501–6.
62. Green M, Williamson P, Samalova M, Davis JJ, Brovelli S, Dobson P, et al. Synthesis of type II/type I CdTe/CdS/ZnS quantum dots and their use in cellular imaging. *J Mater Chem* 2009;**19**:8341–6.
63. Zhang Y, Li Y, Yan XP. Aqueous layer-by-layer epitaxy of type-II CdTe/CdSe quantum dots with near-infrared fluorescence for bioimaging applications. *Small* 2009;**5**:185–9.
64. Zeng RS, Zhang TT, Liu JC, Hu S, Wan Q, Liu XM, et al. Aqueous synthesis of type-II CdTe/CdSe core-shell quantum dots for fluorescent probe labeling tumor cells. *Nanotechnology* 2009;**20**(9):095102–(8pp).
65. Taniguchi S, Green M, Rizvi SB, Seifalian A. The one-pot synthesis of core/shell/shell CdTe/CdSe/ZnSe quantum dots in aqueous media for in vivo deep tissue imaging. *J Mater Chem* 2011;**21**:2877–82.
66. Ayele DW, Chen HM, Su WN, Pan CJ, Chen LY, Chou HL, et al. Controlled synthesis of CdSe quantum dots by a microwave-enhanced process: a green approach for mass production. *Chemistry* 2011;**17**:5737–44.
67. Kalasad MN, Rabinal AK, Mulimani BG. Ambient synthesis and characterization of high-quality CdSe quantum dots by an aqueous route. *Langmuir* 2009;**25**:12729–35.
68. Li RF, Lee J, Kang DF, Luo ZT, Aindow M, Papadimitrakopoulos F. Band-edge photoluminescence recovery from zinc-blende CdSe nanocrystals synthesized at room temperature. *Adv Funct Mater* 2006;**16**:345–50.
69. Zhai CX, Zhang H, Du N, Chen BD, Huang H, Wu YL, et al. One-pot synthesis of biocompatible CdSe/CdS quantum dots and their applications as fluorescent biological labels. *Nanoscale Res Lett* 2011;**6**(1):31–(5pp).
70. Pan D, Wang Q, Jiang S, Ji X, An L. Low-temperature synthesis of oil-soluble CdSe, CdS, and CdSe/CdS core-shell nanocrystals by using various water-soluble anion precursors. *J Phys Chem C* 2007;**111**:5661–6.
71. Pan DC, Wang Q, Jiang SC, Ji XL, An LJ. Synthesis of extremely small CdSe and highly luminescent CdSe/CdS core-shell nanocrystals via a novel two-phase thermal approach. *Adv Mater* 2005;**17**:176.
72. Piven N, Susha AS, Doeblinger M, Rogach AL. Aqueous synthesis of alloyed CdSe(x)Te(1-x) nanocrystals. *J Phys Chem C* 2008;**112**:15253–9.
73. Liang GX, Li LL, Liu HY, Zhang JR, Burda C, Zhu JJ. Fabrication of near-infrared-emitting CdSeTe/ZnS core/shell quantum dots and their electrogenerated chemiluminescence. *Chem Commun* 2010;**46**:2974–6.
74. Mao WY, Guo J, Yang WL, Wang CC, He J, Chen JY. Synthesis of high-quality near-infrared-emitting CdTeS alloyed quantum dots via the hydrothermal method. *Nanotechnology* 2007;**18**(48):485611–(7pp).
75. Lesnyak V, Plotnikov A, Gaponik N, Eychmüller A. Toward efficient blue-emitting thiol-capped Zn(1-x)Cd(x)Se nanocrystals. *J Mater Chem* 2008;**18**:5142–6.
76. Lesnyak V, Lutich A, Gaponik N, Grabolle M, Plotnikov A, Resch-Genger U, et al. One-pot aqueous synthesis of high quality near infrared emitting Cd(1-x)Hg(x)Te nanocrystals. *J Mater Chem* 2009;**19**:9147–52.
77. Tang B, Yang F, Lin Y, Zhuo L, Ge J, Cao L. Synthesis and characterization of wavelength-tunable, water-soluble, and near-infrared-emitting CdHgTe nanorods. *Chem Mater* 2007;**19**:1212–4.

78. Kovalenko MV, Kaufmann E, Pachinger D, Roither J, Huber M, Stangl J, et al. Colloidal HgTe nanocrystals with widely tunable narrow band gap energies: from telecommunications to molecular vibrations. *J Am Chem Soc* 2006;**128**:3516–7.
79. Harrison MT, Kershaw SV, Burt MG, Rogach AL, Kornowski A, Eychmüller A, et al. Colloidal nanocrystals for telecommunications. Complete coverage of the low-loss fiber windows by mercury telluride quantum dots. *Pure Appl Chem* 2000;**72**:295–307.
80. Harrison MT, Kershaw SV, Rogach AL, Kornowski A, Eychmüller A, Weller H. Wet chemical synthesis of highly luminescent HgTe/CdS core/shell nanocrystals. *Adv Mater* 2000; **12**:123–5.
81. Rogach AL, Harrison MT, Kershaw SV, Kornowski A, Burt MG, Eychmüller A, et al. Colloidally prepared CdHgTe and HgTe quantum dots with strong near-infrared luminescence. *Phys Stat Solid B Basic Res* 2001;**224**:153–8.
82. Deng DW, Zhang WH, Chen XY, Liu F, Zhang J, Gu YQ, et al. Facile synthesis of high-quality, water-soluble, near-infrared-emitting PbS quantum dots. *Eur J Inorg Chem* 2009; **23**:3440–6.
83. Ung TDT, Reiss P, Nguyen QL. Luminescence properties of In(Zn)P alloy core/ZnS shell quantum dots. *Appl Phys Lett* 2010;**97**:193104.
84. Xie RG, Peng XG. Synthesis of Cu-doped InP nanocrystals (d-dots) with ZnSe diffusion barrier as efficient and color-tunable NIR emitters. *J Am Chem Soc* 2009;**131**:10645–51.
85. Xu S, Ziegler J, Nann T. Rapid synthesis of highly luminescent InP and InP/ZnS nanocrystals. *J Mater Chem* 2008;**18**:2653–6.
86. Xie R, Battaglia D, Peng X. Colloidal InP nanocrystals as efficient emitters covering blue to near-infrared. *J Am Chem Soc* 2007;**129**:15432–3.
87. Li L, Reiss P. One-pot synthesis of highly luminescent InP/ZnS nanocrystals without precursor injection. *J Am Chem Soc* 2008;**130**:11588–9.
88. Xie RG, Peng XG. Synthetic scheme for high-quality InAs nanocrystals based on self-focusing and one-pot synthesis of InAs-based core-shell nanocrystals. *Angew Chem Int Ed* 2008; **47**:7677–80.
89. Liu WH, Greytak AB, Lee J, Wong CR, Park J, Marshall LF, et al. Compact biocompatible quantum dots via RAFT-mediated synthesis of imidazole-based random copolymer ligand. *J Am Chem Soc* 2010;**132**:472–83.
90. Allen PM, Liu W, Chauhan VP, Lee J, Ting AY, Fukumura D, et al. InAs(ZnCdS) quantum dots optimized for biological imaging in the near-infrared. *J Am Chem Soc* 2009;**132**:470–1.
91. Nam DE, Song WS, Yang H. Noninjection, one-pot synthesis of Cu-deficient CuInS(2)/ZnS core/shell quantum dots and their fluorescent properties. *J Colloid Interface Sci* 2011; **361**:491–6.
92. Yong KT, Roy I, Hu R, Ding H, Cai HX, Zhu J, et al. Synthesis of ternary CuInS(2)/ZnS quantum dot bioconjugates and their applications for targeted cancer bioimaging. *Integr Biol* 2010;**2**:121–9.
93. Xie RG, Rutherford M, Peng XG. Formation of high-quality I-III-VI semiconductor nanocrystals by tuning relative reactivity of cationic precursors. *J Am Chem Soc* 2009;**131**: 5691–7.
94. Pons T, Pic E, Lequeux N, Cassette E, Bezdetnaya L, Guillemin F, et al. Cadmium-free CuInS(2)/ZnS quantum dots for sentinel lymph node imaging with reduced toxicity. *ACS Nano* 2010;**4**:2531–8.
95. Li L, Daou T, Texier I, Tran T, Liem N, Reiss P. Highly luminescent CuInS2/ZnS core/shell nanocrystals: cadmium-free quantum dots for in vivo imaging. *Chem Mater* 2009;**21**: 2422–9.

96. Allen PM, Bawendi MG. Ternary I-III-VI quantum dots luminescent in the red to near-infrared. *J Am Chem Soc* 2008;**130**(29):9240–1.
97. Cassette E, Pons T, Bouet C, Helle M, Bezdetnaya L, Marchal F, et al. Synthesis and characterization of near-infrared Cu-In-Se/ZnS core/shell quantum dots for in vivo imaging. *Chem Mater* 2010;**22**:6117–24.
98. Zhong H, Wang Z, Bovero E, Lu Z, van Veggel FCJM, Scholes GD. Colloidal CuInSe2 nanocrystals in the quantum confinement regime: synthesis, optical properties, and electroluminescence. *J Phys Chem C* 2011;**115**:12396–402.
99. Miao SD, Hickey SG, Rellinghaus B, Waurisch C, Eychmüller A. Synthesis and characterization of cadmium phosphide quantum dots emitting in the visible red to near-infrared. *J Am Chem Soc* 2010;**132**:5613–5.
100. Wang R, Ratcliffe CI, Wu X, Voznyy O, Tao Y, Yu K. Magic-sized Cd3P2 II-V nanoparticles exhibiting bandgap photoemission. *J Phys Chem C* 2009;**113**:17979–82.
101. Xie R, Zhang J, Zhao F, Yang W, Peng X. Synthesis of monodisperse, highly emissive, and size-tunable Cd3P2 nanocrystals. *Chem Mater* 2010;**22**:3820–2.
102. Harris DK, Allen PM, Han HS, Walker BJ, Lee JM, Bawendi MG. Synthesis of cadmium arsenide quantum dots luminescent in the infrared. *J Am Chem Soc* 2011;**133**:4676–9.
103. Ouyang JY, Ratcliffe CI, Kingston D, Wilkinson B, Kuijper J, Wu XH, et al. Gradiently alloyed ZnxCd1-xS colloidal photoluminescent quantum dots synthesized via a noninjection one-pot approach. *J Phys Chem C* 2008;**112**:4908–19.
104. Chen HS, Lo B, Hwang JY, Chang GY, Chen CM, Tasi SJ, et al. Colloidal ZnSe, ZnSe/ZnS, and ZnSe/ZnSeS quantum dots synthesized from ZnO. *J Phys Chem B* 2004;**108**:17119–23.
105. Lomascolo M, Creti A, Leo G, Vasanelli L, Manna L. Exciton relaxation processes in colloidal core/shell ZnSe/ZnS nanocrystals. *Appl Phys Lett* 2003;**82**:418–20.
106. Bae WK, Char K, Hur H, Lee S. Single-step synthesis of quantum dots with chemical composition gradients. *Chem Mater* 2008;**20**:531–9.
107. Dabbousi BO, RodriguezViejo J, Mikulec FV, Heine JR, Mattoussi H, Ober R, et al. (CdSe) ZnS core-shell quantum dots: synthesis and characterization of a size series of highly luminescent nanocrystallites. *J Phys Chem B* 1997;**101**:9463–75.
108. Talapin DV, Rogach AL, Kornowski A, Haase M, Weller H. Highly luminescent monodisperse CdSe and CdSe/ZnS nanocrystals synthesized in a hexadecylamine-trioctylphosphine oxide-trioctylphospine mixture. *Nano Lett* 2001;**1**:207–11.
109. Reiss P, Bleuse J, Pron A. Highly luminescent CdSe/ZnSe core/shell nanocrystals of low size dispersion. *Nano Lett* 2002;**2**:781–4.
110. Carbone L, Nobile C, De Giorg M, Sala FD, Morello G, Pompa P, et al. Synthesis and micrometer-scale assembly of colloidal CdSe/CdS nanorods prepared by a seeded growth approach. *Nano Lett* 2007;**7**:2942–50.
111. Li JJ, Wang YA, Guo WZ, Keay JC, Mishima TD, Johnson MB, et al. Large-scale synthesis of nearly monodisperse CdSe/CdS core/shell nanocrystals using air-stable reagents via successive ion layer adsorption and reaction. *J Am Chem Soc* 2003;**125**:12567–75.
112. Lin YW, Hsieh MM, Liu CP, Chang HT. Photoassisted synthesis of CdSe and core-shell CdSe/CdS quantum dots. *Langmuir* 2005;**21**:728–34.
113. Mekis I, Talapin DV, Kornowski A, Haase M, Weller H. One-pot synthesis of highly luminescent CdSe/CdS core-shell nanocrystals via organometallic and “greener” chemical approaches. *J Phys Chem B* 2003;**107**:7454–62.
114. Sarkar P, Springborg M, Seifert G. A theoretical study of the structural and electronic properties of CdSe/CdS and CdS/CdSe core/shell nanoparticles. *Chem Phys Lett* 2005;**405**:103–7.

115. Talapin DV, Koeppe R, Gotzinger S, Kornowski A, Lupton JM, Rogach AL, et al. Highly emissive colloidal CdSe/CdS heterostructures of mixed dimensionality. *Nano Lett* 2003;**3**:1677–81.
116. Talapin DV, Nelson JH, Shevchenko EV, Aloni S, Sadtler B, Alivisatos AP. Seeded growth of highly luminescent CdSe/CdS nanoheterostructures with rod and tetrapod morphologies. *Nano Lett* 2007;**7**:2951–9.
117. Lim SJ, Chon B, Joo T, Shin SK. Synthesis and characterization of zinc-blende CdSe-based core/shell nanocrystals and their luminescence in water. *J Phys Chem C* 2008;**112**:1744–7.
118. Manna L, Scher EC, Li LS, Alivisatos AP. Epitaxial growth and photochemical annealing of graded CdS/ZnS shells on colloidal CdSe nanorods. *J Am Chem Soc* 2002;**124**:7136–45.
119. Protiere M, Nerambourg N, Renard O, Reiss P. Rational design of the gram-scale synthesis of nearly monodisperse semiconductor nanocrystals. *Nanoscale Res Lett* 2011;**6**:472.
120. Jang E, Jun S, Jang H, Lim J, Kim B, Kim Y. White-light-emitting diodes with quantum dot color converters for display backlights. *Adv Mater* 2010;**22**:3076–80.
121. Chin PTK, Donega CDM, Bavel SS, Meskers SCJ, Sommerdijk N, Janssen RAJ. Highly luminescent CdTe/CdSe colloidal heteronanocrystals with temperature-dependent emission color. *J Am Chem Soc* 2007;**129**:14880–6.
122. Oron D, Kazes M, Banin U. Multiexcitons in type-II colloidal semiconductor quantum dots. *Phys Rev B* 2007;**75**(3):035330–(7pp).
123. Rao G. Sequential synthesis of colloidal type-II core/shell CdTe/CdSe semiconductor nanocrystals demonstrated. *MRS Bull* 2005;**30**:420.
124. Yu K, Zaman B, Romanova S, Wang DS, Ripmeester JA. Sequential synthesis of type II colloidal CdTe/CdSe core-shell nanocrystals. *Small* 2005;**1**:332–8.
125. Zhang WJ, Chen GJ, Wang J, Ye BC, Zhong XH. Design and synthesis of highly luminescent near-infrared-emitting water-soluble CdTe/CdSe/ZnS core/shell/shell quantum dots. *Inorg Chem* 2009;**48**:9723–31.
126. Xing B, Li WW, Sun K. A novel synthesis of high quality CdTe quantum dots with good thermal stability. *Mater Lett* 2008;**62**:3178–80.
127. Liao LF, Zhang H, Zhong XH. Facile synthesis of red- to near-infrared-emitting CdTe(x)Se(1-x) alloyed quantum dots via a noninjection one-pot route. *J Lumin* 2011;**131**:322–7.
128. Liang GX, Gu MM, Zhang JR, Zhu JJ. Preparation and bioapplication of high-quality, water-soluble, biocompatible, and near-infrared-emitting CdSeTe alloyed quantum dots. *Nanotechnology* 2009;**20**(41):415103–(9pp).
129. Bailey RE, Nie SM. Alloyed semiconductor quantum dots: tuning the optical properties without changing the particle size. *J Am Chem Soc* 2003;**125**:7100–6.
130. Taniguchi S, Green M, Lim T. The room-temperature synthesis of anisotropic CdHgTe quantum dot alloys: a "molecular welding" effect. *J Am Chem Soc* 2011;**133**:3328–31.
131. Hines MA, Scholes GD. Colloidal PbS nanocrystals with size-tunable near-infrared emission: observation of post-synthesis self-narrowing of the particle size distribution. *Adv Mater* 2003;**15**:1844–9.
132. Zhao H, Chaker M, Ma D. Effect of CdS shell thickness on the optical properties of water-soluble, amphiphilic polymer-encapsulated PbS/CdS core/shell quantum dots. *J Mater Chem* 2011;**21**:17483–91.
133. Fojtik A, Weller H, Koch U, Henglein A. Photo-chemistry of colloidal metal sulfides. 8. Photo-physics of extremely small CdS particles—Q-state CdS and magic agglomeration numbers. *Berichte Der Bunsen-Gesellschaft-Physical Chemistry Chemical Physics* 1984;**88**:969–77.
134. Lianos P, Thomas JK. Cadmium-sulfide of small dimensions produced in inverted micelles. *Chem Phys Lett* 1986;**125**:299–302.

135. Spanhel L, Haase M, Weller H, Henglein A. Photochemistry of colloidal semiconductors. 20. Surface modification and stability of strong luminescing CdS particles. *J Am Chem Soc* 1987;**109**:5649–55.
136. Xia Y-S, Zhu C-Q. Interaction of CdTe nanocrystals with thiol-containing amino acids at different pH: a fluorimetric study. *Microchim Acta* 2009;**164**:29–34.
137. Tamang S, Beaune G, Texier I, Reiss P. Aqueous phase transfer of InP/ZnS nanocrystals conserving fluorescence and high colloidal stability. *ACS Nano* 2011;**5**(12):9392–402. doi:10.1021/nn203598c.
138. Reiss P. ZnSe based colloidal nanocrystals: synthesis, shape control, core/shell, alloy and doped systems. *New J Chem* 2007;**31**:1843–52.
139. Shavel A, Gaponik N, Eychmüller A. Efficient UV-blue photoluminescing thiol-stabilized water-soluble alloyed ZnSe(S) nanocrystals. *J Phys Chem B* 2004;**108**:5905–8.
140. Murray CB, Norris DJ, Bawendi MG. Synthesis and characterization of nearly monodisperse CdE (E = S, SE, TE) semiconductor nanocrystallites. *J Am Chem Soc* 1993;**115**:8706–15.
141. Peng ZA, Peng XG. Formation of high-quality CdTe, CdSe, and CdS nanocrystals using CdO as precursor. *J Am Chem Soc* 2001;**123**:183–4.
142. Qu LH, Peng ZA, Peng XG. Alternative routes toward high quality CdSe nanocrystals. *Nano Lett* 2001;**1**:333–7.
143. Joo J, Na HB, Yu T, Yu JH, Kim YW, Wu FX, et al. Generalized and facile synthesis of semiconducting metal sulfide nanocrystals. *J Am Chem Soc* 2003;**125**:11100–5.
144. Pradhan N, Katz B, Efrima S. Synthesis of high-quality metal sulfide nanoparticles from alkyl xanthate single precursors in alkylamine solvents. *J Phys Chem B* 2003;**107**: 13843–54.
145. Pradhan N, Efrima S. Single-precursor, one-pot versatile synthesis under near ambient conditions of tunable, single and dual band fluorescing metal sulfide nanoparticles. *J Am Chem Soc* 2003;**125**:2050–1.
146. Chen Y, Vela J, Htoon H, Casson JL, Werder DJ, Bussian DA, et al. "Giant" multishell CdSe nanocrystal quantum dots with suppressed blinking. *J Am Chem Soc* 2008;**130**(15):5026–7.
147. Mahler B, Spinicelli P, Buil S, Quelin X, Hermier J-P, Dubertret B. Towards non-blinking colloidal quantum dots. *Nat Mater* 2008;**7**:659–64.
148. Furdyna JK. Diluted magnetic semiconductors. *J Appl Phys* 1988;**64**:R29–R64.
149. Xue J, Ye Y, Medina F, Martinez L, Lopez-Rivera SA, Giriat W. Temperature evolution of the 2.1 eV band in the Zn1-xMnxSe system for low concentration. *J Lumin* 1998;**78**:173–8.
150. Erwin SC, Zu LJ, Haftel MI, Efros AL, Kennedy TA, Norris DJ. Doping semiconductor nanocrystals. *Nature* 2005;**436**:91–4.
151. Pradhan N, Peng XG. Efficient and color-tunable Mn-doped ZnSe nanocrystal emitters: control of optical performance via greener synthetic chemistry. *J Am Chem Soc* 2007;**129**: 3339–47.
152. Pradhan N, Goorskey D, Thessing J, Peng XG. An alternative of CdSe nanocrystal emitters: pure and tunable impurity emissions in ZnSe nanocrystals. *J Am Chem Soc* 2005;**127**: 17586–7.
153. Pradhan N, Battaglia DM, Liu YC, Peng XG. Efficient, stable, small, and water-soluble doped ZnSe nanocrystal emitters as non-cadmium biomedical labels. *Nano Lett* 2007;**7**: 312–7.
154. Micic OI, Curtis CJ, Jones KM, Sprague JR, Nozik AJ. Synthesis and characterization of InP quantum dots. *J Phys Chem* 1994;**98**:4966–9.
155. Guzelian AA, Katari JEB, Kadavanich AV, Banin U, Hamad K, Juban E, et al. Synthesis of size-selected, surface-passivated InP nanocrystals. *J Phys Chem* 1996;**100**:7212–9.

156. Guzelian AA, Banin U, Kadavanich AV, Peng X, Alivisatos AP. Colloidal chemical synthesis and characterization of InAs nanocrystal quantum dots. *Appl Phys Lett* 1996;**69**:1432–4.
157. Battaglia D, Peng XG. Formation of high quality InP and InAs nanocrystals in a noncoordinating solvent. *Nano Lett* 2002;**2**:1027–30.
158. Xu S, Kumar S, Nann T. Rapid synthesis of high-quality InP nanocrystals. *J Am Chem Soc* 2006;**128**:1054–5.
159. Li L, Protière M, Reiss P. Economic synthesis of high quality InP nanocrystals using calcium phosphide as the phosphorus precursor. *Chem Mater* 2008;**20**:2621–3.
160. Byun HJ, Lee JC, Yang H. Solvothermal synthesis of InP quantum dots and their enhanced luminescent efficiency by post-synthetic treatments. *J Colloid Interface Sci* 2011;**355**:35–41.
161. Cao YW, Banin U. Growth and properties of semiconductor core/shell nanocrystals with InAs cores. *J Am Chem Soc* 2000;**122**:9692–702.
162. Aharoni A, Mokari T, Popov I, Banin U. Synthesis of InAs/CdSe/ZnSe core/shell1/shell2 structures with bright and stable near-infrared fluorescence. *J Am Chem Soc* 2006;**128**:257–64
163. Nairn JJ, Shapiro PJ, Twamley B, Pounds T, von Wandruszka R, Fletcher TR, et al. Preparation of ultrafine chalcopyrite nanoparticles via the photochemical decomposition of molecular single-source precursors. *Nano Lett* 2006;**6**:1218–23.
164. Czekelius C, Hilgendorff M, Spanhel L, Bedja I, Lerch M, Muller G, et al. A simple colloidal route to nanocrystalline ZnO/CuInS2 bilayers. *Adv Mater* 1999;**11**(8):643–6.
165. Castro SL, Bailey SG, Raffaelle RP, Banger KK, Hepp AF. Nanocrystalline chalcopyrite materials (CuInS2 and CuInSe2) via low-temperature pyrolysis of molecular single-source precursors. *Chem Mater* 2003;**15**:3142–7.
166. Castro SL, Bailey SG, Raffaelle RP, Banger KK, Hepp AF. Synthesis and characterization of colloidal CuInS2 nanoparticles from a molecular single-source precursor. *J Phys Chem B* 2004;**108**:12429–35.
167. Nakamura H, Kato W, Uehara M, Nose K, Omata T, Otsuka-Yao-Matsuo S, et al. Tunable photoluminescence wavelength of chalcopyrite CuInS2-based semiconductor nanocrystals synthesized in a colloidal system. *Chem Mater* 2006;**18**:3330–5.
168. Torimoto T, Adachi T, Okazaki K, Sakuraoka M, Shibayama T, Ohtani B, et al. Facile synthesis of ZnS-AgInS2 solid solution nanoparticles for a color-adjustable luminophore. *J Am Chem Soc* 2007;**129**(41):12388–9.
169. Li L, Pandey A, Werder DJ, Khanal BP, Pietryga JM, Klimov VI. Efficient synthesis of highly luminescent copper indium sulfide-based core/shell nanocrystals with surprisingly long-lived emission. *J Am Chem Soc* 2011;**133**:1176–9.
170. Malik MA, O'Brien P, Revaprasadu N. A novel route for the preparation of CuSe and CuInSe2 nanoparticles. *Adv Mater* 1999;**11**:1441–4.
171. Zhong HZ, Li YC, Ye MF, Zhu ZZ, Zhou Y, Yang CH, et al. A facile route to synthesize chalcopyrite CuInSe2 nanocrystals in non-coordinating solvent. *Nanotechnology* 2007;**18**(2):025602–(6pp).
172. Gill R, Zayats M, Willner I. Semiconductor quantum dots for bioanalysis. *Angew Chem Int Ed* 2008;**47**:7602–25.
173. Medintz IL, Uyeda HT, Goldman ER, Mattoussi H. Quantum dot bioconjugates for imaging, labelling and sensing. *Nat Mater* 2005;**4**:435–46.
174. Mattoussi H, Mauro JM, Goldman ER, Anderson GP, Sundar VC, Mikulec FV, et al. Self-assembly of CdSe-ZnS quantum dot bioconjugates using an engineered recombinant protein. *J Am Chem Soc* 2000;**122**:12142–50.
175. Clapp AR, Goldman ER, Mattoussi H. Capping of CdSe-ZnS quantum dots with DHLA and subsequent conjugation with proteins. *Nat Protoc* 2006;**1**:1258–66.

176. Liu D, Snee PT. Water-soluble semiconductor nanocrystals cap exchanged with metalated ligands. *ACS Nano* 2010;**5**:546–50.
177. Pinaud F, King D, Moore H-P, Weiss S. Bioactivation and cell targeting of semiconductor CdSe/ZnS nanocrystals with phytochelatin-related peptides. *J Am Chem Soc* 2004;**126**: 6115–23.
178. Pinaud F, Clarke S, Sittner A, Dahan M. Probing cellular events, one quantum dot at a time. *Nat Methods* 2010;**7**:275–85.
179. Clarke S, Tamang S, Reiss P, Dahan M. A simple and general route for monofunctionalization of fluorescent and magnetic nanoparticles using peptides. *Nanotechnology* 2011;**22**: 175103.
180. Michalet X, Pinaud FF, Bentolila LA, Tsay JM, Doose S, Li JJ, et al. Quantum dots for live cells, in vivo imaging, and diagnostics. *Science* 2005;**307**:538–44.
181. Wang C, Gao X, Su X. In vitro and in vivo imaging with quantum dots. *Anal Bioanal Chem* 2010;**397**:1397–415.
182. Smith AM, Duan H, Mohs AM, Nie S. Bioconjugated quantum dots for in vivo molecular and cellular imaging. *Adv Drug Deliv Rev* 2008;**60**:1226–40.
183. Jamieson T, Bakhshi R, Petrova D, Pocock R, Imani M, Seifalian AM. Biological applications of quantum dots. *Biomaterials* 2007;**28**:4717–32.
184. Somers RC, Bawendi MG, Nocera DG. CdSe nanocrystal based chem-/bio-sensors. *Chem Soc Rev* 2007;**36**:579–91.
185. Klostranec JM, Chan WCW. Quantum dots in biological and biomedical research: recent progress and present challenges. *Adv Mater* 2006;**18**:1953–64.
186. Gao X, Yang L, Petros JA, Marshall FF, Simons JW, Nie S. In vivo molecular and cellular imaging with quantum dots. *Curr Opin Biotechnol* 2005;**16**:63–72.

Chapter 4

Synthesis of Organic Nanoparticles

Gabriela Romero and Sergio E. Moya
CIC biomaGUNE, Paseo Miramón, San Sebastian, Gipuzkoa, Spain

1. INTRODUCTION

Organic nanoparticles (NPs) are templated upon natural or synthetic organic molecules. Nature provides a wide range of examples of organic NPs such as protein aggregates, lipid bodies, milk emulsions, or more complex organized structures such as viruses, to name a few.

Organic NPs also form part of many industrial products, mostly in food and cosmetics. Many food products such as creams, chocolate, and cakes present nanoemulsions in their formulation and that is also the case in the cosmetic industry. Organic NPs are also used in pharmaceutical formulations, that is, liposome vectors, polymersomes, polymer–protein, or polymer–drug conjugates.

One of the most important features of organic NPs is that they offer relatively simple routes for the encapsulation of materials. This together with the fact that the molecules used for the fabrication of the organic NPs can be biodegradable makes organic NPs the most appealing systems for drug delivery and biomedical applications.

1.1. Organic Versus Inorganic

Organic NPs differ conceptually from inorganic NPs in terms of the principles of fabrication. Inorganic NPs are normally formed by the precipitation of inorganic salts, which are linked in a matrix. The nature of the binding among atoms can be different: covalent, metallic, etc., but in any case, the inorganic structure forms a three-dimensional arrangement with linked atoms. In this regard, the closest organic NPs to an inorganic one would be a dendrimer, which is a single-molecule NP. Most of the organic NPs are formed by several organic molecules which are driven together by self-organization or chemical binding. For the synthesis of organic NPs, with the exception of dendrimers and some other polymer NPs, we first need organic molecules that can arrange themselves three-dimensionally. Interactions among the molecules

Frontiers of Nanoscience, Vol. 4. DOI: 10.1016/B978-0-12-415769-9.00004-2

are sometimes not strong like for micelles and vesicles. Self-assembly and the presence of zwitterionic molecules, with polar and nonpolar regions, as the main components of NPs or the means to encapsulate other organic molecules are key elements for the fabrication of many of the organic NPs as we show in this chapter.

Also another important difference from inorganic NPs is that many of the organic NPs, due to the week nature of the interactions holding them together, have a dynamic character. Micelles and vesicles can fuse and generate larger particles. The continuous addition of surfactant to a micellar solution can indeed increase the size of the micelles and even force it to change shape from spherical to cylindrical over the second critical micellar concentration (CMC). Sizes can also not always be truly nano for organic NPs as is the case for inorganic NPs. The formation of inorganic NPs sometimes needs only the clustering of a few atoms. In the case of the organic NPs, the size and geometry of the organic molecules impose in many cases the limits to going nano. In micelles, which are among the smallest organic NPs, size is controlled by the length of the surfactant and the smallest sizes to be achieved are that of a sphere with a radius equal to the length of the surfactant in the micellar phase, typically around 15–20 Å. The lipids forming a vesicle assemble into bilayers because of the relation of their hydrophilic and hydrophobic volume. The bilayers must close in on themselves to make the vesicle, which not only results in an internal volume that can be used for encapsulation but also establishes limits for the smallest reachable size. For organic NPs such as polymer latexes, it is even more difficult to get below a certain size threshold. For latexes, the polymeric core is stabilized by a surfactant. The size of the NP is then that of the core and the stabilizing surfactant. Again, dendrimers, as single-molecule NPs, are an exception and can have a controllable size below 10 nm.

Organic NPs can be fabricated both by "top-down" and by "bottom-up" approaches. In the "top-down" techniques, the most common is the mechanical milling,[1] although, recently, other more complicated techniques involving microfluidics and lithography[2,3] have been used to produce organic nanomaterials.

In the "bottom-up" techniques, organic NPs are typically produced via precipitation and condensation methods,[4,5] using synthetic methodologies[6,7] or applying self-assembly principles.

2. "TOP-DOWN" TECHNIQUES

The techniques that involve the synthesis of nanosized materials from their source material in the macro- or microrange are known as "top-down" techniques, indicating their fabrication via size reduction techniques.

In the "top-down" techniques, probably milling is the most used technique. Milling processes are based on the mechanical crushing of the bulk material to obtain nanostructures with wide size distribution, and in most of the cases, an unavoidable contamination of the produced nanomaterials is due to the abrasion. These processes are mostly used to fabricate drug NPs from poor water-soluble compounds, and almost all the products in the market fabricated by this technique are done by using the pearl mill that was developed by Liversidge and coworkers[8,9] in 1995, today known as the NanoCrystal™ technology. One example of this technique is shown in Ref. 10, where Sigfridsson and coworkers prepared drug NPs of ticagrelor with diameter around 230 nm from poor water-soluble drug crystals by wet-milling technique. The resulting formulation was chemically and physically stable during long periods of time. They also proved in animal models that this nanosuspension has no effect on the hemodynamic parameters.

Other "top-down" techniques include the use of microfluidics and lithography. These have been used recently to produce organic nanomaterials for nanomedicine and electronic applications. This kind of methodologies offers more control over the size, shape, cargo, and composition of the NPs. For example, Rolland *et al.*[3] make use of the particle replication in nonwetting templates (PRINT), a lithography technique that consist in the printing of NPs in a nonwet substrate using a stone or metal pattern with smooth surface; they fabricate isolated monodispersed polymer NPs of poly(ethylene glycol diacrylate), triacrylate resin, poly(lactic acid), and poly(pyrrole), all of them with a size below 200 nm. This approach allows the fabrication of NPs of any shape and the encapsulation of biomaterials such as DNA and proteins.

3. "BOTTOM-UP" TECHNIQUES

"Bottom-up" techniques are based on the building up of NPs via physicochemical processes, starting from single molecules. In contrast to the "top-down" approaches, in "bottom-up" techniques, NPs are fabricated from atoms or molecules via synthetic chemistry and self-organization. The "bottom-up" techniques are used for the generation of a wide range of nanomaterials such as micelles,[11–25] vesicles and liposomes,[26–44] polymersomes,[45–51] polymer conjugates,[52–66] dendrimers,[67–77] capsules,[78–122] and polymeric NPs.[123–146,92,147–151] We will provide a general description of the different organic NPs produced by bottom-up techniques. Indeed, this is the main scope of this chapter, to visualize the rich variety of materials that can be achieved by these techniques.

3.1. Micelles, Vesicles, and Liposomes

Liposomes, vesicles, and micelles are NPs formed by supramolecular self-assembly of surfactants and lipids.[6] The driven force of the assembly is the low water solubility of the amphiphilic molecules.

The term "micelles" was introduced by McBain in 1913[11] to describe the colloidal particles formed by detergents and soaps. Basically, micelles are colloidal aggregates of amphiphilic molecules.[12] When amphiphilic molecules are added into water, the nonpolar chain "tail" of the molecule aggregates, letting the polar side "head" in direct contact with the solvent molecules and a spherical or cylindrical structure that protects the hydrophobic tails in the center. Inverse micelles are obtained when the amphiphilic molecule is added into a nonpolar solvent. In this case, the particle formed has the head group in the center and the tail chains in the outside.

The shape and size of the micelles can be controlled by the nature of the amphiphilic molecule and the solution conditions such as concentration, temperature, pH, and ionic strength.[13] The concentration of the amphiphilic molecule (amphiphilie) plays probably the most important role in micelle formation. Micelles are truly dynamic systems. Normally, micelle formation does not take place below a certain concentration of surfactant, called the CMC. Initially, micelles display spherical shapes and their size increases with the addition of more surfactant, up to a point where they change shape and start to grow as cylinders.

Typical surfactants forming micelles are the sodium dodecyl sulfate and cetyl trimethylammonium bromide, just to give an example. These molecules are formed by a hydrophobic chain with a length of 10–20 carbons and sulfate and ammonium hydrophilic head, respectively. Some proteins and lipids, as well, can aggregate in micelles such as lipoxygenase 3 that by inducing heat form soluble spherical aggregates of around 26 nm in carbonate buffer pH 8.[25]

Vesicles, unlike micelles, are hollow spheres enclosed by a bilayer made of different amphiphilic molecules.[26] When the hollow sphere presents only one amphiphilie bilayer, they are called unilamellar vesicles (UVs), and if it is composed by more than one amphiphilie bilayer, the vesicles are known as mutilamellar vesicles (MLVs) (Figure 4.1).

Depending on their composition, two kinds of vesicles can be distinguished: first, vesicles composed of glycolipids, either natural or synthetic,[27] which are very important because the cell membrane is highly composed of glycolipids and can be formed naturally or can be prepared, and second,

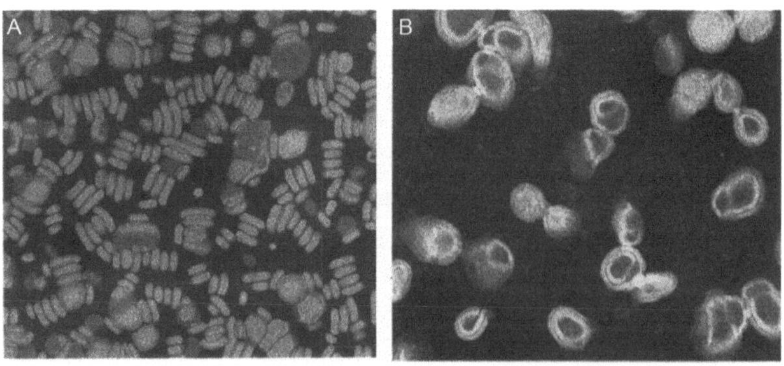

FIGURE 4.1 Examples of electron microscope images of unilamellar (A) and multilamellar (B) liposomes made of egg phosphatidyl choline containing 45 mol% cholesterol and fabricated with French pressure cell.[36]

vesicles made of phospholipids and their mixtures, discovered by Bangham in 1961,[28] also called "liposomes" or "lipid vesicles."

Vesicle properties such as size and polydispersity, surface potential, degree of ionization, phase behavior, permeability, and physical stability depend on the method employed for their preparation as well as on the nature of the component molecules.[29] The methods for vesicle formation can be classified in two groups. The first one is the spontaneous formation of vesicles,[30,40] in which it is just necessary to apply certain stress in order to homogenize the system but no external energy is required for the vesicle formation. The second method of vesicle fabrication is the induced vesicle formation by shearing,[31,40] and this shearing can be via sonication, extrusion, and other methods; all of those techniques require the input of external energy, reason for what vesicles are not thermodynamically stable though they can have high kinetic stability. For the formation of MLVs,[32] usually lipids are dissolved in an organic solvent. Later, the solvent is eliminated via vacuum evaporation or convection to form a dry lipid film; this step is followed by the hydration of the film with an aqueous buffer and finally gentle agitation to form a MLV mixture. The size and the distribution of the MLVs depend on the lipid concentration and composition and on the input energy during agitation. Indeed, vigorous shaking will form vesicles with better size distribution. Another well-known method to fabricate MLVs is the solvent spherule evaporation proposed in 1984 by Kim and coworkers.[33] This method is a reverse-phase evaporation method that consists in the aqueous dispersion of small spherules of volatile hydrophobic solvents in which amphiphilic lipids are dissolved, after evaporation of the organic phase MLVs in aqueous solution are obtained. MLVs can also be formed from UVs by controlled fusion, freeze–thawing, or simple dehydration–hydration cycles. The best size distribution of MLVs is obtained with the extrusion method.[34] In this method, the lipid suspension is forced to pass through a polycarbonate filter with a defined pore size. The extruded vesicles have a diameter near the pore size.

If phospholipids are dispersed in aqueous solution, they form large multilamellar structures as we discussed above. Nevertheless, several methods have been developed to form UVs from multilamellar structures. Sonication is the most common method to produce UVs from MLVs. The ultrasound applied during sonication induces collisions between vesicles, which, in turn, disrupt the lipid multilamellars. This results in the formation of lipid bilayer fragments and other small aggregates, which rearrange into vesicles.[35] There are several drawbacks of this technique such as the nonuniform size distribution of the formed vesicles and the lack of stability of the formed vesicles that tend to aggregate over time. Also, in the case of bath sonication, the process cannot be scaled up and the probe sonication generates some metal contamination that induces lipid degradation. Another way to prepare UVs from MLVs is by the French press technique.[36] This consists in passing the MLV suspension through a small outlet orifice under extreme pressure to form UVs. Finally, the extrusion method described above can be used to produce UVs from MLVs[34] through the sequential extrusion of the MLVs generating small and large UVs.

There are different methods, not based on MLVs, to produce UVs such as the dry lipid hydration methods.[32] This method consists in the creation of a thin film via evaporation of the amphiphile in chloroform, when the thin film in contact with water is dissolved and forms large UVs. Another way to produce UVs is by injection methods[37] that mainly consist of injecting a lipid organic solution into water. The speed of the injection will determine the stability of the system over time. The reverse-phase evaporation method is very well accepted to produce large UVs. This method is based on the dispersion of the aqueous phase in an organic solvent in order to form an emulsion by sonication. The solvent is evaporated under reduced pressure, and finally, the suspension is filtered to obtain vesicles smaller than the filter pore size.[38] There are also methods based on the reconstitution from lipid bilayers, of which detergent depletion technique[39] is the most useful; dry lipids or preformed vesicles are dissolved in presence of detergents that later can be removed by different separation methods. The lipids coalesce and adopt the bilayer configurations that results in seal UVs.

Today, vesicles and liposomes play an important role in research as model systems for permeable biological membranes.[40] They also have significance for the industry, especially in pharmaceutical, cosmetics, and food, as it is known that vesicles can entrap poor water-soluble materials with the possibility to control the action of the embedded molecules. In the pharmaceutical field, significant research is focused on trying to mimic different permeable membranes for drug delivery using vesicles and liposomes.[35,41,44] Some liposome-based delivery systems are already in the market such as Doxil®, Ambisome®, and Daunoxome®. In food industry, vesicles are normally used for encapsulation and controlled release of flavors and nutrients in order to protect them against degradation or contamination and to enhance bioavailability, stability, and shelf-life of sensitive materials.[42] For example, the encapsulation of enzymes in liposomes protects them from degradation. It is also a means to control the release of the enzyme, like for cheese fabrication that encapsulation of enzyme in vesicles is carried out to decelerate cheese ripening.[43] In the cosmetic industry, vesicles are commercially used since 1987 when hydration creams made of liposomes where introduced by C. Dior (Capture) and L'Oréal (Niosomes).[44] Nowadays, we can find several cosmetic formulations of liposomes with encapsulated antibiotics, moisturizers, and different extracts and proteins, such as creams, sunscreens, perfumes, hair conditioners, aftershaves, and shampoos.

3.2. Polymer Micelles, Polymersomes, and Polymer Conjugates

Analogous to amphiphilic molecules, diblock copolymers with a hydrophilic and a hydrophobic block form micelles in certain solvents. When a solvent is thermodynamically good for one of the blocks and generates precipitation in the other block[15]—in other words, when one of the blocks of the copolymers is soluble in the solvent and the other block is insoluble—the copolymer chains associate and form micellar aggregates with a core formed of insoluble copolymer blocks and a shell composed by the soluble copolymer blocks. The micelles of block copolymers have a wide practical interest because they can be used as dispersants,[16] emulsifiers,[17] wetting agents,[18] foam stabilizers, flocculants, demulsifiers,[19] viscosity modifiers,[20] and among other applications in different industrial and pharmaceutical areas such as nanocarriers for drug delivery systems. A very good example of the use of block copolymers micelles as targeted nanocarriers is shown in Ref. 21 in which Zhu *et al.* obtained 146 nm in diameter micelles formed by folate-conjugated amphiphilic star-shaped block copolymers of poly(L-lactide), methoxy poly (ethylene glycol), and folate-poly(ethylene glycol) (Figure 4.2). They loaded the micelles with the anti-inflammatory drug indomethacin via dialysis, and from the release studies, they found a sustained release time of 40 h. They demonstrate as well specific targeting in HeLa cells, where the micelles were taken up by the cells via receptor-mediated endocytosis.

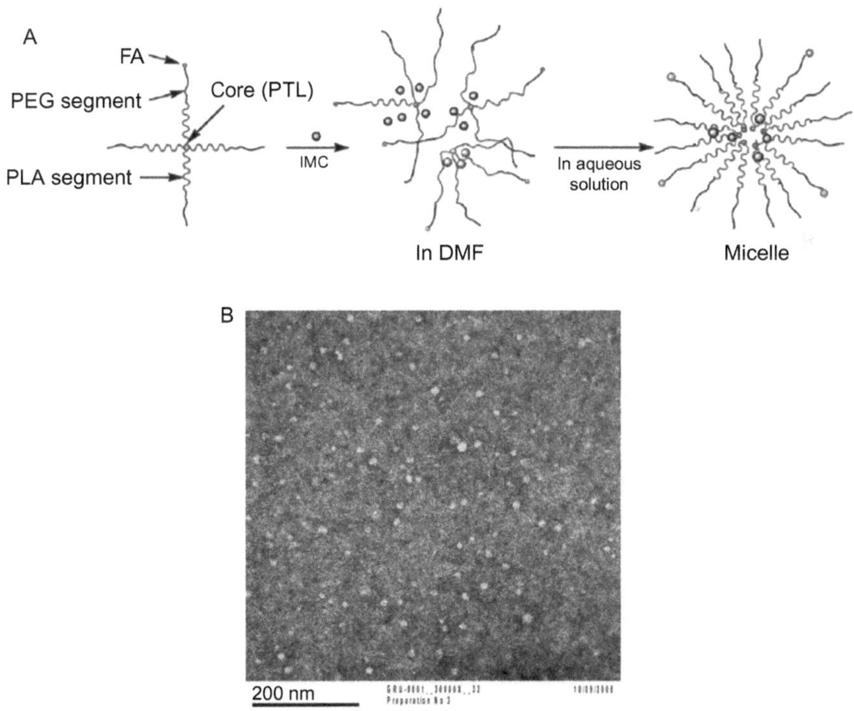

FIGURE 4.2 Diagram of the formation of a star-shaped polymeric micelle (A)[21] and TEM image of polymeric micelles (B) (Romero G., unpublished). (For color version of this figure, the reader is referred to the online version of this chapter.)

Polymersomes are polymer-based artificial vesicles that are formed during hydration of high-molecular-weight amphiphilic block copolymers.[45] A block copolymer without presence of solvent displays a wide range of ordered morphologies including lamellar phase; the hydration of this lamellae by aqueous solution form a stable dispersion of block copolymer aggregates or polymersomes. The difference between the amphiphilic character of a lipid and the one in a block copolymer is that, in the last hydrophobic polymer, chains are covalently linked as a series of two or more hydrophilic blocks (Figure 4.3).

Polymersomes can be prepared by the same methods used for vesicles and liposomes. Basically, the preparation methods can be summarized in two groups: solvent-free techniques and solvent displacement techniques.[46] The solvent-free techniques broadly consist of hydrating the block copolymer in the dry state to form vesicles, such as film rehydration[32] or electroformation.[47] The last method consists in the rehydration of the polymers spread on a pair of electrodes; once the solvent is added, an electric field is applied to facilitate hydration and vesicle formation.

Solvent displacement techniques are those methods which require first to dissolve the block copolymer in an organic solvent before mixing with water. After the mixture, the organic solvent is removed by different procedures.

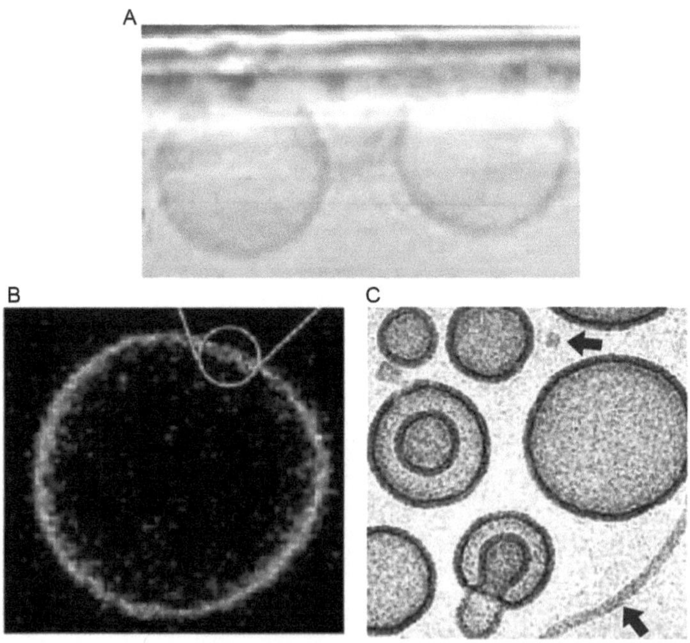

FIGURE 4.3 Illustration of self-directed assembly of polymersomes from hydrated films (A), fluoro-polymersome (B), and cryogenic transmission electron microscopy of 100 nm polymersomes (C).[45] (For color version of this figure, the reader is referred to the online version of this chapter.)

The injection method,[37] reverse-phase evaporation,[33,38] and detergent depletion[39] are examples of solvent displacement techniques used for polymers. All of them described for the vesicle formation.

Polymersomes are very attractive models to mimic biological membranes, and as advantage over vesicles and liposomes, polymersomes have the ability to form versatile structures with high mechanical stability and resistance to external stimulus.[48] Polymersomes can encapsulate either hydrophobic or hydrophilic molecules, and this is the principal reason of their application in medical, pharmaceutical, and environmental fields. In the case of the medical and pharmaceutical field, polymersomes are used as carriers for drug delivery to encapsulate highly toxic therapeutic agents, and they can specifically tune their pharmacokinetics. An example is shown in Ref. 49 where Ghoroghchian *et al.* have prepared polymersomes of poly(ethylene oxide)-*b*-polycaprolactone loaded with doxorubicin and studied the release profiles. The release *in vivo* was probed in tumor-bearing mice and a retard in the tumor growth was successfully shown. In addition to applications in drug delivery, polymersomes have also been used in diagnosis for *ex vivo* cellular labeling and *in vivo* cellular tracking. Christian and coworkers have probed in Ref. 50 the *ex vivo* efficiency for labeling dendritic cells with polymersomes loaded with a porphyrin-based near-infrared agent. Finally, in the environmental area, polymersomes are commonly used to entrap toxic molecules, like in Ref. 51 where Kapakoglou and coworkers present a system in which polymersomes are used to selectively entrap metal ion species in aqueous solutions.

Other types of polymer-based NPs are the polymer conjugates. Polymer conjugates are water-soluble hybrid structures formed by polymer chains covalently linked to a molecule such as a drug or a protein. The concept of polymer conjugate was proposed by Ringsdorf in 1975[52] with a polymer–drug conjugate model based on synthetic polymers. He proposed that certain number of drug molecules or bioactive molecules can be bounded to a macromolecule through a spacer molecule (linker),[56] which can incorporate a breaking point to ensure the release of thc drug at the site of interested; the polymer conjugate can additionally contain specific recognition functions (targeted molecules) or other molecules to enhance bioavailability (Figure 4.4).

In all the methods for synthesis of polymer conjugates, the goal is to obtain well-defined structures with narrow molecular weight distribution and high retention of drug or protein molecules. Controlled radical polymerization methods, such as atom transfer radical polymerization (ATRP) and reversible addition-fragmentation chain transfer (RAFT),[53] are very suitable for the synthesis of polymer conjugates, as they provide a high-level control of the polymer chain size distribution. The chemical synthesis of polymer conjugates via controlled radical polymerization methods has followed three different strategies: grafting to, grafting from, and grafting through. In the "grafting to,"[53] the most typical strategy for the fabrication of polymer conjugates, the ends of the polymer chain are modified after the polymerization to

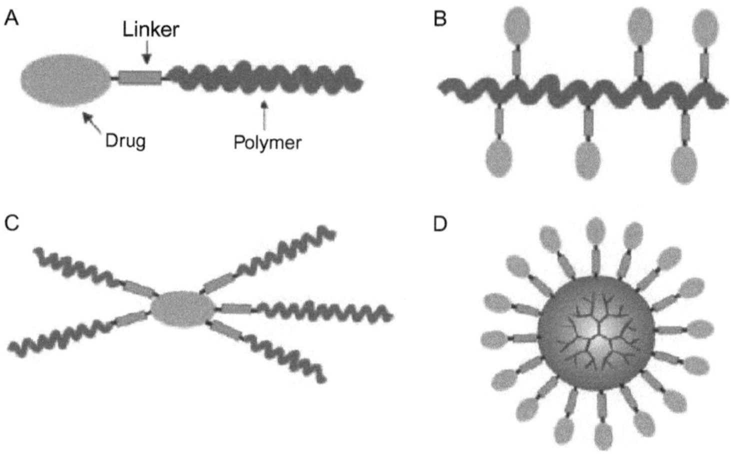

FIGURE 4.4 Different types of structures and architectures of drug–polymer conjugates: monofunctional linear (A), polyfunctional linear (B), starlike (C), and dendritic (D).[57] (For color version of this figure, the reader is referred to the online version of this chapter.)

form reactive groups for covalent linking of drugs or proteins. The "grafting from"[54] consists in the functionalization of the drug or protein with a monomer, which is used as an initiator for the polymer synthesis. In the "grafting through,"[55] the drug or protein to conjugate is employed as monomer for the polymerization.

The majority of the applications of polymer conjugates are in the pharmaceutical and nanomedical field.[56–58] In the clinical area, there are already some polymer–protein conjugates in the market such as Zinostatin Stimalmer® (SMANCS) for hepatocellular carcinoma treatment[60]; Oncaspar® PEG-L-asparaginase for leukemia treatment[61]; PEG-intron™ to treat hepatitis C, multiple sclerosis, and HIV/AIDS[62]; Neulasta™ against neutropenia associated with cancer chemotherapy[63]; and others. Regarding polymer–drug conjugates, most of the successful formulations are under clinical approval evaluations such as CT-2103 XYOTAX™ for treating lung and ovarian cancer,[64] HPMA copolymer–doxorubicin to treat hepatocellular carcinoma,[65] PROTHECAN™ for cancer therapies,[66] etc.

3.3. Dendrimers

Dendrimers are highly perfectly branched monodispersed macromolecules that emanate radially from a central core.[67] This high branching of the polymer molecules gives to the dendrimers chemical and physical properties different from linear polymer molecules that are of enormous interest for biomedical and industrial applications. In this section, we just want to briefly mention some of the methods to synthesize dendrimers and their applications since there is a chapter dedicated to dendrimers within this book.

Dendrimers can be synthesized by divergent and convergent methods. In the divergent methods,[68] the dendrimer grows outward from the core molecule. The monomer that contains one reactive group and two dormant groups reacts first with the core forming the first-generation dendrimer, and successively the new periphery of the molecule is activated and reacts with more monomers until forming a big macromolecular structure. This method has the inconveniences that secondary products are formed which are not easy to purify. In the convergent methods,[69] the dendrimer is synthesized stepwise from the end groups to the inward. When the branched molecules are big enough, they are attached to a core molecule. Dendrimers generated by these methods are easy to purify and have a controlled structure. On the other hand, it is not appropriate to form high-generation dendrimers because of steric impediments of the reaction.

Because of their unique properties and the easiness to tailor their surface and core with different molecules, dendrimers are suitable for many biomedical and industrial applications. In the biomedical area, dendrimers had been used for *in vitro* diagnosis in cardiac testing,[70] as contrast agents for magnetic resonance imaging,[71] as drug delivery systems,[72] as therapeutic agents for boron neutron capture therapy, for example,[73] in gene therapy as vectors to transfer genes through the cell membrane,[74] and in regenerative medicine.[75] Moreover, dendrimers have been used to improve many industrial processes such as nanocatalysis[76] or in purification processes to entrap insoluble materials.[77]

3.4. Nanocapsules

Nanocapsules are hollow polymer shells with an internal cavity where a desired substance can be placed. The best examples of natural nanocapsules are the viral capsids in which the shell is composed of identical protein subunits. Capsids are responsible for the protection of the genetic material of the virus and also play an important role in the infection of the host cell.[78] Inspired by virus capsids, scientists have developed synthetic polymer nanocapsules that offer many interesting applications for encapsulation, delivery, imaging, and catalysis.

Nanocapsules can be obtained by interfacial polymerization or by direct deposition of a preformed polymer. The interfacial polymerization reactions roughly consist in the interfacial reaction between two monomers.[79] First, an emulsion of two immiscible phases is formed. Each phase contains a monomer able to react within each other to form an initial wall. After formation of this initial wall, the polymerization is diffusion controlled, leading the formation of capsules with size dependent on the initial emulsion. In Ref. 80, Bouchemal *et al.* prepared polyurethane and poly(ether urethane) nanocapsules via the interfacial polycondensation reaction. The capsules were loaded with α-tocopherol, which is a strong antioxidant used in many medical and cosmetic formulations. They characterize the nanocapsules as a function of the polymer molecular weight, finding smaller diameters with narrow distributions as the molecular weight decreases. On the other hand, the yield of encapsulation increases when the molecular weight increases. Redox-responsive nanocapsules for protein delivery were recently reported by Zhao and coworkers.[81] The protein caspase 3 was encapsulated via *in situ* interfacial polymerization into a positively charged polymeric shell of polyacrylamide and polymethacrylamide. The described nanocapsules were internalized into cells and the protein was released in the reducing cytosol, proving an effective intracellular delivery strategy for therapeutic, diagnosis, and reprogramming applications. Some advantages of the interfacial polymerization methods are the high efficiency in encapsulation and the possibility of *in situ* formation of the polymer that allows the formation of nanocapsules without using a template. The main disadvantage of this process is the elimination of the organic solvents used to form the emulsions.

Several methods such as nanoprecipitation, emulsion–diffusion, double emulsification, emulsion–coacervation, and layer-by-layer (LbL) technique have been proposed up to now for the preparation of nanocapsules by direct deposition of a preformed polymer.

The nanoprecipitation method is a solvent displacement method in which both organic and aqueous phases are involved.[82] Through this method, nanocapsules are obtained as a colloidal suspension formed when the organic phase (normally containing the polymer) is added slowly to the aqueous phase under stirring. The size and shape of the formed nanocapsules will depend on the organic phase injection rate, aqueous phase agitation rate, organic phase/aqueous phase ratio, concentration, and nature of the polymers. The polymers normally used for preparing nanocapsules by nanoprecipitation are biodegradable polyesters such as poly-*e*-caprolactone (PCL), poly(lactide) (PLA), and poly(lactide-*co*-glycolide) (PLGA). Recently, Moraes *et al.*[84] have optimized a method of nanoprecipitation to prepare PCL nanocapsules as carriers of the anesthetic benzocaine. The optimization of the capsule formation was carried out for different independent variables, resulting in nanocapsules of 188 nm of diameter with negative surface charge and 87% efficiency in drug loading. The same optimization was done by the same group 2 years before to encapsulate benzocaine into 123 nm PLGA nanocapsules.[83]

The emulsion–diffusion method requires three phases: an organic phase (polymer solution), an aqueous phase (active substance in aqueous solution), and a solution of an organic solvent partially miscible in water.[85] First, the organic phase is emulsified in the aqueous phase by vigorous agitation followed by subsequent addition of the aqueous solution that causes the diffusion of the solvent into the external phase resulting in nanocapsules formation. The organic solvent can be eliminated by distillation or crossflow filtration. The size and shape of the nanocapsules depend on the emulsification shear rate, polymer concentration, and dilution ratio. Olvera-Martínez and coworkers[86] prepared cellulose acetate phthalate nanocapsules loaded with the sunscreen octyl methoxycinnamate via emulsion–diffusion method, in order to evaluate the *in vivo* distribution of these nanocapsules and compare them with commercial nanoemulsions. As in this example, several biodegradable polymers are used to prepare nanocapsules via emulsion–diffusion methods such as 4-nitroanisole loaded in PLA[87] or Sudan III encapsulated in Eudragit E nanocapsules.[88]

The double emulsification method is a complex heterodisperse system of water–oil–water emulsion (w/o/w) or oil–water–oil emulsion (o/w/o).[89] Double emulsion nanocapsules are prepared by two-step emulsification process: the first emulsion is formed by ultrasound in which w/o surfactant stabilizes the interface of the w/o internal emulsion; here, the stabilizer of the internal emulsion is the polymer. The second emulsion is also formed by ultrasound, and the nanocapsules dispersion is stabilized by a surfactant present in the last added phase. The organic solvents can be removed by evaporation or extraction under vacuum. An example for nanocapsules preparation by double emulsification method is shown in Ref. 90 where Zhu and coworkers studied the influence of the surfactants on the synthesis via double emulsion of insulin-loaded PLA nanocapsules. The produced capsules had diameters of 150–320 nm depending on the concentration and nature of the emulsifier used. Khoee and Yaghoobian[91] reported the preparation via double emulsion of polybutylene adipate nanocapsules loaded with penicillin-G and studied the influence of the emulsifiers in the nanocapsules size that varies from 75 to 638 nm. More recently, Liu *et al.*[92] prepared daunorubicin-loaded PLGA and PLA nanocapsules via double emulsion with diameters between 200 and 300 nm depending on the ratio of organic solvent, type or surfactant, nature, and molecular weight of polymer. The drug loading efficiency was around 80%, while the *in vitro* release could be controlled over 2 weeks.

The emulsion–coacervation technique involves, first, the o/w emulsification of an organic phase to form a template phase. Then, a simple coacervation method using polyelectrolytes, dehydration agents, or temperature changes is used to precipitate the polymer templates. Finally, the coacervated nanocapsules are stabilized by cross-linking to obtain a rigid shell structure.[82] Some natural polymers had been used to prepare nanocapsules by emulsion–coacervation method such as atelocollagen/xanthane gum that was used to encapsulate porphyrin for photodynamic therapies.[93] The synthesized nanocapsule formulation with size about 300–400 nm was found to be more phototoxic than the typical formulation on the market, showing high efficiency for photodynamic therapies. Zhu and coworkers also reported the use of the natural polymers arabic gum and gelatin in Ref. 94 and acacia and gelatin in Ref. 95 to prepare 30–100 nm capsules via emulsion–coacervation; in both cases, cypermethrin is loaded with efficiencies above 60%.

The LbL process to fabricate capsules was introduced in 1998 by Donath *et al.*[96] This process is based on the alternative assembly of oppositely charged polyelectrolytes on colloidal substrates that can be later destroyed. The assembly results in a thin polymer film of a few nanometers thick. In the LBL method, polyelectrolyte assembly is triggered by the electrostatic attraction between the subsequent polymer layers. After polymer assembly, the colloidal sacrificial template can be removed[97] by dissolution, leading to the formation of polyelectrolyte hollow nanocapsules. The capsules can be later loaded with different materials. Besides loading the capsules, this technique offers the possibility to assemble some delivery materials like proteins or DNA between the layers that will be eventually peeled off and liberated.[98] The thickness of the polyelectrolyte multilayers (PEMs) can be controlled by the number of layers deposited, while the shape and size of the nanocapsules can be engineered by choosing the appropriated template. Then, the stability and surface functionality of the capsules can be controlled by the nature of the polyelectrolytes.

Different combinations of polymers for PEMs and colloids as templates have been used to fabricate nanocapsules via LbL technique like in Ref. 99 where Moya and coworkers fabricated capsules upon PEMs employing polystyrene sulfonate (PSS) and polyallylamine hydrochloride (PAH) on human red blood cells and melamine formaldehyde followed by the template removal by dissolving in hypochlorite solution (Figure 4.5). Also in Ref. 102, Caruso *et al.* coated pyrene and fluorescein diacetate crystals with PAH/PSS PEMs to form capsules after the crystal removal by mild organic solvent. Typical colloids used as templates in this technique are particles made of polystyrene,[103] silica,[98,104,112] calcium carbonate,[105,111,116] cadmium carbonate,[106] or gold.[107] Recently, the use of surfactant vesicles as template has also been reported.[108] For each type of particle used as a template, the route to remove it without dissembling the polyelectrolyte layer is different. Silica particles require relative harsh conditions, employing hydrofluoric acid. Melamine particles, prepared from weakly polymerized melamine monomers, are dissolved by a pH reduction, which breaks the ether linkages between the monomers.[100] Carbonate particles are dissolved by an increase of pH toward basic values that hydrolyze

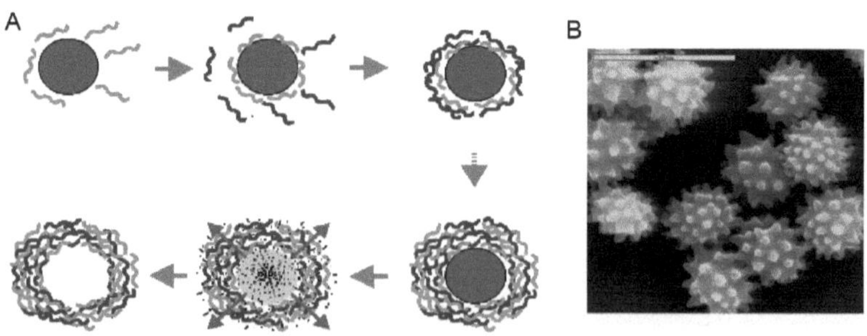

FIGURE 4.5 Scheme of the preparation of hollow polyelectrolyte capsules from spherical particles (A) and SEM micrographs of polyelectrolyte capsules templated on echinocytes (B).[99] (For color version of this figure, the reader is referred to the online version of this chapter.)

the carbonates. Both for silica and carbonate, the core is dissolved into an inorganic salt, that is, ions are released into solution in a progressive way. For the melamine particles, the depolymerization of the melamine core generates melamine oligomers of variable size that have to permeate through the capsule wall.[101] Since the rate of core dissolution and oligomer release through the polyelectrolyte wall may not be the same, the oligomers can be retained in the capsule, soaking up water and causing the capsule to swell osmotically. When the oligomers have been released, the capsule retains its original size. In the case of the inorganic cores, the polyelectrolyte layers have been reported to be almost not affected by the dissolution process. For the melamine particles, the process of swelling can induce the formation of holes and imperfections in the polymer structure. Besides, it is reported that melamine rests are normally retained in the capsule walls. The use of cells for capsule templating with hypochlorite oxidation deserves a special consideration. The hypochlorite degrades the cell proteins through the oxidation of their amine groups. Typical polyelectrolytes employed for the capsule fabrication like PAH and PSS are oxidized by the hypochlorite.[99] Indeed, capsules fabricated with this polymer no longer show amine groups, which are probably cross-linked. The amount of PSS in the capsule walls is also significantly diminished. Cell template capsules show mechanical and structural properties quite different from other LBL template capsules. In general, we can say that the methodology of core dissolution will have a strong impact on the fabricated capsules.

Once the capsule template is removed, it can be loaded with both small- and high-molecular-weight molecules. Different procedures have been reported for the loading of the nanocapsules depending on the size and nature of the molecules to load and the physicochemical properties of the PEMs. If the PEMs are formed by materials responsive to salt concentration or pH, the material can be loaded via diffusion by opening the pores within the shell as shown by Georgieva *et al.*,[109] where they prepared capsules employing PSS, PAH, and dextran on red blood cells. After core removal, the loading of human serum albumin into the capsules was studied as a function of the ionic strength and pH. When the desired cargo forms crystals, like drugs[102] or proteins,[110] the PEMs can be directly assembled using the crystals as template, providing capsules with high concentration of the encapsulated molecule. For high loading efficiency, the encapsulation can be done by the incorporation of the cargo within the sacrificial template porous followed by the LbL assembly and core removal, as is shown by Yufor high-molecular-weight cargos in porous silica particles,[104] or by Volodkin in calcium carbonate particles[111] and for low-molecular-weight cargos in silica particles[112] by Zhu. Finally, in the specific case of high-molecular-weight molecules, that is, RNA, plasmids, or proteins, these can also be encapsulated within the PEMs, like Reibetanz *et al.* show in Ref. 98. In this work, protamine, dextran sulfate, and plasmid DNA capsules were prepared. The defoliation of the PEMs for the intracellular delivery of plasmid DNA was also studied in this chapter.

LbL capsules can also be surface modified to achieve specific targeting within the cells by coating the surface of the capsules with specific recognition molecules via electrostatic interactions or by covalent binding, such as polyethylene glycol (PEG) that has unfolding properties and improve the circulation time of the capsules,[113,114] or lipids that have been proved to protect the capsules from nonspecific interactions.[115] Moreover, to decrease toxicity, biocompatible polyelectrolytes are used to fabricate capsules such as chitosan/alginate,[116] dextran sulfate/protamine,[98] chondroitin sulfate/poly(L-arginine),[117] and poly(L-lysine)/poly(glutamic acid).[104]

LbL capsules can be designed to release the cargo either for burst or for sustained release. Burst release can be achieved by capsule degradation due to an external stimulus, like in Ref. 118, where near-infrared laser pulses are applied on capsules that have metallic NPs incorporated in the PEM wall, promoting burst intracellular release. Additionally, capsule degradation inside the cell can take place when hydrogen-bonded PEMs are cross-linked with disulfide; this linker is retained in the blood stream promoting disruption and burst release of the content in the capsules.[119] In contrast, sustained release can be achieved by slow diffusion of the cargo through the PEM wall[120–123] or by slow degradation of the PEM wall[98,121,124]; in both cases, the release can be environmental stimuli with pH,[120–122] ionic strength,[121–123] or temperature[121,123] changes.

LbL capsules are promising and versatile nanocarriers for drug delivery that allow the specific design of carriers by tuning the surface chemistry, size, permeability, and degradability of the capsules. Different and multiple cargos can be incorporated inside the capsules or within the PEMs, and the release of the cargo and the degradation of the capsules can be easily controlled by different mechanisms.

3.5. Polymeric NPs

Polymeric NPs, also known as polymeric nanospheres, are usually defined as submicron-sized solid polymer particles with matrix type structure,[79] in which a cargo can be encapsulated within the polymer matrix or absorbed in the surface.

Like polymer capsules, polymeric NPs can also be prepared via *in situ* polymerization techniques or directly by formed polymers. The most common reaction used to prepare NPs by polymerization is the emulsion polymerization either o/w or w/o. This reaction consists in disperse the emulsionable monomer in the phase in which is soluble; then, the reaction is initiated by physical or chemical agents forming free radicals that collide with the monomer molecules. The reaction stops when there is no more free available monomer.[125] When encapsulation of a material is desired, the cargo is dissolved in the reaction solution, at the beginning of the polymerization before adding the monomer. The synthesized product is a stable dispersion of polymer colloids in aqueous medium that is called "latex."[126] Latexes have been used for several years in the fabrication of paints, polishes, paper, carpeting, textiles, and adhesives. There is plenty of research about the use of latexes in biotechnology, molecular biology, and biomedicine. Polymethacrylates and polyacrylamides nanolatexes, for example, have been used for analytical detection of soluble metal ions,[127] retention aids in papermaking processes,[130] in protein separation and purification,[128] and for drug delivery and diagnosis.[129,131] Also, polystyrene nanolatexes have been used for protein absorption in biosensor,[134] as a chemical mechanical polishing for oxides,[135] to fabricate paper filter with selective adsorption,[136] and for several drug delivery systems.[137] Due to their excellent adhesive properties, polyalkylcyanoacrylates NPs have been used in several biomedical applications such as substrates for living tissues and skin, surgical glue,[132] and more recently, as carriers for drug delivery.[133]

The emulsion-solvent evaporation method is probably the most used to fabricate NPs. Here, the polymer is dissolved in an organic solvent and emulsified in an aqueous solution containing a surfactant. Then, the organic solvent is evaporated inducing the precipitation of the polymer NPs. Cargo molecules are usually added into the polymer solution for encapsulation. The particle size can be controlled by the emulsification rate, time, and the nature of the polymer and surfactant. The most typical polymers to prepare NPs by this method are PLA,[138] PLGA,[139–142] PCL,[143] ethyl cellulose (EC), cellulose acetate phthalate,[144] and poly(h-hydroxybutyrate).[145]

In the past years, our group has been focused in the fabrication and surface modification of PLGA NPs fabricated by the O/W emulsion-solvent evaporation method (Figure 4.6A). The size of the NPs is controlled by the nature of the stabilizer and by the ratio between organic and aqueous phase. We prepare PLGA NPs using BSA and PEI as stabilizers at different ratios with aqueous phase finding that PEI-stabilized NP size is highly affected by PEI concentration while BSA-stabilized NPs are not affected by the surfactant concentration due to the amphiphilic character of the BSA.[142]

The NPs fabricated have been surface modified via the LbL technique in order to control release and enhance specific cell targeting. The use of the LbL technique offers the possibility of simple routes for covalent functionalization without altering the stability of the NPs (Figure 4.6B). Polyelectrolytes can be chosen to have functional groups like amines or carboxylates that are easy to link covalently[139] with molecules bearing specific recognition or protective like PEG and folic acid (FA) with simple carbodiimide chemistry. The coating with PEG significantly reduces NP uptake, while the addition of FA conjugated with PEG increases it for cancer cell lines, which have overexpressed folate receptors.[140,141] Moreover, a chitosan and alginate coating on PLGA NPs was designed for antifouling protection as alternative to PEGylation, followed by FA immobilization to achieve selective targeting in cancer cells.[141]

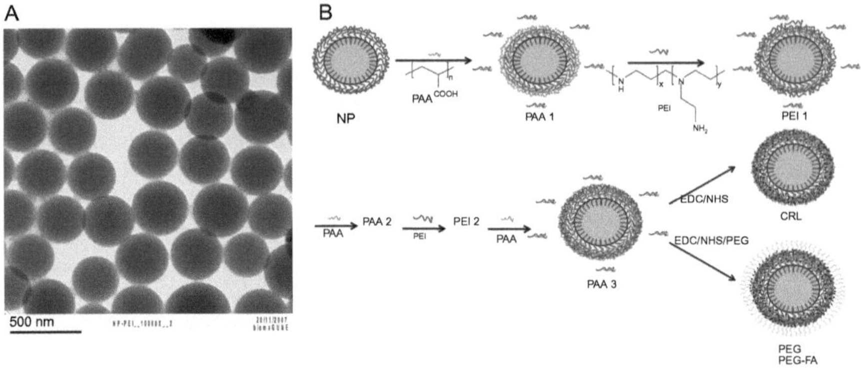

FIGURE 4.6 TEM image of PLGA NPs prepared by o/w emulsion-solvent evaporation method (A) and scheme showing the LbL coating on the PEI-coated PLGA NPs, cross-linking, and PEGylation (B).[137–140] (For color version of this figure, the reader is referred to the online version of this chapter.)

The emulsification solvent diffusion method is a modified version of the emulsification solvent evaporation method in which the particle formation is the result of an interfacial turbulence created by the spontaneous diffusion of a mixture of solvents composed by a water-miscible solvent in presence of a small amount of immiscible organic solvent. The particle size can be controlled by the ratio between water-miscible and -immiscible solvents. Polymers like PLA,[146] PLGA,[92] and gelatin[147] have been used to prepare NPs by this method.

Solvent displacement technique, also called spontaneous emulsification or nanoprecipitation because the particle formation is the result of the precipitation of the polymer from an organic solvent that diffuses in an aqueous medium in presence or absence of surfactant. This method is commonly used to form PLA,[148] PLGA,[148,149] PCL,[149] and PMMA[150] NPs.

Finally, the salting-out method is based on the separation of a water-miscible solvent from aqueous solution by salting-out effect. The polymer is dissolved in a water-miscible solvent and emulsified in aqueous solution containing salting-out agents and stabilizers; then the emulsion is diluted in a large volume of water to enhance the NP formation as a result of the diffusion of the water-miscible solvent. The salting-out agent and the water-miscible solvent can be further removed. This method has been used to prepare NPs from poly(alkyl methacrylate),[151] EC,[152] PLA, and PLGA.[153]

To resume, we have presented a general view of the different organic NPs available. The picture is not complete but gives an idea of the complexity and variety of NPs that can be made from organic molecules. The main features of the bottom-up templated NPs have been summarized in Table 4.1.

A key phenomenon linking most of the NPs presented is the self-assembly as the basis for the organization of the NPs as we have seen for vesicles,

TABLE 4.1 Summary of the Organic NPs Fabricated via "Bottom-Up" Techniques

Nanoparticle Type	Composition	Size	Fabrication Techniques	Applications
Micelles	Amphiphilic molecules	Since 15–20 Å	Self-assembly	Drug delivery
Vesicles and liposomes	Natural and synthetic glycolipids and phospholipids	30–400 nm	Self-assembly, solvent-free techniques and solvent displacement techniques	Nanocarriers for drug delivery, food industry, and cosmetics
Polymersomes	Amphiphilic block copolymers	50–200 nm	Self-assembly, solvent-free techniques and solvent displacement techniques	Nanocarriers for drug delivery, labeling, and tracking and in environmental field to entrap selective toxic molecules
Polymer conjugates	Polymers, proteins, and drugs	5–20 nm	"Grafting to," "grafting from," and "grafting through" using ATRP and RAFT	Pharmaceutics as polymer therapeutics
Dendrimers	Polymers	From 10 nm	Divergent and convergent methods	In the biomedical field for *in vitro* diagnosis, as contrast agents, as drug delivery systems, as therapeutic agents, in gene therapy, and in regenerative medicine. In the industry to improve processes like nanocatalysis or in purification
Nanocapsules	Polymers	From 20 nm	Interfacial polymerization and by direct deposition of preformed polymers	Nanocarriers for drug delivery
Polymeric nanoparticles	Polymers	From 20 nm	Interfacial polymerization and by direct deposition of preformed polymers	Nanocarriers for drug delivery

micelles, polymersomes, polymer micelles, and nanocapsules, or as template for the ulterior formation of the NPs as we have seen for the polymeric NPs and also for nanocapsules. Most of the organic NPs involve several organic or polymeric molecules. Polymer conjugates and dendrimers constitute special cases for organic NPs and have the particularity to be formed on single-molecule basis. The fact that organic NPs are formed by several molecules imposes sometimes limitations to control size below a certain value. This is not the case for polymer conjugates or dendrimers, whose size can be controlled below 10 nm.

A remarkable issue about organic NPs is their capacity to load molecules either by conjugation on the surface or in the core or by physical encapsulation. This property makes them appealing systems for the delivery of molecules and more specifically for drug delivery and biomedical applications.

REFERENCES

1. Müller RH, Gohla S, Keck CM. State of the art of nanocrystals—special features, production, nanotoxicology aspects and intracellular delivery. *Eur J Pharm Biopharm* 2011;**78**:1.
2. Rolland JP, Harberg EC, Denison GM, Carter KR, DeSimone JM. High-resolution soft lithography: enabling materials for nanotechnologies. *Angew Chem Int Ed* 2004;**43**:5796.
3. Rolland JP, Maynor BW, Euliss LE, Exner AE, Denison GM, DeSimone JM. Direct fabrication and harvesting of monodisperse, shape-specific nanobiomaterials. *J Am Chem Soc* 2005;**127**:10096.
4. Texter J. Precipitation and condensation of organic particles. *J Disp Sci Technol* 2001;**22**:499.
5. Zhang H, Wang D, Butler R, Campbell NL, Long J, Tan B, et al. Formation and enhanced biocidal activity of water-dispersable organic nanoparticles. *Nat Nanotechnol* 2008;**3**:506.
6. Euliss LE, DuPont JA, Gratton S, DeSimone J. Imparting size, shape, and composition control of materials for nanomedicine. *Chem Soc Rev* 2006;**35**:1095.
7. Horn D, Rieger J. Organic nanoparticles in aqueous phase—theory, experiment, and use. *Angew Chem Int Ed* 2001;**40**:4330.
8. Liversidge GG, Cundy KC. Particle size reduction for improvement of oral bioavailability of hydrophobic drugs: I. Absolute oral bioavailability of nanocrystalline danazol in beagle dogs. *Int J Pharm* 1995;**125**:91.
9. Merisko-Liversidge E, Liversidge GG, Cooper ER. Nanosizing: a formulation approach for poorly-water-soluble compounds. *Eur J Pharm Sci* 2003;**18**:113.
10. Sigfridsson K, Björkman J, Skantze P, Zachrisson H. Usefulness of a nanoparticle formulation to investigate some hemodynamics parameters of a poorly soluble compound. *J Pharm Sci* 2011;**100**:2194.
11. Schryver SB, Ramsden W, Cross CF, Schidrowitz P, Dreaper WP, McBain JW, Turner T, Worley FP, Martin CJ, Bousfield WR, Morse HN, Henri V, Freundlich H, The C, Ostwald W, Chapman C, Senter G. Discussion. *Trans Faraday Soc* 1913;**9**:93.
12. Wennerström H, Linman B. Micelles. Physical chemistry of surfactant association. *Rev Sec Phys Lett* 1979;**52**:1.
13. Nelson PH, Rutledge GC, Hatton TA. On the size and shape of self-assembled micelles. *J Chem Phys* 1997;**107**:10777.

14. Nagarajan R, Ruckenstein E. Critical micelle concentration: a transition point for micellar size distribution: a statistical thermodynamical approach. *J Colloid Interface Sci* 1977;**60**:221.
15. Riess G. Micellization of block copolymers. *Prog Polym Sci* 2003;**28**:1107.
16. Hariri K, Ruch T, Riess G. Block copolymers as dispersants and migration inhibitors: incorporation of fluorescent dyes in polyethylene. *E-Polymers* 2009; article number 063.
17. Zhang J, Dubay MR, Houtman CJ, Severtson SJ. Sulfonated amphiphilic block copolymers: syntheses, self-assembly in water, and applications as stabilizer in emulsion polymerization. *Macromolecules* 2009;**42**:5080.
18. Steitz R, Schemmel S, Shi H, Findenegg GH. Boundary layers of aqueous surfactants and block copolymers solutions against hydrophobic and hydrophilic solid surfaces. *J Phys Condens Matter* 2005;**17**:S665.
19. AL-Sabagh AM, Maysour NE, El-Din MRN. Investigate the demulsification efficiency of some novel demulsifiers in relation to their surface active properties. *J Dispers Sci Technol* 2007;**28**:547.
20. Candau F, Biggs S, Hill A, Selb J. Synthesis, structure and properties of hydrophobically associating polymers. *Prog Org Coat* 1994;**24**:11.
21. Zhu J, Zhou Z, Yang C, Kong D, Wan Y, Wang Z. Folate-conjugated amphiphilic star-shaped block copolymers as targeted nanocarriers. *J Biomed Mater Res A* 2011;**97A**:498.
22. Kedar U, Phutane P, Shidhaye S, Kadam V. Advances in polymeric micelles for drug delivery and tumor targeting. *Nanomedicine* 2010;**6**:714.
23. Hassan PA, Raghavan SR, Kaler EW. Microstructural changes in SDS micelles induced by hydrotropic salt. *Langmuir* 2002;**18**:2543.
24. Deutschle T, Porkert U, Reiter R, Keck T, Riechelmann H. In vitro genotoxicity of benzalkonium chloride. *Toxicol In Vitro* 2006;**20**:1472.
25. Brault PA, Kariapper MS, Pham CV, Flowers RA, Gunning WT, Shah P, et al. Protein micelles from lipoxygenase 3. *Biomacromolecules* 2002;**3**:649.
26. Davies TS, Ketner AM, Raghavan SR. Self-assembly of surfactant vesicles that transform into viscoelastic wormlike micelles upon heating. *J Am Chem Soc* 2006;**128**:6669.
27. Imura T, Yanagishita H, Ohira J, Skai H, Abe M, Kitamoto D. Thermodynamically stable vesicle formation from glycolipid biosurfactant sponge phase. *Colloids Surf B Biointerfaces* 2005;**43**:115.
28. Bangham AD. A correlation between surface charge and coagulant action of phospholipids. *Nature* 1961;**192**:1197.
29. Hope MJ, Bally MB, Mayer LD, Janoff AS, Cullis PR. Generation of multilamellar and unilamellar phospholipid vesicles. *Chem Phys Lipids* 1986;**40**:89.
30. Jung HT, Coldren B, Zasadzinski JA, Iampietro DJ, Kaler EW. The origins of stability of spontaneous vesicles. *Proc Natl Acad Sci USA* 2001;**98**:1353.
31. Courbin L, Panizza P. Shear-induced formation of vesicles in membrane phases: kinetics and size selection mechanism, elasticity versus surface tension. *Phys Rev E* 2004;**69**:021504.
32. Bangham AD, Standish MM, Watkins JC. Diffusion of univalent ions across the lamellae of swollen phospholipids. *J Mol Biol* 1965;**13**:238.
33. Kim S, Jacobs RE, White SH. Preparation of multilamellar vesicle of defined size-distribution by solvent-spherule evaporation. *Biochim Biophys Acta* 1985;**812**:793.
34. Hunter DG, Frisken BJ. Effect of extrusion pressure and lipid properties on the size and polydispersity of lipid vesicles. *Biophys J* 1998;**74**:2996.
35. Maguire LA, Zhang H, Shamlou PA. Preparation of small unilamellar vesicles (SUV) and biophysical characterization of their complexes with poly-L-lysine-condense plasmid DNA. *Biotechnol Appl Biochem* 2003;**37**:73.

36. Hamilton RL, Goerke JJ, Guo LSS, Williams MC, Havel RJ. Unilamellar liposomes made with French pressure cell: a simple preparative and semiquantitative technique. *J Lipid Res* 1980;**21**:981.
37. Batzri S, Korn E. Single bilayer liposomes prepared without sonication. *Biochim Biophys Acta* 1973;**298**:1015.
38. Pidgeon C, McNeely S, Schmidt TS, Johnson JE. Multilayered vesicles prepared by reverse-phase evaporation: liposome structure and optimum solute entrapment. *Biochemistry* 1987;**26**:17.
39. Schubert R. Liposome preparation by detergent removal. *Methods Enzymol* 2003;**367**:46.
40. Šegota S, Težak Đ. Spontaneous formation of vesicles. *Adv Colloid Interface Sci* 2006;**121**:51.
41. Camelo S, Lajavardi L, Bochot A, Goldenberg B, Naud MC, Brunel N. Protective effect of intravitreal injection of vasoactive intestinal peptide-loaded liposomes on experimental autoimmune uveoretinitis. *J Ocul Pharmacol Ther* 2009;**25**:9.
42. Mozafari MR, Johnson C, Hatziantoniou S, Demetzos C. Nanoliposomes and their applications in food nanotechnology. *J Liposome Res* 2008;**18**:309.
43. Vuillemard JC. Recent advances in the large-scale production of lipid vesicles for use in food products: microfluidization. *J Microencapsul* 1991;**8**:547.
44. Lasic DD. Chapter 10: applications of liposomes. *Handbook Biol Phys* 1995;**1**:491.
45. Discher DE. Polymersomes. *Annu Rev Biomed Eng* 2006;**8**:323.
46. Christian DA, Cai S, Bowen DM, Kim Y, Pajerowski JD, Discher DE. Polymersome carriers: from self assembly to siRNA and protein therapeutics. *Eur J Pharm Biopharm* 2009;**71**:463.
47. Discher MB, Won Y, Ege DS, Lee JC, Bates F, Discher DE, et al. Polymersomes: tough vesicles made from diblock copolymers. *Science* 1999;**284**:1143.
48. Tokarczyk KK, Grumelard J, Haefele T, Meier W. Block copolymers vesicles-using concepts from polymer chemistry to mimic biomembranes. *Polymers* 2005;**46**:3540.
49. Ghoroghchian PP, Li G, Levine DH, Davis KP, Bates FS, Hammer DA, et al. Bioresorbable vesicles formed through spontaneous self-assembly of amphiphilic poly (ethylene oxide)-block-polycaprolactone. *Macromolecules* 2006;**39**:1673.
50. Christian NA, Milone MC, Ranka SS, Li G, Frail PR, Davis KP, et al. Tat-functionalized near-infrared emissive polymersomes for dendritic cell labelling. *Bioconjug Chem* 2007;**18**:31.
51. Kapakoglou NI, Giokas DL, Tsogas GZ, Vlessidis AG. Coacervation of surface-functionalized polymerized vesicles derived from ammonium bromide surfactants. Applications to the selective speciation of chromium in environmental samples. *Anal Chem* 2008;**80**:9787.
52. Ringsdorf H. Structure and properties of pharmacologically active polymers. *J Polym Sci Polym Symp* 1975;**51**:135.
53. Grover GN, Maynard HD. Protein-polymer conjugates: synthetic approaches by controlled radical polymerizations and interesting applications. *Curr Opin Chem Biol* 2010;**14**:818.
54. Kochendoerfer GG. Site-specific polymer modification of therapeutic proteins. *Curr Opin Chem Biol* 2005;**9**:555.
55. Canalle LA, Lowik D, Van-Hest JCM. Polypeptide-polymer bioconjugates. *Chem Soc Rev* 2010;**39**:329.
56. Duncan R, Vicent MJ, Greco F, Nicholson RI. Polymer-drug conjugates: towards a novel approach for the treatment of endocrine-related cancer. *Endocr Relat Cancer* 2005;**12**:S189.

57. Haag R, Kratz F. Polymer therapeutics: concepts and applications. *Angew Chem Int Ed* 2006;**45**:1198.
58. Vicent MJ, Duncan R. Polymer conjugates: nanosized medicines for targeting cancer. *Trends Biotechnol* 2006;**24**:39.
59. Duncan R, Ringsdorf H, Satchi-Fainaro R. Polymer therapeutics: polymer as drugs, drug and protein conjugates and gene delivery: past, present and future opportunities. *Adv Polym Sci* 2006;**192**:1.
60. Matsumura Y, Maeda H. A new concept for macromolecular therapies in cancer chemotherapy: mechanism of tumouritropic accumulation of proteins and the antitumour agent SMANCS. *Cancer Res* 1986;**6**:6387.
61. Graham ML. Pegaspargase: a review of clinical studies. *Adv Drug Deliv Res* 2003;**55**:1293.
62. Wang YS, Youngster S, Grace M, Bausch J, Bordens R, Wyss D. Structural and biological characterization of pegylated recombinant interferon-α-2b and its therapeutic implications. *Adv Drug Deliv Rev* 2002;**54**:547.
63. Molineux G. The design and development of pegfilgrastim (PEG-rmetHuG-CSF, Neulasta). *Curr Pharm Des* 2004;**10**:1235.
64. Langer CJ, Socinski MA, O'Byrne KJ. Paclitaxel polyglumex (PPX)/carboplatin vs paclitaxel/carboplatin for treatment of PS2 patients with chemotherapy-naïve advanced non-small cell lung cancer (NSCLC): a phase III study. *Proc Am Soc Clin Oncol* 2005;**96**:7011.
65. Seymour LW, Ferry DR, Anderson D, Hesslewood S, Julyan PJ, Poyner R, et al. Hepatic drug targeting: phase I evaluation of polymer-bound doxorubicin. *J Clin Oncol* 2002;**20**:1668.
66. Rowinsky EK, Ochoa L, Takimoto CH, Forouzesh B, Schwartz G, Hammond LA, et al. A phase I and pharmacokinetic study of pegylated camptothecin as a 1-hour infusion every 3 weeks in patients with advanced solid malignancies. *J Clin Oncol* 2003;**21**:148.
67. Klajnert B, Bryszewska M. Dendrimers: properties and applications. *Acta Biochim Pol* 2001;**48**:199.
68. Hodge P. Polymer science branches out. *Nature* 1993;**362**:18.
69. Hawker CJ, Fréchet JMJ. Preparation of polymers with controlled molecular architecture. A new convergent approach to dendritic macromolecules. *J Am Chem Soc* 1990;**112**:7638.
70. Sigh P. Dendrimers and their applications in immunoassays and clinical diagnosis. *Biotechnol Appl Biochem* 2007;**48**:1.
71. Kobayashi H, Kawamoto S, Jo SK, Bryant HL, Brechbiel MW, Star RA. Macromolecular MRI contrast agents with small dendrimers: pharmacokinetic differences between sizes and cores. *Bioconjug Chem* 2003;**14**:388.
72. Gillies ER, Fréchet JM. Dendrimers and dendritic polymers in drug delivery. *Drug Discov Today* 2005;**10**:35.
73. Wu G, Barth RF, Yang W, Chatterjee M, Tjarks W, Ciesielski MJ, et al. Site-specific conjugation of boron-containing dendrimers to anti-EGF receptor monoclonal antibody cetuximab (IMC-C225) and its evaluation as potential delivery agent for neutron capture therapy. *Bioconjug Chem* 2004;**15**:185.
74. Dufès C, Uchegbu IF, Schätzlein AG. Dendrimers in gene delivery. *Adv Drug Deliv Rev* 2005;**57**:2177.
75. Oliveira JM, Salgado AJ, Sousa N, Mano JF, Reis RL. Dendrimers and derivatives as potential therapeutic tool in regenerative medicine strategies—a review. *Prog Polym Sci* 2010;**35**:1163.
76. Karakhanov EA, Maximov AL, Skorkin VA, Zolotukhina AV, Smerdov AS, Tereshchenko AY. Nanocatalysis based on dendrimers. *Pure Appl Chem* 2013;**2009**:81.

77. Tanaka K, Inafuku K, Naka K, Chujo Y. Enhancement of entrapping ability of dendrimers by a cubic silsesquioxane core. *Org Biomol Chem* 2008;**6**:3899.
78. Kim D, Kim E, Lee J, Hong S, Sung W, Lim N, et al. Direct syntheses of polymer nanocapsules: self-assembly of polymer hollow spheres through irreversible covalent bond formation. *J Am Chem Soc* 2010;**132**:9908.
79. Reis CP, Neufeld RJ, Ribeiro AJ, Veiga F. Nanoencapsulation I. Methods for preparation of drug-loaded polymeric nanoparticles. *Nanomedicine* 2006;**2**:8.
80. Bouchemal K, Briançon S, Perrier E, Fessi H, Bonnet I, Zydowicz N. Synthesis and characterization of polyurethane and poly(ether urethane) nanocapsules using a new technique of interfacial polycondensation combined to spontaneous emulsification. *Int J Pharm* 2004;**269**:89.
81. Zhao M, Biswas A, Hu B, Joo K, Wang P, Gu Z, et al. Redox-responsive nanocapsules for intracellular protein delivery. *Biomaterials* 2011;**32**:5223.
82. Mora-Huertas CE, Fessi H, Elaissari A. Polymer-based nanocapsules for drug delivery. *Int J Pharm* 2010;**385**:113.
83. Moraes CM, de Matos AP, de Paula E, Rosa AH, Fraceto LF. Benzocaine loaded biodegradable poly-(d, l-lactide-co-glycolide) nanocapsules: factorial design and characterization. *Mater Sci Eng B* 2009;**165**:243.
84. Moraes CM, de Matos AP, Grillo R, de Melo NF, de Paula E, Dias Fiho NL, et al. Screening of formulations variables for the preparation of poly(epsilon-caprolactone) nanocapsules containing the local anesthetic benzocaine. *J Nanosci Nanotechnol* 2011;**11**:2450.
85. Quintanar-Guerrero D, Alléman E, Doelker E, Fessi H. Preparation and characterization of nanocapsules from preformed polymers by new process based on emulsification-diffusion technique. *Pharm Res* 1998;**15**:1056.
86. Olvera-Martínez BI, Cázares-Delgadillo J, Calderilla-Fajardo SB, Villalobos-García R, Ganem-Quintanar A, Quintanar-Guerrero D. Penetration of polymeric nanocapsules containing octyl methoxycinnamate by the emulsification-diffusion technique: penetration across the stratum corneum. *J Pharm Sci* 2005;**94**:1552.
87. Romcro-Cano MS, Vincent B. Controlled release of 4-nitroanisole from poly(lactic acid) nanoparticles. *J Control Release* 2002;**82**:127.
88. Moinard-Chécot D, Chevalier Y, Briançon S, Beney L, Fessi H. Mechanism of nanocapsules formation by the emulsion–diffusion process. *J Colloid Interface Sci* 2008;**317**:458.
89. Garti N, Bisperink C. Double emulsions: progress and applications. *Curr Opin Colloid Interface Sci* 1998;**3**:657.
90. Zhu Y, Zhang G, Yang H, Hong X. Influence of surfactants on the parameters of polylactide nanocapsules containing insulin. *J Surfactants Deterg* 2005;**8**:353.
91. Khoee S, Yaghoobian M. An investigation into the role of surfactants in controlling particle size of polymeric nanocapsules containing penicillin-G in double emulsion. *Eur J Med Chem* 2009;**44**:2392.
92. Liu J, Qiu Z, Wang S, Zhou L, Zhang S. A modified double-emulsion method for the preparation of daunorubicin-loaded polymeric nanoparticles with enhanced in vitro anti-tumor activity. *Biomed Mater* 2010;**5**:065002.
93. Deda DK, Uchoa AF, Caritá E, Baptista MS, Toma HE, Araki K. A new micro/nanoencapsulated porphyrin formulation for PDT treatment. *Int J Pharm* 2009;**376**:76.
94. Zhu Y, Zhang G, Hong X, Dong J, Zhang X, Zeng H. Synthesis of nanocapsules by complex coacervation in microemulsion. *Acta Chim Sin* 2005;**63**:1505.
95. Zhu Y, An X, Li S, Yu S. Nanoencapsulation of β-cypermethrin by complex coacervation in a microemulsion. *J Surfactants Deterg* 2009;12–305.

96. Donath E, Sukhorukov GB, Caruso F, Davis SA, Möhwald H. Novel hollow polymer shells by colloid-templated assembly of polyelectrolytes. *Angew Chem Int* 1998;**37**:2202.
97. Caruso F, Caruso RA, Möhwald H. Nanoengineering of inorganic and hybrid hollow spheres by colloidal templating. *Science* 1998;**282**:1111.
98. Reibetanz U, Claus C, Typlt E, Hofmann J, Donath E. Defoliation and plasmid delivery with layer-by-layer coated colloids. *Macromol Biosci* 2006;**6**:153.
99. Moya S, Dähne L, Voigt A, Leporatti S, Donath E, Möhwald H. Polyelectrolyte multilayer capsules template on biological cells: core oxidation influences layer chemistry. *Colloids Surf A* 2001;**183**:27.
100. Gao C, Moya S, Lichtenfeld H, Casoli A, Fiedler H, Donath E, et al. The decomposition process of melamine formaldehyde cores: the key step in the fabrication of ultrathin polyelectrolyte multilayer capsules. *Macromol Mater Eng* 2001;**286**:355.
101. Gao C, Moya S, Donath E, Möhwald H. Melamine formaldehyde core decomposition as the key step controlling capsule integrity: optimizing the polyelectrolyte capsule fabrication. *Macromol Chem Phys* 2002;**203**:953.
102. Caruso F, Yang W, Trau D, Renneberg R. Microencapsulation of uncharged low molecular weight organic materials by electrolyte multilayer self-assembly. *Langmuir* 2000;**16**:8932.
103. Park MK, Xia C, Advincula RC, Schütz P, Caruso F. Cross-linked, luminescent spherical colloidal and hollow-shell particles. *Langmuir* 2001;**17**:7670.
104. Yu AM, Wang Y, Barlow E, Caruso F. Mesoporous silica particles as templates for preparing enzyme-loaded biocompatible microcapsules. *Adv Mater* 2005;**17**:1737.
105. Delcea M, Yashchenok A, Videnova K, Kreft O, Möhwald H, Skirtach AG. Multicompartmental micro- and nanocapsules: hierarchy and applications in biosciences. *Macromol Biosci* 2010;**10**:465.
106. Shchepelina O, Kozlovskaya V, Kharlampieva E, Mao W, Alexeev A, Tsukruk VV. Anisotropic micro- and nano-capsules. *Macromol Rapid Commun* 2010;**31**:2041.
107. Gittins DI, Caruso F. Multilayers polymer nanocapsules derived from gold nanoparticles templates. *Adv Mater* 1947;**2000**:12.
108. Cuomo F, Lopez F, Miguel MG, Lindman B. Vesicle-templated layer-by-layer assembly for the production of nanocapsules. *Langmuir* 2010;**26**:10555.
109. Georgieva R, Moya S, Hin M, Mitlöhner R, Donath E, Kiesewetter H, et al. Permeation of macromolecules into polyelectrolyte microcapsules. *Biomacromolecules* 2002;**3**:517.
110. Balabushevitch NG, Sukhorukov GB, Moroz NA, Volodkin DV, Larionova NI, Donath E, et al. Encapsulation of proteins by layer-by-layer adsorption of polyelectrolytes onto protein aggregates: factors regulating the protein release. *Biotechnol Bioeng* 2001;**76**:207.
111. Volodkin DV, Petrov AI, Prevot M, Sukhorukov GB. Matrix polyelectrolyte microcapsules: new system for macromolecule encapsulation. *Langmuir* 2004;**20**:3398.
112. Zhu Y, Shi J, Shen W, Dong X, Feng J, Ruan M, et al. Stimuli-responsive controlled drug release from a hollow mesoporous silica sphere/polyelectrolyte multilayer core-shell structure. *Angew Chem Int Ed* 2005;**44**:5083.
113. Khopade AJ, Caruso F. Surface-modification of polyelectrolyte multilayer coated particles for biological applications. *Langmuir* 2003;**19**:6219.
114. Heuberger R, Sukhorukov GB, Vörös J, Textor M, Möhwald H. Biofunctional polyelectrolyte multilayers and micro capsules: control of non-specific and biospecific protein adsorption. *Adv Funct Mater* 2005;**15**:357.

115. Moya S, Donath E, Sukhorukov GB, Auch M, Bäumler H, Lichtenfeld H, et al. Lipid coating on polyelectrolyte surface modified colloidal particles and polyelectrolyte capsules. *Macromolecules* 2000;**33**:4538.
116. Zhao Q, Mao Z, Gao C, Shen J. Assembly of multilayer microcapsules on $CaCO_3$ particles from biocompatible polysaccharides. *J Biomater Sci Polym Ed* 2006;**17**:997.
117. Balkundi SS, Veerabadran NG, Matthew ED, Johnson GR, Lvov YM. Encapsulation of bacterial spores in nanoorganized polyelectrolyte shells. *Langmuir* 2009;**25**:14011.
118. Angelatos AS, Radt B, Caruso F. Light responsive polyelectrolyte/gold nanoparticle microcapsules. *J Phys Chem B* 2005;**109**:3071.
119. Zelikin A, Quinn JF, Caruso F. Disulfide cross-linked polymer capsules: en route to biodeconstructible systems. *Biomacromolecules* 2006;**7**:27.
120. Antipov AA, Sukhorukov GB, Donath E, Möhwald H. Sustained release properties of polyelectrolyte multilayer capsules. *J Phys Chem B* 2001;**105**:2281.
121. Delcea M, Möhwald H, Skirtach AG. Stimuli responsive LbL capsules and nanoshells for drug delivery. *Adv Drug Deliv Rev* 2011;**63**:730–47.
122. Wang Z, Feng Z, Gao C. Stepwise assembly of the same poly electrolytes using host-guest interaction to obtain microcapsules with multiresponsive properties. *Chem Mater* 2008;**20**:4194.
123. Liu X, Gao C, Shen J, Möhwald H. Multilayer microcapsules as anti-cancer drug delivery vehicle. Deposition, sustained release, and in vitro bioactivity. *Macromol Biosci* 2005;**5**:1209.
124. Itoh Y, Matsusaki M, Kida T, Akashi M. Enzyme-responsive release of encapsulated proteins from biodegradable hollow capsules. *Biomacromolecules* 2006;**7**:2715.
125. Cismaru L, Popa M. Polymeric nanoparticles with biomedical applications. *Rev Roum Chim* 2010;**55**:433.
126. Anderson CD, Daniels ES. Emulsion polymerisation and latex applications. *Rapra Rev Rep* 2003;**14**:3.
127. Chen J, Zeng F, Wu S, Su J, Zhao J, Tong Z. A facile approach for cupric ion detection in aqueous media using polyethyleneimine/PMMA core-shell fluorescent nanoparticles. *Nanotechnology* 2009;**20**:365502.
128. Qu F, Guan Y, Ma Z, Zhang Q. Synthesis of Cibacron Blue F3GA-coupled magnetic PMMA nanospheres and their use for protein affinity separation. *Polym Int* 2009;**58**:888.
129. Urbina MC, Zinoveva S, Miller T, Sabilov CM, Monroe WT, Kumer CSSR. Investigation of magnetic nanoparticles-polymer composites for multiple-controlled drug delivery. *J Phys Chem C* 2008;**112**:11102.
130. Lu S, Liu R, Sun X. A study on the synthesis and applications of inverse emulsion of amphoteric polyacrylamide as retention aid in papermaking. *J Appl Polym Sci* 2002;**84**:343.
131. Barari M, Abdollahi M, Hemmati M. Synthesis and characterization of high molecular weight polyacrylamide nanoparticles by inverse-emulsion polymerization. *Iran Polym J* 2011;**20**:65.
132. Reece TB, Maxey TS, Kron IL. A prospectus on tissue adhesives. *Am J Surg* 2001;**182**:40S.
133. Vauthier C, Dubernet C, Fattal E, Pinto-Alphandary H, Couvreur P. Poly(alkylcyanoacrylates) as biodegradable materials for biomedical applications. *Adv Drug Deliv Rev* 2003;**55**:519.
134. Kim S, Kim CA, Pyo HB, Jung MY. Synthesis of anionic sulfate functionalized polystyrene latex and protein adsorption for application as biosensor. *J Nanosci Nanotechnol* 2009;**9**:7171.
135. Chen Y, Long RW, Chen ZG, Ding JN. Synthesis of PS-core ceria-shell composite abrasive and its applications in oxide CMP. *Tribology* 2010;**30**:9.
136. Wang S, Li M, Lu Q. Filter paper with selective absorption and separation of liquids that differ in surface tension. *ACS Appl Mater Interf* 2010;**2**:677.

137. Fritz H, Maier M, Bayer E. Cationic polystyrene nanoparticles: preparation and characterization of a model drug carrier system for antisense oligonucleotides. *J Colloid Interf Sci* 1997;**195**:272.
138. Zambaux MF, Bonneaux F, Gref R, Maincent P, Dellacherie E, Alonso MJ, et al. Influence of experimental parameters of the characteristics of poly(lactic acid) nanoparticles prepared by a double emulsion method. *J Control Release* 1998;**50**:31.
139. Zhou J, Moya S, Ma L, Gao C, Shen J. Polyelectrolyte coated PLGA nanoparticles: templation and release behavior. *Macromol Biosci* 2009;**9**:326.
140. Zhou J, Romero G, Rojas E, Moya S, Ma L, Gao C. Folic acid modified poly(lactide-co-glycolide) nanoparticles, layer-by-layer surface engineers for targeted delivery. *Macromol Chem Phys* 2010;**211**:404.
141. Zhou J, Romero G, Rojas E, Ma L, Moya S, Gao C. Layer by Layer chitosan/alginate coatings on poly(lactide-co-glycolide) nanoparticles for antifouling protection and Folic acid binding to achieve selective cell targeting. *J Colloid Interf Sci* 2010;**345**:241.
142. Romero G, Estrela-Lopis I, Zhou J, Rojas E, Franco A, Sanchez C, et al. Surface engineered poly(lactide-co-glycolide) nanoparticles for intracellular delivery: uptake and cytotoxicity—a confocal Raman microscopic study. *Biomacromolecules* 2010;**11**:2993.
143. Kim BK, Lee JS, Oh JK, Park DJ. Preparation of resveratrol-loaded poly(ε-caprolactone) nanoparticles by oil-in-water emulsion solvent evaporation method. *Food Sci Biotechnol* 2009;**18**:157.
144. Rahman M, Islam S, Sharmin N, Chowdhury JA, Jalil R. Preparation and evaluation of cellulose acetate phthalate and ethyl cellulose based microcapsules of diclofenac sodium using emulsification and solvent evaporation method. *J Pharm Sci* 2010;**9**:39.
145. Koosha F, Muller RH, Washington C. Production of polyhydroxybutyrate (PHB) nanoparticles for drug targeting. *J Pharm Pharmacol* 1987;**39**:136.
146. Trimaille T, Pichot C, Elaïssari A, Fessi H, Briançon S, Delair T. Poly(D, L-lactic acid) nanoparticle preparation and colloidal characterization. *Colloid Polym Sci* 2003;**281**:1184.
147. Cascone MG, Lazzeri L, Carmignani C, Zhu Z. Gelatin nanoparticles produced by simple W/O emulsion as delivery system for methotrexate. *J Mater Sci Mater Med* 2002;**13**:523.
148. Schubert S, Delaney JT, Schubert US. Nanoprecipitation and nanoformulation of polymers: from history to powerful possibilities beyond poly(lactide acid). *Soft Matter* 2011;**7**:1581.
149. Cirpanli Y, Yerlikaya F, Oyturk K, Erdogar N, Launay M, Gegu C, et al. Comparative evaluation of in vitro parameters of tamoxifen citrate loaded poly(lactide-co-glycolide), poly(ε-caprolactone) and chitosan nanoparticles. *Pharmazie* 2010;**65**:867.
150. Aubry J, Ganachaud F, Addad PC, Cabane B. Nanoprecipitation of polymethylmethacrylate by solvent shifting: 1. Boundaries. *Langmuir* 1970;**2009**:25.
151. Perugini P, Simeoni S, Scalia S, Genta I, Modena T, Conti B, et al. Effect of nanoparticle encapsulation on the photostability of the sunscreen agent, 2-ethylhexyl-p-methoxycinnamate. *Int J Pharm* 2002;**246**:37.
152. Allémann E, Leroux JC, Gurny R, Doelker E. In vitro extended-release properties of drug-loaded poly(DL-lactic acid) nanoparticles produced by a salting-out procedure. *Pharm Res* 1993;**10**:1732.
153. Galindo Rodríguez SA, Puel F, Briançon S, Alléman E, Doelker E, Fessi H. Comparative scale-up of three methods for producing ibuprofen-loaded nanoparticles. *Eur J Pharm Sci* 2005;**25**:357.

Chapter 5

Synthetic Strategies to Create Dendrimers: Advantages and Drawbacks

Macarena Sánchez-Navarro and Javier Rojo
Glycosystems Laboratory, Instituto de Investigaciones Químicas, CSIC — Universidad de Sevilla, Seville, Spain

1. INTRODUCTION

Chemists, inspired by nature, have been able to create artificial entities by means of the arsenal of synthetic tools available. In this context, the creation of new chemical bonds to build up any type of architectures is unlimited, at least in theory. In nature, it is possible to find many examples of a specific type of complex and beautiful topologies denoted as dendritic structures. Ordinary examples such as the tree branching, roots, snow crystals, fern leaves, dendrites in neurons, capillaries in the vascular system, etc., can be considered representative of these structures at different scales. The reason why nature makes use of these dendritic topologies in spite of their complexity lies in the advantages inherent to these structures. Dendritic structures allow the maximization of a biological function using a minimum space.

Dendrimers are hyperbranched macromolecules with a well-defined three-dimensional structure.[1–3] In contrast to dendritic polymers, dendrimers are monodisperse compounds. The term dendrimer derives from the Greek words *dendri* (branch tree-like) and *meros* (part of). The interest of these structures lies in their potential applications in fields such as catalysis, molecular electronics, photonics, and nanomedicine.[4] In fact, the broad spectra of functionalities that can be incorporated into these structures as well as the flexibility and modulation of the physical and chemical properties by changing the building blocks used for their construction open an interesting scenario for new applications. Few years ago, the first approval of the FDA for a clinical use of a dendrimer (SLP7013, VivaGel™) has been granted for its application as microbiocide.[5] Many excellent reviews describing the biological applications of functionalized dendrimers are already published.[4,6–19]

Frontiers of Nanoscience, Vol. 4. DOI: 10.1016/B978-0-12-415769-9.00005-4

Dendrimers can be divided into three well-differentiated domains: the core, the inner shells, and the outer shell (Figure 5.1).

The core is normally a small molecule with a variable number of functional groups that define the number of branches (dendrons) in the final structure. Inner shells (interior layers) composed of repeating units (monomers) define the generation (G). The outer shell (periphery or surface groups) contains a large number of functional groups (depending on the generation) that can be manipulated to modulate dendrimer properties and applications.

The potentiality of dendrimers in many fields has pushed the development of more efficient and simple synthetic methodologies for the preparation of these complex molecules in a reasonable and competitive manner doing feasible their industrial applications.

Historically, the first synthesis described, considered as the starting point of this field, goes back to 1978 when Vögtle and coworkers[20] published the synthesis of branched compounds described as cascade-like molecules using a strategy that was the inspiration for synthesizing dendrimers. Since then, dendritic structures with different valency, properties, and applications have been prepared using any kind of cores, building blocks, and functional groups.[21–27]

Addressing the synthesis of dendrimers in a complete monodisperse way is a challenge due to the number of covalent bonds to be created in only one synthetic step and also due to the steric hindrance that can occurs. Since the first article, more than 1500 papers related to dendrimers and hundreds of

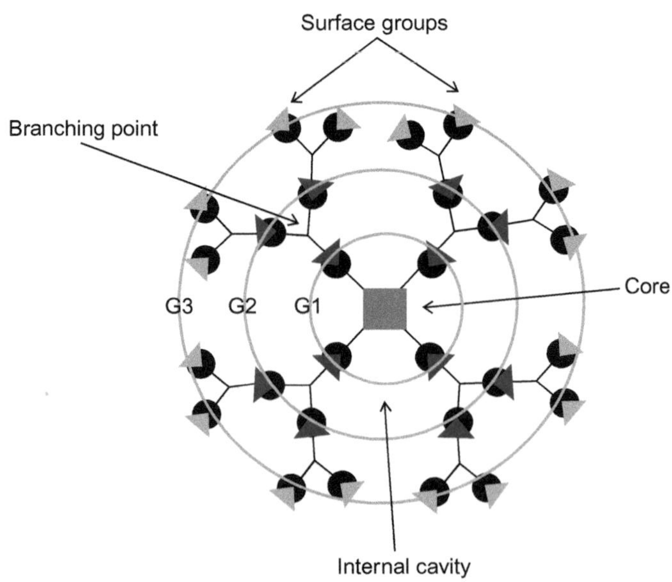

FIGURE 5.1 Domains of a general dendrimeric structure.

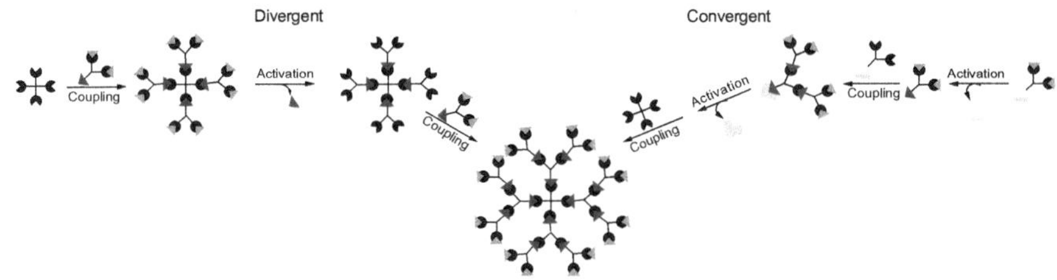

FIGURE 5.2 Schematic representation of the divergent and convergent synthetic strategies.

different dendrimers have been described in the literature. Therefore, due to the limit of space, this review will not be an exhaustive list of all the methodologies described. In this chapter, we would like to summarize some synthetic approaches used to prepare dendrimers and highlight the advances that these approaches have produced. We have selected a few representative examples to illustrate the synthesis of interesting molecules from the early work to more recent achievements. This selection should provide an overview of the state of the art in the field of dendrimers from the point of view of the available synthetic methodologies.

In general, all the synthetic approaches can be classified into two main strategies: the divergent and the convergent approaches (Figure 5.2).

2. DIVERGENT SYNTHETIC APPROACHES

The divergent approach is characterized by the generation growth from the core toward the periphery (inside-out approach) increasing the number of end-groups in each step. This approach basically requires two actions: (1) monomer coupling and (2) deprotection or activation of end-groups. The number of covalent bonds created in each step increases exponentially with the generation. The congestion at the periphery during the generation growth is only a problem for the higher ones because the monomers used for the synthesis are normally small molecules. However, for these high generations, the number of new bonds created is notable and the possibilities to present structural defect increase with every generation. Moreover, these imperfections make the purification and characterization of these molecules complicated. To overcome this problem, the divergent synthesis of dendrimers has been addressed using a large excess of monomer and reagents to improve the reaction yields. Although this strategy works in some cases, in others fail leading to structural polydisperse molecules reducing the efficiency of purification and making the synthesis less attractive and suitable.

The first divergent synthesis of a dendrimer was described in 1985 independently by Tomalia[28] and Newkome.[29] Both groups used different terminology to define these molecules, dendrimer coined for the first time by Tomalia and arborols by Newkome. The synthesis of poly(amidoamine) (PAMAM) dendrimers described by Tomalia is so efficient that nowadays it is the preferred approached of industry for the preparation of these molecules. This classical approach consisted of a two-step synthesis: a Michael addition and an amidation reaction. Up to generation 7 (G7) was prepared using this strategy, where a large excess of reagents was required to achieve a high monodispersity.

Any kind of synthetic route and chemistry can be envisaged to prepare dendrimers using a divergent strategy. Biocompatible polyester dendrimers were prepared by condensation reaction in high yields and purity up to G6.[30–32] Aromatic polyamide dendrimers were prepared up to G4 in yields around 95% using precipitation in alkaline water as purification method in each step.[33] Organometallic chemistry has been applied to prepare high generation (G6) platinum-based dendrimers with diameters larger than 10 nm as described by Takahashi and coworkers[34] (Scheme 5.1).

These are representative example of conventional synthetic methodologies to prepare dendrimers.

The success of the divergent strategy depends on the efficiency of the synthetic steps. The synthesis requires a repetitive stepwise growth based on a coupling reaction of a monomer following by a deprotection or activation step. In 2001, Caminade and Majoral[35] published for the first time a divergent synthesis of a dendrimer using an orthogonal coupling strategy, an approach that does not require the activation step. To address this, a selection of two types of monomers (AB_2 and CD_2) presenting orthogonal functional groups where A reacts only with D and C with B was needed. Using phosphorhydrazides (A) and aldehydes (D) that react under condensation conditions and phosphines (B) and azides (C) that react under Staudinger conditions, a G4 dendrimer was prepared in one-pot synthesis. The dendrimer was produced in quantitative yield without activation reagents, without purification steps, and generating only N_2 and H_2O as side products. This accelerate chemical strategy was used later by Hult and Malkoch to prepare a couple of examples of G4 dendrimers.[36] A Fréchet-type dendrimer (poly(aryl ethers)) was synthesized in only four steps in multigram scale and with an overall yield of 70%. Classical route based on coupling and protection/activation pathway should require eight synthetic steps and several purifications. In a second example, another G4 dendrimer was prepared with high purity and in excellent yields using click chemistry and ester coupling reactions based on bis-MPA as monomer.

The election of adequate synthetic methodology to prepare dendrimers is a key point to achieve an easy synthesis of monodisperse materials. In 2008, Simanek[37] described the preparation of a second generation dendrimer based on cyanuric chloride using a high efficient synthetic strategy. This simple approach allowed the preparation of these dendrimers in kilogram scale using environmentally friendly methods which established their potential use in industry.

A combination of monomers for the preparation of a dendrimer provides the means to modulate the physical and chemical properties of the molecules,

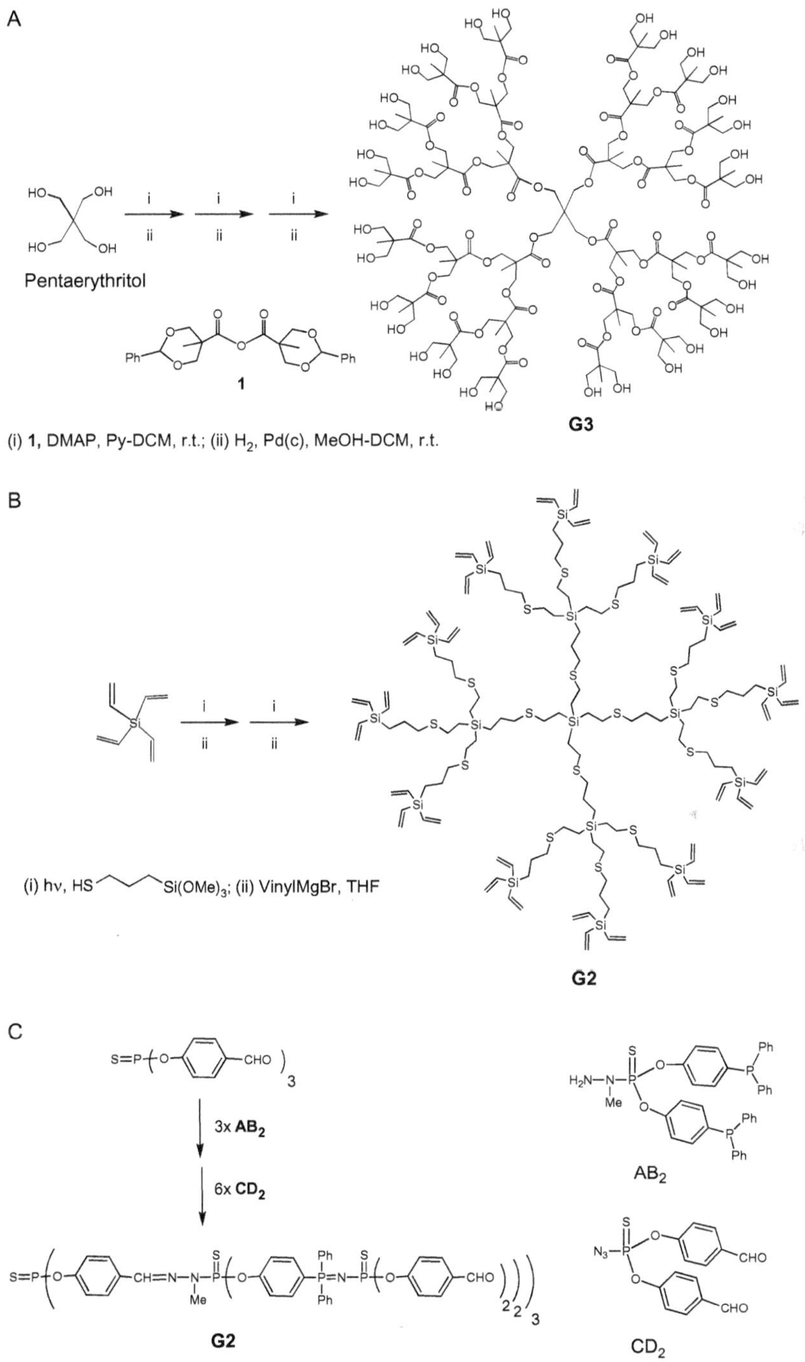

SCHEME 5.1 Three examples of divergent synthesis of dendrimers using: (A) classical approach, (B) thiol–ene click chemistry, and (C) orthogonal monomers AB_2 and CD_2.

and it is an interesting way to produce new dendrimers. In 2008, Tomalia and Backer created POMAM hybrid dendrimers by combination of a core dendrimer based on poly(propylenimine) and shells based on PAMAM.[38]

The advent of click chemistry has marked an inflection point in the development of synthetic methodologies to prepare dendrimers and other macromolecules.[26]

One of the first examples of this methodology was described even before the term click chemistry was coined. Müllen and coworkers used the well-known Diels–Alder (DA) reaction to build up G2 dendrimers contained phenyl rings.[39] A more recent work of the same group demonstrated the scope of the DA strategy to generate polyphenylene dendrimers of higher generation (G4) in yields above 80% using, in this particular case, an iridium complex as core.[40] The DA reaction between maleimide and furan derivatives was used to combine dendrons divergently synthesized.[41] This interesting strategy allows the preparation of unsymmetrical dendrimers when nonidentical dendrons are combined. An interesting behavior of these molecules is the thermoreversibility of the DA reaction to generate the initial components.

However, the reaction that has attracted more attention of chemists working in the macromolecule field is the copper-catalyzed azide-alkyne 1,3-dipolar cycloaddition (CuAAC). This reaction is characterized by the features required to prepare monodisperse dendrimers in a very efficient manner.[26,42–45] The application of this reaction to the divergent synthesis of dendrimers was described for the first time in 2005 by Hawker and Wooley.[46] The application of the CuAAC reaction in each growth step allowed the preparation of G2 and G3 dendrimers in high yields with little or no purification. This example demonstrated the efficiency of these synthetic methodologies to the preparation of dendrimers, which demand facile and cost–effective synthetic methods like click chemistry.

A recent example combines two types of click chemistry reactions, CuAAC and DA.[47] Orthogonal functional groups in the monomers were chosen to apply sequentially both reactions. Also, the use of microwaves to reduce the reaction time makes very attractive the synthetic methodology described in this chapter. This is a very interesting example where some synthetic solutions are applied to improve the preparation of these monodisperse macromolecules.

Together with CuAAC and DA, the thiol–ene reaction is the other important click chemistry reaction that has been developed more recently. This reaction is considered greener than the others because it does not require the use of metals, it can be performed without solvent under controlled conditions, and there are no side products. One of the first examples of the thiol–ene reaction applied to a dendrimer synthesis was described in 2008 by Hawker to prepare a G4 dendrimer functionalized with hydroxyl groups.[48] In this example, a combination of thiol–ene reaction with and ester condensation by the selection of orthogonal terminal groups in the monomers avoids the use of activation or deprotection steps in the synthesis. The environmentally friendly process without solvent, at room temperature, using short reaction times and purification steps by precipitation makes this process very suitable for industrial applications. A similar strategy using thiol–ene reaction and Grignard substitution sequence was applied to prepare carbosilane–thioether dendrimers up to G5 in good to excellent yields.[49] A thio-Michael addition was used recently to prepare up to G4 dendrimers in good yields using a synthetic sequence based on a Grob/Eschenmoser condensation and a thio-Michael addition.[50]

Although most of the monomers used to prepare dendrimers consist of molecules of the AB_2 type, the use of macromonomers or hypermonomers such as AB_3, AB_4, or AB_8 has been considered as an attractive alternative to generate dendrimers in few steps with a high number of terminal functional groups.[51] See, for instance, the work of Fréchet[52] and, more recently, the synthesis of G5 dendrimers based on triazine by Simanek.[53] Another example where the complexity of monomers has been increased was described by Hult and Malkoch in 2009.[54] A monomer of type AB_2C, where A reacts with B but not with C, displaying a high functionality in order to obtain bifunctional dendrimers was prepared. Using this monomer, a G4 dendrimer with 48 (24 + 24) functional groups was obtained in only eight steps. This example demonstrated the versatility of this approach to prepare bifunctional dendrimers with two different types of functional groups on the surface increasing the structural complexity in comparison with those dendrimers constructed with classical AB_2 monomers.

From the first examples described in mid-1980s to the new approaches, the synthetic methodologies have been improved remarkably with the aim to make the production of dendrimers more efficient and easier. The divergent strategy is still an interesting option to prepare these macromolecules. In fact, all commercially available dendrimers are prepared in the industry via divergent approaches.

3. CONVERGENT SYNTHETIC APPROACHES

Synthesizing dendrimers using a convergent approach involves the construction of the dendrimers from the surface to the nucleus (core). Initially, every dendron, which constitutes the dendrimer body, is prepared being the last step its connection to a polyfunctional core.

This segment coupling strategy shows unique characteristics as the low number of bonds built, that remains constant in each step, avoids the use of large excess of reagents and facilitates an easier purification. As a matter of fact, size exclusion chromatography becomes a very useful tool in the purification of these molecules. The rapid increase of the molecular weight at each generation growth step produces high level of monodispersity in the prepared compounds.

Besides, this strategy is a powerful tool for preparing dendrimers with different morphology, joining two or more different segments in the final coupling step.

The main disadvantage of this methodology resides in the steric hindrance that can occur in high generations producing a decrease of the reactivity between the focal point of the dendrons and the core functional groups.

The first convergent synthesis of dendrimers was published by Hawker and Fréchet[55,56] followed by Miller and Neenan[57] in 1990. The former authors prepared polyether dendrimers using 3,5-dihydroxybenzyl alcohol as monomer which was coupled to benzyl bromide forming the repetition unit. The synthetic sequence was repeated generating the G3 dendron, which was finally coupled to the selected core (Scheme 5.2). With this first example, the authors proved the versatility and scope of this methodology which has been widely applied in recent years.

Different chemical alternatives have been used to prepare a vast amount of dendrimers with different properties, size, solubility, or function.

One of the most representative methods developed is based on click chemistry reactions, in particular, CuAAC. The first example was published by Hawker and Sharpless in 2004.[58] Using different cores and AB_2 monomers, they prepared different dendrimers up to G4, probing the value of this methodology due to the high selectivity and yields, and mild conditions required. Encouraged by this result, several groups have achieved the preparation of different type of dendrimers by using this click approach, one of the most efficient to date.[26,44,45]

DA cycloaddition has been used as well to prepare dendrimers by a convergent approach. In 1999, Müllen used the DA reaction to prepare convergently polyphenylene dendrons with more than 60 benzene moieties via [2+4] cycloadditions and Knoevenagel condensations producing dendrimers with different shapes by modulating the core used.[59]

Since the study of self-assembly of molecular species is becoming a very important field in material sciences, this property has been used to prepare dendrimers by noncovalent interactions.[60] One of the first example, reported by Zimmerman,[61] employed hydrogen-bonding interactions between two isophthalic acid units. This units bearing different Fréchet dendrons and coupled to a rigid aromatic ring as binding motif formed hexameric units which are stable in most of the organic solvents but not in polar solvent. Later on, DNA was used for the preparation of more stable noncovalent bonded dendrimers. Tomalia[62] prepared PAMAM dendrons functionalized with single-strand DNA molecules. The assembly of the DNA strands allowed the preparation of different dendrons in a precise manner.[63] Recently, Gothelf and coworkers have used an approach for DNA-templated covalent coupling of modified PAMAM dendrimers. The authors claimed that the concept can be applied to a wide range of macromolecules leading to new applications.[64]

The formation of self-assembled dendrimers has also been studied by metal complexation.[65] Polyether dendrons containing a carboxylate anion at the focal point can trap different lanthanide cations (Er^{3+}, Eu^{3+}, Tb^{3+})

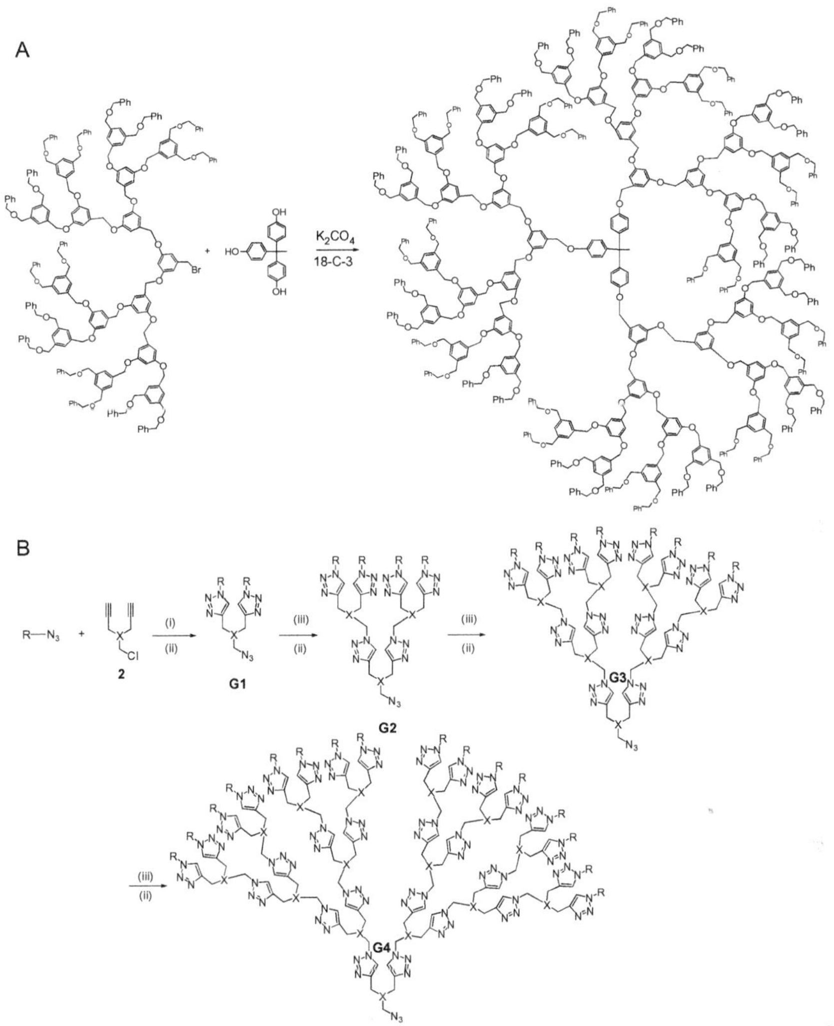

(i) $CuSO_4$(5 mol%), sodium ascorbate (10 mol%), H_2O:*t*BuOH; (ii) 1.5 equiv. NaH, CH_3COCH_3/H_2O (4:1), 60 °C, 1^{-3} h; (iii) **2**, $CuSO_4$(5 mol%), sodium ascorbate (10 mol%), H_2O:*t*BuOH

SCHEME 5.2 Two representative examples of convergent synthesis of dendrimers: (A) first convergent synthesis reported[56] and (B) earliest convergent approach toward triazol dendrimers.

forming dendrimers with enhanced photoluminescence.[66] This process has also been studied with bis-MPA carboxylates bearing pentafluorophenyl groups at the surface. Both complexes have been investigated due to its unique spectroscopic properties.

Since the use of a convergent strategy involves the preparation of the dendrons separately to being finally attached to a central core, multifunctional molecules can be prepared in an easy way. The first unsymmetrical dendrimer was reported by Fréchet in 1993[67] by the preparation of two different benzylether-based dendrons bearing cyano (electron withdrawing) and hydroxyl (electron donating) groups, respectively, and its subsequent attachment to a lineal core. The authors found an enhanced dipolar moment on these compounds in comparison with the corresponding symmetrical dendrimers. Later on, the same author reported the design of a new class of dendrimer shape, "bow tie," characterized by the capability of controlling the loading of two different sorts of ligands.[68]

The combination of an unsymmetrical design and the versatility of "click chemistry," discussed above, were applied to the preparation of bifunctional dendrimers containing both mannose moieties and coumarin fluorescent entity.[69] This bifunctional system joined in one molecule two functions: recognition and detection.

The possibility of developing fancy structures with tunable properties has opened a wide field in the preparation of multiblock dendrimers with two or more functionalities.[27]

Organometallic dendrimers play an important role in the development of the synthetic strategies for dendrimer construction. For instance, siloxano-linked dendrons and dendrimer have been prepared via hydrosilylation as analogues of PAMAM and polyether systems.[70] One interesting example is the preparation of platinum-acetylide dendrimers bearing up to 60 Pt moieties. The authors showed the formation of the desired dendrimers by reaction of previously prepared acetylide dendrons with tri- and tetraplatinum cores, finding this strategy restricted to the preparation of G3.[71]

In recent years, other strategies are emerging for the preparation of dendrimers in order to reduce the number of synthetic and purification steps, and increasing yields. Among the most important approaches are the dual-stage convergent growth,[71–75] the double exponential,[76] and the orthogonal coupling.[77,78] The first approach is the coupling of dendrons previously prepared on a dendritic core, which has a large number of functional groups on the surface. The double exponential growth strategy accelerates the synthesis of dendrimers by the use of orthogonally protected branched monomers, thus allowing the growth of the dendron along the focal point and the periphery. The orthogonal coupling uses two different monomers, in order to eliminate the protection or activation steps, each monomer giving rise to a new generation.

4. CONCLUSIONS

Dendrimer chemistry has experienced extensive development in recent years. The structural and functional versatility of dendrimers and their potential application in the fields of material science and biomedicine have attracted the interest of many research groups, leading to an enormous development of synthetic methodologies to prepare them.

Current synthetic strategies, highlighted herein, afford the preparation of dendrimers with controlled size and shape in multigram scale. More importantly, its properties can be tuned easily by the modification of surface groups, branching groups, or the core. In this sense, different aspects like solubility, biocompatibility, or spectroscopic properties can be modulated. One of the most important achievements in the field is the preparation of a large range of multifunctional dendrimers bearing different groups in the same molecule.

The main advantage of dendrimers over other multivalent systems, such as nanoparticles or polymers, is the monodispersity with a total control over the shape, size, and number of ligands that allows the full characterization of the synthesized molecules, accordingly, the enhancement of its unique properties. The main drawback is always the complexity and difficulty to generate these molecules in a total monodisperse fashion, but the development of very efficient synthetic methods has improved remarkably this aspect; therefore, it has converted these monodisperse macromolecules in a very interesting alternative to other systems. In fact, their biological applications are becoming real due to the development of biocompatible systems such as polyester or poly(L-lysine) dendrimers. The SLP7013, VivaGel™ is a clear example of this fact but hopefully will not be the only one.

From the earliest work at the end of the 1970s, a long road has been traveled in the dendrimer field. Now, we face a very interesting and exciting future for these molecules where new opportunities to develop original and innovative synthetic strategies to prepare dendrimers will be welcome.

REFERENCES

1. Fréchet JMJ, Tomalia DA, editors. *Dendrimers and other dendritic polymers*. Chichester: Wiley; 2001.
2. Newkome GR, Moorefield CN, Vögtle F, editors. *Dendrimers and dendrons, Concepts, synthesis, applications*. Weinheim: Wiley VCH; 2001.
3. Tomalia DA, Fréchet JMJ. Discovery of dendrimers and dendritic polymers: a brief historical perspective. *J Polym Sci A Polym Chem* 2002;**40**:2719.
4. Astruc D, Boisselier E, Ornelas C. Dendrimers designed for functions: from physical, photophysical, and supramolecular properties to applications in sensing, catalysis, molecular electronics, photonics, and nanomedicine. *Chem Rev* 1857;**2010**:110.
5. McCarthy TD, Karellas P, Henderson SA, Giannis M, O'Keefe DF, Heery G, et al. Dendrimers as drugs: discovery and preclinical and clinical development of dendrimer-based microbicides for HIV and STI prevention. *Mol Pharm* 2005;**2**:312.
6. Bosman AW, Janssen HM, Meijer EW. About dendrimers: structure, physical properties, and applications. *Chem Rev* 1999;**99**:1665.
7. Patri AK, Majoros IJ, Baker Jr. JR. Dendritic polymer macromolecular carriers for drug delivery. *Curr Opin Chem Biol* 2002;**6**:466.
8. Cloninger MJ. Biological applications of dendrimers. *Curr Opin Chem Biol* 2002;**6**:742.
9. Stiriba S-E, Frey H, Haag R. Dendritic polymers in biomedical applications: from potential to clinical use in diagnostics and therapy. *Angew Chem Int Ed* 2002;**41**:1329.
10. Boas U, Heegaard PMH. Dendrimers in drug research. *Chem Soc Rev* 2004;**33**:43.
11. Gillies ER, Fréchet JMJ. Dendrimers and dendritic polymers in drug delivery. *Drug Discov Today* 2005;**10**:35.
12. Lee CC, MacKay JA, Fréchet JMJ, Szoka FC. Designing dendrimers for biological applications. *Nat Biotechnol* 2005;**23**:1517.

13. Svenson S, Tomalia DA. Dendrimers in biomedical applications—reflections on the field. *Adv Drug Deliv Rev* 2005;**57**:2106.
14. Rojo J, Delgado R. Dendrimers and dendritic polymers as anti-infective agents: new antimicrobial strategies for therapeutic drugs. *Antiinfective Agents Med Chem* 2007;**6**:151.
15. Wolinsky JB, Grinstaff MW. Therapeutic and diagnostic applications of dendrimers for cancer treatment. *Adv Drug Deliv Rev* 2008;**60**:1037.
16. Rojo J. Dendritic compounds as immune response modulators. New approaches for vaccine development. *Antiinfective Agents Med Chem* 2009;**8**:50.
17. Tekade RK, Kumar PV, Jain NK. Dendrimers in oncology: an expanding horizon. *Chem Rev* 2009;**109**:49.
18. Medina SH, El-Sayed MEH. Dendrimers as carriers for delivery of chemotherapeutic agents. *Chem Rev* 2009;**109**:3141.
19. Mintzer MA, Grinstaff MW. Biomedical applications of dendrimers: a tutorial. *Chem Soc Rev* 2011;**40**:173.
20. Buhleier E, Wehner W, Vögtle F. "Cascade"- and "nonskid-chain-like" syntheses of molecular cavity topologies. *Synthesis* 1978;155.
21. Matthews OA, Shipway AN, Stoddart JF. Dendrimers—branching out from curiosities into new technologies. *Prog Polym Sci* 1998;**23**:1.
22. Klajnert B, Bryszewska M. Dendrimers: properties and applications. *Acta Biochim Pol* 2001;**48**:199.
23. Maraval V, Laurent R, Marchand P, Caminade A-M, Majoral J-M. Accelerated methods of synthesis of phosphorus-containing dendrimers. *J Organomet Chem* 2005;**690**:2458.
24. Boas U, Christensen JB, Heegard PMH. Dendrimers: design, synthesis and chemical properties. *J Mater Chem* 2006;**16**:3785.
25. Carlmark A, Hawker C, Hult A, Malkoch M. New methodologies in the construction of dendritic materials. *Chem Soc Rev* 2009;**38**:352.
26. Franc G, Kakkar AK. "Click" methodologies: efficient, simple and greener routes to design dendrimers. *Chem Soc Rev* 2010;**39**:1536.
27. Hourani R, Kakkar A. Advances in the elegance of chemistry in designing dendrimers. *Macromol Rapid Commun* 2010;**31**:947.
28. Tomalia DA, Baker H, Dewald J, Hall M, Kallos G, Martin S, et al. A new class of polymers: starburst-dendritic macromolecules. *Polym J* 1985;**17**:117.
29. Newkome GR, Yao Z-q, Baker GR, Gupta VK. Cascade molecules: a new approach to micelles. A[27]-arborol. *J Org Chem* 1985;**50**:2003.
30. Ihre H, Padilla De Jesús OL, Fréchet JMJ. Fast and convenient divergent synthesis of aliphatic ester dendrimers by anhydride coupling. *J Am Chem Soc* 2001;**123**:5908.
31. Malkoch M, Malmström E, Hult A. Rapid and efficient synthesis of aliphatic ester dendrons and dendrimers. *Macromolecules* 2002;**35**:8307.
32. Luczkowiack J, Sattin S, Sutkevičiūtė I, Reina JJ, Sánchez-Navarro M, Thépaut M, et al. Pseudosaccharide functionalized dendrimers as potent inhibitors of DC-SIGN dependent Ebola pseudotyped viral infection. *Bioconjug Chem* 2011;**22**:1354.
33. Washio I, Shibasaki Y, Ueda M. Facile synthesis of amine-terminated aromatic polyamide dendrimers via a divergent method. *Org Lett* 2007;**9**:1363.
34. Onitsuka K, Shimizu A, Takahashi S. A divergent approach to the precise synthesis of giant organometallic dendrimers using platinum-acetylides as building blocks. *Chem Commun* 2003;280.
35. Brauge L, Magro G, Caminade A-M, Majoral J-P. First divergent strategy using two AB_2 unprotected monomers for the rapid synthesis of dendrimers. *J Am Chem Soc* 2001;**123**:6698.
36. Antoni P, Nyström D, Hawker CJ, Hult A, Malkoch M. A chemoselective approach for the accelerated synthesis of well-defined dendritic architectures. *Chem Commun* 2007;2249.

37. Chouai A, Simanek EE. Kilogram-scale synthesis of a second-generation dendrimer based on 1,3,5-triazine using green and industrially compatible methods with a single chromatographic step. *J Org Chem* 2008;**73**:2357.
38. Majoros IJ, Williams CR, Tomalia DA, Baker Jr. JR. New dendrimers: synthesis and characterization of POPAM-PAMAM hybrid dendrimers. *Macromolecules* 2008;**41**:8372.
39. Morgenroth F, Reuther E, Müllen K. Polyphenylene dendrimers: from three-dimensional to two-dimensional structures. *Angew Chem Int Ed* 1997;**36**:631.
40. Qin T, Ding J, Wang L, Baumgarten M, Zhou G, Müllen K. A divergent synthesis of very large polyphenylene dendrimers with iridium(III) cores: molecular size effect on the performance of phosphorescent organic light-emitting diodes. *J Am Chem Soc* 2009; **131**:14329.
41. Kose MM, Yesilbag G, Sanyal A. Segment block dendrimers via Diels-Alder cycloaddition. *Org Lett* 2008;**10**:2353.
42. Kolb HC, Finn MG, Sharpless KB. Click chemistry: diverse chemical function from a few good reactions. *Angew Chem Int Ed* 2004;**2001**:40.
43. Rostovtsev VV, Green LG, Fokin VV, Sharpless KB. A stepwise Huisgen cycloaddition process: copper(I)-catalyzed regioselective "ligation" of azides and terminal alkynes. *Angew Chem Int Ed* 2002;**41**:2596.
44. McNerny DQ, Mullen DG, Majoros IJ, Banaszak Holl MM, Baker JR. Dendrimer synthesis and functionalization by click chemistry for biomedical applications. In: Lahann J, editor. *Click Chemistry for biotechnology and materials science*. Chichester, UK: Wiley; 2009.
45. Fournier D, Hoogenboom R, Schubert US. Clicking polymers: a straightforward approach to novel macromolecular architectures. *Chem Soc Rev* 2007;**36**:1369.
46. Joralemon MJ, O'Reilly RK, Matson JB, Nugent AK, Hawker CJ, Wooley KL. Dendrimers clicked together divergently. *Macromolecules* 2005;**38**:5436.
47. Vieyres A, Lam T, Gillet R, Franc G, Castonguay A, Kakkar A. Combined Cu^{I}-catalysed alkyne-azide cycloaddition and furan-maleimide Diels-Alder "click" chemistry approach to thermoresponsive dendrimers. *Chem Commun* 1875;**2010**:46.
48. Killops KL, Campos LM, Hawker CJ. Robust, efficient, and orthogonal synthesis of dendrimers via thiol-ene "click" chemistry. *J Am Chem Soc* 2008;**130**:5062.
49. Rissing C, Son DY. Application of thiol-ene chemistry to the preparation of carbosilane-thioether dendrimers. *Organometallics* 2009;**28**:3167.
50. Hierold J, Gray-Weale A, Lupton DW. A divergent synthesis of modular dendrimers via sequential C-C bond fragmentation thio-Michael addition. *Chem Commun* 2010;**46**:6789.
51. For a review on AB3 monomers see:Newkome GR, Shreiner C. Dendrimers derived from $1 \rightarrow 3$ branching motifs. *Chem Rev* 2010;**110**:6338.
52. Wooley KL, Hawker CJ, Frechet JMJ. A "branched monomer approach" for rapid synthesis of dendrimers. *Angew Chem Int Ed* 1994;**33**:82.
53. Lim J, Mintzer MA, Perez LM, Simanek EE. Synthesis of odd generation triazine dendrimers using a divergent, macromonomer approach. *Org Lett* 2010;**12**:1148.
54. Antoni P, Hed Y, Nordberg A, Nyström D, von Holst H, Hult A, et al. Bifunctional dendrimers: from robust synthesis and accelerated one-pot postfunctionalization strategy to potential applications. *Angew Chem Int Ed* 2009;**48**:2126.
55. Hawker CJ, Fréchet JMJ. A new convergent approach to dendritic macromolecules. *Chem Commun* 1990;1010.
56. Hawker CJ, Fréchet JMJ. Preparation of polymers with controlled molecular architecture. A new convergent approach to dendritic macromolecules. *J Am Chem Soc* 1990;**112**:7638.
57. Miller TM, Neenan TX. Convergent synthesis of monodisperse dendrimers based upon 1,3,5-trisubstituted benzenes. *Chem Mater* 1990;**2**:346.

58. Wu P, Feldman AK, Nugent AK, Hawker CJ, Scheel A, Voit B, et al. Efficiency and fidelity in a click-chemistry route to triazole dendrimers by the copper(I)-catalyzed ligation of azides and alkynes. *Angew Chem Int Ed* 2004;**43**:3928.
59. Wiesler U-M, Müllen K. Polyphenylene dendrimers via Diels–Alder reactions: the convergent approach. *Chem Commun* 1999;2293.
60. Rosen BM, Wilson CJ, Wilson DA, Peterca M, Imam MR, Percec V. Dendron-mediated self-assembly, disassembly, and self-organization of complex systems. *Chem Rev* 2009;**109**:6275.
61. Zimmerman SC, Zeng F, Reichert DEC, Kolotuchin SV. Self-assembling dendrimers. *Science* 1996;**271**:1095.
62. DeMattei CR, Huang B, Tomalia DA. Designed dendrimer syntheses by self-assembly of single-site, ssDNA functionalized dendrons. *Nano Lett* 2004;**4**:771.
63. Choi Y, Thomas T, Kotlyar A, Islam MT, Baker JR. Synthesis and functional evaluation of DNA-assembled polyamidoamine dendrimer clusters for cancer cell-specific targeting. *Chem Biol* 2005;**12**:35.
64. Liu H, Tørring T, Dong M, Rosen CB, Besenbacher F, Gothelf KV. DNA-templated covalent coupling of G4 PAMAM dendrimers. *J Am Chem Soc* 2010;**132**:18054.
65. Kawa M, Fréchet JMJ. Self-assembled lanthanide-cored dendrimer complexes: enhancement of the fluorescence properties of lanthanide cations through site-isolation and antenna effects. *Chem Mater* 1998;**10**:286.
66. Pitois C, Hult A, Lindgren M. Lanthanide-cored fluorinated dendrimer complexes: synthesis and luminescence characterization. *J Lumin* 2005;**111**:265.
67. Wooley KL, Hawker CJ, Fréchet JMJ. Unsymmetrical three-dimensional macromolecules: preparation and characterization of strongly dipolar dendritic macromolecules. *J Am Chem Soc* 1993;**115**:11496.
68. Gillies ER, Fréchet JMJ. Designing macromolecules for therapeutic applications: polyester dendrimer-poly(ethylene oxide) "Bow-Tie" hybrids with tunable molecular weight and architecture. *J Am Chem Soc* 2002;**124**:14137.
69. Wu P, Malkoch M, Hunt JN, Vestberg R, Katgrad E, Finn MG, et al. Multivalent, bifunctional dendrimers prepared by click chemistry. *Chem Commun* 2005;**43**:5775.
70. Morikawa A, Kakimoto M, Imai Y. Convergent synthesis of siloxane starburst dendrons and dendrimers via hydrosilylation. *Macromolecules* 1992;**25**:3247.
71. Onitsuka K, Fujimoto M, Kitajama H, Ohshiro N, Takei F, Takahashi S. Convergent synthesis of platinum-acetylide dendrimers. *Chem Eur J* 2004;**10**:6433.
72. Wooley KL, Hawker CJ, Fréchet JMJ. Hyperbranched macromolecules via a novel double-stage convergent growth approach. *J Am Chem Soc* 1991;**113**:4252.
73. Ihre H, Hult A, Fréchet JMJ, Gitsov I. Double-stage convergent approach for the synthesis of functionalized dendritic aliphatic polyesters based on 2,2-bis(hydroxymethyl)propionic acid. *Macromolecules* 1998;**31**:4061.
74. Labbe G, Forier B, Dehaen W. A fast double-stage convergent synthesis of dendritc polyethers. *Chem Commun* 1996;**18**:2143.
75. Shi Z-F, Chai W-Y, An P, Cao X-P. Novel synthesis of amphiphilic dendrons by the double-stage convergent method. *Org Lett* 2009;**11**:4394.
76. Kawaguchi T, Walker KL, Wilkins CL, Moore JS. Double exponential dendrimer growth. *J Am Chem Soc* 1995;**117**:2159.
77. Spindler R, Fréchet JMJ. Two-step approach towards the accelerated synthesis of dendritic macromolecules. *J Chem Soc Perkin Trans I* 1993;913.
78. Zeng F, Zimmerman SC. Rapid synthesis of dendrimers by an orthogonal coupling strategy. *J Am Chem Soc* 1996;**118**:5326.

Part II

Biotechnological Applications

Chapter 6

Applications of Inorganic Nanoparticles for Biotechnology

Catherine C. Berry
Centre for Cell Engineering, Glasgow University, Glasgow, United Kingdom

In cell biology, many molecules are measured in the nanoscale range, such as proteins and DNA. It is known that nanomaterials less than several hundred nanometer can easily enter cells (with less than 50 nm entering most cells), whilst those under 20 nm can move through the blood vessels and permeate adjacent tissue, and also cross the blood–brain barrier.[1] Nanotechnology has now reached the stage whereby we can create and design nanoparticles, 1–100 nm in size, for use in biomedicine, in particular diagnostics (imaging) and therapeutics (drug/gene delivery), what is termed a "theranostic" approach. Their usefulness is mainly due to two beneficial factors: (1) a large surface area to volume ratio, which alongside ease of chemical functionalisation, permits attachment of a multifunctional cargo payload (such as a fluorescent moiety, targeting molecules, etc.) (Figure 6.1) and (2) the possibility of ubiquitous tissue accessibility.[2]

Organic nanoparticles have been widely investigated to date, with liposomes, polymersomes, polymer constructs and micelles, all being employed for imaging or drug and gene delivery techniques.[3,4] However, inorganic nanoparticles have also gained attention in recent years due in part to their unique material- and size-dependent physiochemical properties, which are not possible with traditional lipid- or polymer-based nanoparticles. In particular, optical, magnetic and other physical properties, in addition to inertness, stability and ease of functionalisation, make inorganic nanoparticles attractive alternatives.[5] Thus, inorganic nanoparticles, such as magnetic, gold, quantum dots (QDs) and carbon nanotubes, have vast potential in modern biomedical applications.

The following body of text describes these four more commonly used inorganic nanoparticles and details their various successes in the biomedical arena.

1. MAGNETIC NANOPARTICLES

Typical magnetic nanoparticles (mNPs) comprise a magnetic core, usually magnetite (Fe^3O^4) or maghemite (γ-Fe^2O^3). Other metals such as cobalt and nickel are also employed; however, these tend to be toxic and susceptible to

Frontiers of Nanoscience, Vol. 4. DOI: 10.1016/B978-0-12-415769-9.00006-6

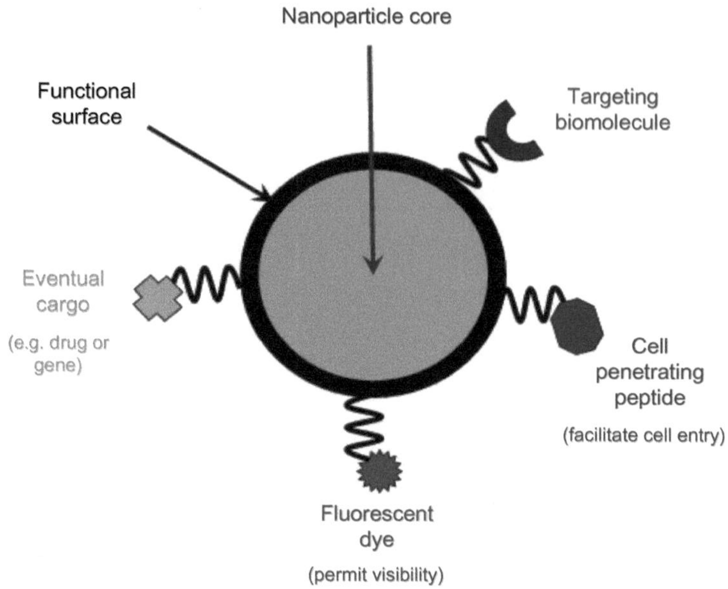

FIGURE 6.1 Schematic of a potential multifunctional approach to disease monitoring and treatment using nanoparticle systems. (For colour version of this figure, the reader is referred to the online version of this chapter.)

oxidation.[6] The adoption of iron oxide mNPs has the added advantage that the body is designed to process excess iron. In the human body, iron is stored primarily in the core of the iron storage protein ferritin. Cationic mNPs enter cells efficiently and remain localised in the endosomes for relatively long periods of time.[7] Iron that is contained in endosomes and lysosomes (i.e. post-cellular uptake) is metabolised into elemental iron and oxygen by hydrolytic enzymes, where the iron joins normal body stores.[8] Iron homeostasis is well controlled by adsorption, excretion and storage, and thus it is postulated that, following the administration of iron oxide nanoparticles, any excess iron in the body can be processed. However, it should also be noted that although iron plays an important role in virtually all living tissues, it has a rather limited bioavailability, and in some situations, it can be toxic to cells in the form of free iron (as opposed to being associated with haemoglobin).[9]

Superparamagnetic iron oxide (SPIO) nanoparticles possess unique magnetic properties that make them attractive candidates as advanced biomedical materials. This superparamagnetic property originates from the large magnetic moment they acquire in the presence of an external magnetic field; removing the field eliminates the paramagnetism. This large moment results in a higher signal change per unit of particles, thus only smaller quantities are needed to produce good signal feedback.[5] A biocompatible surface coating that provides stability under physiological conditions (i.e. inhibits aggregation) is usually surrounding the inorganic iron core. A wide variety of substances can be used as coating materials; these include synthetic[10] or natural (e.g. proteins or dextran)[11,12] polymers and amphiphilic molecules such as fatty acids or phospholipids.[13] Further controlled surface functionalisations may then be made, such as coupling with fragments of DNA strands, proteins, peptides or antibodies. Over the past two decades, mNPs have been increasingly exploited as efficient delivery vectors, leading to opportunities in magnetic resonance imaging (MRI) contrast enhancement, mediators of hyperthermia cancer treatment and targeted therapies (e.g. drug and gene delivery).

1.1. Magnetic Resonance Imaging

MRI is a powerful imaging tool due to its non-invasive nature, high spatial resolution and tomographic capabilities. The applications of MRI have steadily increased over the past decade, offering the advantage of high spatial resolution of contrast differences between tissues. In comparison to conventional gadolinium chelates, mNP-based contrast agents not only offer excellent image enhancement (due to their large magnetic moment) but also improved cellular internalisation and slower clearance from the target site.[14,15] Iron oxide mNPs have received U.S. Food and Drug Administration approval for use as MRI signal enhancers. For MRI purposes, iron oxide cores are commonly used as so-called T_2 contrast agents, due to their ability to shorten T_2 relaxation times. These agents can be divided into either SPIOs, with diameters of >50 nm, or ultrasmall SPIOs (USPIOs), with diameters of <50 nm, the later tending to have longer plasma half-lives of 14–30 h.[16,17] Currently, there are several formulations for clinical applications such as bowel (lumiren and gastromark) and liver or spleen imaging (endoderm and feridex). As with all nanoparticles, the tissue distribution is heavily influenced by size, thus the larger SPIOs tend to rely on passive targeting, such as uptake by the cells of the reticuloendothelial systems (RES), rather than direct labelling, and the USPIOs benefit from slower opsonisation and RES clearance.[18] The next generation of active targeting contrast agents is currently being researched, which provide exciting new opportunities for imaging, diagnosis and treatment. Monoclonal antibodies were the first targeting agents to exploit molecular recognition to deliver mNPs.[19,20] For example, Suzuki *et al.* developed an MRI contrast agent, which was prepared by covalently linking polyethylene glycol (PEG)-coated magnetite to an antibody specific for a human glioma cell surface antigen.[21] However, one drawback of using monoclonal antibodies is their large size ($\sim$12 nm), which can cause poor diffusion through typical biological barriers. Following this, mNPs have been examined intensively as MRI agents to improve detection and diagnosis of solid tumours.[22,23] This is an active area of research, since clinicians need a contrast agent that can be specifically targeted to malignant tumours allowing for more accurate diagnosis of various stages of disease. Direct tumour targeting using antibodies has been successful for rectal carcinoma[24] and breast cancer,[25] as despite the size issue, antibodies remain beneficial in their high specificity. More recent applications utilising antibody-directed targeting tend to focus on using shorter, smaller single-chain fragments, which consist of antibody heavy- and light-chain variable domains connected with a flexible peptide linker.[26] The resulting fragment is smaller than 20% of an intact antibody, but maintains a high binding affinity and specificity.[27]

The signal enhancement caused by conventional iron oxides, however, is still unsatisfactory compared to that obtained with other imaging modalities such as fluorescence and positron emission tomography. A further area of research is to thus develop magnetism-engineered iron oxide nanoparticles, which possess exceptionally high and tunable nanomagnetism.[28] A study by Lee *et al.* demonstrated such engineered particles enhanced sensitivity for cancer cell detection and also made the *in vivo* imaging of small tumours possible.[29]

There are reasons to expect that MRI will become far more sensitive in the next few years. As its sensitivity approaches and perhaps exceeds that of fluorochromes (fluorescent substances used to stain biological specimens), the need for nanoparticles in MRI molecular imaging is increasingly important. However, there is a separate reason why nanoparticles will be required even in more sensitive versions of MRI, leading back to the notion of theranostics. This combination of an image signal generating mNP alongside a conjugated drug/gene payload allows the nanoparticle to be located within the living subject and at the same time to release drugs into a target site, enabling molecular imaging to be combined with therapy. This "therapy" plus "diagnostics", or "theranostics" holds the promise in future of monitoring the effectiveness of therapy simultaneously with delivering the therapy.[30]

1.2. Hyperthermia

It is well recognised that tumour cells are more sensitive to temperature increases than normal healthy cells, thus intracellular hyperthermia methods have been postulated as potentially treating cancers.[31,32] Such a treatment requires a delivery agent to essentially act as a nanoscale heater inside cancer cells, to heat up the cell and instigate necrosis.[33] When subjected to an alternating magnetic field, mNPs convert electromagnetic energy into heat, which can be dissipated to the surrounding medium. Magnetic materials with Curie temperatures between 42 and 60 °C are excellent candidates for effective treatment, as they can potentially act as *in vivo* temperature control switches, preventing overheating, and thus avoiding damage to neighbouring healthy tissue.[34] Intracellular hyperthermia methods have thus been developed using mNPs (in particular SPIOs), whereby the mNPs are concentrated at the tumour site and remotely heated using an applied magnetic field to required hyperthermic temperatures.[35,36]

Tumours can be targeted passively *via* general biodistribution, as tumour tissue tends to have "leaky" vasculature, which allows nanoparticles to accumulate. Alternatively, mNPs can be injected directly into solid tumours, followed by exposure to an alternating magnetic field. This technique has been proven to be capable of inducing tumour regression.[37] Several studies have highlighted the potential of mNPs in human cancer models, such as breast cancer, whereby 10–30 nm iron oxide particles were heated effectively[38]; however, it is noted that further studies are needed to optimise the delivery so as to control heat distribution.[39]

The first clinical experiences with mNP-mediated thermotherapy on prostate carcinoma[40] and other tumour entities[41] have been published. In addition, Van Landeghem *et al.* have published the first report of post-mortem neuropathological findings on glioblastoma multiforme patients, the most common brain tumour in adults, who had undergone injection of mNPs into the tumour and subsequent heating.[42] Results indicated that mNPs were observed as aggregates, restricted to the site of injection, highlighting the need for multiple trajectories on administration. Furthermore, whilst particles and aggregates were phagocytosed by macrophages, with minor uptake from glioblastoma cells, aggregates were preferentially located in necrotic sites in the tumour.

An obvious enhancement of hyperthermia research is the use of tumour-targeted mNPs, with the added potential of drug payload.[43] For example, the use of dextran-coated mNPs further conjugated to breast cancer targeting chimeric L6 monoclonal antibody has demonstrated the feasibility of targeting specific cancer cells.[44,45] Targeted hyperthermia research has also extended to encompass targeted gene expression. The control of gene expression in tumours is important in gene therapy, and the use of an inducible promotor to control expression even more desirable. For example, potentially useful cytotoxic cytokines including TNF-α, IL-1 and IL-2 which possess potent tumouricidal activities (i.e. capable of destroying a tumour) but are currently limited by systemic toxicity would benefit. Hyperthermia has thus been suggested as a non-invasive and non-toxic method of inducing gene expression locally using a thermoinducible promotor.[46] There have been studies describing inducible gene expression systems in which thermal energy was used to activate gene expression. Ito *et al.* combined TNF-α gene therapy driven by a heat-inducible promotor with hyperthermia produced by irradiation of magnetite cationic liposomes.[47] In this context, attempts were made to develop a heat-inducible gene expression system with mNPs. A study developed a method using Hsp70 as a promotor (LacZ gene as a reporter gene) and demonstrated that expression could be elevated 10- to 500-fold by moderate hyperthermia.[48]

1.3. Magnetic Targeted Therapies/Delivery

1.3.1. Magnetic Drug Targeting

When considering tumour therapies, passive targeting is generally arrived at *via* general biodistribution and particle accumulation, as alluded to previously. However, active drug targeting with mNPs should significantly enhance cargo delivery. The technique proposes the targeting of a drug immobilised on mNPs under the action of an external magnetic field. To enhance target specificity, the drug is associated with another molecule capable of target recognition and binding. The effectiveness of this method is dependent on many factors, including the properties of the cargo-loaded mNPs, field strength, depth of the target tissue, rate of blood flow and vascular supply.[49] Most mNPs can be directed to the cell surface *in vitro* using an external magnetic field,[50,51] thus studies and trials are being carried out to determine the efficacy of magnetic targeting *in vivo*.

A highly publicised example of magnetic drug delivery is as a replacement or to augment chemotherapy/radiotherapy. The development of techniques that could selectively deliver drugs to the diseased site without a concurrent increase in adjacent healthy tissue is an active area of cancer research. The first clinical trials in humans were reported by Lubbe *et al.*, who used a ferrofluid (particle size 100 nm) to which the drug epirubicin was bound. Early clinical trials demonstrated successful accumulation in the target site in approximately half the patients.[52,53] Unfortunately, there were several problems associated with the technique, such as the possibility of embolisation of the blood vessels, difficulty in scaling up from animal models due to limited field penetration and control of drug diffusion after release from the mNPs.[54] Subsequently, a mathematical model incorporating some of these practical hurdles was formed, which concluded that magnetic drug targeting could only be used effectively for targets close to the surface of the body.[55]

With regards to targeted drug delivery, mNPs have been evaluated for conjugated with a variety of chemotherapeutic agents including etoposide, doxorubicin and methotrexate.[56,57] In addition to drugs, mNPs have also been assessed as carriers for therapeutic proteins and peptides. For example, Herceptin™ (known as trastuzumab), which has been demonstrated to cause cell arrest during G1 phase of the cell cycle, was successfully incorporated into mNP-loaded liposomes and has been shown to reduce breast cancer cell proliferation.[37]

1.3.2. Magnetic Transfection

The potential of mNPs in antisense and gene therapy is currently being assessed, the aim being the binding of a large amount of negatively charged DNA to the mNPs, ideally through electrostatic interactions, and subsequent release after cell internalisation.[58] However, these therapies are hampered due to limited half-life *in vivo*, lack of specificity and poor diffusion across the cell membrane, resulting in poor transfection efficacies. It was proposed that the use of mNPs as carriers may overcome some of these problems. Several research groups have employed mNPs and developed transfection methods, and the generic term “magnetofection” is widely used in the literature to describe nucleic acid delivery under the influence of a magnetic field acting on nucleic acid vectors that are conjugated to mNPs.[59,60] In a typical *in vitro* magnetofection system, target cells are located at the bottom of a fluidic chamber (e.g. well of a culture plate), and a rare-earth magnet beneath the chamber provides a magnetic force that attracts biofunctionalised mNPs toward the cells (Figure 6.2).

Magnetofection has significant advantages over traditional transfection methods, including reduced process time (in the order of 10 min as opposed to 2–4 h), high transfection rates with lower vector doses and an increase in efficacy.[61] Magnetofection is now well established and used for *in vitro* applications, which include overexpressing a transduced/transfected gene using almost any vector type.[62]

Furthermore, the technique can be used for both "large" nucleic acids (e.g. plasmids) and small constructs (e.g. siRNA).[63] RNA interference is a natural cellular process used to silence gene expression,[64] which can be exploited artificially.[65] The use of wide-scale therapeutic RNA silencing requires the development of a suitable transfection method that can administer siRNA molecules into human cells *in vivo*. Nanoparticles are an attractive delivery vehicle for siRNA due to their multifunctionality, allowing them to overcome many problems associated with systemic siRNA application for human cell transfection. For example, nanoparticles can prevent rapid excretion of siRNA by the kidney.[66] This is a fate that succumbs naked siRNA molecules; however,

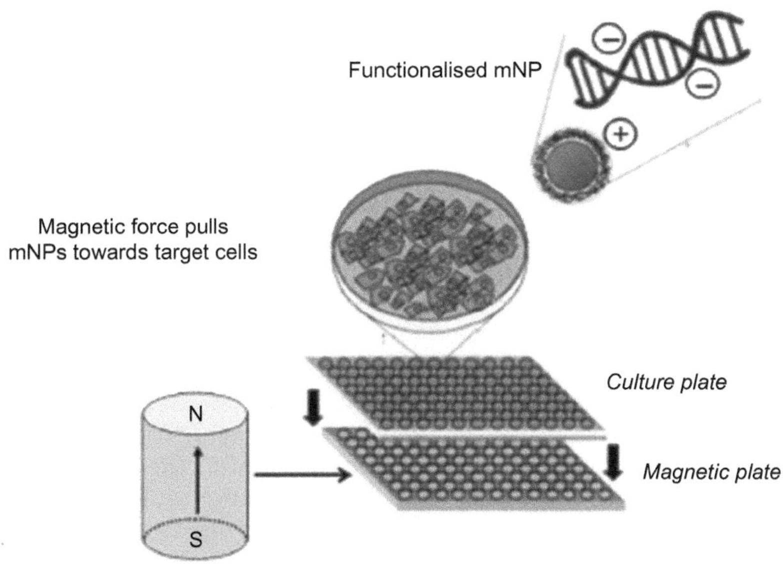

FIGURE 6.2 Schematic of magnetofection technique. Magnetic nanoparticles are electrostatically conjugated to specific genetic material and added to cell medium. The particles are subsequently physically introduced to cells via the attractive force of a magnetic field, provided by a static magnet under the cell culture plate (adapted with permission from Furlani and Ng[61] copyright 2008 American Physical Society). (For colour version of this figure, the reader is referred to the online version of this chapter.)

attachment to nanoparticles results in minimal renal filtration,[67] extending the unmodified siRNA half-life within the circulation.[68] In 2007, Medarova *et al.* demonstrated cancer cell transfection by nanoparticle-delivered siRNA, reporting the uptake of mNPs into tumours. In this case, membrane translocation was facilitated by an attached myristoyl-coupled polyarginine peptide.[69] In addition to delivering siRNA treatment, these nanoparticles were further functionalised with a near-infrared dye (Cy5.5) to allow non-invasive imaging of their localisation within the body. This imaging was performed by both MRI and near-infrared *in vivo* optical imaging (NIRF). Imaging results from this study showed nanoparticle accumulation in tumours, most likely passive due to the enhanced permeability and retention effect associated with the leaky vasculature of tumours. The study concluded that nanoparticles further functionalised with tumour-targeting moieties would increase tumour localisation in order to deliver siRNA treatment.

A subsequent study involving tumour targeting used mNPs attached to a cyclic Arg-Gly-Asp (RGD) peptide, which binds to $\alpha_v\beta_3$ integrin expressed on the surface of cancer cells.[70,71] PEGylation of this targeting peptide increased nanoparticle stability. Similar imaging techniques were employed (MRI and by the attachment of a Cy5 near-infrared dye, NIRF). Treatment delivery was facilitated by the attachment of siRNA to silence green fluorescent protein. Interestingly, the authors state that their mNPs provide better targeting and reduced toxicity compared to the popular delivery vehicle polyethyleneimine, famous for its "proton sponge" method of endosomal escape. This study concluded that joining the tasks of diagnosis and treatment in future *in vivo* and clinical applications will reduce invasiveness and side effects.

2. GOLD NANOPARTICLES

Gold nanoparticles (AuNPs), which are commonly referred to as colloidal gold, are easily prepared and have been used for hundreds of years due to their intense colours which vary depending on particle size; smaller ($\leq$100 nm) being red and larger being yellow, for example, in stained glass. While AuNPs share most of their attractive qualities with regard to bioapplications with other nanoparticles (e.g. size, inertness, ease of synthesis, biocompatible), AuNPs are a very attractive candidate for biological imaging techniques as they can be visualised based on the interaction between the nanoparticles and light, whereby the particles strongly absorb and scatter visible light.[72] Upon light absorption, the light energy excites the free electrons in the gold particles to a collective oscillation, the so-called surface plasmon.[73] This absorption lies in the visible region for gold, silver and copper,[74] with the surface plasmon resonance (SPR) of AuNPs being visible down to 3 nm.[75] AuNPs give rise to both absorption and scattering, the size of which depends on AuNP size. Particles smaller than 20 nm essentially show absorption, but larger sizes of 80 nm increase the ratio of scattering to absorption.

2.1. Biological Imaging

Colloidal gold has been employed as a contrast agent in electron microscopy for several decades. Owing to the high atomic number of gold, the colloidal gold particles are electron dense, which renders them amongst the best for electron microscopy. However, AuNPs are involved in several other imaging techniques that rely on the plasmon band. AuNPs greater than 20 nm can be used in optical microscopy under phase contrast or differential interference contrast mode. Other techniques include fluorescence microscopy, photothermal coherence tomography (similar to ultrasound with good depth penetration),[76] multiphoton SPR microscopy[77] and X-ray scattering.[78]

The perhaps most referenced application of AuNPs in bioimaging is in immunostaining. Essentially, antibody-conjugated AuNPs are designed to bind against antigens on fixed and permeabilised cells, being subsequently visualised via TEM or light microscopy. Typically, an excess of gold is used so that virtually all entities are labelled to improve contrast. As cells are fixed and permeabilised, targets outside as well as inside the cells can be labelled. As an ultrastructural marker for detection of proteins, peptides or amino acids, gold can be used for immunostaining thick or thin sections prior to embedding, or for immunostaining ultrathin sections after embedding. By virtue of its particulate nature, gold as an immunolabel facilitates a semiquantitative analysis of antigen densities on ultrathin sections.[79] Various combinations of different size gold particles, or dual immunolabelling with enzymatic immunolabels together with colloidal gold or silver-intensified gold, serve well for ultrastructural immunocytochemical localisation of two antigens in the same tissue section.[80] Compared to fluorescence microscopy, AuNPs are more stable as they do not suffer photobleaching and, in the case of TEM imaging, offer excellent lateral resolution with high contrast (indeed individual cell receptors can be resolved by visualising the bound AuNPs with TEM).

2.2. Cell Delivery Vehicles

Most delivery strategies, such as using cancer-targeting moieties conjugated to NPs for delivery into cancer tissue, are very similar to those used for magnetic and other types of NP. Indeed, gold has been used for many years to deliver molecules into cells. For delivery applications, AuNPs are employed for their small size, colloidal stability, ease of synthesis and conjugation and their inert, biocompatible nature. Introduction into cells can either be forced, for example, with gene guns, or be achieved *via* cellular uptake. With regard to the gene gun technique, DNA is adsorbed onto the surface of AuNPs, which are then essentially shot into cells.[78] The force required is typically provided via gas pressure or electric discharge.[81] While gene guns were more commonly employed in plant cell biology (to breach the plant cell wall), they have also been used for DNA delivery into animal cells.[82] Alternatively, specific or non-specific endocytosis cellular uptake can be relied upon for AuNP delivery. Specific uptake depends on receptor–ligand binding, for example, using transferrin adsorbed onto the AuNPs as a means of instigating cellular uptake,[83] and is far more effective than passive non-specific uptake. Following uptake, the NPs are stored in endosomal/lysosomal vesicles inside the cells (Figure 6.3).[84]

There have been efforts to avoid endosomal uptake and subsequent vesicle storage. Pioneering work by Feldherr *et al.* demonstrated nuclear translocation of AuNPs (10–35 nm) coated with nucleoplasmin (a *Xenopus* oocyte protein that contains a well characterised nuclear localisation sequence[85]), although it was acknowledged that their work was performed using microinjection on chemically modified cells, thus bypassing plasma membrane entry. Further work showed nuclear targeting was only achieved when exposing cells in culture to AuNPs (20 nm) derivatised with bovine serum albumin and functionalised with a variety of short peptide sequences that exhibit nuclear localisation sequences, when the whole peptide was present, thus including the sequence for plasma membrane translocation (otherwise the particles were trapped in the endosome).[86] Following the advent of cell-penetrating peptide research, whereby direct plasma membrane translocation was achieved, conjugating tat peptide to AuNPs also achieved excellent cellular uptake levels.[87]

2.3. Biosensors

Where imaging and delivery above tend to use passive methods, plasmon-related sensing uses AuNPs in a more active role. Basically, the NPs are required to (i) specifically register the presence of analyte molecules (e.g. antigens, nucleic acids, enzymatic reactions) and (ii) provide a concentration readout. This is typically achieved by changes in the optical properties of the

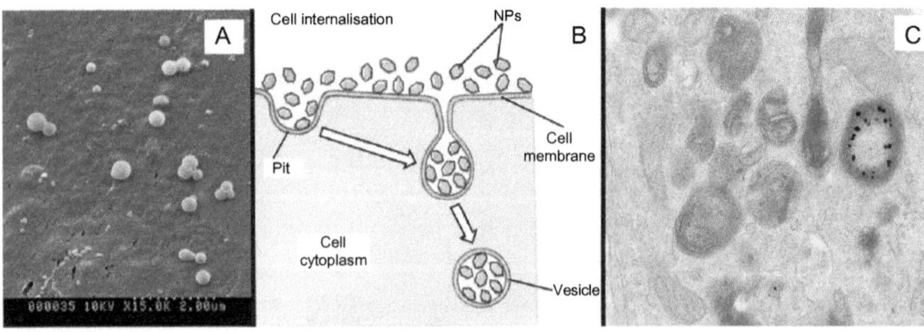

FIGURE 6.3 Conjugated particles bind to receptors on the cell surface, as shown in the SEM image (A), are typically internalised via endocytosis by the cell (B) and stored in vesicles, as illustrated by the TEM cell section (C) (personal images). (For colour version of this figure, the reader is referred to the online version of this chapter.)

AuNPs.[88,89] The plasmon resonance frequency is a very reliable intrinsic feature of AuNPs that can be used for sensing. The binding of molecules to the particle surface can change the plasmon resonance frequency directly, which is visible by scattered light. In addition, the influence of the inter-AuNP distance on the plasmon resonance, when this distance is reduced to less than the particle diameter, is the crucial factor in the sensor application and linking the NPs with a biological analyte results in a colour change that makes the basis of sensing. The first colourimetric sensing of nucleic acids was reported by Mirkin *et al.* and is now the most recognised example of a gold-based biosensor.[90] The DNA linked the AuNPs together with an interparticle distance of 0.34 nm, which caused a red-to-purple colour change (depending on whether the gold particles are free in colloidal suspension (red) or attached to the DNA target (blue-violet)). The colour change is temperature reversible, thus when the sample is heated, even a single sequence mismatch will result in a different melting temperature, which will cause a different colour change. Further studies have lead to detection of DNA sequences at very low concentrations.[91]

AuNPs possess excellent fluorescent properties and display antiphotobleaching behaviour under strong light illumination.[74] AuNPs exhibit strong native fluorescence under relatively high excitation power. For example, if cells stained with AuNPs are illuminated with strong light, AuNPs fluorescence can be recorded for cell imaging.[92] A further phenomenon is that the fluorescence of many fluorophores is quenched when they are in close proximity to gold,[93,94] and this effect can be used for several sensing strategies.

3. QUANTUM DOTS

QDs are colloidal nanometre-sized crystals, comprising atoms of elements from groups II to VI (e.g. Cd, Zn, Se, Te) or III to V (e.g. In, P, As) in the periodic table.[95] In terms of bioapplications, the most typical QD employed is composed of CdSe and encased in a ZnS shell (to protect from the highly toxic cadmium).[96] By confining the electrons in variable sizes, the energy band gap, absorption spectra allows their corresponding emission wavelengths to be tuned from the ultraviolet to near-infrared (NIR) region.[95] Smaller QDs ($\sim$2 nm) demonstrate blue fluorescence emission (380–440 nm), while larger particles ($\sim$5 nm) demonstrate red fluorescence emission (605–630 nm).[97] QDs have further advantages over organic dye molecules, aside from their tunable fluorescence, that they are robust and stable light emitters due to their inorganic makeup and are less susceptible to photobleaching than organic dye molecules. This photostability makes them extremely useful in observing cells over longer periods of time.[97,98]

Bruchez *et al.* were one of the first to report the use of QD conjugates for labelling biological specimens.[97] Subsequently, several authors have shown labelling of whole cells and tissue sections using different surface modifications of QDs.[99,100]

3.1. Biological Imaging

Due to their high photostability and limited cytotoxicity, QDs appear as a very promising probe for longer term experiments in living cells, providing that several issues are addressed. The first is water solubility, an important requirement for *in vitro* and *in vivo* imaging. Thiol groups (SH) are generally anchored to the ZnS shell with terminal carboxyl (COOH) in order to increase the hydrophilicity of QDs.[5]

The second issue is cell internalisation. As with both magnetic and gold NPs, cell uptake is typically enhanced *via* particle surface functionalisation. Several methods have been proposed to deliver QDs to the cell cytoplasm. Aside from the physical methods, such as microinjection and electroporation, many attempts have been made using lipid- or polymer-mediated endocytosis (e.g. FuGENE, lipofectamine, etc.)[101] and peptide-mediated endocytosis (e.g. tat peptide, nuclear localisation sequences).[102] The main issue here is that the QDs remain trapped in the endosomes for several days. An alternative pathway is to exploit pinocytosis.[103] This is activated by an increase in osmotic pressure in cell culture medium. To counterbalance this, cells uptake medium in pinosomes, together with QDs in the medium. Pinosomes can subsequently be disrupted by a second osmotic shock, releasing the QDs into the cell cytoplasm.[104]

3.1.1. *Single Cell Imaging*

Tracking of single cells was initially developed to study membrane receptor dynamics. The first studies included diffusion of transmembrane proteins using micrometre-sized beads of AuNPs.[105] More recently, similar experiments have been done using QDs to target membrane proteins, and study the mobility and kinetics of receptors, transmembrane proteins (e.g. integrins and channels) and synapses.[106]

QDs can be detected and tracked with the same approach as for traditional organic dyes. In the cellular context, tracking ODs is technically difficult. First, the background noise due to cell autofluorescence and to the surrounding QDs diminishes precision. To reduce this noise, QDs emitting in the red field are used, where the cell autofluorescence disappears.[107] Furthermore, using small levels of QDs helps improve the signal-to-noise ratio. The second difficulty arises from the need to track in a three-dimensional space. In recent years, techniques have been developed to acquire image stacks, thus reconstructing the three-dimensional structure of the cell, for example, total internal refection fluorescence microscopy, where only particles closest to the glass coverslip are excited (typically 100–200 nm of a cell).[108] The most popular reconstruction of a *z*-series, however, is the scanning confocal microscope, which limits excitation to a particular volume.

3.1.2. In Vivo *Imaging*

QD–peptide conjugates were first used *in vivo* to target tumour vasculature in mice.[109] Histology indicated that the tissue-specific peptide coating on (CdSe)

ZnS QDs increased nanoparticle accumulation at vascular sites following intravascular injection. Although this did not describe QD imaging in a live animal, it demonstrated the potential of using QDs for molecular level detection. Several subsequent studies, typically on mice, used tail vein injection as a means of entry. This indicated that untargeted or passive targeting of QD probes demonstrated weak or no signal, but antibody-conjugated QDs resulted in intense fluorescent signals.[110,111]

Despite using targeting moieties with QDs, such as antibodies or peptides, targeted to tumours in live animals for cancer imaging, light penetration and autofluorescence of deep tissue remain a major hurdle. As with single cell imaging above, the use of red field emitting QDs has minimised light absorption by blood and water and improved tissue depth.[112]

3.2. Targeted Therapies

Similar to mNPs employed in transfection therapies *in vivo* (Section 1.3.2), QDs are also used as delivery and reporter systems. A big advantage of nanoparticle transfection, compared to other types of delivery vehicles, is that they can be functionalised with many different oligonucleotides and cell-binding ligands at once, potentially allowing multiple gene knockdowns and higher affinity for the target cell simultaneously.[66] Derfus *et al.* reported that one siRNA per particle in conjunction with >15 peptides, or two siRNA per particle in conjunction with <10 peptides, gave optimal knockdown and targeting.[67]

Derfus *et al.* furthered their studies by using QDs designed with tumour targeting, facilitated by the attachment of a tumour-homing peptide (F3), which binds to nucleolin expressed on the surface of cancer cells.[113] This addition of F3 increased QD tumour cell uptake by two orders of magnitude compared to nontargeting QDs. Diagnostic imaging was facilitated *via* the QD core with emission in the NIR, and treatment delivery was facilitated by the attachment of siRNA. This siRNA silenced the enhanced green fluorescent protein gene, but the authors noted that their design could also be applied to silence oncogenes for cancer treatment. Results showed that siRNA attached to the QD by a disulphide bond increased gene knockdown compared to siRNA attached to the QD by a covalent bond. This example demonstrates the importance of such acute attention to molecular design in the development of an efficient nanoparticle-based RNAi platform.

4. CARBON NANOTUBES

Hollow and porous nanoparticles, such as nanotubes, nanoshells and hollow spheres, can be loaded with a large amount of cargo, thus enhancing signal and sensitivity. Carbon nanotubes are cylindrical graphene sheets. Although most applications of carbon nanotubes have focused on microelectronic devices, due to their unique electronic and physical properties,[114] carbon nanotubes have shown some attractive properties for biomedical use, including easy translocation across the cell membrane and relatively low toxicity.[115] Both single-walled carbon nanotubes (SWCNTs; 1–3 nm in diameter and 5–30 nm in length) and multiwalled CNTs (MWCNTs; 10–150 nm in diameter and 200 nm to several microns in length) have been investigated for various bioapplications. As with QDs, it has been indicated that SWCNTs have strong optical absorbance in the NIR region where biological systems are known to be highly transparent.[116]

4.1. Neuronal Tissue Engineering

CNTs have been rapidly developing as a technology platform for designing novel neuroimplantable devices. They combine incredible strength with extreme flexibility, in addition to exhibiting physical and chemical properties that allow them to efficiently conduit electrical current in electrochemical interfaces. Thus, CNTs can be employed in tissue engineering scaffolds, organised as fibres or tubes, with diameters similar to those in neuronal processes such as axons and dendrites.[117] To date, both SWCNTs and MWCNTs have attracted attention as potential scaffolds for reestablishing the intricate connections between neurons (Figure 6.4).[119–121]

The first application of nanotubes to neuroscience research by Mattson *et al.* employed MWCNTs for growth of rat brain neurons.[119] It was noted that on unmodified tubes neurons extended only one or two neurites, which exhibited very few branches. In contrast, neurons grown on tubes coated with a bioactive molecule (4-hydroxynonenal) demonstrated multiple neurites with extensive branching. Further studies have shown CNTs to be ideal for long-term neural implants as they enhance electrical recording of neurons in culture and in living animals[122,123] by reducing the impedance between devices and cell membranes.[124]

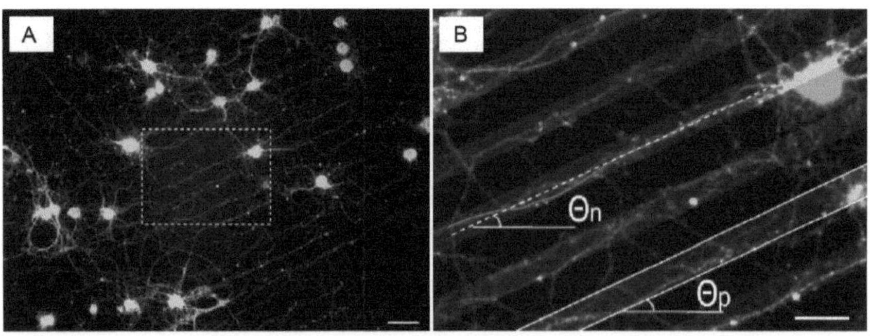

FIGURE 6.4 Neuronal networks cultured on a carbon nanotube patterned substrate (A). Patterned neuronal growth over long 14 days cultivation resulted in aligned neurites (note dotted annotated line in B) (modified and reproduced with permission from Ref. 118). (For colour version of this figure, the reader is referred to the online version of this chapter.)

4.2. Imaging and Cancer Treatment

Similar to mNPs and AuNPs, SWCNTs have been utilised in thermal necrosis of cancer cells. Intratumoural injection of tubes, alongside NIR irradiation resulted in thermal death of human epidermoid mouth carcinoma KB tumour cells in xenografted mice with minimal side effects up to 6 months after treatment, with excretion via urine in 3 months.[125] Similarly, thermal cancer therapy was applied to mice bearing kidney tumours using MWCNTs, and complete tumour regression was observed with no recurrence within 3 months.[126] Tumour targeting has also been employed, using antibodies[127] and folate receptors (which are known to be highly expressed in a range of tumour cells).[128]

When considering targeting, CNTs have also been studied for their use in delivering therapeutic drugs to tumours. SWCNTs exhibiting paclitaxel conjugated to PEG chains were injected into breast tumour xenografts in mice.[129] Particles were well dispersed, with inhibition of tumour growth by almost 60%. Similar studies using other drugs, such as doxorubicin-loaded SWCNTs,[130] RGD peptides to target-specific integrins,[131] anti-cancer agent cisplatin,[132] all had varying effects on tumour growth and some retention in the liver, kidney, and spleen.

There are other types of inorganic nanoparticles, such as silica, nanoshells, nanorods, hybrid particles, etc., but the particle systems described above are the four most commonly used when considering bioapplications. Such systems, due to their multifunctional approach, hold great promise in diagnostics, drug and gene delivery, sensing and biosensing, and both *in vitro* and *in vivo* imaging.

While each of the particles described exhibits some features that are original to them (e.g. magnetism, optical properties of gold, tunable fluorescence of QDs and electrical properties of carbon nanotubes), the bioapplications do overlap, with many sharing functionalities and targeting groups. Each particle type is designed with the view to boosting cellular uptake efficiency, for image/signal enhancement or cargo delivery, and also to target-specific tissues/cells (e.g. cancer treatments). It is envisaged that diseases may be managed by multifunctional nanoparticles which encompass both imaging and therapeutic capabilities, thus allowing simultaneous disease monitoring and treatment.

REFERENCES

1. Bhupender S, Sekhon S, Seema R, Kamboj R. Inorganic nanomedicine—part 1. *Nanomed Nanotechnol Biol Med* 2010;**6**:516–22.
2. Berry CC. Progress in functionalization of magnetic nanoparticles for applications in biomedicine. *J Phys D Appl Phys* 2009;**42**:224003.
3. Koo H, Huh MS, Sun IC, Yuk SH, Choi K, Kim K, et al. In vivo targeted delivery of nanoparticles for theranosis. *Acc Chem Res* 2011;**44**:1018–28.
4. Bae KH, Chung HJ, Park TG. Nanomaterials for cancer therapy and imaging. *Mol Cells* 2011;**31**:295–302.

5. Huang HC, Barua S, Sharma G, Dey SK, Rege K. Inorganic nanoparticles for cancer imaging and therapy. *J Control Release* 2011;**155**:344–57.
6. Connolly J, St. Pierre TG, Rutnakornpituk M, Riffle JS. Cobalt nanoparticles formed in polysiloxane copolymer micelles: effect of production methods on magnetic properties. *J Phys D Appl Phys* 2004;**37**:2475.
7. Xu C, Sun S. Superparamagnetic nanoparticles as targeted probes for diagnostic and therapeutic applications. *Dalton Trans* 2009;**29**:5583–91.
8. Brigger I, Dubernet C, Couvreue P. Nanoparticles in cancer therapy and diagnosis. *Adv Drug Deliv Rev* 2002;**13**:631–51.
9. Dobson J. Nanoscale biogenic iron oxides and neurodegenerative disease. *FEBS Lett* 2001;**496**:1–5.
10. Bulte JW, Douglas T, Witwer B. Magnetodendrimers allow endosomal magnetic labelling and in vivo tracking of stem cells. *Nat Biotechnol* 2001;**19**:1141–7.
11. Berry CC, Wells S, Charles S, Curtis ASG. Dextran and albumin derivatised nanoparticles: influence on fibroblasts in vitro. *Biomaterials* 2003;**24**:4551–7.
12. Berry CC, Wells S, Charles S, Atchinson G, Curtis ASG. *Biomaterials* 2004;**25**:5405–541.
13. Soenan SJH, Vercauteren D, Braeckmans K, Noppe W, de Smelt S, de Cuyper M. Stable long-term intracellular labelling with fluorescently tagged cationic magnetic liposomes. *Chembiochem* 2009;**10**:257–67.
14. Enochs WS, Harsh F, Hochberg F, Weissledher R. Improved delineation of human brain tumors on MR images using a long-circulating, superparamagnetic iron oxide agent. *J Magn Reson Imaging* 1999;**9**:228–32.
15. Varallyay P, Nesbit G, Muldoon LL, Nixon RR, Delashaw JI, Cohen JI, et al. Comparison of two superparamagnetic viral-sized iron oxide particles ferumoxides and ferumoxtran-10 with a gadolinium chelate in imaging intracranial tumors. *AJNR Am J Neuroradiol* 2002;**23**:510–9.
16. Varallyay P, Nesbit G, Muldoon LL, Nixon RR, Delashaw J, Cohen JI. Comparison of two superparamagnetic viral-sized nanoparticles ferumoxides and ferumoxtran-10 with a gadolinium chelate in imaging intracranial tumors. *Am J Neuroradiol* 2002;**23**:510–9.
17. Babincova M, Altanerova V, Altaner C, Cicmanex P, Babinec P. In vivo heating of magnetic nanoparticles in an alternating magnetic field. *Med Phys* 2004;**31**:2219–21.
18. Barry SE. Challenges in the development of magnetic particles for therapeutic applications. *Int J Hyperthermia* 2008;**24**:451–566.
19. Cerdan S, Lotscher HR, Kunnecke B, Seeling J. Monoclonal antibody-coated magnetite particles as contrast agents in magnetic resonance imaging of tumors. *Magn Reson Med* 1989;**12**:151–63.
20. Bulte JW, Hoekstra Y, Kamman RL, Magin RL, Webb AG, Briggs RW, et al. Specific MR imaging of human lymphocytes by monoclonal antibody-guided dextran-magnetite particles. *Magn Reson Med* 1992;**25**:148–57.
21. Suzuki M, Honda H, Kobayashi T, Wakabayashi T, Yoshida J, Takahashi M. Development of a target directed magnetic resonance contrast agent using monoclonal antibody-conjugated magnetic nanoparticles. *Brain Tumor Pathol* 1996;**13**:127–32.
22. Huh YM. In vivo magnetic resonance detection of cancer by using multifunctional magnetic nanocrystals. *J Am Chem Soc* 2006;**127**:12387–91.
23. Arternov D, Mori N, Okollie B, Bhujwalla AM. MR imaging of the HER-2/neu receptor in breast cancer cells using targeted iron oxide nanoparticles. *Magn Reson Med* 2003;**49**:403–8.
24. Toma A, Otsuji E, Kuriu Y, Okamoto K, Ichikawa D, Hagiwara A, et al. Monoclonal antibody A7-superparamagnetic iron oxide as contrast agent of MR imaging of rectal carcinoma. *Br J Cancer* 2005;**93**:131–6.

25. Funovics MA, Kapeller B, Hoeller C, Su HS, Kunstfield R, Puig S, et al. MR imaging of the her2/neu and 9.2.27 tumor antigens using immunospecific contrast agents. *Magn Reson Imaging* 2004;**22**:843–50.
26. Zhou Y, Drummond DC, Zou H, Hayes ME, Adams GP, Kirpotin DB, et al. Impact of single-chain Fv antibody fragment affinity on nanoparticle targeting of epidermal growth factor receptor-expressing tumor cells. *J Mol Biol* 2007;**371**:934–47.
27. Yang L, Mao H, Want A, Cao Z, Peng X, Wang X, et al. Single chain epidermal growth factor receptor antibody conjugated nanoparticles for in vivo tumor targeting and imaging. *Small* 2009;**2**:235–43.
28. Jun YW, Huh YM, Choi JS, Lee JH, Song HT, Kim S, et al. Nanoscale size effect of magnetic nanocrystals and their utilization for cancer diagnosis via magnetic resonance imaging. *J Am Chem Soc* 2005;**127**:5732–3.
29. Lee JH, Huh YM, Jun Y, Seo J, Jang J, Song HT, et al. Artificially engineered magnetic nanoparticles for ultra sensitive molecular imaging. *Nat Med* 2007;**13**:95–9.
30. McCarthy JR, Weissleder R. Multifunctional magnetic nanoparticles for targeted imaging and therapy. *Adv Drug Deliv Rev* 2008;**60**:1241–51.
31. Masashige S. Functional magnetic particles for medical application. *J Biosci Bioeng* 2002;**94**:606–13.
32. Fortin JP, Gaxeau G, Wilhelm C. Intracellular heating of living cells through Neel relaxation of magnetic nanoparticles. *Eur Biophys J* 2008;**37**:223–8.
33. Gazeau F, Levy M, Wilhelm C. Optimizing magnetic nanoparticle design for nanothermotherapy. *Nanomedicine* 2008;**3**:831–44.
34. Giri J, Ray A, Dasgupta S, Datta D, Bahadur D. Investigation of T_c tuned nanoparticles of magnetic oxides for hyperthermia applications. *Biomed Mater Eng* 2003;**13**:387–99.
35. Jordan A, Scholz R, Wust P, Fahling H, Felix R. Magnetic fluid hyperthermia (MFH): cancer treatment with AC magnetic field inducted excitation of biocompatible superparamagnetic nanoparticles. *J Magn Magn Mater* 1999;**201**:413–9.
36. Jordan A, Scholz R, Maier-Hauff K, Johannsen M, Wust P, Nadobny J. Presentation of a new magnetic field therapy system for the treatment of human solid tumors with magnetic fluid hyperthermia. *J Magn Magn Mater* 2001;**225**:118–26.
37. Ito A, Kuga Y, Honda H, Kikkawa H, Horiuchi A, Watanabe Y, et al. Magnetite nanoparticle-loaded anti-HER2 immunoliposomes for combination of antibody therapy with hyperthermia. *Cancer Lett* 2004;**212**:167–75.
38. Jin H, Kang KA. Application of novel metal nanoparticles as optical/thermal agents in optical mammography and hyperthermic treatment for breast cancer. *Adv Exp Med Biol* 2007;**599**:45–52.
39. Salloum M, Ma RH, Weeks D, Zhu L. Controlling nanoparticle delivery in magnetic nanoparticle hyperthermia for cancer treatment: experimental study in agarose gel. *Int J Hyperthermia* 2008;**24**:337–45.
40. Johannsen M, Gneveckow U, Thiesen B, Taymoorian K, Cho CH, Waldofner N. Thermotherapy of prostate cancer using magnetic nanoparticles: feasibility, imaging, and three-dimensional temperature distribution. *Eur Urol* 2007;**52**:1653–62.
41. Wust P, Gneveckow U, Johannsen M, Bohmer D, Henkel T, Kahmann F. Magnetic nanoparticles for interstitial thermotherapy—feasibility, tolerance and achieved temperatures. *Int J Hyperthermia* 2006;**22**:673–85.
42. Van Landeghem FKH, Maier-Hauff K, Jordan A, Hoffmann KT, Gneveckow R, Scholz R, et al. Post-mortem studies in glioblastoma patients treated with thermotherapy using magnetic nanoparticles. *Biomaterials* 2009;**30**:52–7.

43. Kim DH, Kim KN, Kim KM, Lee YK. Targeting to carcinoma cells with chitosan and starch coated magnetic nanoparticles for magnetic hyperthermia. *J Biomed Mater Res* 2008;**88A**:1–11.
44. DeNardo SJ, DeNArdo GL, Miers LA, Natarajan A, Foreman AR, Gruettner C, et al. Development of tumor targeting bioprobes (^{111}In-chimeric L6 monoclonal antibody nanoparticles) for alternating magnetic field cancer therapy. *Clin Cancer Res* 2005;**11**:7087–92.
45. Naratajan A, Gruettner R, Ivkov GL, DeNardo G, Mirick A, Yuan A, et al. Nanoferrite particle based radioimmunonanoparticles: binding affinity and in vivo pharmacokinetics. *Bioconjug Chem* 2008;**19**:1211–8.
46. Walther W, Stein U, Schlag P. Use of the human MRD1 promotor for heat-inducible expression of therapeutic genes. *Int J Cancer* 2002;**98**:291–6.
47. Ito A, Shinkai M, Honda H, Kobayashi T. Heat-inducible TNF-alpha gene therapy combined with hyperthermia using magnetic nanoparticles as a novel tumor-targeted therapy. *Cancer Gene Ther* 2001;**8**:649–54.
48. Tang QS, Zhang DS, Cong XM, Wan ML, Jin LQ. Using thermal energy produced by irradiation of Mn-Zn ferrite magnetic nanoparticles (MZF-NPs) for heat-inducible gene expression. *Biomaterials* 2008;**29**:2673–9.
49. Dobson J. Magnetic nanoparticles for drug delivery. *Drug Dev Res* 2006;**67**:55–60.
50. Dejardin T, de la Fuente J, Pino P, Furlani EP, Mullin M, Smith CA, et al. The influence of both a static magnetic field and penetratin on magnetic nanoparticle delivery into fibroblasts. *Nanomedicine* 2011;**6**:1719–31.
51. Smith CA, de la Fuente J, Pelaz B, Furlani EP, Mullin M, Berry CC. The effect of static magnetic fields and tat peptides on cellular and nuclear uptake of magnetic nanoparticles. *Biomaterials* 2010;**15**:4392–400.
52. Lubbe AS, Alexiou C, Bergmann C. Clinical applications of magnetic drug targeting. *J Surg Res* 1999;**95**:200–6.
53. Lubbe AS, Bergmann C, Brock J, McClure DG. Physiological aspects in magnetic drug targeting. *J Magn Magn Mater* 2000;**194**:149–55.
54. Briley-Saebo K, Bjornerud A, Grant D, Ahlstrom H, Berg T, Kindberg GM. Hepatic cellular distribution and degradation of iron oxide nanoparticles following single intravenous injection in rats: implications for magnetic resonance imaging. *Cell Tissue Res* 2004;**316**:315–23.
55. Grief AD, Richardson G. Mathematical modelling of magnetically targeted drug delivery. *J Magn Magn Mater* 2005;**293**:455–63.
56. Jain TK, Morales MA, Sahoo SK, Leslie-Pelecky DL, Labhasetwar V. Iron oxide nanoparticles for sustained delivery of anticancer agents. *Mol Pharm* 2005;**2**:194–205.
57. Kohler N, Sun C, Fichtenholz A, Gunn J, Fang C, Zhang MQ. Methotrexate-immobilised poly(ethylene glycol) magnetic nanoparticles for MR imaging and drug delivery. *Small* 2006;**2**:785–92.
58. McBain SC, Yiu HH, Dobson J. Magnetic nanoparticles for gene and drug delivery. *Int J Nanomed* 2008;**3**:169–80.
59. Hirao K, Sugita T, Kubo T, Igarashi K, Tanimoto K, Murakami T, et al. Targeted gene delivery to human osteocarcinoma cells with magnetic cationic liposomes under a magnetic field. *Int J Oncol* 2003;**22**:1065–71.
60. Haim H, Steiner I, Panet A. Synchronised infection of cell cultures by magnetically controlled virus. *J Virol* 2005;**79**:622–5.
61. Furlani EP, Ng KC. Nanoscale magnetic biotransport with application to magnetofection. *Phys Rev* 2008;**77**:0619141–8.
62. Dobson J. Gene therapy progress and prospectives: magnetic nanoparticle-based gene delivery. *Gene Ther* 2006;**13**:283–7.

63. Mykhaylyk O, Antequera YS, Vlaskou D, Plank C. Generation of magnetic nonviral gene transfer agents and magnetofection in vitro. *Nat Protoc* 2007;**2**:2391–411.
64. Fire A, Xu SQ, Montgomery MK, Kostas SA, Driver SE, Mello CC. Potent and specific genetic interference by double-stranded RNA in Caenorhabditis elegans. *Nature* 1998;**391**:806–11.
65. Gondi CS, Rao JS. Concepts in in vivo siRNA delivery for cancer therapy. *J Cell Physiol* 2009;**220**:285–91.
66. Juliano R, Alam MR, Dixit V, Kang H. Mechanisms and strategies for effective delivery of antisense and siRNA oligonucleotides. *Nucleic Acids Res* 2008;**36**:4158–71.
67. Derfus AM, Chen AA, Min DH, Ruoslahti E, Bhatia SN. Targeted quantum dot conjugates for siRNA delivery. *Bioconjug Chem* 2007;**18**:1391–6.
68. Soutschek J, Akinc A, Bramlage B, Charisse K, Constien R, Donoghue M. Therapeutic silencing of an endogenous gene by systemic administration of modified siRNAs. *Nature* 2004;**432**:173–8.
69. Medarova Z, Pham W, Farrar C, Petkova V, Moore A. In vivo imaging of siRNA delivery and silencing in tumors. *Nat Med* 2007;**13**:372–7.
70. Lee JH, Lee K, Moon SH, Lee Y, Park TG, Cheon J. All-in-one target-cell-specific magnetic nanoparticles for simultaneous molecular imaging and siRNA delivery. *Angew Chem Int Ed* 2009;**48**:4174–9.
71. Arap W, Pasqualini R, Ruoslahti E. Cancer treatment by targeted drug delivery to tumor vasculature in a mouse model. *Science* 1998;**279**:377–80.
72. Huang XH, Jain PK, El-Sayed IH, El-Sayed MA. *Nanomedicine* 2007;**2**:681–93.
73. Sonnichsen C, Franzl T, Wilk T, von Plessen G, Feldmann J, Wilson O, et al. Drastic reduction of plasmon damping in gold nanorods. *Phys Rev Lett* 2002;**88**:077402.
74. Boisselier E, Astruc D. Gold nanoparticles in nanomedicine: preparations, imaging, diagnostics, therapies and toxicity. *Chem Soc Rev* 2009;**38**:1759–82.
75. Souza GR, Christianson DR, Staquicini FI, Ozawa MG, Snyder EY, Sidman RL, et al. Networks of gold nanoparticles and bacteriophage as biological sensors and cell-targeting agents. *Proc Natl Acad Sci USA* 2006;**103**:1215–20.
76. Skala MC, Crow MJ, Wax A, Izatt JA. Photothermal optical coherence tomography of epidermal growth factor receptor in live cells using immunotargeted gold nanospheres. *Nano Lett* 2008;**8**:3461–7.
77. Yelin D, Oron D, Thiberge S, Moses E, Silberberg Y. Multiphoton plasmon-resonance microscopy. *Opt Express* 2003;**16**:1385–91.
78. Sperling RA, Rivera Gil P, Zhang F, Zanella M, Parak WJ. Biological applications of gold nanoparticles. *Chem Soc Rev* 2008;**37**:1896–908.
79. Brendayan M. A review of the potential and versatility of colloidal gold cytochemical labelling for molecular morphology. *Biotech Histochem* 2000;**75**:203–42.
80. van del Pol AN. Neuronal imaging with colloidal gold. *J Microsc* 1980;**155**:27–59.
81. Yang NS, Burkholder J, MArtinell B, McCabe D. In vivo and in vitro gene transfer to mammalian somatic cells by particle bombardment. *Proc Natl Acad Sci USA* 1990;**87**:9568–72.
82. Kuriyama S, Mitoro A, Tsujinoue H, Nakatani T, Yoshiji H, Tsujimoto T, et al. Particle-mediated gene transfer into murine livers using a newly developed gene gun. *Gene Ther* 2000;**7**:1132–6.
83. Yang PH, Sun X, Chiu JF, Sun H, Je QY. Transferring-mediated gold nanoparticle cellular uptake. *Bioconjug Chem* 2005;**16**:494–6.
84. Chithrani BD, Chan WC. Elucidating the mechanism of cellular uptake and removal of protein-coated gold nanoparticles of different sizes and shapes. *Nano Lett* 2007;**7**:1542–50.

85. Feldherr CM, Lanford RE, Akin D. Signal-mediated nuclear transport in simian virus 40-transformed cells is regulated by large tumor antigen. *Proc Natl Acad Sci USA* 1992;**89**: 11002–5.
86. Tkachenko AG, Xie H, Coleman D, Glomm W, Ryan J, Anderson MF, et al. Multifunctional gold nanoparticle-peptide complexes for nuclear targeting. *J Am Chem Soc* 2003;**125**:4700–1.
87. Berry CC, de la Fuente J, Mullin M, Wai Ling Chu S, Curtis A. Nuclear localisation of HIV-1 tat functionalised gold nanoparticles. *IEEE TNB* 2007;**6**:262–9.
88. Kumar S, Harrison N, Richards-Kortum R, Sokolov K. Plasmonic nanosensors for imaging intracellular biomarkers in live cells. *Nano Lett* 2007;**7**:1338–43.
89. Doering WE, Piotti ME, NAtan MJ, Freeman RG. SERS as a foundation for nanoscale optically detected biological labels. *Adv Mater* 2007;**19**:3100–8.
90. Mirkin CA, Letsinger RL, Mucic RC, Storhoff JJ. A DNA-based method for rationally assembling nanoparticles into macroscopic materials. *Nature* 1996;**382**:607–9.
91. Man JM, Stoeva SI, Mirkin CA. Bio-bar-code-based DNA detection with PCR-like sensitivity. *J Am Chem Soc* 2004;**126**:5932–3.
92. He H, Xie C, Ren J. Nonbleaching fluorescence of gold nanoparticles and its applications in cancer cell imaging. *Anal Chem* 2008;**80**:5951–7.
93. Dulkeith E, Morteani AC, Niedereichholz T, Klar TA, Feldmann J, Levi SA, et al. Fluorescence quenching f dye molecules near gold nanoparticles: radiative and nonradiative effects. *Phys Rev Lett* 2002;**89**:2003002.
94. Dulkeith R, Ringler M, Klar TA, Feldmann J, Munoz JA, Parak WJ. Gold nanoparticles quench fluorescence by phase induced radiative rate suppression. *Nano Lett* 2005;**5**:585–9.
95. Alivisatos AP. Semiconductor clusters, nanocrystals, and quantum dots. *Science* 1996;**271**: 933–7.
96. Alivisatos AP, Gu W, Larabell C. Quantum dots as cellular probes. *Annu Rev Biomed Eng* 2005;**7**:55–76.
97. Bruchez M, Moronne M, Gin P, Weiss S, Alivisatos AP. Semiconductor nanocrystals as fluorescent biological labels. *Science* 1988;**281**:2013–6.
98. Michalet X, Pinaud F, Lacoste TD, Dahan M, Bruchez M. Properties of fluorescent semiconductor nanocrystals and their applications to biological labelling. *Single Mol* 2001;**2**:261–76.
99. Rosenthal SJ, Tomlinson I, Adkins EM, Schroeter S, Adams S, Swafford L, et al. Targeting cell surface receptors with ligand-conjugated nanocrystals. *J Am Chem Soc* 2002;**124**: 4586–94.
100. Wu X, Liu H, Liu J, Haley KN, Treadway JA, Larson JP, et al. Immunofluorescent labeling of cancer marker Her2 and other cellular targets with semiconductor quantum dots. *Nat Biotechnol* 2003;**21**:41–6.
101. Yoo J, Kambara T, Gonda K, Higuchi H. Intracellular imaging of targeted proteins labeled with quantum dots. *Exp Cell Res* 2008;**314**:3563–9.
102. Berry CC, Harianawala H, Loebus J, Oreffo ROC, de la Fuente J. Enhancement of human bone marrow cell uptake of quantum dots using tat peptide. *Curr Nanosci* 2009;**5**:390–5.
103. Okada CY, Rechsteiner M. Introduction of macromolecules into cultured mammalian cells by osmotic lysis of pinocytic vesicles. *Cell* 1982;**29**:33–41.
104. Courty S, Luccardini C, Bellaïche Y, Cappello G, Dahan M. Tracking individual kinesin motors in living cells using single quantum dot imaging. *Nano Lett* 2006;**6**:1491–5.
105. Kusumi A, Sako Y, Yamamoto M. Conned lateral diffusion of membrane receptors as studied by single particle tracking (nanovid microscopy). Effects of calcium-induced differentiation in cultured epithelial cells. *Biophys J* 1993;**65**:2021–40.

106. Chen H, Titushkin I, Stroscio M, Cho M. Altered membrane dynamics of quantum dot-conjugated integrins during osteogenic differentiation of human bone marrow derived progenitor cells. *Biophys J* 2007;**92**:1399–408.
107. Dubertret B, Skourides P, Norris DJ, Noireaux V, Brivanlou AH, Libchaber A. In vivo imaging of quantum dots encapsulated in phospholipid micelles. *Science* 2002; **298**:1759–62.
108. Axelrod D. Cell surface contacts illuminated by total internal reflection fluorescence. *J Cell Biol* 1981;**89**:141–5.
109. Mkerman MA, Chan WCW, Laakkonen P, Chatia SN, Ruoslahti E. Nanocrystal targeting in vivo. *Proc Natl Acad Sci USA* 2002;**99**:12617–21.
110. Gao X, Yang L, Petros JA, Marshall FF, Simons J, Nie S. In vivo molecular and cellular imaging with quantum dots. *Curr Opin Biotechnol* 2004;**16**:63–72.
111. Nair A, Shen JH, Thevenot P, Zou L, Cai T, Hu ZB, et al. Enhanced intratumoral uptake of quantum dots concealed within hydrogel nanoparticles. *Nanotechnology* 2008;**19**:48.
112. Hu R, Yong KT, Roy I, Ding H, Law WC, Cai HX, et al. Functionalized near-infrared quantum dots for in vitro tumor vasculature imaging. *Nanotechnology* 2010;**21**:14.
113. Christian S, Pilch J, Akerman ME, Porkka K, Laakkonen P, Ruoslahti E. Nucleolin expressed at the cell surface is a marker of endothelial cells in angiogenic blood vessels. *J Cell Biol* 2003;**163**:871–8.
114. Kong J, Soh HT, Cassell AM, Quate CF, Dai H. Synthesis of individual single-walled carbon nanotubes on patterned silicon wafers. *Nature* 1998;**395**:878–81.
115. Blanco A, Kostarelos K, Partidos CD, Prato M. Biomedical applications of functionalised carbon nanotubes. *Chem Commun* 2005;**5**:571–7.
116. O'Connell MJ, Bachilo SM, Huffman CB, Moore VC, Strano MS, Haroz EH, et al. Band gap fluorescence from individual single walled carbon nanotubes. *Science* 2002;**297**:593–6.
117. Fabbro A, Cellot G, Prato M, Ballerini L. Interfacing neurons with carbon nanotubes: (re) engineering neuronal signalling. *Prog Brain Res* 2011;**194**:241–52.
118. Jang MJ, Namgung S, Hong S, Nam Y. Directional neurite growth using carbon nanotube patterned substrates as a biomimetic cue. *Nanotechnology* 2010;**21**:235102–8.
119. Mattson MP, Haddon RC, Rao AM. Molecular functionalization of carbon nanotubes and use as substrates for neuronal growth. *J Mol Neurosci* 2000;**14**:175–82.
120. Ni Y, Hu H, Malarkey EB, Zhao B, Montana V, Haddon RC, et al. Chemically functionalized water soluble single-walled carbon nanotubes modulate neurite outgrowth. *J Nanosci Nanotechnol* 2005;**5**:1707–12.
121. Cellot G, Cilia E, Cipollone S, Rancic V, Sucapane A, Giordani S, et al. Carbon nanotubes might improve neuronal performance by favouring electrical shortcuts. *Nat Nanotechnol* 2009;**4**:126–33.
122. Keefer EW, Botterman BR, Romero MI, Rossi AF, Gross GW. Carbon nanotube coating improves neuronal recordings. *Nat Nanotechnol* 2008;**3**:434–9.
123. Shoval A, Adams C, David-Pur M, Shein M, Hanein Y, Sernagor E. Carbon nanotube electrodes for effective interfacing with retinal tissue. *Front Neuroeng* 2009;**2**:1–8.
124. Kotov NA, Winter JO, Clements IP, Jan E, Timko BP, Campidelli S, et al. Nanomaterials for neural interfaces. *Adv Mater* 2009;**21**:3970–4004.
125. Moon HK, Lee SH, Choi HC. In vivo near-infrared mediated tumor destruction by photothermal effect of carbon nanotubes. *ACS Nano* 2009;**3**:370.7–370.3713.
126. Burke A, Ding X, Singh R, Kraft RA, Levi-Polyachenko N, Rylander MN, et al. Long-term survival following a single treatment of kidney tumors with multiwalled carbon nanotubes and near-infrared radiation. *Proc Natl Acad Sci USA* 2009;**106**:12897–902.

127. McDevitt MR, Chattopadhyay D, Kappel BJ, Jaggi JS, Schiffman SR, Antczak C, et al. Tumor targeting with antibody-functionalized, radiolabeled carbon nanotubes. *J Nucl Med* 2007;**48**:1180–9.
128. Zhou FF, Xing D, Ou ZM, Wu BY, Resasco DE, Chen WR. Cancer photothermal therapy in the near-infrared region using single walled carbon nanotubes. *J Biomed Opt* 2009;**14**:021009.
129. Liu Z, Chen K, Davis C, Sherlock S, Cao QZ, Chen XY, et al. Drug delivery with carbon nanotubes for in vivo cancer treatment. *Cancer Res* 2008;**68**:6652–60.
130. Liu Z, Fan AC, Rakhra K, Sherlock S, Goodwin A, Chen XY, et al. Supramolecular stacking of doxorubicin on carbon nanotubes for in vivo cancer therapy. *Angew Chem Int Ed* 2009;**48**:7668–72.
131. Xiang LZ, Yuan Y, Xing D, Ou ZM, Yang SH, Zhou FF. Photoacoustic molecular imaging with antibody functionalized single walled carbon nanotubes for early diagnosis of tumor. *J Biomed Opt* 2009;**14**:021008.
132. Bhirde AA, Patel V, Gavard J, Zhang GF, Sousa AA, Masedunskas A, et al. Targeted killing of cancer cells in vivo and in vitro with EGF-directed carbon nanotube-based drug delivery. *ACS Nano* 2009;**3**:307–16.

Chapter 7

Investigating Nanoparticle Internalization Patterns by Quantitative Correlation Analysis of Microscopy Imaging Data

Raimo Hartmann, Susana Carregal-Romero, Wolfgang J. Parak and Pilar Rivera_Gil
Fachbereich Physik, Philipps Universität Marburg, Marburg, Germany

1. INTRODUCTION

The fate of nanoparticles (NPs) taken up by cells depends on various nanoscale parameters such as size, surface charge/chemistry, or colloidal stability.[1–4] Although great effort has been put into finding and characterizing dependencies between NP parameters and internalization pathways, most of the results are insufficient for a deeper understanding or even inconsistent when comparing the findings of different research institutions.[5] Especially for drug delivery based on nanocarriers, the knowledge of the subcellular distribution of the pharmaceuticals is essential for exploring and understanding therapeutic mechanisms.

Intracellular trafficking is typically studied with fluorescent microscopy. By comparing the fluorescent pattern of labeled and internalized NPs with the distribution of cellular organelles, possible intracellular pathways can be derived for the material. It is mandatory to analyze the kinetics of the uptake progress and further transport of the nanosized material within the cells as conclusions cannot be drawn from a snapshot taken at a fixed point in time from one single cell. Following endocytic uptake, NPs are generally trapped in vesicular compartments. Vesicle migration governs the intracellular pathways of embedded nanomaterials and finally, their fate. The detection and imaging of typical proteins associated to those enclosed structures allows their identification and allocation in for example, endosomes or lysosomes. If such image material is superimposed with signal gained from fluorescent NPs,

Frontiers of Nanoscience, Vol. 4. DOI: 10.1016/B978-0-12-415769-9.00007-8

structures associated with NP uptake, transport, and processing can be identified.

To analyze the uptake and enrichment of NPs inside a certain organelle, fluorescent labeling of both the nanomaterial (in case the NP is not fluorescent itself) and the organelle is typically performed. The uptake study is based on the correlation of the emission of the labeled nanomaterial with fluorescence signal of the organelle. If both structures are colocalizing within the detection volume, the overlay of the corresponding two fluorescence image channels (e.g., red and green) would result in a new color value (yellow). In a qualitative manner, the degree of colocalization can be estimated by looking at the superimposed image.

As a matter of fact, any processing having impact on the image's histogram influences the "amount of yellow" in the overlay and alters the subjective impression of the degree of colocalization.

For a more sophisticated correlation of the image material of both structures, several approaches to perform a quantitative colocalization analysis exist. In intensity-based methods, voxel or pixel intensities in both fluorescence channels are correlated by calculating, for example, Pearson's or Manders' colocalization coefficients.[6,7] In Li's approach, the correlation between the variations of the intensity distributions within both channels is analyzed.[8] In object-based approaches, the imaged structures are transformed into binary objects and the overlap is quantified.[7] In live cell imaging also, methods for trajectory correlation of those binary objects have been introduced.[9] Nevertheless, as long as single NP detection and tracking is hard to realize by conventional confocal microscopy, the relevance of trajectory correlation is quite low, although the results seem to bear good prospects due to the discrimination of false colocalization caused by low axial resolution.

To understand the kinetics of different nanomaterials, uptake and trafficking, images of NPs, and organelles have to be acquired over time, and structures of interest playing part in intracellular trafficking have to be highlighted. This can be done either by preparing fixated samples using immunocytochemistry or by live cell imaging.

Considering the first case, specific endocytic vesicles are stained by employing fluorescently labeled antibodies (immunoglobulins) that bind selected proteins only present in those structures. Antibodies are available for almost all relevant proteins known to be involved in cellular uptake and trafficking. The disadvantages of this method are the need of fixation and permeabilization of the cellular material, which can cause artifacts,[10] and the time-consuming staining procedure. To obtain time-resolved images prior to any colocalization analysis, various individual samples have to be prepared to cover the desired time frame. In addition, many cells have to be imaged to decrease the deviations in the degree of colocalization for a given time point and to provide reproducibility of the experiment. An example for possible results is given in Figure 7.1 where the kinetics of the enrichment of polymer-coated superparamagnetic iron oxide NPs (SPIONs) inside lysosomal structure of A549 lung carcinoma cells was

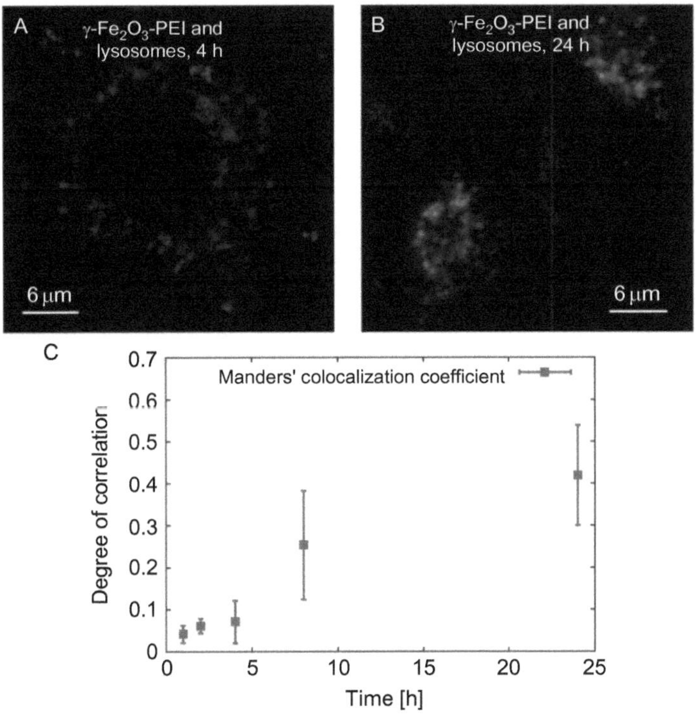

FIGURE 7.1 Comparison: qualitative versus quantitative colocalization analysis. Degree of colocalization between poly(ethyleneimine) (PEI)-coated iron oxide nanoparticles and lysosomes. Nanoparticles are labeled with FITC while lysosomal structures were detected using fluorescent antibodies against LAMP1. (For interpretation of the references to color in this figure legend, the reader is referred to the online version of this chapter.)

analyzed. The lysosomes were detected using immunoglobulin with high affinity to the lysosomal-associated protein 1 (LAMP1).

In live cell imaging, growth conditions for cells are maintained during imaging. In contrast to the last method, a fluorescent staining of intracellular structures in living material requires different techniques, because the intact cell membrane acts as a barrier for most staining agents. The most common way is the application of markers with known internalization pathways to label certain structures prior to NP exposure. For a more sophisticated and more specific labeling, cells can be transfected with DNA constructs encoding a modified, fluorescent version of a protein, residing in specific endocytic vesicles. Due to the availability of several such modified proteins playing part in intracellular trafficking, time-lapse imaging of live cells is gaining more and more importance and the idea of live colocalization experiments is promising.[11]

In the present chapter, we provide an approach to analyze intracellular pathways by fluorescence microscopy based on time-resolved colocalization analysis in a quantitative and reproducible manner.

2. TIME-RESOLVED QUANTITATIVE CORRELATION ANALYSIS

2.1. Image Acquisition

The conclusions which can be drawn from a colocalization analysis are strongly influenced by the capabilities of the optical system used for imaging. For a meaningful appraisal of the obtained results, one has to be aware of the capabilities and limits of the image acquisition system. Unfortunately, a perfect copy of the distribution of fluorophores within a probe cannot be imaged. In every optical system without super-resolution capabilities, the image information is blurred due to the system's characteristic point spread function (PSF). The PSF describes how a single point source is seen by the detector due to the diffraction-limited nature of photon propagation.

Due to the relatively large spatial distribution of the PSF for any fluorophore in the probe, an image acquired with a widefield fluorescence microscope is blurred because many photons are collected originating from unfocussed optical sections. Therefore, the detection volume in such a system cannot be quantified. To circumvent this problem, the image information of several optical slices around the desired axial position has to be acquired. By means of a numerical deconvolution of the data, blur can then be reassessed to inverse the effects of the PSF. The resulting resolution of the widefield image material after an appropriate deconvolution is comparable to that obtained with a confocal system, which is introduced in the next section.

In contrast to a standard widefield microscope, in a confocal system, light which is not originating from the focus is suppressed. First, due to higher detection sensitivity, the illuminating light intensity can be decreased to minimize fluorophore-excitation outside of the focus. Second, photons originating from other axial positions than defined through the focal plane are diminished by a small pinhole within the emission light path. This allows for "cropping" only the central part of every fluorophore's PSF and therefore providing an additional axial resolution. In such a system, the detected signal behind the pinhole is influenced by the illuminating and the detection pathway. In other words, the image information is blurred due to the effects of the PSF_{ex}, characterizing the illuminating light path of the instrument and the PSF_{em} of the optical path of the detection setup. If the dimensions or the full width at half maximum of the PSF_{ex} is larger, but in the same range as the PSF_{em}, the resolving capacity is governed by the size of the PSF_{ex} and can be stated as[12]

$$r_{\text{lateral}} = 0.51 \frac{\lambda_{\text{ex}}}{\text{NA}} \quad (7.1)$$

$$r_{\text{axial}} = 0.88 \frac{\lambda_{\text{ex}}}{n - \sqrt{n^2 - \text{NA}^2}} \quad (7.2)$$

where λ_{ex} is the excitation wavelength, n the refractive index of the surrounding medium, and NA the numerical aperture of the used objective. The thickness of a single optical slice (d_{slice}) is defined as the axial dimension, that is, the full width at half maximum of the PSF_{em} given by[13]

$$d_{slice} = \sqrt{\left(\frac{0.88\lambda_{em}}{n-\sqrt{n^2-NA^2}}\right)^2 + \left(\frac{\sqrt{2}nd_{pinhole}}{NA}\right)^2} \tag{7.3}$$

Here, λ_{em} is the emission wavelength, $\lambda_{pinhole}$ the back-projected, object-related, diameter of the confocal pinhole (the size of the pinhole as seen in the focal plane), n the refractive index of the surrounding medium, and NA the numerical aperture of the used objective.

Due to the pointwise measurement nature of a confocal setup, the probe has to be sampled in a serial manner to obtain 2D or 3D image data, respectively. To discriminate objects in the size range of the optical resolution, the lateral and axial sampling frequency should fulfill the Nyquist–Shannon sampling criterion.[14] Furthermore, the lateral and axial resolution depends on the excitation wavelength. For a quantitative correlation analysis of structures excited with different wavelengths, the resolution for each channel is therefore different. By varying the pinhole diameter $\lambda_{pinhole}$, the optical detection volume for each dye species can be adjusted by controlling the size of the PSF for the corresponding imaging light path.

The image quality and the signal-to-noise ratio are influenced by various factors. Quality and reproducibility of correlations analyses suffer from background signal and intrinsic, Poisson-distributed shot noise which goes along with photon detection at low count rates. In addition, any spectral overlap between the used fluorophores distorts the results. Furthermore, the detection of two different dye species requires the use of different filter settings for each channel. Any misalignment of the optical system or aberrations can have serious impact on the results of a quantitative correlation analysis.

If signal-to-noise ratio of the image acquisition system is already maximized, various methods are available to enhance the quality of the image material after recording prior to analysis and to convert the intensity variations within the images into desired binary-like structures.

In conventional fluorescent widefield microscopy, the image information is blurred due to convolution with the characteristic PSF of the optical system (refer to Figure 7.2). In addition to the lateral lack of definition, the axial resolution cannot be quantified in a proper way. Here, deconvolution techniques try to inverse the impact of the PSF. By re-assessing blur (lateral and axial), the sharpness and the dynamic range of an image are enhanced and the signal-to-noise ratio is maximized.[15] Beside this, shot noise is also eliminated effectively.[16] Unfortunately, most software packages which include easy-to-use deconvolution algorithms are quite expensive.

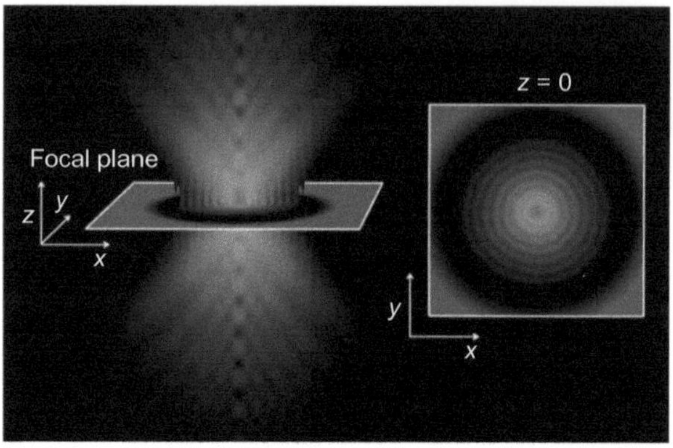

FIGURE 7.2 Point spread function (PSF). Cross-section of a 3D PSF at $y=0$ (left) and top view of the same slice in the focal plane ($z=0$). By adjusting the diameter of the confocal pinhole, ideally, only the first main maximum (red) of the PSF is detected with a confocal laser scanning microscope to obtain a high depth of sharpness. The detection volume is shaped as a rotational ellipsoid. (For interpretation of the references to color in this figure legend, the reader is referred to the online version of this chapter.)

If the image material is not obtained with a confocal microscope (or devices offering an even better axial resolution), proper deconvolution of the image material is mandatory. To suppress shot noise and eliminate background signal in confocal data, in most cases, low pass, Gaussian, or median filtering followed by thresholding is sufficient[17] (for examples regarding the effects of image processing refer to Figure 7.5). All mentioned filtering techniques are quite fast but result in a loss of resolution because the image is smoothed. This loss can be reduced by oversampling (i.e., acquiring images at a higher physical resolution than necessary) during image acquisition. To remove background signal, typically a threshold is applied to the image material. Thereby, a certain intensity value is subtracted from every pixel or voxel. Attention has to be put on the choice of the threshold if the chosen colocalization technique is very sensitive to background noise as it is the case for object-based colocalization analyses.

Misalignment between different channels of the optical system can be identified by imaging a pattern of dual-color microspheres. If a hardware-based correction is not possible, a distortion map can be created to derive the relative drift for every image pixel between the two channels of interest. Often, the effect of minor aberrations is reduced by shifting the image information of the second channel for a few pixels in reference to the first channel. If two structures are present, which are known to colocalize, their images should be used to maximize their degree of positive correlation by introducing a small pixel shift between both channels. For this purpose, either Pearson's or Manders' colocalization coefficient (introduced in the next section) can be calculated. The lateral displacement of the second channel should be noted and applied to the entire image material.

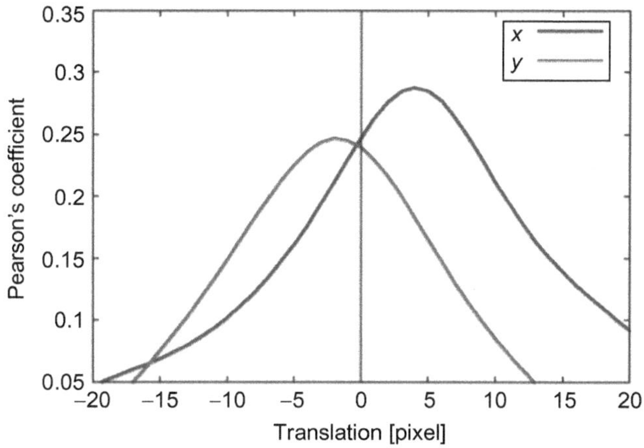

FIGURE 7.3 Pixel-shift. Example for a global, lateral displacement of two images of the same sample acquired with different setups. Pearson's correlation coefficients of early endosomes-stained cellular structures and the image pattern of fluorescent nanoparticles incorporated by these cells after 4 h of exposure time are plotted against a small lateral displacement between both channels in *x*- and *y*-direction. The image was sampled at a lateral resolution of 50 nm/pixel. For further analysis, the image information of the second channel may be shifted 4 pixel in negative *x*-direction and 2 pixels in *x*-direction relative to the first channel. (For color version of this figure, the reader is referred to the online version of this chapter.)

2.2. Quantitative Analysis of Colocalization

Several approaches exist to quantify the degree of colocalization of different patterns being imaged in two different channels.

2.2.1. Intensity-Based Analysis

i) Pearson's Colocalization Coefficient

Pearson's correlation coefficient R_r is defined as the covariance of two variables divided by the product of their standard deviations and is widely used in pattern recognition.[6]

$$R_r = \frac{\sum(R_i - \bar{R})(G_i - \bar{G})}{\sqrt{\sum(R_i - \bar{R})^2 \sum(G_i - \bar{G})^2}} \in [-1, 1] \tag{7.4}$$

R_i or G_i, respectively, is the intensity of the ith voxel in channel R or G while $\bar{R}$ and $\bar{G}$ are the mean values of all voxel intensities in each channel. Positive values for R_r indicate a high degree of colocalization while negative values may represent exclusion. Due to the inclusion of the average image intensities, this coefficient is only slightly biased by different background levels of the two images.

ii) Manders' Overlap Coefficients

To compare objects/patterns of different signal intensities caused by different labeling efficiencies of amplifier settings, Manders *et al.* introduced their colocalization coefficient R in 1993.[7]

$$R = \frac{\sum(R_i G_i)}{\sqrt{\sum R_i^2 \sum G_i^2}} \in [0,1] \tag{7.5}$$

Due to fact that both channels contribute equally to R_r or R, it is not possible to indentify situations where the objects in two channels colocalize asymmetrically as it is shown exemplary in Figure 7.4. Therefore, it is mandatory to quantify the degree of correlation using two independent parameters k_1 and k_2 defined as

$$k_1 = \frac{\sum R_i G_i}{\sum R_i^2} \in \mathbb{R}_+ \text{ and } k_2 = \frac{\sum R_i G_i}{\sum G_i^2} \in \mathbb{R}_+ \tag{7.6}$$

2.2.2. *Object-Based Analysis*

In the last case, Equation (7.6), colocalizing objects are weighted by their intensities. To quantify simply the spatial overlap of two patterns (object-based correlation), the distinct colocalization coefficients M_1 and M_2 are widely used:

$$M_1 = \frac{\sum R_{i,\text{coloc}}}{\sum R_i} \in [0,1] \text{ and } M_2 = \frac{\sum G_{i,\text{coloc}}}{\sum G_i} \in [0,1] \tag{7.7}$$

In comparison to M_1 and M_2, the coefficients k_1 and k_2 are weighted by the voxel intensities and therefore less affected by noise (refer also to Figure 7.5).

2.2.3. *Correlated Intensity Variations*

The Intensity Correlation Quotient (ICQ) calculates the dependency of the intensity variation of two channels around their mean intensity value.[8] If the intensity in both channels at a certain position within the image is above their

Example		R_r	R	k_1	k_2	M_1	M_2	ICQ
1		–0.34	0	0	0	0	0	0.00
2		–0.03	0.23	0.23	0.23	0.42	0.42	0.10
3		1.00	1.00	1.00	1.00	1.00	1.00	0.50
4		0.46	0.52	0.30	0.91	0.42	1.00	0.29

FIGURE 7.4 Resulting correlation coefficients calculated for selected examples. (For color version of this figure, the reader is referred to the online version of this chapter.)

Endosomes (EEA1) **Iron oxide nanoparticles (FITC)** **Correlation**

Raw data (8 bit)

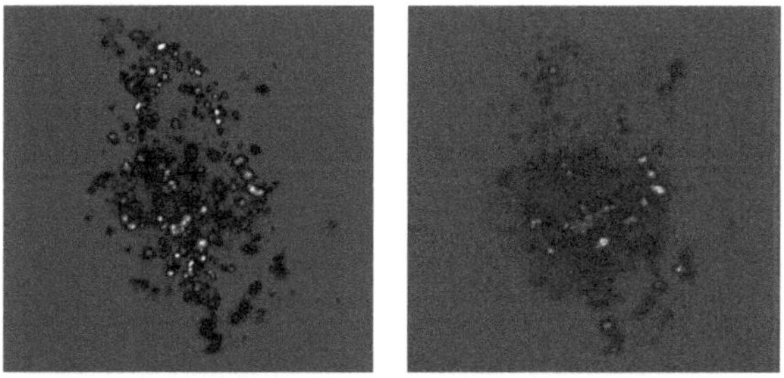

R_r	0.43
R	0.45
k_1	0.23
k_2	0.86
M_1	0.69
M_2	0.69
ICQ	0.37

Median filtering (3×3)

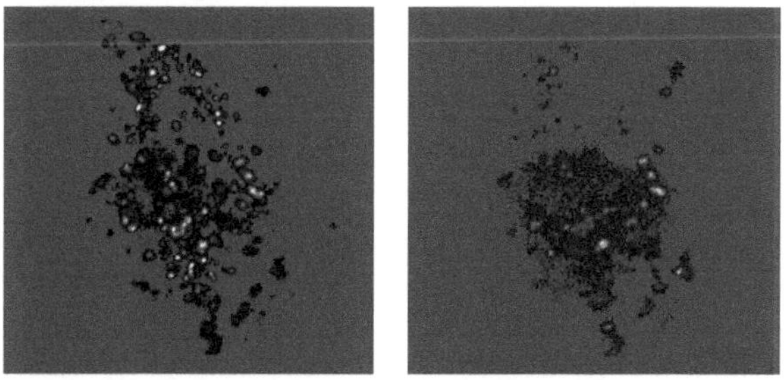

R_r	0.51
R	0.53
k_1	0.23
k_2	1.24
M_1	0.72
M_2	0.85
ICQ	0.42

Median filtering (3×3)+thresholding (20 intensity values)

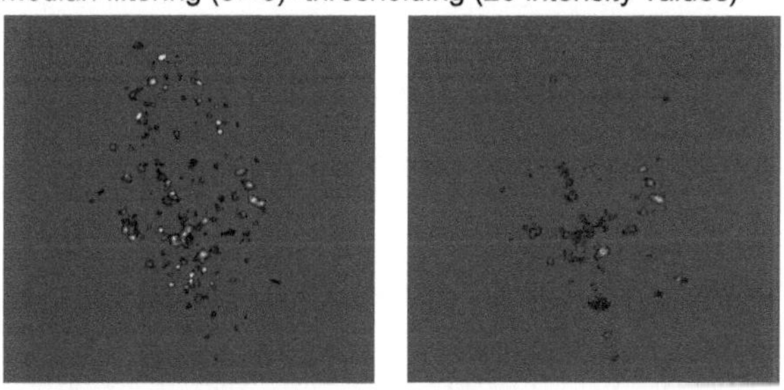

R_r	0.39
R	0.40
k_1	0.13
k_2	1.28
M_1	0.42
M_2	0.74
ICQ	0.47

2D Deconvolution with theoretical PSF

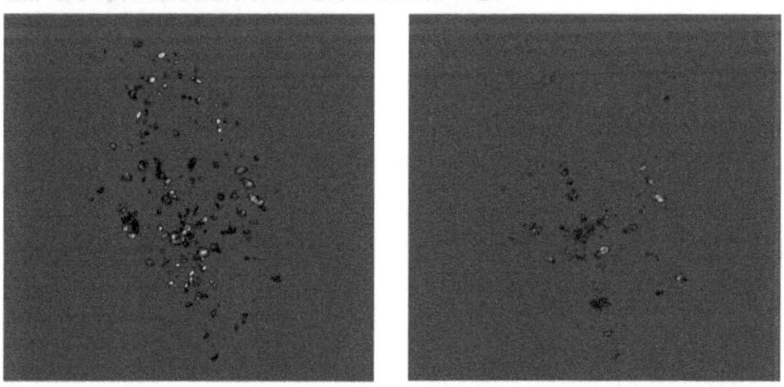

R_r	0.31
R	0.32
k_1	0.17
k_2	0.61
M_1	0.34
M_2	0.77
ICQ	0.47

FIGURE 7.5 Image processing. The effects of median filtering, thresholding, and deconvolution on the image raw material are shown exemplary. (For color version of this figure, the reader is referred to the online version of this chapter.)

mean value, the value of the product of the differences from the mean (PDM), which is defined for each voxel, becomes positive. Asynchronous intensity variations are resulting in negative PDM values.

$$\mathrm{PDM}_i = (R_i - \bar{R})(G_i - \bar{G}) \tag{7.8}$$

The ICQ is based on a sign-test analysis of all PDM values as stated below.

$$\mathrm{ICQ} = \frac{N_{\text{positive}}}{N_{\text{total}}} - 0.5 \in [0.5, 0.5] \tag{7.9}$$

N_{positive} is the number of positive PDM values while N_{total} is the number of voxel intensities which are nonequal to zero in both channels.

3. EXAMPLE STUDY: INTERNALIZATION PATHWAYS OF SUPERPARAMAGNETIC IRON OXIDE NPs

An exemplary quantitative and time-resolved colocalization analysis was performed to understand the impact of surface charge and chemistry of different SPIONs model systems on their internalization mechanisms. The localization of these systems within A549 lung carcinoma cells was analyzed with respect to endosomes and lysosomes. In detail, A549 cells were exposed to positively charged poly(ethyleneimine) (PEI)-coated and negatively charged polymaleic anhydride derivative (PMA)-coated γ-Fe_2O_3 NPs.

3.1. Preparation and Characterization of SPIONs

Within the framework of this example analysis, two SPIONs species were used keeping a similar size distribution, but differing in synthesis procedures, surface charge, and surface chemistry.

γ-Fe_2O_3-PEI NPs were prepared via aqueous coprecipitation[18] and embedded into a polymeric PEI matrix providing a positive zeta-potential of $+53.0 \pm 11.0$ mV. Through conjugation with fluorescein isothiocyanate (FITC), the particles could be detected with a fluorescence microscope. In contrast, negatively charged, polymer-coated γ-Fe_2O_3-PMA NPs were obtained by thermal decomposition of organometallic precursor molecules and coated with Dy636-conjugated PMA, which render the NPs hydrophilicity (resulting zeta-potential: -38.0 ± 5.6 mV).[19,20] The size distribution obtained by TEM image analysis was 10.4 ± 2.4 nm for positively charged particles and 10.8 ± 0.12 nm for negatively charged NPs, respectively. The absorption and emission spectra of both NP systems are shown in Figure 7.6B and C.

3.1.1. Cell Culture and Sample Preparation

A549 cells were exposed to γ-Fe_2O_3-PEI NPs and γ-Fe_2O_3-PMA NPs simultaneously at an iron concentration of 1 μg/ml for 30 min, 1, 2, 4, 8, and 24 h, each. Finally, the samples were prepared for immunostaining of cellular

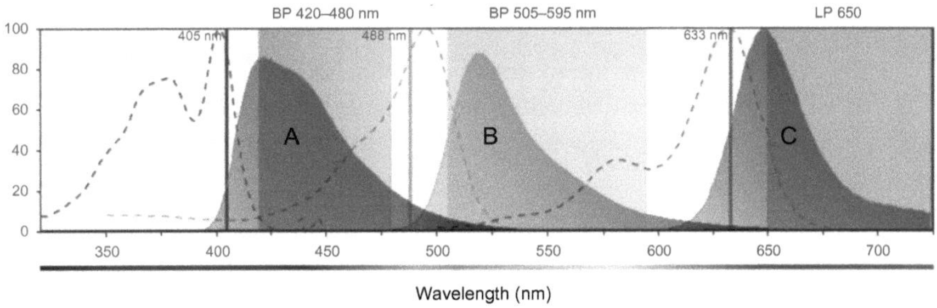

FIGURE 7.6 Absorption and emission spectra of used dyes. Absorption spectra are drawn with dots, emission spectra as solid lines. (A) DyLight405 was excited at $\lambda=405$ nm, and the emission was detected between 420 and 480 nm. (B) γ-Fe_2O_3-PEI nanoparticles labeled with FITC were excited at $\lambda=488$ nm and the emission were detected between 505 and 550 nm. (C) γ-Fe_2O_3-PMA nanoparticles labeled with Dy636 were excited at $\lambda=633$ nm. In this case, the emitting photons were filtered spectrally using a long-pass at $\lambda=650$ nm. (For color version of this figure, the reader is referred to the online version of this chapter.)

organelles. Either lysosomes were stained with monoclonal mouse anti-human LAMP1/CD107a antibodies (Developmental Studies Hybridoma Bank), or a labeling of the early endosomes with polyclonal rabbit anti-human EEA1 immunoglobulin (Cell Signaling) was performed. Donkey anti-mouse DyLight405 antibodies (Jackson ImmunoResearch) were used to detect the LAMP1-specific primary antibodies while goat anti-rabbit, DyLight405-conjugated, immunoglobulin (Invitrogen) was used as a secondary antibody for the early endosomes. The absorption and emission spectra of the different fluorescent molecules used are shown in Figure 7.6.

3.1.2. Fluorescence Microscopy

The stained samples were examined with a confocal laser scanning microscope (510 Meta; Zeiss). The device was equipped with a laser diode emitting at 405 nm, an argon laser with a line at 488 nm, and a HeNe laser emitting at 633 nm (the excitation wavelengths are also shown in Figure 7.6). The emission originating from early endosomes and lysosomes (stained with DyLight405) was selected using a band-pass (BP) filter passing signal covering a spectral range of 420–480 nm. γ-Fe_2O_3-PEI-FITC NPs were detected using a 420–480 nm BP, while the photons emitted by γ-Fe_2O_3-PMA-Dy636 particles were selected with a 650 nm long-pass (LP) filter (refer to Figure 7.6). To determine the localization of γ-Fe_2O_3-PEI and γ-Fe_2O_3-PMA NPs in a subsequent step with respect to cellular organelles, 20 cells from different positions of each sample were imaged in randomly chosen sectional planes. The thickness of the optical section was adjusted to 0.9 μm for each type of fluorophore. A Plan-Apochromat 63 ×/1.40 Oil DIC M27 objective was used, and for each image, the fluorescent signal of a 102.75 μm × 102.75 μm-sized square was sampled with an lateral resolution of 50.2 nm (resulting image size: 2048 × 2048 pixels, 8 bit).

3.2. Results

To understand the cellular internalization of the two types of SPIONs which differed in their synthetic procedure and surface chemistry, their degree of colocalization between endosomes and lysosomes was analyzed for different exposure times.

Looking at the superimposed fluorescent pattern of NPs with endosomes, both NP species colocalize with these cellular structures within the observed period of time (Figure 7.7). In contrast, the results from a quantitative analysis give a much better impression of the internalization process. In Figure 7.8, Manders' distinct colocalization coefficients were calculated. It can be seen that the fraction of SPIONs overlapping spatially with early endosomes has a maximum at 4 h (PEI-coated NPs, green) and 2 h (PMA-coated NPs, magenta), that is, the intracellular transport of γ-Fe_2O_3-PMA-Dy636 NPs toward the endosomes takes place faster than that of γ-Fe_2O_3-PEI-FITC NPs. If the signal of the NPs is correlated with lysosomal structures (see Figure 7.10), it becomes clear that a major fraction of particles are either processed from the endosomal toward lysosomal compartments or the endosomes itself mature to early lysosomes. The values of the coefficient, representing the degree of colocalization "as seen by the endosomes" (yellow), differ strongly from the NP correlation coefficients. An explanation could be that the cells are exposed to "fresh" extracellular NPs for the whole period of time, and the endosomes are not assumed to be the final destination of internalized material. If the uptake rate is not decreasing significantly over time, NPs are endocytosed and traverse endosomes steadily.

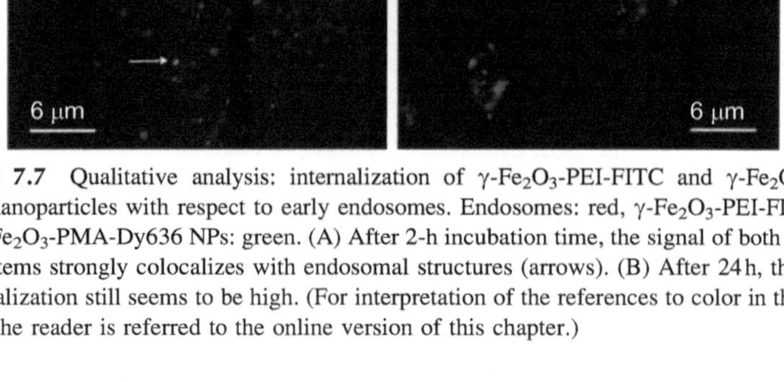

FIGURE 7.7 Qualitative analysis: internalization of γ-Fe_2O_3-PEI-FITC and γ-Fe_2O_3-PMA-Dy636 nanoparticles with respect to early endosomes. Endosomes: red, γ-Fe_2O_3-PEI-FITC NPs: blue, γ-Fe_2O_3-PMA-Dy636 NPs: green. (A) After 2-h incubation time, the signal of both nanoparticle systems strongly colocalizes with endosomal structures (arrows). (B) After 24 h, the degree of colocalization still seems to be high. (For interpretation of the references to color in this figure legend, the reader is referred to the online version of this chapter.)

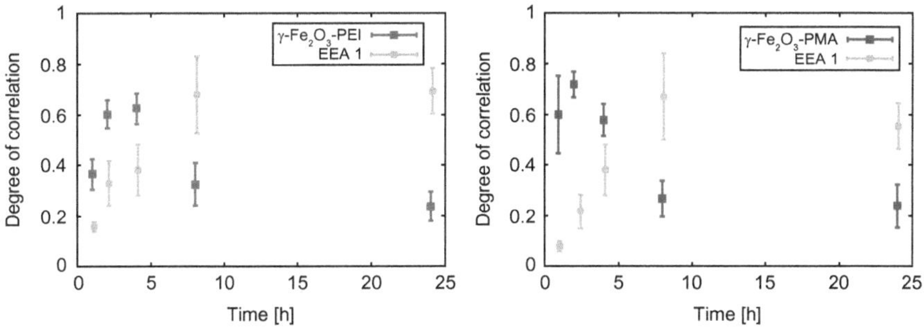

FIGURE 7.8 Quantitative analysis: internalization of γ-Fe_2O_3-PEI-FITC and γ-Fe_2O_3-PMA-Dy636 nanoparticles with respect to early endosomes. Shown are Manders' distinct colocalization coefficients (M_1: green/magenta, M_2: yellow). (For interpretation of the references to color in this figure legend, the reader is referred to the online version of this chapter.)

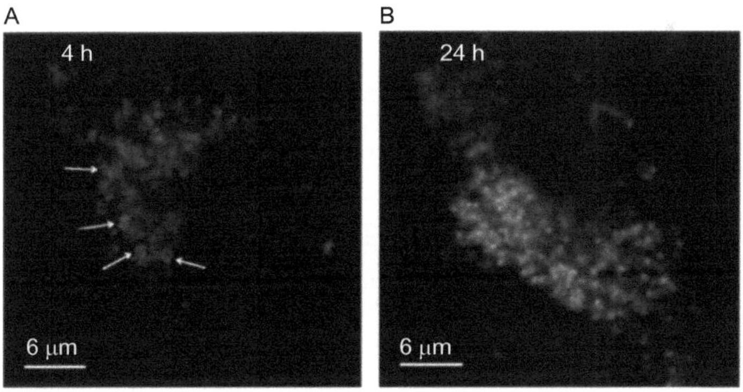

FIGURE 7.9 Qualitative analysis: internalization of γ-Fe_2O_3-PEI-FITC and γ-Fe_2O_3-PMA-Dy636 nanoparticles with respect to lysosomes. Lysosomes: red, γ-Fe_2O_3-PEI-FITC NPs: blue, γ-Fe_2O_3-PMA-Dy636 NPs: green. (A) After 4-h incubation time, no nanoparticles can be found inside lysosomal structures. (B) In contrast, after 24 h, the degree of colocalization is very high. (For interpretation of the references to color in this figure legend, the reader is referred to the online version of this chapter.)

The fraction of both SPIONs species enriched in lysosomal structures is increasing steadily as it can be seen qualitatively in Figure 7.9 and quantitatively in Figure 7.10. The higher degree of colocalization for PMA-coated particles after 4 h of exposure time in comparison to PEI-coated NPs is also consistent with the finding that these particles are processed faster in early endosomes (refer to Figure 7.8, right part).

Finally, the colocalization between both SPIONs after concomitant exposure was evaluated by calculating the ICQ values and Manders' overlap coefficient as shown in Figure 7.11. The uptake pattern differs for $t < 4$ h. This might be caused by the earlier detection of γ-Fe_2O_3-PMA-Dy636 NPs inside endosomal structures in comparison to γ-Fe_2O_3-PEI-FITC NPs (Figure 7.8).

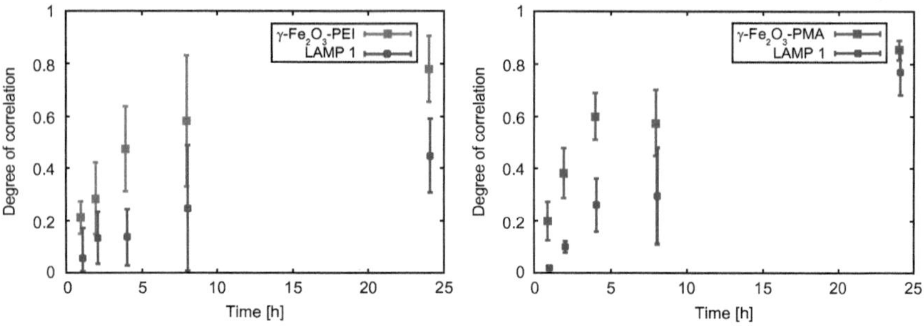

FIGURE 7.10 Quantitative analysis: internalization of γ-Fe_2O_3-PEI-FITC and γ-Fe_2O_3-PMA-Dy636 nanoparticles with respect to lysosomes. Shown are Manders' distinct colocalization coefficients (M_1: green/magenta, M_2: red). (For interpretation of the references to color in this figure legend, the reader is referred to the online version of this chapter.)

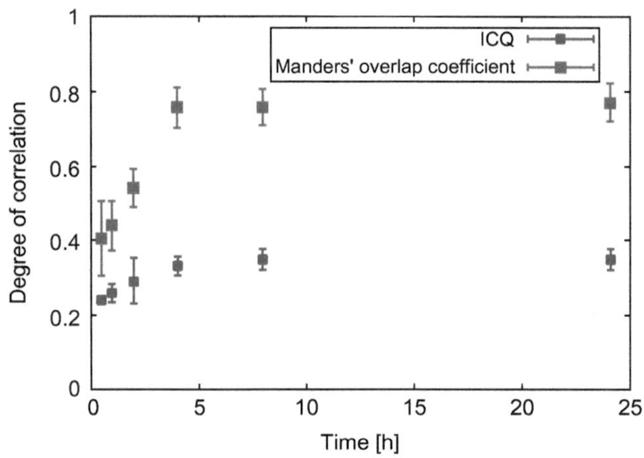

FIGURE 7.11 Concomitant exposure: spatial correlation between fluorescence signal from γ-Fe_2O_3-PEI-FITC and γ-Fe_2O_3-PMA-Dy636 nanoparticles. Shown are the intensity correlation coefficient (blue) and Manders' overlap coefficient (green). (For interpretation of the references to color in this figure legend, the reader is referred to the online version of this chapter.)

The uptake patterns of both SPIONs are very similar despite different surface coating and opposite charge. Probably, PEI-coated particles form aggregates with PMA-coated ones due to strong electrostatic interactions and these agglomerates are internalized.

4. LIMITATIONS AND CONCLUSION

Concerning the characterization of NPs, knowledge about their intracellular pathways, such as their endosomal sorting after cellular uptake, is mandatory, especially if the particles are designed as cargos for drug delivery purposes.

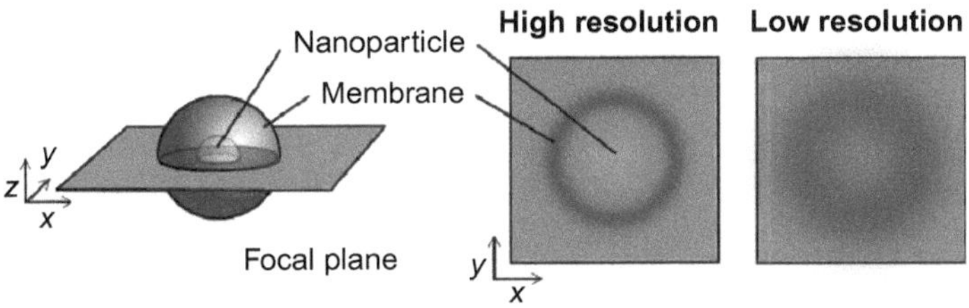

FIGURE 7.12 Organelle with enclosed particle. This scheme shows a fluorescent particle encapsulated by an intracellular vesicle. If only the membrane of the vesicle is stained, no colocalization is resulting at very high resolutions and high contrast. If the definition of the image is lowered, both structures become positively correlated. (For color version of this figure, the reader is referred to the online version of this chapter.)

The conclusions drawn by eye-based interpretation of superimposed, fluorescent images presenting the distribution of NPs and certain cellular structures are strongly biased by any acquisition parameters and image processing. For a first impression or a proof of principle, this method may be sufficient but the generalization of any observation has to be taken with a grain of salt.

The averaging over colocalization data of several individual experiments and the imaging of various cells for each data point is needed. For quantification, the described procedure of time-resolved colocalization analysis is a well-suited tool that may help to retrace NP internalization in a reproducible manner.

Attention has to be put on the choice of the optical and physical resolution of the microscope. If the resolution is too low, spatially separated structures are correlated positively. In contrast, the resolution can also be too high. If the transport in relatively large organelles shall be quantified using a membrane-associated fluorescent marker and a cellular structure is "cut" by an optical section, only colocalization with the fluorescent membrane is accounting, especially if a confocal image is deconvolved prior to analysis. In those special cases, deconvolution may be omitted to enhance the level of blur and, hence, the degree of colocalization (refer to Figure 7.12).

REFERENCES

1. Mailander V, Landfester K. Interaction of nanoparticles with cells. *Biomacromolecules* 2009;**10**(9):2379–400.
2. Harush-Frenkel O, Rozentur E, Benita S, Altschuler Y. Surface charge of nanoparticles determines their endocytic and transcytotic pathway in polarized MDCK cells. *Biomacromolecules* 2008;**9**(2):435–43.
3. Chung YI, Kim JC, Kim YH, Tae G, Lee SY, Kim K, et al. The effect of surface functionalization of PLGA nanoparticles by heparin- or chitosan-conjugated Pluronic on tumor targeting. *J Control Release* 2010;**143**(3):374–82.

4. Ge YQ, Zhang Y, Xia JG, Ma M, He SY, Nie F, et al. Effect of surface charge and agglomerate degree of magnetic iron oxide nanoparticles on KB cellular uptake in vitro. *Colloids Surf B Biointerfaces* 2009;**73**(2):294–301.
5. Zhao F, Zhao Y, Liu Y, Chang XL, Chen CY, Zhao YL. Cellular uptake, intracellular trafficking, and cytotoxicity of nanomaterials. *Small* 2011;**7**(10):1322–37.
6. Gonzalez RC, Woods RE. *Digital image processing*. Upper Saddle River, New Jersey 07458: Pearson Education, Inc. Pearson Prentice Hall Pearson Education, Inc.; 2008.
7. Manders EMM, Verbeek FJ, Aten JA. Measurement of colocalization of objects in dual-color confocal images. *J Microsc (Oxford)* 1993;**169**:375–82.
8. Li Q, Lau A, Morris TJ, Guo L, Fordyce CB, Stanley EF. A syntaxin 1, Galpha(o), and N-type calcium channel complex at a presynaptic nerve terminal: analysis by quantitative immunocolocalization. *J Neurosci* 2004;**24**(16):4070–81.
9. Vercauteren D, Deschout H, Remaut K, Engbersen JFJ, Jones AT, Demeester J, et al. Dynamic colocalization microscopy to characterize intracellular trafficking of nanomedicines. *ACS Nano* 2011;**5**(10):7874–84.
10. Bacallao R, Sohrab S, Phillips C. Pawley JB, editor. *Handbook of biological confocal microscopy*. 3rd ed. New York: Springer Science and Business Media LLC; 2006. p. 368.
11. Watson P. Live cell imaging for target and drug discovery. *Drug News Perspect* 2009;**22**(2):69–79.
12. Cole RW, Jinadasa T, Brown CM. Measuring and interpreting point spread functions to determine confocal microscope resolution and ensure quality control. *Nat Protoc* 2011;**6**(12): 1929–41.
13. Wilhelm S, Gröbler B, Gluch M, Heinz H. *Confocal laser scanning microscopy: principles*. Carl Zeiss Publication, Jena, Germany; 2000 40-617e.
14. Nyquist H. Certain topics in telegraph transmission theory. *Proc IEEE* 2002;**90**(2):280–305 (Reprinted from Trans AIEE, February 1928;617–644).
15. Nasse MJ, Woehl JC. Realistic modeling of the illumination point spread function in confocal scanning optical microscopy. *J Opt Soc Am A Opt Image Sci Vis* 2010;**27**(2):295–302.
16. Landmann L, Marbet P. Colocalization analysis yields superior results after image restoration. *Microsc Res Tech* 2004;**64**(2):103–12.
17. Demandolx D, Davoust J. Multicolour analysis and local image correlation in confocal microscopy. *J Microsc (Oxford)* 1997;**185**:21–36.
18. Bee A, Massart R, Neveu S. Synthesis of very fine maghemite particles. *J Magn Magn Mater* 1995;**149**(1–2):6–9.
19. Hyeon T, Lee SS, Park J, Chung Y, Na HB. Synthesis of highly crystalline and monodisperse maghemite nanocrystallites without a size-selection process. *J Am Chem Soc* 2001;**123**(51): 12798–801.
20. Lin C-AJ, Sperling RA, Li JK, Yang T-Y, Li P-Y, Zanella M, et al. Design of an amphiphilic polymer for nanoparticle coating and functionalization. *Small* 2008;**4**(3):334–41.

Chapter 8

Organic Nanoparticles

Helene Feracci*
*Centre de Recherche Paul Pascal, CNRS UPR 8641, Université Bordeaux 1, Pessac, France

Berta Saez Gutierrez[†,‡]
[†]Instituto de Nanociencia de Aragón, Universidad de Zaragoza, Edificio I+D, Mariano Esquillor s/n, Campus Rio Ebro, Zaragoza, Spain
[‡]Instituto Aragones de Ciencias de la Salud, Avenida de Alcalde Gómez Laguna, 25, Zaragoza, Spain

William Hempel[§]
[§]Biosystems International, 1 rue Pierre Fontaine, Evry, France

Isabel Segura Gil[†]
[†]Instituto de Nanociencia de Aragón, Universidad de Zaragoza, Edificio I+D, Mariano Esquillor s/n, Campus Rio Ebro, Zaragoza, Spain

1. INTRODUCTION

Organic nanoparticles/nanobeads are of major interest in the material and life sciences. Their small size provides them with unique properties putting them at the forefront of emerging technologies. They are highly useful in bioanalysis and basic science as model systems to study single molecule behavior, molecular recognition, signal transduction, and protein-mediated membrane fusion processes once functionalized/associated with biological entities (such as DNA or antibodies) as catalytic interfaces, for sensor applications, for drug delivery, as well as other applications. There are a multitude of possible applications of fabricated nanoparticles in biotechnology, fundamental and clinical research, and in medicine. While nanobiotechnology deals with the development of electronic circuits, molecular "switches," biosensors, and nanosized microchips, nanomedicine focuses on disease treatment, diagnosis/monitoring, and rational delivery and targeting of pharmaceutical, therapeutic, and diagnostic agents. Rather than using inorganic matrices, biopolymer nanoparticles may provide some significant advantages. They have been primarily developed for drug delivery to overcome the risk of chronic toxicity due to the intracellular and/or tissue overloading of nondegradable polymers. The need for developing such nanoparticles (liposome, virus-like particle, protein, polysaccharide, etc.) not only as effective drug

Frontiers of Nanoscience, Vol. 4. DOI: 10.1016/B978-0-12-415769-9.00008-X

delivery devices but also for immunoassays was realized years ago.[1] Indeed, in addition to the general advantages of nanoparticles, biopolymer nanoparticles in particular offer several advantages, which include the ease of their preparation from well-understood biodegradable polymers and their high stability in biological fluids and during storage.[2]

The aim of this chapter is to give examples from different fields in order to provide a prospective of the applications of nanomaterials to biotechnology and biomedicine, a field that is rapidly expanding and will continue to do so for the foreseeable future.

2. USE OF NANOPARTICLES IN NUCLEIC ACID DELIVERY

2.1. Background

Safe and effective gene delivery systems are tremendously important in gene therapy for tackling various genetic diseases, viral infections, and cardiovascular disorders. Gene therapy is a process by which genetic material in the form of oligonucleotides or plasmids is introduced into specific target cells to recover or induce the expression of a normal protein to treat of human disorders. Gene therapy can also deliver antisense oligonucleotides or small interfering RNA to interrupt the function of target genes and trigger silencing. In recent decades, many methods for gene delivery have been developed for a wide range of cells and tissues.[3]

In addition to these therapeutic applications, gene delivery is widely used *in vitro* as a research tool to investigate gene function/regulation within a cellular and physiological context.[4] Initial systems for gene delivery have been developed, with the potential to significantly impact biotechnology, diagnostic applications, and basic research[3c,5]

Gene delivery systems should enable the formation of stable complexes with nucleic acids, provide low cytotoxicity, and disassemble intracellularly to release the nucleic acid. Environmental interactions are manipulated through the incorporation of functional groups that may stabilize the vector in the extracellular milieu, target a specific tissue or cell type, or maintain the vector within the delivery location. Intracellular trafficking addresses the need to deliver the plasmid to the nucleus for expression, thus bioactive groups can be incorporated to facilitate endosomal escape or nuclear localization.

Gene delivery systems include viral and nonviral vectors. Viral vectors are the most effective, but there is an increasing concern about their application, which is limited by their limited DNA loading capactiy, and in therapeutic application by their oncogenicity and immunogenicity. Nonviral vectors are more reproducible, do not present DNA size-limits, and are safer and less costly than viral vectors. Nonviral transfection systems are usually composed of cationic peptides, cationic polymers, or cationic lipids, although the combination of some of them is also possible.[6]

These so-called modular nonviral vectors consist of a vector backbone modified with functional groups that mediate environmental interactions and intracellular trafficking to overcome the multiple barriers to gene transfer.[3c] For both *in vitro* and *in vivo* applications, nonviral vectors must be designed taking into consideration their interactions with serum components, the extracellular milieu within the tissue or cell culture media, and binding to the cell surface. Many of the aforementioned vectors interact with serum proteins that can inactivate the complex or promote clearance from the desired tissue, which limits the opportunity for cellular internalization. Cellular internalization can be improved by the addition of receptor ligands to the nanoparticle, thus increasing binding to cell-surface receptors and allowing the targeting of specific cell types.[7]

Once the DNA is introduced into the cell, it must pass a series of barriers that could damage it, before reaching the nucleus. To avoid this, extensive research is underway to define the ideal carrier to facilitate cell entry and transport to the nucleus of the DNA in a protected manner. The carrier should be sufficiently small to allow cellular internalization and pass through nuclear pores, must have flexible tropism, and should be able to escape the endosome–lysosome process that follows endocytosis.[3b] It must also not be cytotoxic nor elicit an immune response. All these conditions can be accomplished by nonviral nanovectors including cationic molecules such as cationic lipids and synthetic or natural cationic polymers, which have been widely developed to condense DNA and to efficiently deliver therapeutic genes within mammalian cells.[8]

While gene delivery has been tested in many cell types using a number of different conditions and readout systems to test for transfection efficiency and toxicity, the lack of a reference standard makes it difficult to directly compare different approaches. Thus, this section presents selective approaches with illustrative examples, rather than an exhaustive comparison of the different methods that have been developed over the years.

2.2. Engineered Biomaterials for Vector Backbones

Numerous cationic lipids and polymers have been developed to package DNA for cellular internalization and protect it from degradation, leading to the identification of some structure–function design relationships.[9]

Cationic lipids have three basic constituents: a polar headgroup, a linker, and a hydrophobic moiety. The cationic headgroup promotes interaction with DNA, whereas the hydrophobic moiety provides self-association to form either micelles or liposomes in the presence of a helper lipid, such as dioleylphosphatidylethanolamine. Lipoplexes form a multilayered structure consisting of the plasmid sandwiched between the cationic lipids. Several recent reviews have described the design of the cationic headgroup, hydrophobic moiety, and linker for optimized transfection.[10] Despite the fact that cationic lipids have low immunogenicity, they show *in vitro* and *in vivo* cytotoxicity and relatively low gene transfer efficiency.[11]

Cationic polymers can be classified into two groups: natural polymers, such as proteins and peptides, and synthetic polymers, such as polyethylenimine, dendrimers, and polyphosphoesters.[12] They spontaneously associate with plasmids by electrostatic interactions due to protonatable amine residues to form condensed complexes called polyplexes.[13] With the electrostatic forces between the polymer and the nucleic acid, the complex should maintain a stable and condensed nano-size structure, promote cellular endocytosis, and possibly enhance transfection efficiency of therapeutic genes.[13] Furthermore, DNA plasmids can be condensed inside the nanoparticles for protection against nuclease degradation.[14] High-molecular-weight polymers tend to form relatively stable, small complexes compared to low-molecular-weight polymers. Low-molecular-weight polymers, however, can enhance transfection efficiencies, likely due to a decreased cytotoxicity and the increased ability of the plasmid to dissociate from the cationic polymer.[15] Block copolymers can potentially regulate assembly and structure of the complex, while providing for multiple functions due to each component.[16] Poly(2-diethylaminoethyl methacrylate) or poly(2-dimethylaminoethyl methacrylate), which have primary and secondary amines to facilitate complexation and intracellular trafficking, can be combined with poly(ethylene oxide) (PEO) or poly(propylene oxide) to prevent aggregation and reduce toxicity.[17]

Although the condensing vectors seem to be an excellent substitute for viral vectors, some drawbacks inherent in the condensing system limit their application for systemic delivery. These include toxicity of the cationic polymer or lipid, rapid clearance by the reticuloendothelial system, inability of the complex to escape from the endosome/lysosome compartments in the cells, and lack of intracellular unpacking of the nucleic acid construct from the electrostatic complex.[18]

Noncondensing lipids and polymers posses either a neutral or net negative charge. They can be used to engineer nanoscale vectors for tissue and cell specific delivery and allow for enhanced transfection efficiency with significantly less toxicity concerns.[19] Nucleic acids are encapsulated within such vectors (e.g., liposomes or nanoparticles) either by physical entrapment within the matrix or through hydrogen bonds between polymer and nucleic acid bases.[20] Physical encapsulation offers protection from the enzymes and other plasma proteins during its transit from blood to the site of action. Cellular uptake is facilitated since masking the negative charge of DNA prevents electrostatic repulsion with the negatively charged cell surface. Moreover, in contrast to condensing lipids and polymers, the absence of positive charges on noncondensing systems limits (a) their recognition by the mononuclear phagocyte system and hence limits their early clearance and (b) opsonization by IgM and the innate immune response.[21]

Solid lipid nanoparticles (SLNs) have also been tested. They are generated by exchanging the liquid lipid of emulsions for a solid lipid, making them solid at room and body temperature, thus inducing a reduction in particle size.[22] The capacity of SLN:DNA vectors to induce the expression of a foreign protein after intravenous administration has been demonstrated.[23] No significant differences in transfection rate were observed between an SLN and its counterpart liposome. However, addition of chloroquine or monesin, led to an increase in transfection efficiency due to protonation of these weak bases in acidic cellular compartments leading to endosomal buffering. By neutralizing the acidic pH inside the endosome, DNA molecules, protected against acid attack of protons, remain intact inside the cell. The most important difference with liposomes is that the core matrix used for release and delivery of bioactive substances becomes lipophilic instead of aqueous for liposomes. Some advantages of SLN are their low toxicity, good storage stability, and the possibility of steam sterilization and lyophilization.

The use of cholesterol and some of its derivatives for the synthesis of gene delivery carriers has been recently tested. Cationic cholesterol disulfide lipids (CHOSS) were synthesized by binding cholesterol to cationic head groups via disulfide linkage. Headgroups which were used included histidine, pyrimidine, or methyl imidazole which demonstrated low cytotoxicity and high transfection efficiency.[24] CHOSS can bind to DNA with high-affinity forming stable lipoplex nanoparticles.[25] Transfection experiments with COS-7 cells confirmed the low cytotoxicity and particularly high uptake capability of these particles. Interestingly, the particles were specifically localized to the periphery of cell nuclei following uptake. Once inside the cell, release of the DNA may occur via cleavage of the disulphide bond in the reducing intracellular environment.[25]

Another approach has been to create biomimetic nanocapsules. They consist of an oily core made up of a mixture of triglycerides and polyglyceryl-6 dioleate surrounded by a shell of free polyethylene glycol (PEG) and HS-PEG. They have been used for DNA encapsulation to improve the resistance to rapid clearance.[26] Through PEG coating, the positive charges of such complexes are masked and thus prevent opsonization by immune proteins. Furthermore, because PEG leads to destabilization of the lipid endosomal membrane, DNA nanocapsules are better than lipoplexes when the limiting step in transfection of endosomes has to be avoided.[26] Another important consideration is the biodegradability of the nanospheres once they have traversed the required specific sites. Early biopolymer nanoparticle design focused primarily on the use of nonbiodegradable synthetic polymers such as polyacrylamide and poly (methylacrylate).[27] The risks of chronic toxicity due to the intracellular and/or tissue overloading of nondegradable polymers were soon considered as a major limitation for the systemic administration of polyacrylamide and poly (methylacrylate) nanoparticles in humans. As a consequence, the focus has shifted toward nanoparticles designed using synthetic biodegradable polymers including polyalkylcyanoacrylate, poly(lactic-*co*-glycolic acid) and polyanhydride.[28] The therapeutic potential of these biodegradable colloidal systems was investigated for various applications.[29] Despite the very interesting results reported in the literature, these systems may also be associated with toxicological problems.[30]

There is another limitation for the bionanoparticle-based administration of hydrophilic molecules including nucleic acids (oligonucleotides and genes). This limitation is mainly because the polymers forming these nanoparticles are mostly hydrophobic while biomolecules including nucleic acids are hydrophilic. This leads to difficulties for efficient encapsulation and protection against enzymatic degradation. Therefore, the preparation of nanoparticles using more hydrophilic and naturally occurring materials has been explored.[31]

Dendrimers have attracted interest for drug and gene delivery systems because they possess a number of unique and interesting characteristics such as defined structures, inner cavities able to encapsulate guest molecules, and controllable multivalent functionalities in their inner or outer parts. These properties make dendrimers an important option for the development of nanoscale nonviral vectors for nucleic acid delivery.[32] Dendrimers can interact with various forms of nucleic acids: DNA, RNA, and oligonucleotides. These interactions, primarily electrostatic, lead to complexes, which protect the nucleic acid from degradation. The properties of these complexes depend on many factors, such as stoichiometry, the concentration of dendrimer amines and nucleic acids phosphates, solvent properties like pH, salt concentration, etc.[32b] However, the inherent difficulty of synthesizing new dendrimers that are suitable carriers for drug delivery has led researchers to focus primarily on the modification of existing dendrimers, instead of the development of novel dendrimers for gene delivery systems.

Poly(amido amine) (PAMAM) dendrimers have been tested as genetic material carriers and have been also modified with PEG, amino acids, or ligands in order to enhance their gene delivery potency.[33] Numerous reports have been published describing the use of amino-terminated PAMAM or PPI (poly(propileneimine)) dendrimers as nonviral gene transfer agents, enhancing the transfection of DNA by endocytosis and, ultimately, into the cell nucleus.[32a] The observed high transfection efficiency of dendrimers may not only be due to their well-defined structure but may also be caused by the low pK_a of the amines (3.9 and 6.9), which permits the dendrimer to buffer the pH change in the endosomal compartment.[34] Dendrimers hold a promising future for various biomedical applications for the coming years, as they possess unique properties, such as a high degree of branching and multivalency. Also as research progresses, newer applications of dendrimers will emerge and the future should witness an increasing number of commercialized dendrimer-based DNA delivery systems.

The first naturally occurring material used for the preparation of nanoparticles consisted of two proteins, albumin and gelatin.[35] Protein-based colloidal systems are very promising because they are biodegradable, nontoxic, and less immunogenic; they have greater stability *in vivo* and during storage, are relatively easy to prepare and to monitor size distribution (reviewed in Ref. 2a). In addition, because of the defined primary structure of proteins, protein-based nanoparticles offer various possibilities for surface modification and covalent drug attachment. For all these reasons, a number of proteins have been used to develop protein-based nanoparticles for drug delivery. These include albumin, collagen, gelatin, fibroin, sericin, and keratin. As an example, gelatin is one of the most versatile natural biopolymers derived from collagen, and it has been widely used in food products and medicines. Many researchers have used gelatin nanoparticles as a gene delivery vehicle.[36] With solvent displacement, type B gelatin, derived from alkaline hydrolysis of collagen, which has an isoelectric point at around 4.5–5.5, can physically encapsulate nucleic acid constructs at neutral pH. Furthermore, the physical encapsulation in gelatin nanoparticles preserves the supercoiled structure of the plasmid DNA and improves the transfection efficiency upon intracellular delivery.[37]

An approach, which has recently been used for gene delivery, includes genetically engineered protein-based polymers, which incorporate peptide motifs such as elastin, silk, and collagen. This approach has the advantage that the properties of the resulting nanoparticles can be tailored to avoid cytotoxicity and rapid clearance, while insuring delivery of the DNA package to the intended target. Another promising approach, which has primarily been tested *in vitro* combines the properties of lipids and peptides to achieve high efficiency of transfection with minimal toxicity even in the presence of serum. This approach consists of the use of oligoarginine–lipid conjugates. The lipidic part of these complexes consists of 3,5-bis(dodecyloxy)benzamide (BDB) and a PEG spacer which is introduced between the amide group of BDB and the C-terminal of oligo-Arg.[38]

Polysaccharide-derived nanoparticle surfaces help to improve biocompatibility of cell toxic material, together with new immobilization approaches, which are currently in development for novel bionanoparticle-derived pharmaceutical formulations. Nanoparticles from naturally occurring polysaccharides were designed for the administration of peptides and proteins, as well as nucleic acids.[31a,b,39] An important example of this approach is the use of chitosan, a natural biodegradable cationic polysaccharide derived from chitin consisting of D-glucosamine and *N*-acetyl-D-glucosamine. It is produced by deacetylation of chitin extracted from the shells of crabs, shrimp, and krill.[40] This linear polymer has been shown to be biocompatible and nonimmunogenic and to possess mucoadhesive properties,[40,41] making it an excellent biopolymer for the preparation of nanoparticles[2b,42] as vectors for DNA delivery. Indeed chitosan can spontaneously bind to DNA via ionic interactions to form nanoparticles.[39b] The biochemical properties of chitosan make it a suitable vehicle for gene transfection. The amino groups confer to the molecule a high charge density and are readily available for chemical modification and salt formation with acids. *In vivo*, it is degraded by lysozyme[43] and it has been also shown to partially protect DNA from nuclease degradation.[44] Cellular transfection with chitosan-DNA complexes has demonstrated that once inside the cell, complexes of 100–250 nm can be found accumulated in the nucleus. Finally, chitosan-DNA nanospheres have been shown to be nontoxic in both experimental animals and humans.

Although not discussed here, alginate is another polysaccharide that has been extensively studied for the preparation of nanoparticles.[31a]

2.3. Environmental Interactions

As already mentioned, nonspecific interactions need to be taken into account when optimizing gene delivery protocols. Interactions with serum components can be minimized by modifying the surface of polymeric nanoparticles with hydrophilic polymer chains in a process called passive targeting as it allows the modified nanoparticles to avoid the body's clearance mechanisms, increasing the chances that they will reach their intended target.[45] There are several ways to achieve passive targeting such as surface modification of polymeric nanoparticles with hydrophilic polymer chains or incorporation of environmently insensitive polymers into the nanoparticles.[19] PEG or PEO, with a structure of $HO—(CH_2CH_2O)_n—CH_2CH_2—OH$ is a commonly used polymer for surface modification of long-circulating vesicles.[46] A dense, hydrophilic shell of long chains forms that protect the core from interacting with solutes, including nonspecific hydrophobic interactions with the reticuloendothelial system.[46] This polymeric protection is referred to as "steric stabilization."[46,47] Modifications of the cationic polymers, such as PEGylation, have been incorporated to facilitate delivery and enhance transgene expression. At the same time, the terminal hydroxyl groups of PEG can be derivatized, leading to monofunctional, homo-, or hetero-bifunctional and even multi-arm PEG, allowing further conjugation of selected ligands.[20a] To summarize, PEG modification of nanoparticles provides the following advantages: increased circulation time, stabilization of the therapeutic payload during transportation, decreased RES accumulation, and the potential for conjugation of targeting moieties.

Kaul and Amiji[48] were the first to develop type B gelatin-based nanoparticles as a noncondensing gene delivery system for tumor targeting. They prepared unmodified and PEG-modified gelatin nanoparticles by ethanol precipitation leading to particles in the range of 200–500 nm. Tetramethylrhodamine-dextran, a hydrophilic fluorescently labeled molecule, was first used as model drug for *in vitro* cell uptake studies. The results showed that control and PEG-modified type B gelatin nanoparticles were taken up by cells through nonspecific endocytosis.[48] Within 12 h of PEG-modified gelatin nanoparticle internalization by NIH-3T3 murine fibroblasts, the payload was released and accumulated around the perinuclear region. Further experiments using encapsulated plasmid DNA, encoding for enhanced green fluorescence protein

(EGFP-N1), confirmed the long-lasting transgene expression potential of PEG-modified type B gelatin nanoparticles compared to other methods. In addition, neither gelatin nor PEG-modified gelatin nanoparticles demonstrated any toxicity.[49] To complement the *in vitro* evaluations, the authors further examined the biodistribution profiles of unmodified and PEG-modified ^{125}Iodine (^{125}I)-labeled gelatin nanoparticles following intravenous administration through the tail vein in Lewis lung carcinoma-bearing C57BL/6J mice. PEG-modified nanoparticles remained in circulation for an extended period of time and preferentially accumulated in the tumor for up to 24 h post-administration as well as the liver. Conversely, unmodified nanoparticles were rapidly cleared from the circulation and remained mostly in the liver and spleen. These results show that PEG-modified gelatin nanoparticles can be passively targeted to the tumor mass following systemic administration and have the potential to be an effective vector for anticancer gene therapy.[50] In another study, the potential of PEG-modified gelatin nanoparticles for passively targeting gene delivery to tumors was further demonstrated. Interestingly, intravenous administration of an encapsulated reporter plasmid DNA into tumor bearing mice actually led to a higher level of transfection of the tumor than following intratumoral administration.[51]

A more active tumor targeting strategy takes advantage of the fact that tumors express elevated levels of glutathione (GSH) compared to normal cells. GSH is a tripeptide, generally expressed in the cell cytoplasm and functions as an antioxidant to prevent damage related to the reactive oxygen species.[52] The intracellular GSH concentration (5–10 mM) is generally higher than the extracellular concentrations (1–10 mM). During active proliferation of tumor cells, GSH and peroxide levels are further elevated in the cytoplasm.[52] Introduction of thiol (i.e., SH) groups is a common modification that can allow for intracellular delivery through the reduction disulfide cross-links. This approach has been tested *in vitro* and leads to increased transfection efficiency compared to other nanoparticle based systems.[20a]

Chitosan-DNA nanospheres have also been tested for clinical applications. Several delivery routes of chitosan-DNA nanoparticles have been investigated in animal models including intranasal[53] and oral delivery resulting in transduced gene expression in the stomach and intestinal epithelium of a peanut allergen gene providing protection against allergen induced anaphylaxis.[54] In addition, such nanoparticles have also been designed for tissue engineering and local gene delivery in periodontal tissue regeneration.[55]

2.4. Intracellular Trafficking

Following extracellular transport to the target cell, the vector is internalized primarily by endocytosis and then must escape the endosomal compartment to avoid lysosomal degradation before transport to the nucleus and crossing the nuclear membrane for subsequent transcription (reviewed in Ref. 56).

Nuclear localization occurs either following cell division or through transport through the nuclear pore complex. The rate-limiting step for this process is dependent upon the vector.

Lipoplexes exhibit poor cellular uptake, whereas polyplexes exhibit limited nuclear import.[33c,57] Small molecule peptides (NLS, RGD, the HIV protein TAT) and proteins proteins such as transferrin (for cancer cell targeting), antibodies, or galactose (for hepatocyte targeting) have been attached to either the plasmid or to the cationic lipid or polymer to promote endosomal escape and nuclear localization.[1,58]

Thus fusogenic peptides have been incorporated to disrupt membranes, which may facilitate either cell entry without endocytosis or endosomal escape following endocytosis.[59] Nuclear targeting can be enhanced by the attachment of nuclear localization signals (NLS), which are oligopeptides that bind to importins, cytoplasmic receptors responsible for binding and transport through the nuclear pore complex. For example NLS addition to the polysaccharide chitosan improved transfection efficiency.[60] However, success with NLS has been varied: while some papers have reported that inclusion of NLS-containing proteins or peptides increases gene transfer and expression, such as for chitosan, others have found no such enhancement.[61]

Numerous opportunities remain to increase gene transfer by virus mimicking: that is the design of vectors to interface with specific cellular processes, with functional groups derived from the understanding viruses or cellular processes. Nonviral vectors engineered with functional groups that allow for directed motion along the cytoskeleton could increase accumulation at the nucleus and decrease the amount of DNA required.[62] Peptide nucleic acids (PNAs) have been developed that are able to bind tightly to specific DNA sequences. Incorporation of PNAs could potentially target the transgene to a specific chromosomal location as some viruses do.[63] Alternatively, engineered zinc-finger proteins with nuclease activity were developed to recognize a unique chromosomal site and induce a double-strand break. At this break site, the chromosome can recombine with an extrachromosomal sequence of interest.[64] Finally, the nucleic acid itself could be engineered to avoid silencing by methylation, thereby extending the duration of transgene expression.

3. IMMUNOASSAYS

3.1. Background

Immunoassays play a pivotal role in laboratory research, clinical diagnostics, and food and environmental monitoring. They combine immunology and chemistry to create scientific tests such as enzyme immunoassays for the specific and sensitive detection of the analytes of interest. These assays are based on the principle of the specificity of the interaction between the antibody and its cognate antigen. Immunoassays are relatively easy to perform contributing to their widespread use. *R*adio*i*mmuno*a*ssays (RIAs) and enzyme immunoassays such as ELISA (*e*nzyme-*l*inked *i*mmunosorbent *a*ssay), *l*uminescent *i*mmuno*a*ssays, and *f*luorescent *i*mmuno*a*ssays (FIAs) are all currently used. The antibodies can be labeled in several ways including radioisotopes, fluorescent dyes, or enzymes which catalyze fluorogenic or luminogenic reactions, thus allowing visualization of the antibody–antigen interaction.

In the past, RIAs were widely used but have been slowly replaced by assays using fluorescent molecules and enzymes as labels to avoid the obvious disadvantages of radioisotopes. Nevertheless the use of enzymes and fluorescent labels is often less straightforward and do not yet achieve the sensitivity and limits of detection of radioisotope-based assays. When first developed in the 1970s, ELISAs and Western blot assays provided a significant increase in sensitivity over existing detection methods.[65] Today, a wide variety of enzyme-linked antibodies (with choices of chemiluminescent, bioluminescent, chemifluorescent, fluorescent, and more traditional colorimetric detection systems) are commercially available from various suppliers. Despite recent advances, there is an urgent need to improve immunoassay sensitivity for multiple applications such as the deciphering of normal and pathophysiological biological processes, early disease detection or the detection of environmental hazards, all instances where marker levels may be very low.

Antibody–enzyme (Enz-Ab) conjugates are most often prepared by cross-linking enzymes to the antibody via their functional groups such as the primary amines and sulfhydryls, or by cross-linking through sugar moieties attached to one of the proteins. Chemical activation of the target residues on both the enzyme and the antibody tends to be random and difficult to control. Thus, the use of homobifunctional reagents such as glutaraldehyde often results in very low yields, and the resulting Enz-Ab complexes can be heterogeneous and highly polymerized.[66,67] Such Enz-Ab complexes can compromise the sensitivity of the subsequent immunological assay due to steric hindrance, which affects antigen-binding capacity. In addition, large molecular weight oligomers formed by uncontrolled cross-linking may produce insoluble aggregates that are impossible to manipulate in immunoassays. In the case of glycoproteins, some of these difficulties can be circumvented by periodate oxidation. This method has been used to cross-link horseradish peroxidase (HRP), a glycoprotein, to functional groups on an antibody in a more controlled way.[68] However, this approach is limited to enzymes that contain carbohydrate moieties and thus is not applicable to all enzymes including the commonly used reporter enzyme alkaline phosphatase.

In FIAs, signal amplification is typically achieved by coupling fluorophores, such as organic dyes, to the antibody probes. The sensitivity of FIAs is mainly determined by the number of light quanta emitted/analyte molecule. Increasing the fluorescent dye to Ab ratio (i.e., the F/P ratio) results in improved signal amplification and therefore sensitivity. As for Enz-Ab conjugates, labeling antibodies with large numbers of fluorophores usually leads to reduced specificity and binding affinity as well as a reduced quantum yield due to dye self-quenching effects. For these reasons, the F/P ratio is normally kept around 4–8 thus limiting the sensitivity of the assays.

As will be described below, biotinylated antibodies with avidin or streptavidin enzyme conjugates provide an ideal alternative to conventional enzyme-linked antibodies for many applications.[69] One of the most advantageous features of the biotin–streptavidin systems is that they offer the possibility of signal amplification due to the potential of multiple site binding of the streptavidin–enzyme conjugate to several biotinylation sites on the antibody. Obviously the number and location of the biotinylation sites and steric constraints limit the potential of the method.

3.2. Nanotechnology and Improvement of Sensitivity

The arrival of nanotechnology has created new opportunities for the development of highly sensitive bioaffinity and biocatalytic assays.[70] Organic nanoparticles provide the possibility to dramatically increase label loading or the surface available for Ig linkage, without the need to modify the detection system.

3.2.1. Functionalization of Nanoparticles by Biological Entities

The creation of nanoparticles with the desired physiobiochemical properties remains a challenge. First, it is important that the antibody molecules are stably attached to the particle surfaces while maintaining their ability to interact with the target analyte. In addition, nanoparticle aggregation and their nonspecific binding with biological molecules remain a serious issue. For example, proteins can be either hydrophilic or hydrophobic, with either negative or positive charges, making it very difficult to avoid nonspecific interactions. To address these technical challenges, numerous surface modifications and immobilization procedures have been explored and developed.

A common surface modification strategy is post-coating and modification of the nanoparticle surface with different functional groups, including carboxylate, amine, PEG, or combinations of different functionalities.[71] Alternatively, the direct synthesis on the surface of the nanoparticles of a mixed monolayer of PEG, and a functional group has been tested.[72] At the same time that the ethylene glycol chains, which are water soluble and neutral in charge, function as a shielding component to minimize nonspecific binding, the functional group can be used as a capture agent for antibody conjugation acts.

Immobilization of biomolecules on surfaces can be achieved in a number of ways-adsorption, chemical linkage, or by affinity-based interactions.

Adsorption: In this case, molecules are adsorbed at the interface via physical forces such as van der Waals, electrostatic, or hydrophobic interactions depending on the chemical nature of the surfaces and molecules. As a consequence, the conditions used for immobilization are highly sensitive to ionic strength, pH, and temperature, leading under certain conditions to subsequent dissociation of biomolecules. Moreover, this method of immobilization may interfere with proper folding and can lead to multilayer adsorption and loss of orientation or enzymatic activity. However, because adsorption of biomolecular targets is relatively simple to perform, it is widely used.

Chemical linkage: The disadvantages of adsorption can be overcome to some extent by covalently linking biomolecules to the particle surface via amide, ester, ether, or sulfide bonds. Because of the large variety of reactive groups found on organic nanoparticles made of proteins, polysaccharides, lipids, or polymers, the well-defined methods of chemical modification are often not position-specific and therefore lack orientation. In some cases, the reactive groups are located close to the active centre, affecting activity and function.

Affinity-based interactions: Streptavidin–biotin interactions are widely used for biotechnology applications, in particular immunoassays. Although biotin–streptavidin binding is not covalent, its high-affinity constant produces a highly specific and nearly irreversible immobilization. The streptavidin–biotin interaction is the strongest noncovalent receptor–ligand interaction ($K_a = 10^{15}\ M^{-1}$) currently known, with a stronger affinity than in any known antigen–antibody interaction.[73] Bond formation between biotin and streptavidin is very rapid, and once formed, is unaffected by most extremes of pH, temperature, organic solvents, and other denaturing agents. Streptavidin contains four subunits, each with a single biotin-binding site, allowing signal amplification.

A variety of biomolecules can be biotinylated and subsequently used with streptavidin labeled probes.[74] Biotinylated antibodies, which serve as recognition agents, are used in a variety of different assays, and therefore, many conjugates are now commercially available. As such, streptavidin linkage is a very convenient and efficient conjugation method to attach recognition agents onto nanoparticles.

An easy way of immobilizing streptavidin onto a surface is based on electrostatic interaction. The positively charged streptavidin naturally adsorbs to a negatively charged surface and can then be stabilized by cross-linking using a bifunctional reagent such as glutaraldehyde. Resulting streptavidin-coated particles can then be functionalized by conjugation to biotinylated recognition molecules.

An interesting alternative to the streptavidin–biotin system called "protein-assisted nanoassembler" has been recently described.[75] This approach allows the robust self-assembly of multifunctional superstructures consisting of different single-function particles such as labels, carriers, recognition, and targeting agents, including antibodies. This bioengineering method employs two unique proteins from *Bacillus amyloliquefaciens*, barnase and barstar, to rapidly bring together the structural components directly in solution.[76] These two proteins establish a very specific and strong noncovalent interaction as their binding constant is $K_a \sim 10^{14}\ M^{-1}$. That is only slightly smaller than the binding constant of streptavidin and biotin ($10^{15}\ M^{-1}$) but is much stronger than most antibody–antigen binding constants, which normally range from 10^8 to 10^{11} M^{-1}. The properties of the superstructures can be designed on demand by linking different agents of various sizes and chemical nature, as a function of the specific purpose. It has been demonstrated that using barnase and barstar, it is possible to assemble colloidally stable trifunctional structures by binding together magnetic particles, quantum dots, and antibodies.[75] Indeed the bonds between these proteins are strong enough to hold macroscopic (5 nm–3 μm) particles together. Specific interaction of such superstructures with cancer cells resulted in fluorescent labeling of the cells and their responsiveness to a magnetic field. This very recent and versatile method can be used for multiple nanotechnology applications including immunoassays.

Encapsulation of labels: In addition to adsorption and covalent linkage, the physical encapsulation of molecules, that is, entrapment within biodegradable polymers such as lipid vesicles may be a very interesting technique for certain biotechnological, bioanalytical, and biomedical applications.[77]

3.2.2. Bioanalytical Applications

Numerous efforts have been made to optimize antibody labeling in order to further improve the performance of immunoassays. In a careful study, Lee *et al.*[78] have described the optimization of biotin labeling of a mouse IgG by varying the classical parameters of the labeling protocol. The immobilization of biotin-tagged mouse IgGs on avidin-coated plates was then investigated by incubating the bound antibodies with goat anti-mouse IgGs linked to fluorescent beads. The optimum conditions were successfully applied in sandwich immunoassays for two different analytes resulting in the detection of as little as 2 and 5 ng of troponin I and N-terminal probrain natriuretic peptide (BNP), respectively. In an interesting but still little used approach, Simons *et al.* described the covalent *in vacuo* cross-linking of HRP to anti-rabbit immunoglobulin G (IgG).[79] The advantageous feature of this co-lyophilization-based procedure is that the cross-linking to form Enz-Ab conjugates is accomplished without the use of chemical modifying or activating reagents, reducing the potential activity loss due to chemical modification. The resulting soluble multi-enzyme-IgG immunoconjugate exhibited a 100-fold increased sensitivity for antigen detection compared to a commercial conjugate prepared by conventional chemical cross-linking methods.

The use of nanoparticles provides an interesting and powerful avenue to achieve significant signal amplification by allowing the linkage of a single antibody molecule with up to thousands of reporter molecules such as fluorophores (Figure 8.1A) or enzymes (Figure 8.1B). In an interesting example of this strategy, a polypeptide containing 20 lysine residues each conjugated to a HRP molecule was attached to 0.44 μm streptavidin polystyrene spherical nanoparticles introducing hundreds of HRP molecules and making a signal amplifying detector conjugate (Figure 8.1B).[80] Such highly labeled nanoparticles were further functionalized with IgG molecules to achieve a molar ratio of 1 IgG to 10^5 HRP complexes. These immunoconjugates efficiently bound to plasma anti-HIV-1 antibodies that had been captured by HIV antigens

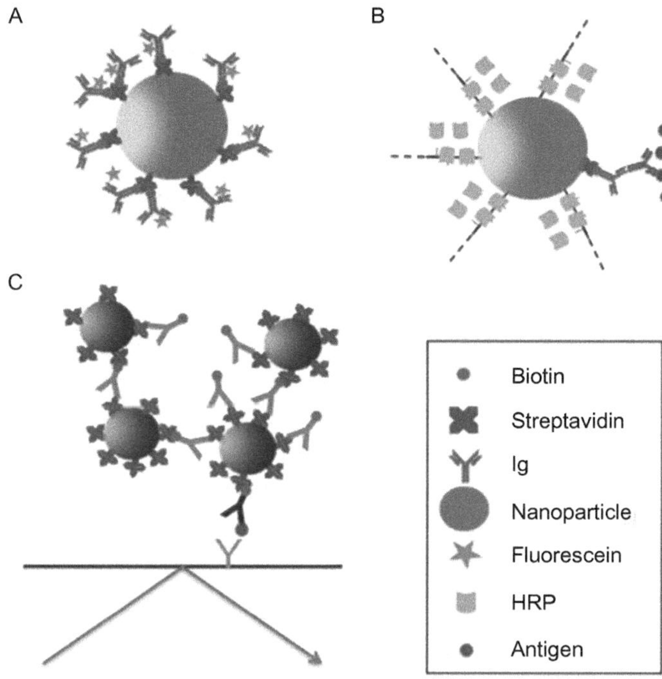

FIGURE 8.1 Examples of signal amplification using nanoparticle technology. (A) Fluorescently labeled antibodies, (B) enzyme conjugate loaded nanoparticles,[80] (C) with no need of label, the immunoconjugates formed are amplified by a nanoparticle network and detected by surface plasmon resonance.[81] (For color version of this figure, the reader is referred to the online version of this chapter.)

on 5 μm carboxyl magnetic microparticles and produced a detection signal with five to eight times more sensitivity as compared to conventional HRP-conjugated goat anti-human IgG.[80]

Similarly, sulfate functionalized polystyrene nanoparticles were used to electrostatically attract positively charged antibodies. Protein immobilization maintained the Y-shaped orientation of the molecules as well as their immunological activity, optimizing the sensitivity of the immunosensor. Such complexes tested in sandwich immunoassays conducted to detect cardiac troponin I showed a fivefold higher activity over the control.[82]

The interaction between antigen–antibody complexes has also been detected by surface plasmon resonance (SPR)-based sensing systems.[83] The principle is based on the detection of a small refractive index change on a thin gold film caused by the complex formation between biomolecules.[84] As shown on Figure 8.2C, streptavidin-conjugated dextran beads of 50 nm were used to develop an SPR-based highly sensing system for the detection of BNP.[81] By this method, the SPR signals were highly intensified leading to an in increase in the detection of BNP from nanogram to picogram levels.

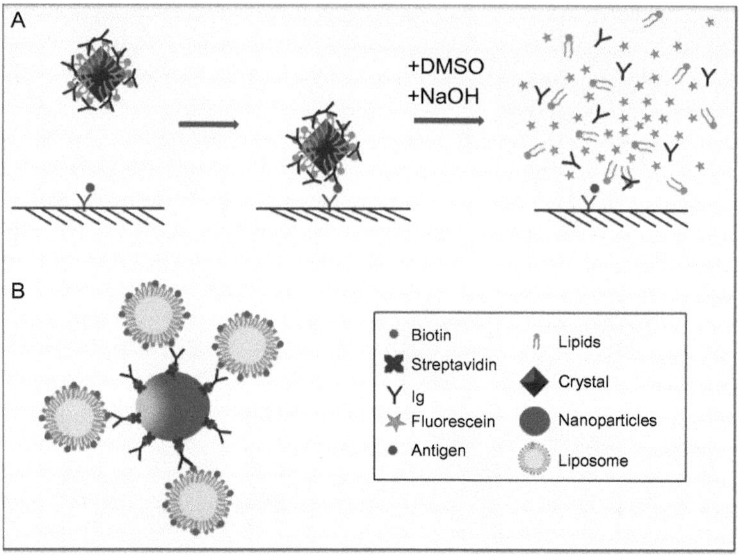

FIGURE 8.2 High load systems for signal amplification. (A) IgGs are labeled with fluorescein diacetate (FDA) nanocrystals, dissolved after formation of the immunocomplex,[87] (B) biotin-tagged, carboxyfluorescein-encapsulated liposomes used as label.[88] (For color version of this figure, the reader is referred to the online version of this chapter.)

Recently, numerous attempts have been made to improve the performance of fluorescence-based immunoassays as well. The sensitivity of these assays is generally limited by the ratio of label (fluorescent or enzyme) molecules per biomolecule (L/P ratio). The L/P ratio is typically 4–8 for a conventional, covalently coupled fluorescent immunolabel, for example, an IgG labeled with fluorescein isothiocyanate (IgG-FITC) conjugate. A higher L/P may lead to a decrease of the specific binding affinity of the biomolecule, and additionally cause self-quenching effects. Increasing the effective dye/biomolecule ratio while minimizing dye self-quenching and maintaining the biomolecule's binding properties is thus an important goal in assay development. Several ways to increase the F/P ratio have been investigated, in particular those which are based on coprecipitation or self-assembly without the formation of covalent bonds. One approach has been the substitution of labeling molecules by micro- or nanocrystalline dyes. For example, perylene, a fluorescent polycyclic aromatic hydrocarbon consisting of two molecules of naphthalene that have been fused together has been used as label. A higher F/P ratio ($\sim$1400) can be obtained by precipitating fluorescent perylene microparticles in the presence of the antibodies.[85] After the immunoreaction, the large number of fluorescent molecules contained in these particles can then be dissolved in a suitable solvent for detection. An analogous route reported by Trau *et al.*[86] was to link antibodies to polyelectrolyte encapsulated microcrystalline fluorescent material. The surface of these particulate structures is typically engineered by the layer-wise assembly of oppositely charged polyelectrolytes, the outer layer consisting of biorecognition molecules, for example, immunoglobulins. Because of the exceptionally high F/P ratio of the detection antibodies (~51,000–190,000), a dramatically (70–2000-fold) amplified immunoassay was achieved. Despite the advantages of these two methods, a key limitation lies in the restricted number of materials that can be precipitated or crystallized for encapsulation. Given the limitations of existing fluorescence-based biochemical assays, the development of new strategies and biolabeling systems will be necessary.

A novel signal amplification technology based on a new class of biofunctional fluorescent nanocrystals has been developed, consisting of a two-step approach to encapsulate the fluorogenic precursor fluorescein diacetate (FDA) nanocrystals (107-nm average size) followed by conjugation of the antibody.[87] Distearoylphosphatidylethanolamine modified with PEG(2000) Amine is coated on the surface of the FDA nanocrystals to provide an interface for antibody coupling (Figure 8.2A). Anti-mouse antibodies are subsequently attached to the nanocrystalline FDA biolabels by adsorption.

A high molar ratio of fluorescent molecules to biomolecules (2.8×10^4) is achieved in this nanocrystal biolabel system. The analytical performance of the nanocrystal-based label system has been evaluated in a model sandwich immunoassay for the detection of mouse IgG. The limit of detection was lowered by a factor of 5–28, and the sensitivity was 400–2700-fold higher compared with a state-of-the-art immunoassay using directly fluorescent-labeled antibodies. The high sensitivity of this assay may be explained by the boosting effect of the high ratio of dye/antibody, but washing out the dye molecules after the affinity reaction may also contribute to improve the signal, the mean dye-to-dye distances being large enough to diminish the quenching effects.

Other high-load systems have been successfully designed and used as immunolabels such as fluorescent conjugated dendrimers, fluorophore-loaded latex beads,[89] and liposome encapsulated fluorophores.[77,90] For example, liposomes used in immunodetection and drug delivery systems can be tagged with antibodies to form immunoliposomes. Numerous procedures for the conjugation of antibodies to liposomes have been developed, falling into four general categories defined by the particular functionality of the antibody being modified, namely, amine modification, carbohydrate modification, disulfide modification, and non-covalent conjugation. Interestingly, Ho *et al.*[88] have inserted a monosialoganglioside (GM1), which exhibits specific affinity toward cholera toxin (CT), into the phospholipid bilayer during the liposome synthesis. These GM1-sensitized, sulforhodamine B dye entrapping liposomes were then used for the determination of CT. Subsequently, this group reported the successful preparation of biotin-tagged, carboxyfluorescein-encapsulated liposomes by using the reversed-phase evaporation method from a lipid mixture containing biotin-X-DHPE (Figure 8.2B).[77] Such liposomes have been successfully used to improve biosensing systems. Carboxy-encapsulated fluorescein biotin-tagged liposomes were used as a novel alternative analytical method for the detection of biotin to levels as low as 5.0 pg of biotin[77] and CT down to ≈1.1 zmol.[88] As suggested by the authors, this method may be useful in the monitoring and surveillance of CT in water and food supplies to ensure the health of humans.

4. NANOPARTICLES AND THE FUNDAMENTAL STUDY OF CELL ADHESION MECHANISMS

4.1. Adhesion Basics

Adhesion is a highly important and fundamental phenomenon in biology. Living cells are endowed with different receptors expressed at the plasma membrane that allow the continuous perception of the extracellular environment. These ubiquitously present receptors are quite diverse in function and include, in particular, receptors that anchor cells to the extracellular matrix (integrins) or those involved in cell–cell interactions (such as selectins or cadherins).

The study of cell adhesion has to be tackled using multiple approaches, from molecular, developmental, or cell biology to biophysics. The field of the study of cell adhesion is thus, by nature, multidisciplinary, involving a large number of research groups producing an ever-increasing number of publications as our understanding of this complex process grows. We have chosen to limit this review to cadherin-mediated interactions and, through some examples, show how liposomes and organic nanobeads can help to decipher the biological mechanism of these molecules.

Intercellular contacts, created by morphologically distinct structures, are made of the clustering of cell-surface transmembrane adhesive receptors into multi-protein assemblies, or junctions, connected to the cytoskeleton and intracellular signaling pathways. Numerous molecules are involved which regulate various mechanisms such as differentiation, migration, or apoptosis.

Adherens junctions consist of complex intercellular structures formed by localized clusters of *trans* dimers between classical cadherins from apposed cells.[91] Cadherins are single-pass transmembrane glycoproteins and signal transducing molecules involved in Ca^{2+}-dependent homophilic cell–cell adhesion.[92] During development, cadherins contribute to the regulation of a large number of processes including tissue morphogenesis such as mesenchymal–epithelial transition, cell sorting and tissue rearrangement through convergence-extension, neurite elongation, and synaptogenesis.[93] In addition, deregulation of cadherin-mediated adhesion has been associated with alterations of tissue homeostasis.[94] Thus, cadherins are key morphoregulatory molecules in developmental processes, as well as essential contributors to cell–cell cohesion within adult tissues and organs (see for reviews Ref. 95). A comprehensive view of cadherin recruitment and dynamics at cell–cell contacts and its regulation is of major importance for the understanding of the control of cell fate in normal as well as pathological situations, such as carcinogenesis.

Classical cadherins (type I and type II) consist of an extracellular segment typically containing five tandem repeats of an approximately 110 amino acid module (numbered EC1 to EC5 from the outermost domain), a transmembrane region and a highly conserved cytoplasmic domain.[93a] Cadherin engagement triggers a series of still only partially understood intracellular signaling events that lead to the reorganization of the actin cytoskeleton via cytoplasmic proteins such as catenins, plakoglobin, and p-120 that, in turn, trigger changes in cell morphology and motility.[91a,96] The formation of adherens junctions is likely to represent the first step in this signaling cascade.

Despite detailed studies of cadherin-mediated adhesion in multicellular organisms, the molecular understanding of the adhesive states of cadherin is less clear.[97] Cadherin expression is cell type or tissue specific, and a cell type may express more than one type of cadherin. Cells expressing cadherins sort out and aggregate only with cells expressing identical cadherins. This is the basis of tissue patterning and architecture in both cell to cell contact and cell migration.[98] How these molecules interact with each other and the mechanisms by which they transfer specific intracellular signals remain poorly understood.

Structural studies have shown that cadherin–cadherin contacts are mediated by the cadherin extracellular domain. A detailed structural description of adherens junctions is emerging from the elucidation of the structure of individual molecular partners.[99] However, despite numerous studies over the past 25 years, the details of *trans* dimerization are still under debate. The crystal structures of the entire (EC15) ectodomain from classical type I C-cadherin[99b] and more recently E- and N-cadherins[99d] reveal a "strand swap" *trans* interface in which the N-terminal β-strand from the EC1 domain of each paired cadherin exchanges with that of the partner molecule. A second functionally important *trans* interface, involving the linker region between the EC1 and EC2 domains, has also been identified and constitutes a kinetic intermediate on the path to the formation of strand swapped dimers.[99d] However, a widely accepted model as summarized in recent review articles[100] is that the functional unit of cadherin adhesion is a *cis* dimer formed by binding of the extracellular domains of two cadherins on the same cell surface. The interplay between *trans* binding and lateral *cis* interactions among proteins on the same membrane theoretically play a crucial role in the clustering of cadherins into junctions, but evidence for these *trans*- and *cis*-cadherin binding states remains controversial.

Unlike the contradictory data on *cis* dimers, there is a better, albeit incomplete, understanding of how *trans* interactions occur between cadherins on opposing cell surfaces. Numerous data have implicated the EC1 domain in cadherin *trans* interactions. However, direct force measurements supported by cell attachment and bead-binding assays revealed that strong cadherin adhesion may also require *trans* binding along additional EC domains.[101]

These data raise questions about the EC domains that participate in *trans* cadherin binding.

4.2. Liposomes as Biomimetic Cell Membranes

Little is known about the mode of cadherin assembly in adherens junctions. Electron tomography has been used to determine the 3D structure of intact desmosomes *in situ*.[102] Desmosomes are adhesive intercellular structures also mediated by interactions between the ectodomains of specialized cadherins, desmocollins, and desmogleins.[103] The known determinants of strand swapping in type I cadherins are conserved in desmosomal cadherins,[104] suggesting a similar mode for *trans* binding between their distal membrane EC1 domains. He *et al.*[105] performed a 3D electron tomography study of dehydrated and plastic-embedded sections of desmosomes and concluded that cadherin molecules were disordered and associated into discrete groups. Then, by cryo-EM of frozen sections, Dubochet and his coworkers imaged desmosomes in their hydrated state.[106] Although desmosomal electron tomography experiments have shown *trans* interactions involving both desmosomal cadherin monomers and *cis* dimers,[105] cryoelectron tomographs suggest that *cis* dimerization is mandatory for *trans* adhesion.[106] No such studies have been carried out on adherens junctions to date.

Adherens junctions have been artificially reconstituted *in vitro* by assembly of cadherin fragments at the surfaces of liposomes and their organization studied by electron microscopy. The liposomes were prepared using phospholipid mixtures that included Ni-$^{2+}$-chelated head groups to bind C-terminally 6$\times$ Histidine-tagged E-cadherin ectodomains on the surface. Immobilized metal chelators, such as nitrilotriacetic acid (NTA), are routinely used for purifying recombinant proteins bearing polyHis tags by metal-ion affinity chromatography. These metal-chelating head groups covalently linked to acyl-like chains or lipids can be used to anchor suitably tagged molecules onto planar lipid bilayer membranes as well as liposomes.[107] Liposomes coated with vascular endothelium cadherin (VE-cadherin) recombinant fragments consisting of four domains (VE-EC14)[108] and with E-cadherin ectodomains (E/EC15)[99d] exhibited junction-like structures in the space between apposed liposomes. The VE-cadherin reconstituted adherens junctions are formed by 2D self-assembly of VE-EC14 hexamers. The inherent symmetry of the junctions implies that each hexamer is consists of three antiparallel cadherin dimers. The hexameric organization in VE-cadherin-mediated junctions may be specific to this cadherin isoform or to the fact that these recombinant fragments comprise four modules and are not glycosylated. Adhesion between liposomes coated with E-cadherin ectodomains is mediated by an ordered structure that is consistent with the lattice structure observed in the cadherin crystals.[99b] Even though *trans*-cadherin binding is a low-affinity interaction, cell–cell adhesion is believed to be enhanced cooperatively by the lateral

FIGURE 8.3 Electron microscopy of artificial junctions between E/EC12 decorated liposomes. Image of frozen-hydrated liposomes decorated with EC1-2 fragments forming artificial junctional structures. Scale bar: 100 nm.

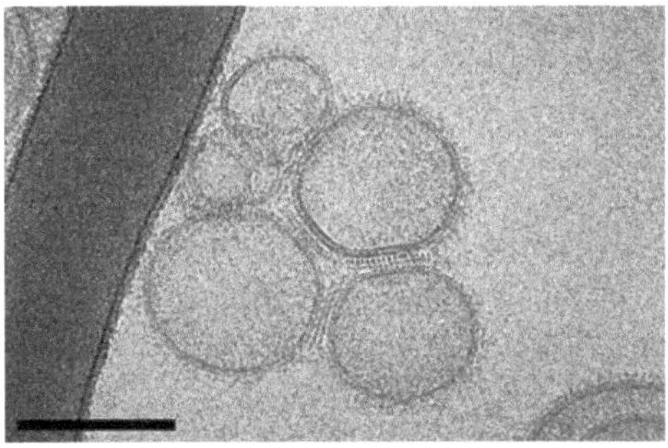

clustering of cadherins, it should be underlined that adhesion also occurs when the *cis* interface is ablated.[99d]

The two outermost modules of E-cadherin (E/EC12) have been shown to represent the minimal segment able to realize the initial events in the E-cadherin recognition step.[101c,109] They have been used to functionalize giant unilamellar vesicles (GUVs) and supported bilayers through the hexahistidine-NTA interaction.[110] GUVs are spherical shells which because of their size (5–200 mm in diameter) can be seen under the light microscope,[111] and various morphological changes such as cytomimetic processes occurring upon adhesion can be viewed and easily monitored.[112] Adhesive foci form between these two surfaces in a specific and calcium-dependant manner and have been quantified.[110]

Interestingly, liposomes coated with wild-type E/EC12 fragments also reveal junction-like structures in the space between apposed liposomes (Figure 8.3; unpublished results). The mechanical strength resulting from the assembly of junction-like structures by cadherin fragments consisting of as little as two modules appears to be sufficient to induce the flattening of the membrane surface of liposomes reminiscent of adherens junction morphology (Patrick Bron and Hélène Feracci, unpublished results).

Altogether, these studies indicate that the information required to build up an adherens junction is contained within the extracellular moiety of cadherin molecules. Still many questions remain open and further studies using this model system will help *in* understanding cadherin structure/function relationships in the formation of cadherin-mediated junctional contacts.

4.3. Polymeric Beads

4.3.1. Bead–Bead and Bead–Cell Assays

Cell aggregation assays were often used in early studies to evaluate the role of cadherins on cell–cell adhesion.[98b] Such studies are challenging in live cells and this approach provides mostly qualitative data as it is difficult to take into account the number of cadherin molecules involved. Moreover, cadherins are multimodular and involved in complex multimolecular structures making it difficult to elucidate their biological properties.

For this reason, glass and polymeric nanoparticles have been often used in standard assays to assess the biological activity of cadherin recombinant fragments.[101b,109a,113] Fluorescent protein A coated 0.9 μM polystyrene beads functionalized with chimeric C-cadherin fragments,[101b] and streptavidin-coated 2.8 μm polystyrene beads decorated with E-cadherin fragments[109a] were used to address the involvement of the different ectodomain modules in bead aggregation assays.

Bead–cell binding assays have also been used to dissect the molecular mechanisms of signal transduction.[113] N (neural)-cadherin chimera loaded latex beads of 6 μm diameter self aggregate and specifically bind, in a Ca^{2+}-dependent manner, to N-cadherin-expressing cells. N-cadherin full-length chimera coated beads fully mimic cadherin-mediated cell–cell interactions, inducing the accumulation of N-cadherin, catenins, and F-actin as well as membrane remodeling at the bead–cell contact. Streptavidin spherical polystyrene beads (2.8 μm in diameter) decorated with E/EC12 have been shown to efficiently interact with HC11 cells and activate membrane dynamics events (Figure 8.4). Indeed, subsequent to these contacts, the beads were rapidly engulfed by the E-cadherin expressing cells,[114] and this internalization appears to be highly specific and sensitive to point mutations (Figure 8.4; Sébastien Chevalier, Elisabeth Sellier, Helene Feracci, unpublished results).

4.3.2. Single Molecule Assays

Most approaches used to study cadherin–cadherin interactions provide information from the behavior of multimolecular systems with an often incompletely defined geometrical organization. Many contradictory results and unanswered questions suggest that it would be a hopeless task to derive clear molecular properties from these data. An understanding of the intrinsic kinetic properties of cadherin interactions requires the measurement of parameters at the single

FIGURE 8.4 Scanning electron microscopy of a bead/cell interaction. HC11 cells expressing E-cadherin at their plasma membrane rapidly internalize E/EC12 loaded polystyrene beads.

molecule level. During the past 15 years, new biophysical approaches have been developed that allow the study of ligand–receptor interactions with unprecedented accuracy, down to the single bond level. Reported results include information on bond mechanical properties, association behavior of surface-attached molecules, and the dissection of energy landscapes and reaction pathways. Indeed, monitoring single bond formation and dissociation has made it possible to bypass difficult problems such as force sharing between multiple bonds or assessing the effect of partial geometrical matches on the kinetics of bond formation. Several of these methodologies have been used to study cadherin interactions at the molecular level. A surface force apparatus has been used to investigate mechanisms of cadherin binding by measuring the force/distance between cadherin ectodomains.[101a] Single molecule atomic force microscopy (AFM) was used to study the mechanical resistance of cad herin interactions.[115] However, interpretation of the data was not straightforward: the relationship between the unbinding force, as measured by AFM, and the dissociation rate is complex and dependent on the cantilever stiffness and rate of sample displacement.[116] At the present time, these experiments have yet to give us a comprehensive view of the mechanisms underlying cadherin interactions.

Two approaches, the flow chamber[117] and the biomembrane force probe[116] (BFP) offer a way to bring into contact two surfaces covered with molecules and a means of measuring the duration of the interaction as well as the rupture force under stress. The molecules to be tested are linked in the case of the BFP approach onto two micron-sized glass beads, whereas in the flow chamber approach, they are linked onto a polystyrene bead and a flat surface. These methodologies allow (i) the analysis of the association of surface-attached rather than soluble molecules reproducing a molecular orientation relevant to physiological conditions, (ii) highly sensitive detection of molecular interactions, and (iii) the kinetic study of bond formation and dissociation, under stress. These two approaches provided the first quantitative data describing the dissociation kinetics of individual E-cadherin[101c,109a,118] and C-cadherin[119] interactions.

5. CONCLUSION

Organic nanoparticles already play an important role in many biotechnology applications and promise to take center stage for many new and emerging applications in the coming years. Interesting future developments include not only biomedical applications such as improved delivery of drugs to tumor cells and the use of dendrimers for regenerative medicine but also fields such as water purification and disinfection, food production, and packaging.

For example, it may be possible to circumvent the anti-apoptotic mechanism of the Bcl-2/Bax system, which contributes to the survival and drug resistance of tumoral cells, by introducing proapoptotic molecules into the cells using nanoparticles, aiding in the escape from lysosomes and thus sensitizing cancer cells to the induction of apoptosis by several therapeutic agents. As an example, lipid nanocapsules used as nanocarriers have been shown to improve the activity of a number of anticancer hydrophobic compounds and to transport drugs in the close vicinity to mitochondria.[120] The capability of obtaining nanocarriers that recognize mitochondria is a major improvement to introduce into cell drugs which have subcellular compartment as a target.

In addition, dendrimers may be useful in the field of regenerative medicine. The use of dendrimers as support biomaterial for the generation of new tissue may play an increasingly important role, whether for the formation of scaffolds to allow for cell growth or the generation of bioartificial tissues.[121]

Organic nanoparticles may also play an increasingly important role in water treatment and disinfection. Although disinfection methods currently used in drinking water treatment can effectively control microbial pathogens, research in the past few decades has revealed a dilemma between effective disinfection and formation of harmful disinfection by products, many of which are carcinogens. Nanomaterials are excellent adsorbents, catalysts, and sensors due to their large specific surface area and high reactivity. Interestingly, naturally occurring chitin and certain peptides have been long recognized for their antimicrobial properties toward bacteria, viruses, and fungi. Chitosan prevents the multiplication of bacteriophages in bacteria and induces resistance toward viral diseases in plants.[122] Recently, these materials have been engineered into nanoparticles.[42] Nanoscale chitosan has potential drinking water disinfection applications as an antimicrobial agent in membranes, sponges, or surface coatings of water storage tanks.

Nanoparticles are now also starting to be used for all stages of food production. Nanotechnology is being applied in areas such as food security and quality, development of new products and packaging. Reducing the particle size of materials to that of NPs has been used to enhance emulsification, gelation, foaming, water-binding capacity. Nanoparticles are also being used to deliver vitamins or other nutrients in food and beverages without affecting the taste or appearance. These nanoparticles actually encapsulate the nutrients and carry them through the stomach into the bloodstream.[123] Nanopackaging has also started to impact food storage by increasing the shelf life of fresh foods and carbonated beverages, and many other food packaging products are currently under development.[124]

As the number of biotechnological applications of organic nanoparticles continues to expand, the use of this technology promises to play an increasingly important role in the biomedical and industrial arenas, affecting the lives of all of us.

ACKNOWLEDGMENTS

We gratefully acknowledge the financial support of the research grants from the SUDOE TRAIN2 program, CNano GSO, Programme Hubert Curien Picasso, Région Aquitaine, and Fondation pour la Recherche Médicale. B. S. G. acknowledges support of a Sara Borrell fellowship. The authors would like to thank Audrey Belloc and J. Paul Salvetat for their help with the chapter.

REFERENCES

1. Roth CM, Sundaram S. Engineering synthetic vectors for improved DNA delivery: insights from intracellular pathways. *Annu Rev Biomed Eng* 2004;**6**:397–426.
2. (a) Sundar S, Kundu J, Kundu SC. Biopolymeric nanoparticles. *Sci Technol Adv Mater* 2010;**11**(1):014104–17. (b) Nayak S, Lyon LA. Soft nanotechnology with soft nanoparticles. *Angew Chem Int Ed* 2005;**44**(47):7686–708.
3. (a) Li S, Huang L. Nonviral gene therapy: promises and challenges. *Gene Ther* 2000;**7** (1):31–4. (b) Mansouri S, Lavigne P, Corsi K, Benderdour M, Beaumont E, Fernandes JC. Chitosan-DNA nanoparticles as non-viral vectors in gene therapy: strategies to improve transfection efficacy. *Eur J Pharm Biopharm* 2004;**57**(1):1–8. (c) De Laporte L, Rea JC, Shea LD. Design of modular non-viral gene therapy vectors. *Biomaterials* 2006;**27**(7): 947–54.
4. Panyam J, Labhasetwar V. Biodegradable nanoparticles for drug and gene delivery to cells and tissue. *Adv Drug Deliv Rev* 2003;**55**(3):329–47.
5. Ziauddin J, Sabatini DM. Microarrays of cells expressing defined cDNAs. *Nature* 2001;**411** (6833):107–10.
6. (a) Hyndman L, Lemoine JL, Huang L, Porteous DJ, Boyd AC, Nan XS. HIV-1 Tat protein transduction domain peptide facilitates gene transfer in combination with cationic liposomes. *J Control Release* 2004;**99**(3):435–44. (b) Tokunaga M, Hazemoto N, Yotsuyanagi T. Effect of oligopeptides on gene expression: comparison of DNA/peptide and DNA/peptide/liposome complexes. *Int J Pharm* 2004;**269**(1):71–80. (c) del Pozo-Rodriguez A, Pujals S, Delgado D, Solinis MA, Gascon AR, Giralt E, Pedraz JL. A proline-rich peptide improves cell transfection of solid lipid nanoparticle-based non-viral vectors. *J Control Release* 2009;**133**(1):52–9.
7. (a) Kim SH, Jeong JH, Joe CO, Park TG. Folate receptor mediated intracellular protein delivery using PLL-PEG-FOL conjugate. *J Control Release* 2005;**103**(3):625–34. (b) Ogris M, Brunner S, Schuller S, Kircheis R, Wagner E. PEGylated DNA/transferrin-PEI complexes: reduced interaction with blood components, extended circulation in blood and potential for systemic gene delivery. *Gene Ther* 1999;**6**(4):595–605. (c) Kunath K, Merdan T, Hegener O, Haberlein H, Kissel T. Integrin targeting using RGD-PEI conjugates for in vitro gene transfer. *J Gene Med* 2003;**5**(7):588–99. (d) Blessing T, Kursa M, Holzhauser R, Kircheis R, Wagner E. Different strategies for formation of PEGylated EGF-conjugated PEI/DNA complexes for targeted gene delivery. *Bioconjug Chem* 2001;**12**(4):529–37. (e) Kircheis R, Kichler A, Wallner G, Kursa M, Ogris M, Felzmann T, Buchberger M, Wagner E. Coupling of cell-binding ligands to polyethylenimine for targeted gene delivery. *Gene Ther* 1997;**4**(5):409–18.
8. Garnett MC. Gene-delivery systems using cationic polymers. *Crit Rev Ther Drug Carrier Syst* 1999;**16**(2):147–207.

9. Schatz C, Lecommandoux S. Polysaccharide-containing block copolymers: synthesis, properties and applications of an emerging family of glycoconjugates. *Macromol Rapid Commun* 2010;**31**(19):1664–84.
10. Hirko A, Tang FX, Hughes JA. Cationic lipid vectors for plasmid DNA delivery. *Curr Med Chem* 2003;**10**(14):1185–93.
11. Zabner J, Fasbender AJ, Moninger T, Poellinger KA, Welsh MJ. Cellular and molecular barriers to gene-transfer by a cationic lipid. *J Biol Chem* 1995;**270**(32):18997–9007.
12. (a) Boussif O, Lezoualch F, Zanta MA, Mergny MD, Scherman D, Demeneix B, Behr JP. A versatile vector for gene and oligonucleotide transfer into cells in culture and in-vivo—polyethylenimine. *Proc Natl Acad Sci U S A* 1995;**92**(16):7297–301. (b) Braun CS, Vetro JA, Tomalia DA, Koe GS, Koe JG, Middaugh CR. Structure/function relationships of polyamidoamine/DNA dendrimers as gene delivery vehicles. *J Pharm Sci* 2005;**94**(2):423–36. (c) Wang J, Zhang PC, Mao HQ, Leong KW. Enhanced gene expression in mouse muscle by sustained release of plasmid DNA using PPE-EA as a carrier. *Gene Ther* 2002;**9**(18):1254–61.
13. Kabanov AV, Felgner PL, Seymour LW. *Self-assembling complexes for gene delivery: from laboratory to clinical trial*. Chichester, NY: Wiley; 1998, p xvii, 442 pp.
14. Gao X, Huang L. Potentiation of cationic liposome-mediated gene delivery by polycations. *Biochemistry* 1996;**35**(3):1027–36.
15. (a) Neu M, Fischer D, Kissel T. Recent advances in rational gene transfer vector design based on poly(ethylene imine) and its derivatives. *J Gene Med* 2005;**7**(8):992–1009. (b) Schaffer DV, Fidelman NA, Dan N, Lauffenburger DA. Vector unpacking as a potential barrier for receptor-mediated polyplex gene delivery. *Biotechnol Bioeng* 2000;**67**(5):598–606.
16. Lomas H, Johnston APR, Such GK, Zhu Z, Liang K, van Koeverden MP, Alongkornchotikul S, Caruso F. Polymersome-loaded capsules for controlled release of DNA. *Small* 2011;**7**(14):2109–19.
17. Agarwal A, Unfer R, Mallapragada SK. Novel cationic pentablock copolymers as non-viral vectors for gene therapy. *J Control Release* 2005;**103**(1):245–58.
18. Merdan T, Kopecek J, Kissel T. Prospects for cationic polymers in gene and oligonucleotide therapy against cancer. *Adv Drug Deliv Rev* 2002;**54**(5):715–58.
19. Xu J, Ganesh S, Amiji M. Non-condensing polymeric nanoparticles for targeted gene and siRNA delivery. *Int J Pharm* 2012;**427**(1):21–34.
20. (a) Kommareddy S, Amiji M. Preparation and evaluation of thiol-modified gelatin nanoparticles for intracellular DNA delivery in response to glutathione. *Bioconjug Chem* 2005;**16**(6):1423–32. (b) Lemieux P, Guerin N, Paradis G, Proulx R, Chistyakova L, Kabanov A, Alakhov V. A combination of poloxamers increases gene expression of plasmid DNA in skeletal muscle. *Gene Ther* 2000;**7**(11):986–91.
21. (a) Bhavsar MD, Amiji MM. Polymeric nano- and microparticle technologies for oral gene delivery. *Expert Opin Drug Deliv* 2007;**4**(3):197–213. (b) Morille M, Passirani C, Vonarbourg A, Clavreul A, Benoit J-P. Progress in developing cationic vectors for non-viral systemic gene therapy against cancer. *Biomaterials* 2008;**29**(24–25):3477–96.
22. Tabatt K, Kneuer C, Sameti M, Olbrich C, Muller RH, Lehr CM, Bakowsky U. Transfection with different colloidal systems: comparison of solid lipid nanoparticles and liposomes. *J Control Release* 2004;**97**(2):321–32.
23. (a) Schwarz C, Mehnert W, Lucks JS, Muller RH. Solid lipid nanoparticles (SLN) for controlled drug-delivery. 1. Production, characterization and sterilization. *J Control Release* 1994;**30**(1):83–96. (b) Muller RH, Mader K, Gohla S. Solid lipid nanoparticles (SLN) for controlled drug delivery—a review of the state of the art. *Eur J Pharm Biopharm* 2000;**50**(1):161–77. (c) Vighi E, Ruozi B, Montanari M, Battini R, Leo E. Re-dispersible cationic

solid lipid nanoparticles (SLNs) freeze-dried without cryoprotectors: characterization and ability to bind the pEGFP-plasmid. *Eur J Pharm Biopharm* 2007;**67**(2):320–8.
24. Medvedeva DA, Maslov MA, Serikov RN, Morozova NG, Serebrenikova GA, Sheglov DV, Latyshev AV, Vlassov VV, Zenkova MA. Novel cholesterol-based cationic lipids for gene delivery. *J Med Chem* 2009;**52**(21):6558–68.
25. Sheng R, Luo T, Zhu Y, Li H, Sun J, Chen S, Sun W, Cao A. The intracellular plasmid DNA localization of cationic reducible cholesterol-disulfide lipids. *Biomaterials* 2011;**32**(13): 3507–19.
26. Vonarbourg A, Passirani C, Desigaux L, Allard E, Saulnier P, Lambert O, Benoit J-P, Pitard B. The encapsulation of DNA molecules within biomimetic lipid nanocapsules. *Biomaterials* 2009;**30**(18):3197–204.
27. (a) Birrenbach G, Speiser PP. Polymerized micelles and their use as adjuvants in immunology. *J Pharm Sci* 1976;**65**(12):1763–6. (b) Kreuter J, Speiser PP. New adjuvants on a polymethylmethacrylate base. *Infect Immun* 1976;**13**(1):204–10.
28. (a) Couvreur P, Kante B, Roland M, Guiot P, Bauduin P, Speiser P. Polycyanoacrylate nanocapsules as potential lysosomotropic carriers—preparation, morphological and sorptive properties. *J Pharm Pharmacol* 1979;**31**(5):331–2. (b) Gurny R, Peppas NA, Harrington DD, Banker GS. Development of biodegradable and injectable lattices for controlled release of potent drugs. *Drug Dev Ind Pharm* 1981;**7**(1):1–25. (c) Allemann E, Leroux JC, Gurny R, Doelker E. In-vitro extended-release properties of drug-loaded poly (DL-lactic acid) nanoparticles produced by a salting-out procedure. *Pharm Res* 1993;**10** (12):1732–7.
29. (a) Mathiowitz E, Jacob JS, Jong YS, Carino GP, Chickering DE, Chaturvedi P, Santos CA, Vijayaraghavan K, Montgomery S, Bassett M, Morrell C. Biologically erodable microsphere as potential oral drug delivery system. *Nature* 1997;**386**(6623):410–4. (b) Couvreur P, Vauthier C. Polyalkylcyanoacrylate nanoparticles as drug carrier—present state and perspectives. *J Control Release* 1991;**17**(2):187–98. (c) De Boer AG. *Drug absorption enhancement: concepts, possibilities, limitations, and trends.* Switzerland; Langhorne, PA: Harwood Academic Publishers; 1994, p xv, 490 pp. (d) Couvreur P, Dubernet C, Puisieux F. Controlled drug-delivery with nanoparticles—current possibilities and future-trends. *Eur J Pharm Biopharm* 1995;**41**(1):2–13. (e) Fattal E, Vauthier C, Aynie I, Nakada Y, Lambert G, Malvy C, Couvreur P. Biodegradable polyalkylcyanoacrylate nanoparticles for the delivery of oligonucleotides. *J Control Release* 1998;**53**(1–3):137–43. (f) Labhasetwar V, Song CX, Levy RJ. Nanoparticle drug delivery system for restenosis. *Adv Drug Deliv Rev* 1997;**24**(1):63–85.
30. Fernandezurrusuno R, Fattal E, Porquet D, Feger J, Couvreur P. Evaluation of liver toxicological effects induced by polyalkylcyanoacrylate nanoparticles. *Toxicol Appl Pharmacol* 1995;**130**(2):272–9.
31. (a) Rajaonarivony M, Vauthier C, Couarraze G, Puisieux F, Couvreur P. Development of a new drug carrier made from alginate. *J Pharm Sci* 1993;**82**(9):912–7. (b) Wang N, Wu XS. Preparation and characterization of agarose hydrogel nanoparticles for protein and peptide drug delivery. *Pharm Dev Technol* 1997;**2**(2):135–42. (c) Calvo P, RemunanLopez C, VilaJato JL, Alonso MJ. Chitosan and chitosan ethylene oxide propylene oxide block copolymer nanoparticles as novel carriers for proteins and vaccines. *Pharm Res* 1997;**14** (10):1431–6.
32. (a) Svenson S, Tomalia DA. Commentary—dendrimers in biomedical applications—reflections on the field. *Adv Drug Deliv Rev* 2005;**57**(15):2106–29. (b) Dufes C, Uchegbu IF, Schatzlein AG. Dendrimers in gene delivery. *Adv Drug Deliv Rev* 2005;**57**(15):2177–202.

33. (a) Luo D, Haverstick K, Belcheva N, Han E, Saltzman WM. Poly(ethylene glycol)-conjugated PAMAM dendrimer for biocompatible, high-efficiency DNA delivery. *Macromolecules* 2002;**35**(9):3456–62. (b) Choi JS, Nam K, Park J, Kim JB, Lee JK. Enhanced transfection efficiency of PAMAM dendrimer by surface modification with L-arginine. *J Control Release* 2004;**99**(3):445–56. (c) Kim T, Seo HJ, Choi JS, Jang HS, Baek J, Kim K, Park JS. PAMAM-PEG-PAMAM: novel triblock copolymer as a biocompatible and efficient gene delivery carrier. *Biomacromolecules* 2004;**5**(6):2487–92.
34. Klajnert B, Bryszewska M. Dendrimers: properties and applications. *Acta Biochim Pol* 2001;**48**(1):199–208.
35. (a) Scheffel U, Wagner HN, Rhodes BA, Nataraja TK. Albumin microspheres for study of reticuloendothelial system. *J Nucl Med* 1972;**13**(7):498–503. (b) Gao ZH, Shukla AJ, Johnson JR, Crowley WR. Controlled-release of a contraceptive steroid from biodegradable and injectable gel formulations—in-vitro evaluation. *Pharm Res* 1995;**12**(6):857–63. (c) Marty JJ, Oppenheim RC, Speiser P. Nanoparticles—new colloidal drug delivery system. *Pharm Acta Helv* 1978;**53**(1):17–23.
36. (a) Truong-Le VL, Walsh SM, Schweibert E, Mao HQ, Guggino WB, August JT, Leong KW. Gene transfer by DNA-gelatin nanospheres. *Arch Biochem Biophys* 1999;**361**(1):47–56. (b) Mladenovska K, Kumbaradzi EF, Dodov GM, Makraduli L, Goracinova K. Biodegradation and drug release studies of BSA loaded gelatin microspheres. *Int J Pharm* 2002;**242**(1–2):247–9.
37. Young S, Wong M, Tabata Y, Mikos AG. Gelatin as a delivery vehicle for the controlled release of bioactive molecules. *J Control Release* 2005;**109**(1–3):256–74.
38. Furuhata M, Kawakami H, Toma K, Hattori Y, Maitani Y. Design, synthesis and gene delivery efficiency of novel oligo-arginine-linked PEG-lipids: effect of oligo-arginine length. *Int J Pharm* 2006;**316**(1–2):109–16.
39. (a) van de Manakker F, Vermonden T, van Nostrum CF, Hennink WE. Cyclodextrin-based polymeric materials: synthesis, properties, and pharmaceutical/biomedical applications. *Biomacromolecules* 2009;**10**(12):3157–75. (b) Calvo P, RemunanLopez C, VilaJato JL, Alonso MJ. Novel hydrophilic chitosan-polyethylene oxide nanoparticles as protein carriers. *J Appl Polym Sci* 1997;**63**(1):125–32.
40. Kas HS. Chitosan: properties, preparations and application to microparticulate systems. *J Microencapsul* 1997;**14**(6):689–711.
41. (a) Berscht PC, Nies B, Liebendorfer A, Kreuter J. Incorporation of basic fibroblast growth-factor into methylpyrrolidinone chitosan fleeces and determination of the in-vitro release characteristics. *Biomaterials* 1994;**15**(8):593–600. (b) Lehr CM, Bouwstra JA, Schacht EH, Junginger HE. In vitro evaluation of mucoadhesive properties of chitosan and some other natural polymers. *Int J Pharm* 1992;**78**(1):43–8. (c) Luessen HL, Lehr CM, Rentel CO, Noach ABJ, Deboer AG, Verhoef JC, Junginger HE. Bioadhesive polymers for the peroral delivery of peptide drugs. *J Control Release* 1994;**29**(3):329–38.
42. Gazit E. Self-assembled peptide nanostructures: the design of molecular building blocks and their technological utilization. *Chem Soc Rev* 2007;**36**(8):1263–9.
43. Nordtveit RJ, Varum KM, Smidsrod O. Degradation of partially N-acetylated chitosans with hen egg white and human lysozyme. *Carbohydr Polym* 1996;**29**(2):163–7.
44. (a) MacLaughlin FC, Mumper RJ, Wang JJ, Tagliaferri JM, Gill I, Hinchcliffe M, Rolland AP. Chitosan and depolymerized chitosan oligomers as condensing carriers for in vivo plasmid delivery. *J Control Release* 1998;**56**(1–3):259–72. (b) Richardson SCW, Kolbe HJV, Duncan R. Potential of low molecular mass chitosan as a DNA delivery system:

biocompatibility, body distribution and ability to complex and protect DNA. *Int J Pharm* 1999;**178**(2):231–43.
45. Kichler A. Gene transfer with modified polyethylenimines. *J Gene Med* 2004;**6**:S3–S10.
46. Napper DH. *Polymeric stabilization of colloidal dispersions*. London, New York: Academic Press; 1983 p xvi, 428 pp.
47. Veronese FM, Pasut G. PEGylation, successful approach to drug delivery. *Drug Discov Today* 2005;**10**(21):1451–8.
48. Kaul G, Amiji M. Long-circulating poly(ethylene glycol)-modified gelatin nanoparticles for intracellular delivery. *Pharm Res* 2002;**19**(7):1061–7.
49. Dong L, Xia S, Wu K, Huang Z, Chen H, Chen J, Zhang J. A pH/enzyme-responsive tumor-specific delivery system for doxorubicin. *Biomaterials* 2010;**31**(24):6309–16.
50. Kaul G, Amiji M. Biodistribution and targeting potential of poly(ethylene glycol)-modified gelatin nanoparticles in subcutaneous murine tumor model. *J Drug Target* 2004;**12**(9–10):585–91.
51. Kaul G, Amiji M. Tumor-targeted gene delivery using poly(ethylene glycol)-modified gelatin nanoparticles: in vitro and in vivo studies. *Pharm Res* 2005;**22**(6):951–61.
52. Navarro J, Obrador E, Carretero J, Petschen I, Avino J, Perez P, Estrela JM. Changes in glutathione status and the antioxidant system in blood and in cancer cells associate with tumour growth in vivo. *Free Radic Biol Med* 1999;**26**(3–4):410–8.
53. (a) Aspden TJ, Mason JDT, Jones NS, Lowe J, Skaugrud O, Illum L. Chitosan as a nasal delivery system: the effect of chitosan solutions on in vitro and in vivo mucociliary transport rates in human turbinates and volunteers. *J Pharm Sci* 1997;**86**(4):509–13. (b) Kumar M, Behera AK, Lockey RF, Zhang J, Bhullar G, De La Cruz CP, Chen LC, Leong KW, Huang SK, Mohapatra SS. Intranasal gene transfer by chitosan-DNA nanospheres protects BALB/c mice against acute respiratory syncytial virus infection. *Hum Gene Ther* 2002;**13**(12):1415–25.
54. (a) Roy K, Mao HQ, Huang SK, Leong KW. Oral gene delivery with chitosan-DNA nanoparticles generates immunologic protection in a murine model of peanut allergy. *Nat Med* 1999;**5**(4):387–91. (b) Sun CJ, Pan SP, Xie QX, Xiao LJ. Preparation of chitosan-plasinid DNA nanoparticles encoding zona pellucida glycoprotein-3 alpha and its expression in mouse. *Mol Reprod Dev* 2004;**68**(2):182–8.
55. Peng L, Cheng X, Zhuo R, Lan J, Wang Y, Shi B, Li S. Novel gene-activated matrix with embedded chitosan/plasmid DNA nanoparticles encoding PDGF for periodontal tissue engineering. *J Biomed Mater Res A* 2009;**90A**(2):564–76.
56. (a) Mailaender V, Landfester K. Interaction of nanoparticles with cells. *Biomacromolecules* 2009;**10**(9):2379–400. (b) Gu FX, Karnik R, Wang AZ, Alexis F, Levy-Nissenbaum E, Hong S, Langer RS, Farokhzad OC. Targeted nanoparticles for cancer therapy. *Nano Today* 2007;**2**(3):14–21. (c) Arruebo M, Fernandez-Pacheco R, Ibarra MR, Santamaria J. Magnetic nanoparticles for drug delivery. *Nano Today* 2007;**2**(3):22–32.
57. Varga CM, Tedford NC, Thomas M, Klibanov AM, Griffith LG, Lauffenburger DA. Quantitative comparison of polyethylenimine formulations and adenoviral vectors in terms of intracellular gene delivery processes. *Gene Ther* 2005;**12**(13):1023–32.
58. (a) Morris MC, Chaloin L, Heitz F, Divita G. Translocating peptides and proteins and their use for gene delivery. *Curr Opin Biotechnol* 2000;**11**(5):461–6. (b) Godeau G, Bernard J, Staedel C, Barthelemy P. Glycosyl-nucleoside-lipid based supramolecular assembly as a nanostructured material with nucleic acid delivery capabilities. *Chem Commun* 2009;**34**:5127–9.
59. (a) Kichler A, Mechtler K, Behr JP, Wagner E. Influence of membrane-active peptides on lipospermine/DNA complex mediated gene transfer. *Bioconjug Chem* 1997;**8**(2):213–21.

(b) Wyman TB, Nicol F, Zelphati O, Scaria PV, Plank C, Szoka FC. Design, synthesis, and characterization of a cationic peptide that binds to nucleic acids and permeabilizes bilayers. *Biochemistry* 1997;**36**(10):3008–17. (c) Baru M, Nahum O, Jaaro H, Sha'anani J, Nur I. Lysosome-disrupting peptide increases the efficiency of in-vivo gene transfer by liposome-encapsulated DNA. *J Drug Target* 1998;**6**(3):191–9.

60. Opanasopit P, Rojanarata T, Apirakaramwong A, Ngawhirunpat T, Ruktanonchai U. Nuclear localization signal peptides enhance transfection efficiency of chitosan/DNA complexes. *Int J Pharm* 2009;**382**(1–2):291–5.
61. Dean DA, Strong DD, Zimmer WE. Nuclear entry of nonviral vectors. *Gene Ther* 2005;**12**(11):881–90.
62. Kelkar SA, Pfister KK, Crystal RG, Leopold PL. Cytoplasmic dynein mediates adenovirus binding to microtubules. *J Virol* 2004;**78**(18):10122–32.
63. Check E. Harmful potential of viral vectors fuels doubts over gene therapy. *Nature* 2003;**423**(6940):573–4.
64. Urnov FD, Miller JC, Lee YL, Beausejour CM, Rock JM, Augustus S, Jamieson AC, Porteus MH, Gregory PD, Holmes MC. Highly efficient endogenous human gene correction using designed zinc-finger nucleases. *Nature* 2005;**435**(7042):646–51.
65. (a) Voller A, Bartlett A, Bidwell DE. Enzyme immunoassays with special reference to ELISA techniques. *J Clin Pathol* 1978;**31**(6):507–20. (b) Towbin H, Staehelin T, Gordon J. Electrophoretic transfer of proteins from polyacrylamide gels to nitrocellulose sheets—procedure and some applications. *Proc Natl Acad Sci U S A* 1979;**76**(9):4350–4.
66. Avrameas S. Coupling of enzymes to proteins with glutaraldehyde use of conjugates for detection of antigens and antibodies. *Immunochemistry* 1969;**6**(1):43–52.
67. Kluger R, Alagic A. Chemical cross-linking and protein-protein interactions—a review with illustrative protocols. *Bioorg Chem* 2004;**32**(6):451–72.
68. Wilson MB, Nakane PK. Covalent coupling of proteins to periodate-oxidized sephadex—new approach to immunoadsorbent preparation. *J Immunol Methods* 1976;**12**(1–2):171–81.
69. Diamandis EP, Christopoulos TK. The biotin (strept)avidin system—principles and applications in biotechnology. *Clin Chem* 1991;**37**(5):625–36.
70. Wang J. Nanomaterial-based amplified transduction of biomolecular interactions. *Small* 2005;**1**(11):1036–43.
71. (a) Zheng M, Davidson F, Huang XY. Ethylene glycol monolayer protected nanoparticles for eliminating nonspecific binding with biological molecules. *J Am Chem Soc* 2003;**125**(26):7790–1. (b) Lin PC, Chou PH, Chen SH, Liao HK, Wang KY, Chen YJ, Lin CC. Ethylene glycol-protected magnetic nanoparticles for a multiplexed immunoassay in human plasma. *Small* 2006;**2**(4):485–9. (c) Bagwe RP, Hilliard LR, Tan WH. Surface modification of silica nanoparticles to reduce aggregation and nonspecific binding. *Langmuir* 2006;**22**(9):4357–62.
72. (a) Zheng M, Li ZG, Huang XY. Ethylene glycol monolayer protected nanoparticles: synthesis, characterization, and interactions with biological molecules. *Langmuir* 2004;**20**(10):4226–35. (b) Gan DJ, Lyon LA. Synthesis and protein adsorption resistance of PEG-modified poly(N-isopropylacrylamide) core/shell microgels. *Macromolecules* 2002;**35**(26):9634–9. (c) Prencipe G, Tabakman SM, Welsher K, Liu Z, Goodwin AP, Zhang L, Henry J, Dai H. PEG branched polymer for functionalization of nanomaterials with ultralong blood circulation. *J Am Chem Soc* 2009;**131**(13):4783–7.
73. Gitlin G, Bayer EA, Wilchek M. Studies on the biotin-binding sites of avidin and streptavidin—tyrosine residues are involved in the binding-site. *Biochem J* 1990;**269**(2):527–30.

74. (a) Jana NR, Earhart C, Ying JY. Synthesis of water-soluble and functionalized nanoparticles by silica coating. *Chem Mater* 2007;**19**(21):5074–82. (b) Barnhill HN, Claudel-Gillet S, Ziessel R, Charbonniere LJ, Wang Q. Prototype protein assembly as scaffold for time-resolved fluoroimmuno assays. *J Am Chem Soc* 2007;**129**(25):7799–806.
75. Nikitin MP, Zdobnova TA, Lukash SV, Stremovskiy OA, Deyev SM. Protein-assisted self-assembly of multifunctional nanoparticles. *Proc Natl Acad Sci U S A* 2010;**107**(13):5827–32.
76. (a) Hartley RW. Barnase-barstar interaction. *Ribonucleases Pt A* 2001;**341**:599–611. (b) Alsallaq R, Zhou H-X. Energy landscape and transition state of protein-protein association. *Biophys J* 2007;**92**(5):1486–502.
77. Ho JA, Hung CH. Using liposomal fluorescent biolabels to develop an immunoaffinity chromatographic biosensing system for biotin. *Anal Chem* 2008;**80**(16):6405–9.
78. Lee JH, Choi HK, Chang JH. Optimization of biotin labeling of antibodies using mouse IgG and goat anti-mouse IgG-conjugated fluorescent beads and their application as capture probes on protein chip. *J Immunol Methods* 2010;**362**(1–2):38–42.
79. Simons B, Kaplan H, Hefford MA. Novel cross-linked enzyme-antibody conjugates for Western blot and ELISA. *J Immunol Methods* 2006;**315**(1–2):88–98.
80. Dhawan S. Design and construction of novel molecular conjugates for signal amplification (II): use of multivalent polystyrene microparticles and lysine peptide chains to generate immunoglobulinhorseradish peroxidase conjugates. *Peptides* 2002;**23**(12):2099–110.
81. Teramura Y, Arima Y, Iwata H. Surface plasmon resonance-based highly sensitive immunosensing for brain natriuretic peptide using nanobeads for signal amplification. *Anal Biochem* 2006;**357**(2):208–15.
82. Kim S, Pyo H-B, Ko SH, Ah CS, Kim A, Kim W-J. Fabrication of anionic sulfate-functionalized nanoparticles as an immunosensor by protein immobilization. *Langmuir* 2010;**26**(10):7355–64.
83. Homola J. Present and future of surface plasmon resonance biosensors. *Anal Bioanal Chem* 2003;**377**(3):528–39.
84. (a) Raether H. Surface-plasmons on smooth and rough surfaces and on gratings. *Springer Tract Modern Phys* 1988;**111**:1–133. (b) Kyo M, Usui-Aoki K, Koga H. Label-free detection of proteins in crude cell lysate with antibody arrays by a surface plasmon resonance imaging technique. *Anal Chem* 2005;**77**(22):7115–21. (c) Wink T, van Zuilen SJ, Bult A, van Bennekom WP. Liposome-mediated enhancement of the sensitivity in immunoassays of proteins and peptides in surface plasmon resonance spectrometry. *Anal Chem* 1998;**70**(5):827–32.
85. Kamyshny A, Magdassi S. Fluorescence immunoassay based on fluorescer microparticles. *Colloids Surf B Biointerfaces* 2000;**18**(1):13–7.
86. Trau D, Yang WJ, Seydack M, Caruso F, Yu NT, Renneberg R. Nanoencapsulated microcrystalline particles for superamplified biochemical assays. *Anal Chem* 2002;**74**(21):5480–6.
87. Chan CPY, Bruemmel Y, Seydack M, Sin KK, Wong LW, Merisko-Liversidge E, Trau D, Renneberg R. Nanocrystal biolabels with releasable fluorophores for immunoassays. *Anal Chem* 2004;**76**(13):3638–45.
88. Ho J-AA, Wu L-C, Huang M-R, Lin Y-J, Baeumner AJ, Durst RA. Application of ganglioside-sensitized liposomes in a flow injection immunoanalytical system for the determination of cholera toxin. *Anal Chem* 2007;**79**(1):246–50.
89. Hall M, Kazakova I, Yao YM. High sensitivity immunoassays using particulate fluorescent labels. *Anal Biochem* 1999;**272**(2):165–70.
90. Schott H, Voncunow D, Langhals H. Labeling of liposomes with intercalating perylene fluorescent dyes. *Biochim Biophys Acta* 1992;**1110**(2):151–7.

91. (a) Troyanovsky S. Cadherin dimers in cell-cell adhesion. *Eur J Cell Biol* 2005;**84**(2–3):225–33. (b) Mareel M, Leroy A. Clinical, cellular, and molecular aspects of cancer invasion. *Physiol Rev* 2003;**83**(2):337–76.
92. (a) Patel SD, Chen CP, Bahna F, Honig B, Shapiro L. Cadherin-mediated cell-cell adhesion: sticking together as a family. *Curr Opin Struct Biol* 2003;**13**(6):690–8. (b) Niessen CM, Gottardi CJ. Molecular components of the adherens junction. *Biochim Biophys Acta—Biomembranes* 2008;**1778**(3):562–71.
93. (a) Nollet F, Kools P, van Roy F. Phylogenetic analysis of the cadherin superfamily allows identification of six major subfamilies besides several solitary members. *J Mol Biol* 2000;**299**(3):551–72. (b) Takeichi M. Morphogenetic roles of classic cadherins. *Curr Opin Cell Biol* 1995;**7**(5):619–27.
94. Mareel M, Boterberg T, Noe V, VanHoorde L, Vermeulen S, Bruyneel E, Bracke M. E-cadherin/catenin/cytoskeleton complex: a regulator of cancer invasion. *J Cell Physiol* 1997;**173**(2):271–4.
95. (a) Takeichi M. The cadherins—cell cell-adhesion molecules controlling animal morphogenesis. *Development* 1988;**102**(4):639–55. (b) Gumbiner BM. Regulation of cadherin-mediated adhesion in morphogenesis. *Nat Rev Mol Cell Biol* 2005;**6**(8):622–34.
96. (a) Delmas V, Pla P, Feracci H, Thiery JP, Kemler R, Larue L. Expression of the cytoplasmic domain of E-cadherin induces precocious mammary epithelial alveolar formation and affects cell polarity and cell-matrix integrity. *Dev Biol* 1999;**216**(2):491–506. (b) Nelson WJ. Regulation of cell-cell adhesion by the cadherin-catenin complex. *Biochem Soc Trans* 2008;**36**:149–55. (c) Raptis L, Arulanandam R, Vultur A, Geletu M, Chevalier S, Feracci H. Beyond structure, to survival: activation of Stat3 by cadherin engagement. *Biochem Cell Biol* 2009;**87**(6):835–43.
97. Pokutta S, Weis WI. Structure and mechanism of cadherins and catenins in cell-cell contacts. *Annu Rev Cell Dev Biol* 2007;**23**:237–61.
98. (a) Steinberg MS, McNutt PM. Cadherins and their connections: adhesion junctions have broader functions. *Curr Opin Cell Biol* 1999;**11**(5):554–60. (b) Steinberg MS, Takeichi M. Experimental specification of cell sorting, tissue spreading, and specifiC spatial patterning by quantitative differences in cadherin expression. *Proc Natl Acad Sci U S A* 1994;**91**(1):206–9.
99. (a) Koch AW, Bozic D, Pertz O, Engel J. Homophilic adhesion by cadherins. *Curr Opin Struct Biol* 1999;**9**(2):275–81. (b) Boggon TJ, Murray J, Chappuis-Flament S, Wong E, Gumbiner BM, Shapiro L. C-cadherin ectodomain structure and implications for cell adhesion mechanisms. *Science* 2002;**296**(5571):1308–13. (c) Huber AH, Weis WI. The structure of the beta-catenin/E-cadherin complex and the molecular basis of diverse ligand recognition by beta-catenin. *Cell* 2001;**105**(3):391–402. (d) Harrison OJ, Jin X, Hong S, Bahna F, Ahlsen G, Brasch J, Wu Y, Vendome J, Felsovalyi K, Hampton CM, Troyanovsky RB, Ben-Shau A, Frank J, Troyanovsky SM, Shapiro L, Honig B. The extracellular architecture of adherens junctions revealed by crystal structures of Type I cadherins. *Structure* 2011;**19**(2):244–56.
100. (a) Stemmler MP. Cadherins in development and cancer. *Mol Biosyst* 2008;**4**(8):835–50. (b) Niessen CM. Tight junctions/adherens junctions: basic structure and function. *J Invest Dermatol* 2007;**127**(11):2525–32.
101. (a) Sivasankar S, Brieher W, Lavrik N, Gumbiner B, Leckband D. Direct molecular force measurements of multiple adhesive interactions between cadherin ectodomains. *Proc Natl Acad Sci U S A* 1999;**96**(21):11820–4. (b) Chappuis-Flament S, Wong E, Hicks LD, Kay CM, Gumbiner BM. Multiple cadherin extracellular repeats mediate homophilic

binding and adhesion. *J Cell Biol* 2001;**154**(1):231–43. (c) Perret E, Leung A, Feracci H, Evans E. Trans-bonded pairs of E-cadherin exhibit a remarkable hierarchy of mechanical strengths. *Proc Natl Acad Sci U S A* 2004;**101**(47):16472–7.

102. Al-Amoudi A, Diez DC, Betts MJ, Frangakis AS. The molecular architecture of cadherins in native epidermal desmosomes. *Nature* 2007;**450**(7171):832–7.
103. Nollet F, Berx G, van Roy F. The role of the E-cadherin/catenin adhesion complex in the development and progression of cancer. *Mol Cell Biol Res Commun* 1999;**2**(2):77–85.
104. Posy S, Shapiro L, Honig B. Sequence and structural determinants of strand swapping in cadherin domains: do all cadherins bind through the same adhesive interface? *J Mol Biol* 2008;**378**(4):954–68.
105. He WZ, Cowin P, Stokes DL. Untangling desmosomal knots with electron tomography. *Science* 2003;**302**(5642):109–13.
106. Al-Amoudi A, Norlen LPO, Dubochet J. Cryo-electron microscopy of vitreous sections of native biological cells and tissues. *J Struct Biol* 2004;**148**(1):131–5.
107. Dorn IT, Pawlitschko K, Pettinger SC, Tampe R. Orientation and two-dimensional organization of proteins at chelator lipid interfaces. *Biol Chem* 1998;**379**(8–9):1151–9.
108. Lambert O, Taveau JC, Him JLK, Al Kurdi R, Gulino-Debrac D, Brisson A. The basic framework of VE-cadherin junctions revealed by cryo-EM. *J Mol Biol* 2005;**346**(5):1193–6.
109. (a) Perret E, Benoliel AM, Nassoy P, Pierres A, Delmas V, Thiery JP, Bongrand P, Feracci H. Fast dissociation kinetics between individual E-cadherin fragments revealed by flow chamber analysis. *EMBO J* 2002;**21**(11):2537–46. (b) Leckband D, Prakasam A. Mechanism and dynamics of cadherin adhesion. *Annu Rev Biomed Eng* 2006;**8**:259–87.
110. Puech PH, Feracci H, Brochard-Wya F. Adhesion between giant vesicles and supported bilayers decorated with chelated E-cadherin fragments. *Langmuir* 2004;**20**(22):9763–8.
111. Menger FM, Keiper JS. Chemistry and physics of plant vesicles as biomembrane models. *Curr Opin Chem Biol* 1998;**2**(6):726–32.
112. (a) Boulbitch A, Guttenberg Z, Sackmann E. Kinetics of membrane adhesion mediated by ligand-receptor interaction studied with a biomimetic system. *Biophys J* 2001;**81**(5):2743–51. (b) Goennenwein S, Tanaka M, Hu B, Moroder L, Sackmann E. Functional incorporation of integrins into solid supported membranes on ultrathin films of cellulose: impact on adhesion. *Biophys J* 2003;**85**(1):646–55. (c) Lorz BG, Smith A-S, Gege C, Sackmann E. Adhesion of giant vesicles mediated by weak binding of Sialyl-Lewis(x) to E-selectin in the presence of repelling poly(ethylene glycol) molecules. *Langmuir* 2007;**23**(24):12293–300. (d) Marchi-Artzner V, Lorz B, Gosse C, Jullien L, Merkel R, Kessler H, Sackmann E. Adhesion of Arg-Gly-Asp (RGD) peptide vesicles onto an integrin surface: visualization of the segregation of RGD ligands into the adhesion plaques by fluorescence. *Langmuir* 2003;**19**(3):835–41. (e) Evans E, Heinrich V, Ludwig F, Rawicz W. Dynamic tension spectroscopy and strength of biomembranes. *Biophys J* 2003;**85**(4):2342–50. (f) Rawicz W, Olbrich KC, McIntosh T, Needham D, Evans E. Effect of chain length and unsaturation on elasticity of lipid bilayers. *Biophys J* 2000;**79**(1):328–39. (g) Rawicz W, Smith BA, McIntosh TJ, Simon SA, Evans E. Elasticity, strength, and water permeability of bilayers that contain raft microdomain-forming lipids. *Biophys J* 2008;**94**(12):4725–36. (h) Fenz SF, Merkel R, Sengupta K. Diffusion and intermembrane distance: case study of avidin and E-cadherin mediated adhesion. *Langmuir* 2009;**25**(2):1074–85.
113. Lambert M, Padilla F, Mege RM. Immobilized dimers of N-cadherin-Fc chimera mimic cadherin-mediated cell contact formation: contribution of both outside-in and inside-out signals. *J Cell Sci* 2000;**113**(12):2207–19.

114. Arulanandam R, Vultur A, Cao J, Carefoot E, Elliott BE, Truesdell PF, Larue L, Feracci H, Raptis L. Cadherin-cadherin engagement promotes cell survival via Rac1/Cdc42 and signal transducer and activator of transcription-3. *Mol Cancer Res* 2009;**7**(8):1310–27.
115. (a) Baumgartner W, Hinterdorfer P, Ness W, Raab A, Vestweber D, Schindler H, Drenckhahn D. Cadherin interaction probed by atomic force microscopy. *Proc Natl Acad Sci U S A* 2000;**97**(8):4005–10. (b) du Roure O, Buguin A, Feracci H, Silberzan P. Homophilic interactions between cadherin fragments at the single molecule level: an AFM study. *Langmuir* 2006;**22**(10):4680–4. (c) Zhang Y, Sivasankar S, Nelson WJ, Chu S. Resolving cadherin interactions and binding cooperativity at the single-molecule level. *Proc Natl Acad Sci U S A* 2009;**106**(1):109–14.
116. Merkel R, Nassoy P, Leung A, Ritchie K, Evans E. Energy landscapes of receptor-ligand bonds explored with dynamic force spectroscopy. *Nature* 1999;**397**(6714):50–3.
117. Pierres A, Benoliel A-M, Bongrand P. Studying molecular interactions at the single bond level with a laminar flow chamber. *Cell Mol Bioeng* 2008;**1**(4):247–62.
118. Pierres A, Feracci H, Delmas V, Benoliel AM, Thiery JP, Bongrand P. Experimental study of the interaction range and association rate of surface-attached cadherin 11. *Proc Natl Acad Sci U S A* 1998;**95**(16):9256–61.
119. (a) Pierres A, Prakasam A, Touchard D, Benoliel A-M, Bongrand P, Leckband D. Dissecting subsecond cadherin bound states reveals an efficient way for cells to achieve ultrafast probing of their environment. *FEBS Lett* 2007;**581**(9):1841–6. (b) Bayas MV, Leung A, Evans E, Leckband D. Lifetime measurements reveal kinetic differences between homophilic cadherin bonds. *Biophys J* 2006;**90**(4):1385–95.
120. (a) Garcion E, Lamprecht A, Heurtault B, Paillard A, Aubert-Pouessel A, Denizot B, Menei P, Benoit J-P. A new generation of anticancer, drug-loaded, colloidal vectors reverses multidrug resistance in glioma and reduces tumor progression in rats. *Mol Cancer Ther* 2006;**5**(7):1710–22. (b) Weyland M, Manero F, Paillard A, Gree D, Viault G, Jarnet D, Menei P, Juin P, Chourpa I, Benoit JP, Gree R, Garcion E. Mitochondrial targeting by use of lipid nanocapsules loaded with SV30, an analogue of the small-molecule Bcl-2 inhibitor HA14-1. *J Control Release* 2011;**151**(1):74–82.
121. (a) Oliveira JM, Salgado AJ, Sousa N, Mano JF, Reis RL. Dendrimers and derivatives as a potential therapeutic tool in regenerative medicine strategies—a review. *Prog Polym Sci* 2010;**35**(9):1163–94. (b) Petite H, Viateau V, Bensaid W, Meunier A, de Pollak C, Bourguignon M, Oudina K, Sedel L, Guillemin G. Tissue-engineered bone regeneration. *Nat Biotechnol* 2000;**18**(9):959–63. (c) Uemura T, Dong J, Wang YC, Kojima H, Saito T, Iejima D, Kikuchi M, Tanaka J, Tateishi T. Transplantation of cultured bone cells using combinations of scaffolds and culture techniques. *Biomaterials* 2003;**24**(13):2277–86.
122. Chirkov SN. The antiviral activity of chitosan (review). *Appl Biochem Microbiol* 2002;**38**(1):1–8.
123. (a) Dutta J, Tripathi S, Dutta PK. Progress in antimicrobial activities of chitin, chitosan and its oligosaccharides: a systematic study needs for food applications. *Food Sci Technol Int* 2012;**18**(1):3–34. (b) Kuan C-Y, Yee-Fung W, Yuen K-H, Liong M-T. Nanotech: propensity in foods and bioactives. *Crit Rev Food Sci Nutr* 2012;**52**(1–3):55–71.
124. (a) Silvestre C, Duraccio D, Cimmino S. Food packaging based on polymer nanomaterials. *Prog Polym Sci* 2011;**36**(12):1766–82. (b) Duncan TV. Applications of nanotechnology in food packaging and food safety: barrier materials, antimicrobials and sensors. *J Colloid Interface Sci* 2011;**363**(1):1–24.

Part III

Applications in Diagnostics and Biosensing

Chapter 9

Application of Inorganic Nanoparticles for Diagnosis Based on MRI

Pedro M. Enriquez-Navas and Maria L. Garcia-Martin
Bionand-Centro Andaluz de Nanomedicina y Biotecnología, Parque Tecnológico de Andalucía, Malaga, Spain

1. INTRODUCTION

Magnetic resonance imaging (MRI) has evolved rapidly over the past few decades to become a fundamental tool for clinical diagnosis. MRI is one of a variety of modern medical imaging technologies, including X-ray, ultrasound, computed tomography, and nuclear medicine. Unlike other image modalities, MRI uses nonionizing radiation that is harmless to human tissue. Additionally, MRI offers the advantage of very good image resolution together with unlimited tissue penetration. Because of this unique combination of qualities, MRI has attracted a great deal of interest in the field of diagnostic imaging. For instance, MRI would be the method of choice to avoid repeated exposure to radiation in follow-up studies or for patients who cannot receive iodinated contrast dye.

MRI signal comes from the magnetization of water protons. The observed contrast in regular morphologic MRI scans reflects difference in signal intensity, which depends on the concentration of water molecules within the tissue, the relaxation times, T_1 and T_2, of the water protons, and the mobility of the water molecules (diffusion, flow).[1]

The observed contrast can be modulated by modifying the parameters of the MRI sequence, such as the repetition time, the echo time, or diffusion factor. In this way, images with different "weighting" can be obtained, namely, T_1-, T_2-, proton-density-, or diffusion-weighted images. This type of images provides morphologic information about the tissue, being merely qualitative. However, all these parameters, T_1, T_2, diffusion rate, and many others, such as fractional anisotropy and cerebral blood flow, etc., can also be quantitatively measured to generate parametric images, that is, maps of a specific parameter.

The above-described sources of contrast correspond to the so-called intrinsic contrast. However, this type of contrast is often not enough to allow sensitive

Frontiers of Nanoscience, Vol. 4. DOI: 10.1016/B978-0-12-415769-9.00009-1

and specific diagnosis and the use of exogenous contrast agents (CAs) becomes necessary. The most commonly used are the Gd-chelates, which largely shorten T_1 relaxation times. They are known as positive CAs because they produce signal enhancement in T_1-weighted images. The second largest group of MRI CAs is the iron-oxide-based nanoparticles (NPs). As opposed to Gd-chelates, they are classified as negative CAs because they are observed as signal voids in T_2-weighted images. Magnetic NPs are especially attractive in the emerging field of molecular imaging, which aims to *in vivo* visualize and measure biological processes at the molecular and cellular level, because they produce an outstanding signal amplification due to their superparamagnetism, as it will be discussed later.

In this chapter, we focus on the use iron oxide NPs as CAs for MRI-based diagnosis.

2. MAGNETIC NPs FOR MRI

Iron-oxide-based NPs were introduced as MRI CAs after the Gd-chelates.[2] The main difference between both of them is the mechanism through which they produce magnetic relaxation of water protons and thus image contrast. Paramagnetic agents, such as Gd-chelates, induce relaxation through the chemical exchange between water molecules coordinated to the paramagnetic complex and bulk water molecules. On the other hand, the relaxation induced by iron oxide NPs is due to the diffusion of water molecules through the magnetic field inhomogeneities created by the magnetic NPs. The core of iron oxide NPs is made of thousands of iron ions, and when these ions are magnetically ordered within the crystal, in the presence of an external magnetic field, the net magnetic moment of the particle exceeds by far that of paramagnetic ions. This effect is known as superparamagnetism and is responsible for the outstanding increase in sensitivity produced by iron oxide NPs, which has made it possible to even detect a single cell *in vivo* by MRI.[3,4] For this reason, superparamagnetic NPs are the best candidates for MRI-based molecular imaging.[5] In addition, the iron ions are much less toxic than the gadolinium ones and can be reused or recycled by cells using normal biochemical pathways for iron metabolism, which makes them more suitable for *in vivo* studies.

2.1. The Superparamagnetism Phenomenon

Superparamagnetism appears in small ferromagnetic or ferrimagnetic NPs. If the size of these NPs is small enough, their magnetization can randomly flip direction under the influence of temperature. The time between two flips is known as the Néel relaxation time. If the time used to measure the magnetization of the NPs is much longer than the Néel relaxation time and no external field is present, their average magnetization seems to be zero, and they are said to be in superparamagnetic state.

The concept of superparamagnetism of magnetic materials in the nanoscale was first proposed by Frenkel and Doefman in 1930.[6] They predicted that NPs made of magnetic materials with a particle size small enough would display superparamagnetic properties.

Superparamagnetic NPs can be dispersed into an aqueous solution and remain stable by coating them with an appropriate layer, thus forming a *ferrofluid*. These ferrofluids can be used in different bioapplications, including *in vivo* and *in vitro* applications. *In vivo* applications hold such specialities as drug delivery, hyperthermia procedures, and MRI CAs.

2.2. Conventional Iron-Oxide-Based NPs for MRI

The first NPs developed for clinical diagnosis based on MRI were particles of magnetite, Fe_3O_4, for the detection of liver and spleen lesions.[2] These NPs would fall within the category of superparamagnetic iron oxides (SPIOs), one of the two in which iron oxide NPs have been classified according to their size, the other one being ultrasmall superparamagnetic iron oxides (USPIOs). The hydrodynamic diameter of SPIO particles ranges from 50 to 500 nm, whereas in the case of USPIOs is less than 50 nm. SPIOs are eliminated by the reticuloendothelial system (RES), which makes them mainly useful for liver or spleen MRI. USPIOs are also taken up by the RES, but because of their smaller size, they remain longer in the blood stream, making them suitable for other MRI applications such as MR angiography. The first USPIO NPs were developed by Weissleder *et al.*[7] They claimed several potential applications of these NPs in the clinics, namely: as intravenous CA for the lymphatic system, as bone marrow CA, as long-half-life perfusion agent for brain and heart, and as the magnetic moiety in organ-targeted superparamagnetic CAs for MRI.

A large variety of superparamagnetic NPs have been reported since then, but very few have been commercialized for clinical use. Some examples are Resovist (SHU-555A, Bayer Schering Pharma AG), Supravist (SHU-555C, Bayer Schering Pharma AG), or Sinerem (AMI-227, Guerbet). Resovist is coated with carboxydextran to render a 60-nm size NPs and has T_2 relaxivities of 150 $mmol^{-1}\ s^{-1}$. It has been used for liver/spleen MRI. Supravist is an improved version of Resovist, coated with carboxydextran and with an average size of 20 nm, similar to Sinerem, which is coated with dextran. Their T_2 relaxivities are 57 and 44 $mmol^{-1}\ s^{-1}$, respectively. Both of them are USPIOs and were proposed as MRI CAs to image lymph nodes, bone marrow, and inflammatory processes. Other NPs of larger size are Endorem and Lumirem (AMAG Pharmaceuticals Inc.) with particle sizes of 150 and 300 nm, respectively. Endorem is used to detect liver lesions and Lumirem for gastrointestinal contrast.

Currently, most of these CAs have been withdrawn from the market mainly due to low demand and high prices.

2.3. Magnetic NPs in the Nanotechnology Era

Over the past two decades, a large variety of NPs have been developed as a result of multidisciplinary scientific teams. For molecular imaging using MRI, iron oxide NPs need to be nontoxic, biocompatible, magnetic, and with high specificity toward the target tissue. To this end, NPs have been functionalized with bioactive materials, such as antibodies, peptides, and oligonucleotides[8] (Figure 9.1).

Various surface modification methods, including ligand exchange with water, dispersible ligands, and encapsulation with biocompatible shells, have been developed to turn organic-dispersible NPs into water-dispersible and biocompatible.[9]

Not only Fe, but also other elements, like Gd or Mn, have been used as the source of contrast in the development of new MRI NPs. NPs of Gd_2O_3,[10–12] GdF_3,[13] and $GdPO_4$[14] have been investigated as T_1 MRI CAs. The gadolinium complexes have been immobilized in different nanostructured materials, including nanoporous silicas, dendrimers, perfluorocarbon NPs, and nanotubes. This new generation of Gd-based NPs exhibits largely improved relaxation properties due to their ability to carry a large number of paramagnetic ions.

Mn-based NPs have also been developed as new T_1 MRI CA for body organs such as brain, liver, and kidney.[15]

Iron, in its metallic form, is the element with the highest saturation magnetization at room temperature and it is also biocompatible. These features make of iron

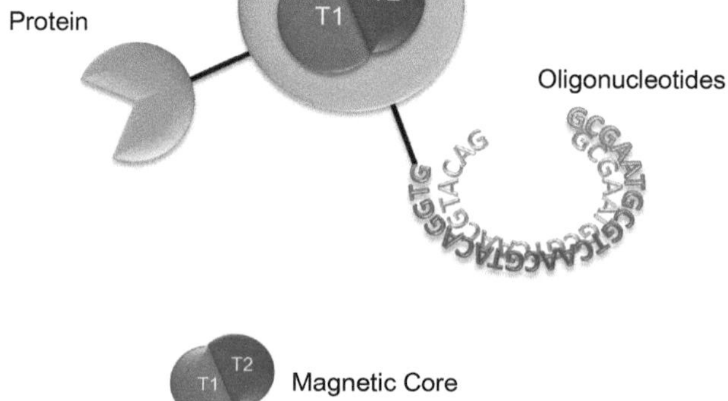

FIGURE 9.1 Functionalized magnetic nanoparticle for MRI applications. (For color version of this figure, the reader is referred to the online version of this chapter.)

the ideal candidate for MRI NPs. However, the use of iron in the development of new NPs has been challenging because it is difficult to obtain stable metallic NPs using conventional synthesis methods. To overcome this problem, Cheong *et al.* have developed a new synthesis protocol to create an oxide shell around the iron core.[16] The authors demonstrated the high effectiveness of these NPs as MRI CAs by injecting them in mice bearing tumors and showing their ability to successfully detect very small tumors (<3 mm) at 1.5 T.

An interesting work has been published by Chow *et al.*, in which gas-filled polymeric microtubules (PMBs) are modified by incorporating iron oxide into microbubble shells. These PMB showed very high susceptibility effects, allowing their detection by MRI at low concentrations. PMB can potentially be used as intravascular MRI CAs or as MRI guidance in microbubble-based drug delivery and therapeutic applications.[17]

Another type of NPs is based on liposomes. Recently, the use of liposomes loaded with metallic ions as MRI CAs has been investigated. Liposomes are specially promising in the area of theranostic (therapy plus diagnosis) because they can easily be loaded with different substances, including CAs and therapeutic drugs. Nonetheless, most of the challenges related to the *in vivo* use of liposomes still need to be addressed.[18]

Most of the NPs reported for MRI applications are still in the early stage of *in vitro* testing or animal studies. Before they reach extensive clinical use, several issues need to be addressed, including toxicological effects, specificity, long-term stability, and biodistribution. Interdisciplinary teams are essential to reach the ultimate goal of using NP-based MRI CAs for molecular imaging that can be applied to personalized diagnosis.[8]

Summing up, the challenge in this area of research is the development of NPs with improved properties, including better magnetic relaxation properties, better functionalization, and higher specificity toward the targeted diseased tissue. New MRI techniques are also needed for the accurate visualization and measurement of magnetic NPs *in vivo*, which is essential for the assessment of NP biodistribution, pharmacokinetics, specificity, and therapeutic efficacy during their developmental stage, and also for more efficient clinical diagnosis.

3. MRI METHODS TO IMAGE NPs

Conventional MRI techniques for the detection of CAs can be summarized in T_1- and T_2*-weighted imaging. T_1-weighted MRI sequences are mostly used in combination with Gd-chelates to detect the signal enhancement produced by T_1 shortening of the water protons that get in contact with the paramagnetic element. T_2*-weighted sequences allow the signal loss produced by a superparamagnetic material to be detected.

To explain the NMR basics underlying these effects, we will focus on two basic sequence schemes: the spin-echo and the gradient-echo sequences (Figure 9.2).

The observed signal (S) is given by Equations (9.1) and (9.2) in spin-echo and gradient-echo sequences, respectively:

$$S_{SE} \approx \rho\left(1 - e^{-TR/T_1}\right)e^{-TE/T_2} \quad (9.1)$$

$$S_{GE} \approx \rho\frac{\left(1 - e^{-TR/T_1}\right)\sin\alpha}{\left(1 - e^{-TR/T_1}\right)\cos\alpha}e^{-TE/T_{2*}} \quad (9.2)$$

where ρ is the proton density and α is the radiofrequency (RF) pulse flip angle, which is typically short for gradient-echo MRI sequences.

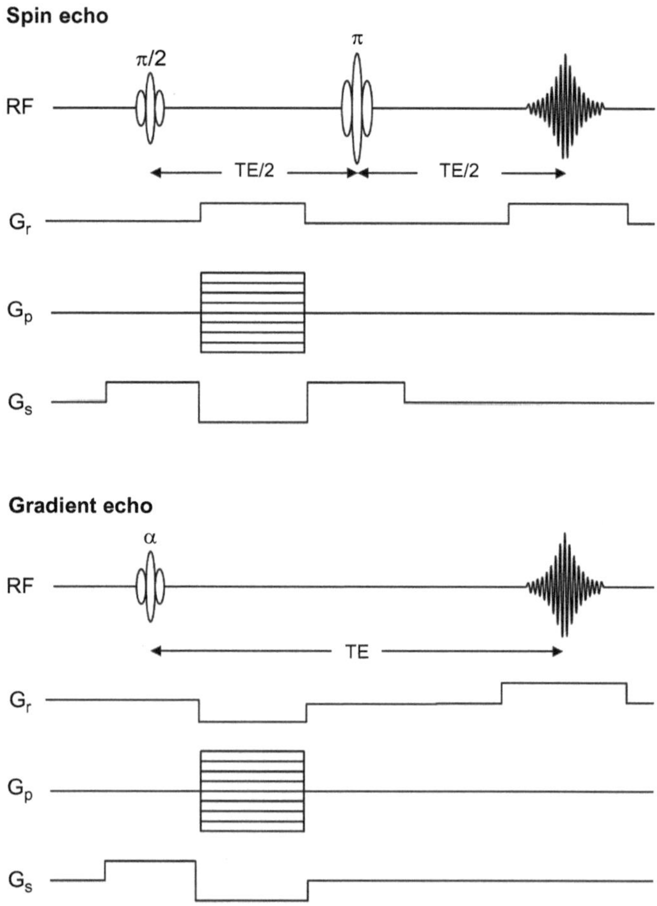

FIGURE 9.2 Spin-echo and gradient-echo MRI acquisition schemes. (For color version of this figure, the reader is referred to the online version of this chapter.)

The signal in spin-echo sequences is dependent on T_1 and T_2, whereas in gradient echo, the signal decay is determined by T_2*. The relationship between these two variables is described in the following equation:

$$\frac{1}{T_2*} = \frac{1}{T_2} + \frac{1}{T_2'} \tag{9.3}$$

where $1/T_2'$ is the T_2 decay due to field contributions. For a gradient-echo sequence, a fair approximation is to assume that $1/T_2'$ is given by $\gamma \Delta B$, being ΔB the field inhomogeneity across a voxel.

From the above, it becomes evident that gradient-echo sequences are more sensitive to field inhomogeneities, such as those produced by superparamagnetic agents.

T_2*-weighted imaging has been widely used to image iron oxide NPs both in human[19–24] and animal[25–30] studies. However, regular gradient-echo sequences are prone to susceptibility artifacts caused by air/tissue interfaces, hemorrhage, metals, etc. In the past decade, new techniques have been developed to suppress these unwanted background susceptibility effects and keep just the local susceptibility effects caused by magnetic NPs. Some of these techniques are discussed in the following sections.

3.1. Susceptibility-Weighted Imaging

There are several intrinsic sources of magnetic susceptibility *in vivo*, among them, deoxygenated hemoglobin or air/tissue interfaces. In addition, these susceptibility differences can be also produced by the administration of CAs, such as iron oxide NPs. Regardless their origin, the difference in susceptibility between neighboring tissues causes phase differences that can be used to generate image contrast. The simplest way is to calculate phase images from gradient-echo sequences. However, the presence of undesired background magnetic fields can obscure the useful information contained in the phase images. That is the reason why phase information has been disregarded for long time. In 1997, Haacke's group developed a method to remove most of the unwanted phase artifacts and keep just the local phase of interest.[31] Later, they proposed the combined use of phase and magnitude images to create a new type of image, susceptibility-weighted magnitude image, which is currently referred to as SWI (susceptibility-weighted imaging).[32] Briefly, the low-frequency components of the background field are removed by applying a high-pass filter, and then the phase-corrected image thus generated is used to create a phase mask, which is designed to suppress pixels with certain phases. The phase mask is then multiplied by the magnitude image resulting from the gradient-echo acquisition to obtain the susceptibility-weighted image.

SWI has been extensively used in clinical neuroradiology.[33] As for biomedical research, SWI has been applied for the detection of iron oxide NPs in animal models.[34,35]

3.2. Positive Contrast Techniques

The susceptibility effects produced by superparamagnetic NPs give rise to large signal voids that can be detected as negative contrast using regular gradient-echo imaging or SWI. These effects extend well beyond the size of the NPs, which is good in terms of sensitivity, but it also has important drawbacks: (i) negative contrast is much less desirable than positive contrast because it cannot always be distinguished from intrinsic low signals in tissues caused by the presence of blood or air and (ii) iron oxide NPs produce "blooming artifacts" in gradient-echo images, which may lead to an overestimation of the area occupied by the biomarker or the targeted lesion. Recently, experimental approaches to overcome these limitations have been reported. On the one hand, negative contrast can be converted into positive by using specific acquisition sequences[36] or by applying postprocessing algorithms to data acquired with regular T_2*-weighted sequences.[37–39] On the other hand, blooming artifacts can be minimized by using new MRI techniques, such as quantitative susceptibility mapping (QSM), which will be commented in the following section.

The new sequence developed by Cunningham *et al.*[36] uses spectrally selective RF pulses to excite and refocus the off-resonance water surrounding the iron-oxide-labeled cells and suppress the on-resonance signal, so that only the fluid and tissue adjacent to the labeled cells are visible in the image. This method has been successfully used *in vivo* to image iron-oxide-labeled stem cells transplanted into the mouse hind limb.[40]

The postprocessing methods use different approaches to generate positive contrast from complex data acquired with standard MRI sequences. Thus, the susceptibility gradient mapping proposed by Dahnke *et al.* measures the strength and direction of the susceptibility gradients in each pixel. Then, positive contrast images can be generated from these data by assigning gray values to the strength of the susceptibility gradient vector. This method has been used *in vivo* to image SPIO-labeled C6 glioma xenografts[38] or atherosclerotic plaques,[41] for example.

Eibofner *et al.*[37] described another postprocessing method to produce positive contrast. The technique is based on the SWI described previously, but it introduces changes in the phase mask so that the tissue surrounding the iron oxide NPs appears highlighted against a homogeneous background. For *in vivo* applications, however, this approach needs to be improved to allow gray scale values on the background instead of the complete suppression. That way anatomical information would be preserved while obtaining positive contrast around the NPs.

3.3. Quantitative Susceptibility Mapping

QSM pursues the actual quantification of local susceptibilities to generate parametric maps. Calculation of susceptibility from phase images is an ill-posed inverse problem for which several approaches have been proposed, including thresholding,[42] oversampling,[43] or regularization[44–47] methods. QSM has been applied *in vivo* to image the human brain[45,48,49] and also iron oxide NPs in mice.[47]

Although QSM is a technique of very recent development and its application *in vivo* is still reduced, it has two crucial advantages that make it very promising for molecular imaging of iron-oxide-based biomarkers: (i) it allows for absolute quantification of the CA and (ii) it removes imaging-parameter-dependent blooming artifacts, allowing for more accurate delineation of the lesion under study or the distribution of the biomarker.

Quantitative analysis of biomarkers is essential to exploit full potential of molecular imaging for diagnosis, disease staging, and therapeutic monitoring. For example, the use of NPs as delivery vehicles for diagnostic and therapeutic agents requires the accurate measurement of NPs accumulated at the disease tissue to make diagnostic decisions and gauge therapeutic dose. The measurement of drug dose at the targeted tissues is essential for monitoring therapy. Also, the count of stem cells homing at the targeted site is essential for optimizing cell therapy protocols, not to mention the importance of accurate quantification and reproducibility to standardize and cross-validate molecular MRI methods. However, current quantification of magnetic CAs in MRI is based on relaxation measurements, which are known to have important limitations: *in vitro* calibrations are needed for absolute quantification, and the assumption needs to be made that free water is uniformly available for relaxation, which cannot longer be sustained when CAs are bound to cells and molecules as it is in present molecular imaging strategies. Consequently, the implementation of new methods, like QSM, that allow absolute quantification of magnetic biomarkers will be very useful for the development of new diagnostic and therapeutic drugs in the emerging field of nanomedicine.

3.4. Ultra-Short TE and Zero TE MRI

The techniques described up to here have in common the use of magnetic field perturbations caused by iron oxide NPs as the source of contrast, that is, they are all based on T_2* weighting. However, iron oxide NPs also produce a significant T_1 shortening that could potentially be exploited to generate positive contrast without relying on magnetic field disturbances and thus avoid the limitations intrinsic to T_2*-weighted sequences (bulk magnetic susceptibility effects). However, regular MRI sequences do not allow echo times short enough to avoid the strong T_1 signal attenuation due to the fast T_2 decay produced by superparamagnetic NPs.

Recently, new acquisition schemes have been developed that allow for ultra-short echo times (UTEs), in the range of microseconds, that open new possibilities to image NPs using T_1-weighted imaging.[50,51]

UTE MRI has been used in the clinics to image tissues such as cortical bone, ligaments, menisci, tendons, or intervertebral disks,[52–54] which contain a predominant proportion of short T_2 components and hence appear as dark background (or very low signal intensity) in conventional T_2-weighted imaging.

UTE sequences do not require the formation of an echo, instead they acquire signal during the free induction decay that follows a short RF pulse. They use center-out radial k-space trajectory combined with a ramp-sampling strategy to allow signal acquisition right after the excitation and thus avoid the effect of rapid signal decay. The multiecho versions of UTE allow for the combination of both T_1 and T_2* effects in a synergistic way by subtracting the second echo image from the first one.[51] This subtraction method, known as SubUTE, produces long-T_2* signal suppression, while reversing the short-T_2* contrast into positive contrast.

Very recently, another technique has been reported that goes a step further and uses zero echo time (ZET).[55] In ZET MRI, the gradients are turn on before the RF pulse, instead of immediately after, as UTE, thus allowing for virtually ZET. In their work, Weiger *et al.*[55] have successfully image iron oxide NPs *in vitro* at ZET.

In summary, there are several techniques that can be used to image superparamagnetic NPs *in vivo*. Some are merely qualitative, some are quantitative, some are based just on T_2* effects, and some others take advantage of both T_2* and T_1 shortening effects. All of them have advantages and disadvantages and the method of choice will depend on the goal of the study and the available technological media.

REFERENCES

1. Torrey HC. Bloch equations with diffusion terms. *Phys Rev* 1956;**104**(3):563–5.
2. Mendonca Dias MH, Lauterbur PC. Ferromagnetic particles as contrast agents for magnetic resonance imaging of liver and spleen. *Magn Reson Med* 1986;**3**(2):328–30.
3. Shapiro EM, et al. In vivo detection of single cells by MRI. *Magn Reson Med* 2006;**55**(2): 242–9.
4. Smirnov P, et al. Single-cell detection by gradient echo 9.4 T MRI: a parametric study. *Contrast Media Mol Imaging* 2006;**1**(4):165–74.
5. Bulte JW, Kraitchman DL. Iron oxide MR contrast agents for molecular and cellular imaging. *NMR Biomed* 2004;**17**(7):484–99.
6. Frenkel J, Doefman J. Spontaneous and induced magnetisation in ferromagnetic bodies. *Nature* 1930;**126**:274–5.
7. Weissleder R, et al. Ultrasmall superparamagnetic iron oxide: characterization of a new class of contrast agents for MR imaging. *Radiology* 1990;**175**(2):489–93.
8. Na HB, Song IC, Hyeon T. Inorganic nanoparticles for MRI contrast agents. *Adv Mater* 2009;**21**(21):2133–48.
9. Laurent S, et al. Magnetic iron oxide nanoparticles: synthesis, stabilization, vectorization, physicochemical characterizations, and biological applications. *Chem Rev* 2008;**108**(6):2064–110.
10. McDonald MA, Watkin KL. Investigations into the physicochemical properties of dextran small particulate gadolinium oxide nanoparticles. *Acad Radiol* 2006;**13**(4):421–7.
11. Bridot JL, et al. Hybrid gadolinium oxide nanoparticles: multimodal contrast agents for in vivo imaging. *J Am Chem Soc* 2007;**129**(16):5076–84.

12. Guay-Begin AA, et al. Surface modification of gadolinium oxide thin films and nanoparticles using poly(ethylene glycol)-phosphate. *Langmuir* 2012;**28**(1):774–82.
13. Evanics F, et al. Water-soluble GdF3 and GdF3/LaF3 nanoparticles—physical characterization and NMR relaxation properties. *Chem Mater* 2006;**18**(10):2499–505.
14. Hifumi H, et al. Gadolinium-based hybrid nanoparticles as a positive MR contrast agent. *J Am Chem Soc* 2006;**128**(47):15090–1.
15. Na HB, et al. Development of a T1 contrast agent for magnetic resonance imaging using MnO nanoparticles. *Angew Chem Int Ed Engl* 2007;**46**(28):5397–401.
16. Cheong S, et al. Simple synthesis and functionalization of iron nanoparticles for magnetic resonance imaging. *Angew Chem Int Ed Engl* 2011;**50**(18):4206–9.
17. Chow AM, et al. Enhancement of gas-filled microbubble R2* by iron oxide nanoparticles for MRI. *Magn Reson Med* 2010;**63**(1):224–9.
18. Terreno E, et al. Challenges for molecular magnetic resonance imaging. *Chem Rev* 2010;**110**(5):3019–42.
19. Harisinghani MG, et al. MR imaging of pelvic lymph nodes in primary pelvic carcinoma with ultrasmall superparamagnetic iron oxide (Combidex): preliminary observations. *J Magn Reson Imaging* 1997;**7**(1):161–3.
20. Schmitz SA, et al. Magnetic resonance imaging of atherosclerotic plaques using superparamagnetic iron oxide particles. *J Magn Reson Imaging* 2001;**14**(4):355–61.
21. McCauley TR, Rifkin MD, Ledet CA. Pelvic lymph node visualization with MR imaging using local administration of ultra-small superparamagnetic iron oxide contrast. *J Magn Reson Imaging* 2002;**15**(4):492–7.
22. Kumano S, et al. Using superparamagnetic iron oxide-enhanced MRI to differentiate metastatic hepatic tumors and nonsolid benign lesions. *AJR Am J Roentgenol* 2003;**181**(5):1335–9.
23. Baghi M, et al. The efficacy of MRI with ultrasmall superparamagnetic iron oxide particles (USPIO) in head and neck cancers. *Anticancer Res* 2005;**25**(5):3665–70.
24. Kim T, et al. Effect of superparamagnetic iron oxide on tumor-to-liver contrast at T2*-weighted gradient-echo MRI: comparison between 3.0T and 1.5T MR systems. *J Magn Reson Imaging* 2009;**29**(3):595–600.
25. Yeh TC, et al. In vivo dynamic MRI tracking of rat T-cells labeled with superparamagnetic iron-oxide particles. *Magn Reson Med* 1995;**33**(2):200–8.
26. Schmitz SA, et al. Superparamagnetic iron oxide-enhanced MRI of atherosclerotic plaques in watanabe hereditable hyperlipidemic rabbits. *Invest Radiol* 2000;**35**(8):460–71.
27. Schmitz SA, et al. Iron-oxide-enhanced magnetic resonance imaging of atherosclerotic plaques: postmortem analysis of accuracy, inter-observer agreement, and pitfalls. *Invest Radiol* 2002;**37**(7):405–11.
28. Kim J, et al. Imaging of the inflammatory response in reperfusion injury after transient cerebral ischemia in rats: correlation of superparamagnetic iron oxide-enhanced magnetic resonance imaging with histopathology. *Acta Radiol* 2008;**49**(5):580–8.
29. Christoforidis GA, et al. High resolution ultra high field magnetic resonance imaging of glioma microvascularity and hypoxia using ultra-small particles of iron oxide. *Invest Radiol* 2009;**44**(7):375–83.
30. Marradi M, et al. Paramagnetic Gd-based gold glyconanoparticles as probes for MRI: tuning relaxivities with sugars. *Chem Commun (Camb)* 2009;(26):3922–4.
31. Reichenbach JR, et al. Small vessels in the human brain: MR venography with deoxyhemoglobin as an intrinsic contrast agent. *Radiology* 1997;**204**(1):272–7.
32. Haacke EM, et al. Susceptibility weighted imaging (SWI). *Magn Reson Med* 2004;**52**(3):612–8.

33. Mittal S, et al. Susceptibility-weighted imaging: technical aspects and clinical applications, part 2. *AJNR Am J Neuroradiol* 2009;**30**(2):232–52.
34. Hamans BC, et al. Contrast enhanced susceptibility weighted imaging (CE-SWI) of the mouse brain using ultrasmall superparamagnetic ironoxide particles (USPIO). *Z Med Phys* 2006;**16**(4):269–74.
35. Cheng JL, et al. In vivo tracing of superparamagnetic iron oxide-labeled bone marrow mesenchymal stem cells transplanted for traumatic brain injury by susceptibility weighted imaging in a rat model. *Chin J Traumatol* 2010;**13**(3):173–7.
36. Cunningham CH, et al. Positive contrast magnetic resonance imaging of cells labeled with magnetic nanoparticles. *Magn Reson Med* 2005;**53**(5):999–1005.
37. Eibofner F, et al. Positive contrast imaging of iron oxide nanoparticles with susceptibility-weighted imaging. *Magn Reson Med* 2010;**64**(4):1027–38.
38. Liu W, et al. In vivo MRI using positive-contrast techniques in detection of cells labeled with superparamagnetic iron oxide nanoparticles. *NMR Biomed* 2008;**21**(3):242–50.
39. Dahnke H, et al. Susceptibility gradient mapping (SGM): a new postprocessing method for positive contrast generation applied to superparamagnetic iron oxide particle (SPIO)-labeled cells. *Magn Reson Med* 2008;**60**(3):595–603.
40. Suzuki Y, et al. In vivo serial evaluation of superparamagnetic iron-oxide labeled stem cells by off-resonance positive contrast. *Magn Reson Med* 2008;**60**(6):1269–75.
41. Makowski MR, et al. Noninvasive assessment of atherosclerotic plaque progression in ApoE-/- mice using susceptibility gradient mapping. *Circ Cardiovasc Imaging* 2011;**4**(3): 295–303.
42. Shmueli K, et al. Magnetic susceptibility mapping of brain tissue in vivo using MRI phase data. *Magn Reson Med* 2009;**62**(6):1510–22.
43. Liu T, et al. Calculation of susceptibility through multiple orientation sampling (COSMOS): a method for conditioning the inverse problem from measured magnetic field map to susceptibility source image in MRI. *Magn Reson Med* 2009;**61**(1):196–204.
44. de Rochefort L, et al. Quantitative MR susceptibility mapping using piece-wise constant regularized inversion of the magnetic field. *Magn Reson Med* 2008;**60**(4):1003–9.
45. de Rochefort L, et al. Quantitative susceptibility map reconstruction from MR phase data using bayesian regularization: validation and application to brain imaging. *Magn Reson Med* 2010;**63**(1):194–206.
46. Kressler B, et al. Nonlinear regularization for per voxel estimation of magnetic susceptibility distributions from MRI field maps. *IEEE Trans Med Imaging* 2010;**29**(2):273–81.
47. Liu T, et al. Unambiguous identification of superparamagnetic iron oxide particles through quantitative susceptibility mapping of the nonlinear response to magnetic fields. *Magn Reson Imaging* 2010;**28**(9):1383–9.
48. Liu C, et al. High-field (9.4 T) MRI of brain dysmyelination by quantitative mapping of magnetic susceptibility. *Neuroimage* 2011;**56**(3):930–8.
49. Li W, Wu B, Liu C. Quantitative susceptibility mapping of human brain reflects spatial variation in tissue composition. *Neuroimage* 2011;**55**(4):1645–56.
50. Zhang L, et al. T-weighted ultrashort echo time method for positive contrast imaging of magnetic nanoparticles and cancer cells bound with the targeted nanoparticles. *J Magn Reson Imaging* 2011;**33**(1):194–202.
51. Girard OM, et al. Optimization of iron oxide nanoparticle detection using ultrashort echo time pulse sequences: comparison of T1, T2*, and synergistic T1–T2* contrast mechanisms. *Magn Reson Med* 2011;**65**(6):1649–60.

52. Robson MD, Bydder GM. Clinical ultrashort echo time imaging of bone and other connective tissues. *NMR Biomed* 2006;**19**(7):765–80.
53. Bae WC, et al. Ultrashort echo time MR imaging of osteochondral junction of the knee at 3 T: identification of anatomic structures contributing to signal intensity. *Radiology* 2010;**254**(3): 837–45.
54. Bae WC, et al. Conventional and ultrashort time-to-echo magnetic resonance imaging of articular cartilage, meniscus, and intervertebral disk. *Top Magn Reson Imaging* 2010;**21**(5): 275–89.
55. Weiger M, Pruessmann KP, Hennel F. MRI with zero echo time: hard versus sweep pulse excitation. *Magn Reson Med* 2011;**66**(2):379–89.

Chapter 10

Biosensors Based on Nanoparticles and Electrochemical Detection

Ester Polo*
**Nanotherapy and Nanodiagnostics Group (GN^2), Instituto de Nanociencia de Aragón, Universidad de Zaragoza, Edificio I+D, Mariano Esquillor s/n, Campus Rio Ebro, Zaragoza, Spain*

Sara Puertas*
**Nanotherapy and Nanodiagnostics Group (GN^2), Instituto de Nanociencia de Aragón, Universidad de Zaragoza, Edificio I+D, Mariano Esquillor s/n, Campus Rio Ebro, Zaragoza, Spain*

Pilar Batalla[†]
[†]Department of Analytical Chemistry and Chemical Engineering, University of Alcalá, Alcalá de Henares, Madrid, Spain

1. INTRODUCTION

Colloidal gold has been used since Ancient Roman times to color glass (yellow, red, or mauve) depending on the concentration of gold, and these applications are still continuing now. Perhaps the most famous example is the Lycurgus Cup that was manufactured in the 5th to 4th century BC. It is ruby red in transmitted light and green in reflected light, due to the presence of gold colloids. However, the beginning of more exhaustive studies started in 1857 with Michael Faraday's work.[1] He investigated the optical properties of thin films prepared from dried colloidal solutions and observed reversible color changes of the films upon mechanical compression. Since then till now, lots of works have been reported showing the exceptional optical and electrical properties of gold nanoparticles (GNPs) (electron dense core, highly resonant particle plasmons, catalytic size enhancement by silver deposition, etc.).[2–8]

In the field of therapeutic medicine, gold compounds were used by Chinese and Indian cultures in the 3th century BC for increasing vital power and curing male impotence.[9] In the 19th and 20th centuries, they were used for the treatment of epilepsy, syphilis, rheumatic, tuberculosis, and various inflammatory skin diseases.[10,11] Nowadays, the research lines are focused on cancer.[12–14] Nevertheless, they are used not only in therapeutic medicine but also in detection and

Frontiers of Nanoscience, Vol. 4. DOI: 10.1016/B978-0-12-415769-9.00010-8

diagnosis. GNPs have great interest in biosensing due to their intrinsic characteristics. This can be shown by the increase in the number of publications in the past decade (only 8 works were reported in 2002 whereas 800 articles have been published until now). Hereby, the authors will dedicate this chapter to analyze the role of GNPs in biosensing.

By definition, a biosensor is a quantitative or semiquantitative detection analytical technique (device) containing a sensing biomolecule which can convert a biological signal into an electrochemical or optical signal. Besides, a biosensor gives shorter analytical times and procedures, smaller sample volumes, higher test efficiencies, and lower cost, making it attractive for companies. Combination of the properties of GNP with the activity of biomolecules bound to them has allowed, and improved, the development of sensor devices. Concerning the biofunctionalization process, several proteins, antibodies, antigens, and nucleic acids are used in order to make them viable for biomedical, environmental, or food sensing.[15–19]

Biosensors could be classified into electrochemical, mass-sensitive, and optical sensors depending on the signal's transduction (Figure 10.1):

i. *Electrochemical sensors:* They are based on changes in electrical properties of a biofunctionalized electrode when the analyte is present in the sample. Among this kind of sensor, it could be distinguished, depending on the electronically amplified signal, between amperometric, impedimetric, potentiometric, and conductometric methods.[20–22]
ii. *Optical sensors:* They are based on optical changes, such appearance/disappearance of color or in surface plasmon resonance (SPR). Among the first ones, the most useful ones are lateral flow immunochromatographic assays (LFIAs).[23,24] They consist of a nitrocellulose membrane where a biological recognition agent has been deposited and traps the targeted analyte if present in the sample. On the other hand, in SPR technique, light

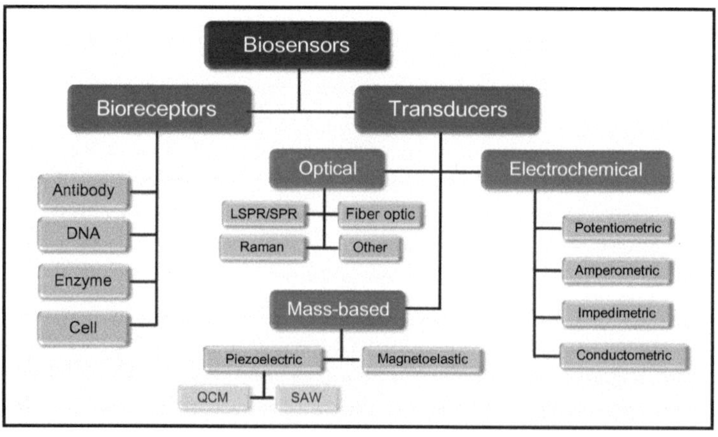

FIGURE 10.1 Classification of biosensors according to their bioreceptors and transducers. (For color version of this figure, the reader is referred to the online version of this chapter.)

photons interact with metal plasmons resulting in a diminished intensity of reflected light. This intensity varies when the analyte is attached.[25,26]

iii. *Mass-sensitive sensors (piezoelectric)*: They are based on resonance frequency variations. A functionalized quartz crystal changes its resonance frequency due to the increase in mass generated by antigen recognition. Their sensitivity depends on the crystal's composition and thickness.[27–30]

Throughout this chapter, the reader can find more details about the basics of each biosensor and several examples showing the diversity of detected analytes.

2. ELECTROCHEMICAL SENSORS

Electrochemical biosensors have been extensively studied because of their specific features: fast response, easy fabrication, high sensitivity, selectivity, quantification of analyte target, and moreover they offer the possibility of portable assays that could be used in point-of-care devices.[31,32]

In order to improve the properties of electrochemical sensors, new concepts of signal amplification are desirable for ultrasensitive biosensors. The design of new electrochemical biosensors combining an electrode with GNPs is recently gaining importance. Therefore, some reviews have covered the main contributions to this technique.[33,34] As they state, GNPs are especially attractive because they are able to immobilize biomolecules retaining their bioactivity, allowing the direct electron transfer between redox proteins and bulk electrode materials and giving a high surface-to-volume ratio.

Nanosized structures provide excellent electron transfer rates, and they enable direct electron transfer when they are used to modify the surface of working electrodes. The electrical contact of the redox-active sites of enzymes with electrodes is one of the most fundamental issues in bioelectrochemistry, providing the basis for the development of biosensors. In many cases, the spatial separation between the enzymes and electrodes hampers these reactions, hence nanoparticles can act as a connector for transporting electrodes between the redox sites of the enzyme and the electrode.[35] The catalytic reaction of an apoenzyme took place when GNPs functionalized with a cofactor were bound to an Au electrode via dithiol bridges.[36] In other cases, nanoparticles are used as a new reagent that enhances the signal and improves the sensitivity and stability of the device. They can be considered a label that amplifies a single recognition event.

Somehow, the application of nanoparticles enhances the sensitivity of the devices, making it feasible to apply them to a broad variety of fields: clinical,[37] environmental,[38] or food[39] are some examples, because they can detect different kind of molecules such as proteins, nucleic acids, toxins, or pesticides.

Due to their versatility, GNPs can play different roles within the sensor. Depending on this, the GNPs can perform as a label to amplify sensor's signal, be directly deposited onto the electrode surface, be mixed with the electrode material forming a composite, or combine their properties with carbon nanotubes (CNTs) (Figure 10.2).

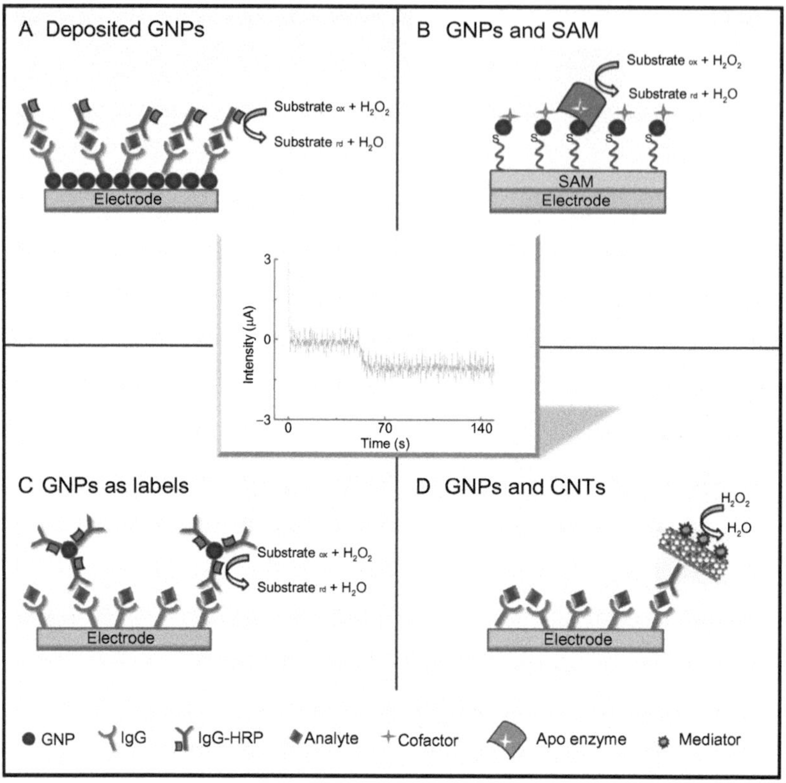

FIGURE 10.2 Scheme of different applications of GNPs in electrochemical devices. (A) Direct deposited GNPs on electrodes, (B) GNPs functionalized SAM layer, (C) GNPs used as markers, and (D) Immobilized GNPs on CNT. In the middle, example of an amperometric signal. (For color version of this figure, the reader is referred to the online version of this chapter.)

2.1. GNPs Mixed with the Electrode Material

It is possible to mix the composite matrix of the electrode with GNPs, obtaining a more intimate interaction that enhances both the enzyme reaction and the electrochemical one.

Pingarron's group has worked with this approach developing a tyrosinase biosensor through the construction of a graphite–Teflon composite electrode matrix in which the enzyme and colloidal GNPs are incorporated by simple physical adsorption.[40] Another example consists in a xanthine oxidase biosensor that mixes glassy carbon microparticles with GNPs.[41]

2.2. Direct Deposition of GNPs onto the Electrode

The direct deposition of nanoparticles onto the electrode surface creates a more favorable surface for the attachment of the enzymes. They generate a roughened surface, increasing its superficial area and enhancing the interaction of the enzyme with the electrode (Figure 10.2A).

One common procedure to reach this requires the direct electrodeposition of nanoparticles onto the electrode. With this technique, myoglobin has been immobilized on a colloidal gold-modified carbon paste electrode,[42] using amperometric measurements. Amperometry was also the technique used to measure the signal of a glucose oxidase (GOx) sensor, enzyme that was covalently attached to one GNP monolayer that modifies an Au electrode.[43] Another example implies a sensor for the detection of the pesticide carbofuran at nanomolar concentrations where acetylcholinesterase is immobilized onto the electrode after electrodeposition of colloidal GNPs.[38] In this case, in the absence of GNPs layer, the sensor response was significantly reduced.

The surface of the electrode can be modified with an ionic reagent, thus the GNPs react with the electrode through an ionic interaction. This is the case of a prostate-specific antigen (PSA) sensor able to detect concentrations of PSA in the range of pg/mL.[44] GNPs protected with glutathione were adsorbed onto a graphite disk electrode activated with a cationic reagent (polydimethyldiallylammonium). Anti-PSA antibody was covalently linked onto the carboxyl groups of the GNP layer using EDC/NHS chemistry. After the PSA recognition, the sandwich reaction takes place with a secondary antibody. In order to amplify the signal, this secondary antibody was bound to magnetic beads with multiple HRP labels, obtaining with such a packed GNP platform an ultrasensitive electrochemical immunosensor.

2.3. GNPs Immobilized onto Self-assembled Monolayers

The modification of electrode surfaces with self-assembled monolayers (SAMs) of thiols provides a simple way to design tailored materials that can be further used as functionalized sites to immobilize GNPs. With this approach, the GNPs are also bound to the electrode, although there is the SAM between them, this layer enables the functionalization of the electrode (Figure 10.2B).

After the functionalization of gold electrodes with cysteamine monolayers, GNPs react with the monolayer, covering the surface. This is the basis of a sensor that can be use with enzymes immobilized onto the GNPs[45] or DNA sensor.[46]

2.4. GNPs as a Label

The use of GNPs as label amplifying the signal from a biorecognition event is probably one of their most common applications. A great variety of biosensors has been built following this approach, particularly in immunosensors and DNA sensors (Figure 10.2C).

Systems including nanoparticle-based electrochemical amplification have been stated to be extremely sensitive being able to detect concentration between 10^{-15} and 10^{-21} M. This technology is well-suited to multiple-target detection with different nanoparticles (in case of using nanoparticles of different nature), but it also presents some drawbacks such as presenting many development steps in the assays.[47]

Among the enzyme immunosensors, it has been reported a magnetically based sandwich immunoassay using GNPs as labels that detect human IgG antibody with very low limits of detection (1.7 pM). The use of magnetic particles makes possible this low detection limit: a magnetic particle collection is a step included in the assay that helps to concentrate the sample.[48]

GNP presence can be recorded with an enzyme or through the redox signal of the Au itself. This last kind of detection was the one applied to an immunosensor where the capture protein was adsorbed to the electrode, while a colloidal gold-labeled antibody performed the sandwich assay. Once the sandwich was ready, the GNPs were oxidized and its signal recorded by adsorptive voltammetry.[46]

The GNPs that participate in an immunoassay can also be enlarged with silver, improving their characteristics as labels. A silver ion-selective microelectrode is used then to detect potentiometrically the silver ions released from the silver.[49]

With the genosensors, one can also find examples of systems that detect the GNPs with a silver deposition technique,[50] where DNA strands were immobilized between two electrodes, while the complementary strand is labeled with GNPs, and the silver deposition facilitated by these GNPs gives readily measurable conductivity changes. Otherwise, the signal can be detected by the oxidation signal of colloidal Au, where an Au electrode is modified with DNA that hybridizes with complementary probes conjugated to GNPs. These DNA sequences were able to detect a mutation of the Factor V Leiden. In this case, the electroactivity of GNPs was used for the detection of the hybridization without any external indicators. The specific sequences of DNA were directly detected by anodic analysis of gold colloids.[37]

2.5. GNPs and Carbon Nanotubes

Hybrid nanoparticle–nanotube materials have shown to have interesting properties, enhancing the electrochemical reaction. These devices can be prepared by selective attachment of GNPs to the CNT surfaces, with a previous functionalization of CNTs to immobilize GNPs. The functionalization is achieved many times with a cationic or anionic reagent that changes the surface properties of CNTs and allows them to be coated successfully with the GNPs through the electrostatic interaction (Figure 10.2D).

In one of the first works that mix these materials, the nanoparticles were made of Pt instead of gold.[51] But more recently, there are other examples in the literature of this technique using GNPs. The composite electrode from a glucose biosensor was prepared mixing GNPs, CNT, and Teflon, showing significant improved responses to hydrogen peroxide, including those sensors based on CNTs.[15] The incorporation of GOx into the matrix allowed the preparation of a mediatorless glucose biosensor.

One application of the CNT as labels can be found in immunosensors based on GNP-decorated MWCNTs (multiwall carbon nanotubes) with multiple hydrazine labels.[39] These immunosensors were able to detect Neomycin (Neo), a powerful antibacterial that has toxicological effects in human beings. The device was fabricated immobilizing a monoclonal antiNeo antibody onto GNPs functionalized with a monomer bearing a benzoic acid group (DPB). The GNPs were electrodeposited onto bare electrode and then the DPB polymer was added. The antibody was immobilized via EDC/NHS reaction. Sandwich immunosensing method was adopted, where the secondary antibody was indirectly labeled with MWCNT decorated with hydrazine-labeled GNP to achieve multiamplification of the response. Using this sensor, they detected concentrations in the range of ng/mL.

3. OPTICAL SENSORS

The unique light-scattering properties of GNPs provide a wide range of opportunities for application of GNPs in interfacing biological recognition events with signal transduction and in designing multiple biosensing devices.

GNPs that are linked to bioreceptors provide labeled conjugates that can be used to optically follow the biorecognition events at biosensor surfaces. The most prominent detection techniques are based on the interaction between GNPs and light, and allow analyzing numerous components in samples directly. Various optical methods have been employed to detect the association of NPs on biochips, including the scanometric detection of light scattering, SPR spectroscopy, resonance-enhanced absorption by NPs, nanoparticle fluorescence, and enhanced Raman scattering.[52–54]

The colorimetric change pointed to a simple and inexpensive way of diagnosing disease. Many solution-based nanoparticle assays take advantage of analyte-induced aggregation events that result in measurable changes and shifts of nanoparticle surface plasmon absorption bands.[32]

3.1. Surface Plasmon

GNPs strongly absorb and scatter visible light. Upon light absorption, freely mobile electrons (6s electrons of the conduction band) in the GNP surface show a characteristic collective oscillation frequency of the plasma resonance, giving rise to the so-called plasmon resonance band SPB (observed near around 510–530 nm wavelengths in the 4–40 nm diameter range).[55] The SPB maximum and bandwidth are also influenced by the particle shape, medium dielectric constant, and temperature. The refractive index of the solvent has been shown to induce a shift of the SPB.[56] Applications of the sensitivity of the position of the SPB are known, especially in the fields of sensors and biology.

The binding of molecules to the particle surface can change the plasmon resonance frequency directly. Besides, the plasmon resonance frequency is dramatically changed when the average distance between gold particles is reduced so that they form small aggregated colloids. Separate GNPs have a red color as the consequence of their plasmon absorbance, but upon aggregation, another absorbance arise at lower energies because of interparticle plasmon coupling. The method was pioneered by Mirkin and co-workers and is nowadays maybe the most well-known example of a gold-based sensor.[57] The original assay was developed for the detection of DNA; aggregation of GNPs linked by the oligonucleotides provokes a red-to-blue color change (red shift from 520 to 600 nm of the SPB) that is most useful for this DNA-sensing method (Figure 10.3A). The effect of the length of the DNA strands that control the interparticle distance was studied, and it was found that the SPB frequency changes are inversely dependent on the oligonucleotide linker length.[58]

Methods based on nucleotide interactions have been the most reported of these particle–particle coupling systems. Limits of detection in the picomolar range for a variety of analytes have been achieved using oligonucleotide-modified GNPs. This effect of plasmon coupling can be used for colorimetric detection of multiple analytes, apart from DNA sensing, such as metal ions (Pb^{2+}),[59] adenosine, amino acids,[60] proteins,[61,62] toxins,[63] or cancer cells by use the excellent selectivity of aptamers or aptazymes.[64,65] The biochemical interactions induce the aggregation of the nanoparticles, which can be easily detected through the corresponding color change, without the use of complex instrumentation. This colorimetric assay can be carried out in complex biological media without sample preparation; immunoglobulin has been detected by aggregated antibody-GNPs, in saline serum and whole blood.[62]

3.2. SPR and Localized SPR

SPR is an optical phenomenon arising from the interaction between an electromagnetic wave and the conduction electrons in a metal and is used for probing and characterizing physicochemical changes of thin films on metal surface. SPR can be excited on thin metal films using prism couplers. This resonance is a coherent oscillation of the surface conduction electrons excited by electromagnetic radiation. The binding of specific molecules onto the surface of metallic films can induce a variation in the dielectric constant. A change in refractive index shifts the plasmon resonance condition, which can be detected as intensity, wavelength, or angle shifts to provide quantitative information about the recognition event. The sensitivity of the method which depends on the refractive index changes can be improved by coupling the analyte recognition at the surface of the metal with another event that leads to larger changes in the SPR signal.[66]

GNPs, whose optical extinction spectrum is highly sensitive to the dielectric constant of the surrounding medium, have been studied intensively in SPR

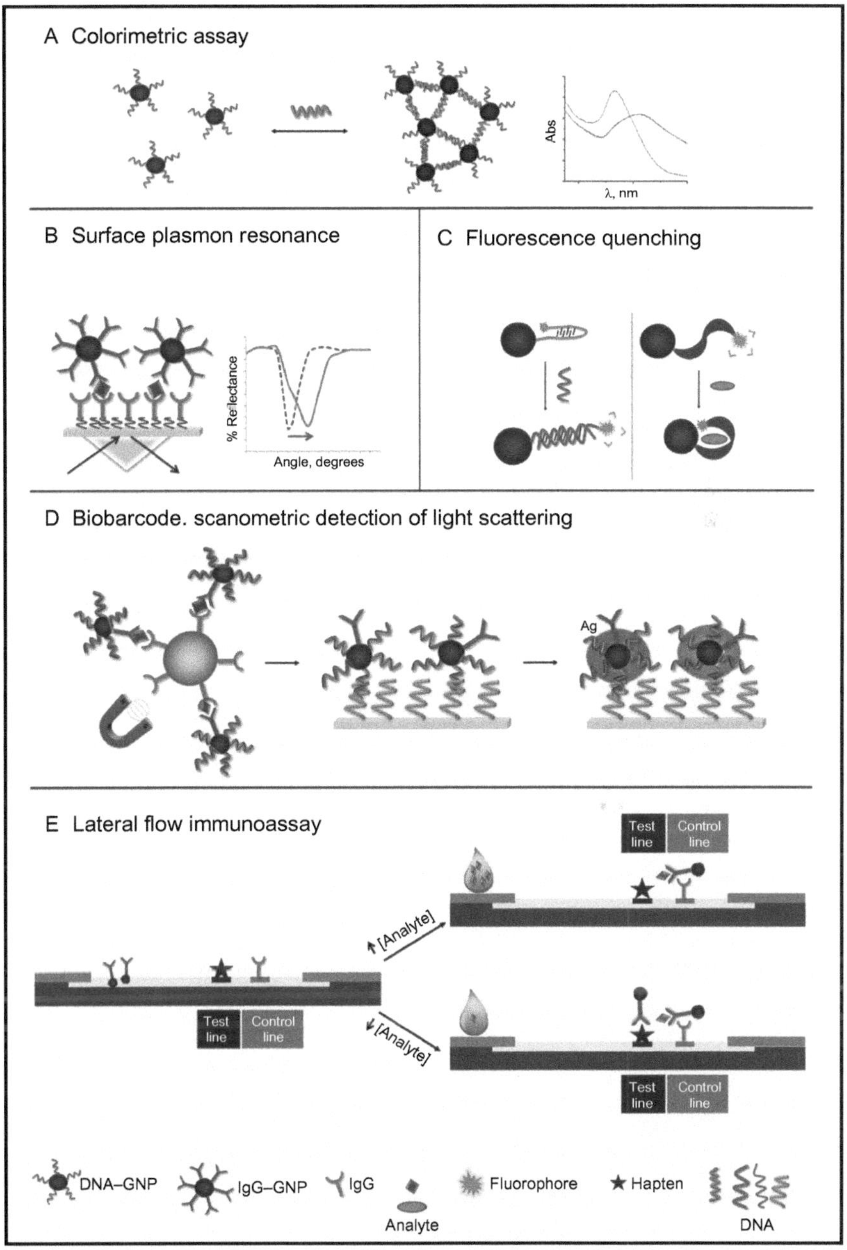

FIGURE 10.3 Representative scheme of the employment of GNPs in optical sensors. (A) Aggregation of the surface plasmon band, (B) SPR methodologies, (C) GNPs in fluorescence techniques, (D) GNP mixed with MNPs (barcode), and (E) GNP in a competitive lateral flow immunoassay. (For color version of this figure, the reader is referred to the online version of this chapter.)

to provide better analytical characteristics.[67] Real-time DNA hybridization on gold surfaces could be detected by SPR with detection limits of 150 nM DNA; however, using oligonucleotide-GNP labels, the detection limit improves 100-fold to less than 10 pM DNA. The signal amplification mechanism of GNPs is due to the electronic coupling between the localized surface plasmon of GNPs and the propagating plasmon on the SPR gold surface. Nanoparticle-enhanced SPR spectroscopy has been extensively used for immunosensing, DNA sensing,[68] detecting protein conformational changes, and more recently, to follow biocatalytic reactions due to the plasmon coupling between the charged GNPs and the metallic film resulted in a shift of the SPR spectrum which was dependent on the charge generated on the nanoparticles[69] (Figure 10.3B).

The LSPR is an optical phenomenon generated by collective oscillations of the electron gas in metal nanostructures surrounded by a dielectric. When metal nanostructures interact with a light beam, part of the incident photons are absorbed and part are scattered in different directions. Both absorption and scattering are greatly enhanced when the LSPR is excited.[19] Consequently, optical spectroscopy is the simplest method to detect the LSPR on metal nanostructures and is generally based on extinction or scattering measurements. Extinction is often employed to characterize systems containing a large amount of nanostructures, such as nanoparticle colloids. In contrast, scattering measurements, based on dark-field or total internal reflection spectroscopy, present a much lower signal-to-background ratio than extinction.[70] Therefore, scattering detection is more interesting in samples with a low density of nanostructures, allowing the optical characterization of even single nano-objects.[71]

The maximum absorbance wavelength of their localized SPR spectrum is sensitive to the size, shape, interparticle spacing, and dielectric properties of the material and local external dielectric environment of the nanoparticle. This latter property has been exploited for developing new nanoscale affinity biosensors that are based on transducing small changes in the refractive index near the nanoparticle surface into a measurable wavelength–shift response.[72] A huge variety of gold nanostructures have been used in solution or immobilized on surfaces for this detection method. GNP immobilized on surfaces can be applied for potential on-chip sensing process, and biorecognition events result in LSPR changes measured with a commercially available UV–Vis spectrometer. GNP-layered substrates and different gold nanostructures such as nanorods,[73] stars,[74] or nanoholes[75] have been investigated as potential LSPR sensing platforms, for the label-free detection of DNA hybridization or antigen–antibody reactions (more recently to detect the interaction between molecules involved in Alzheimer's disease).[76]

3.3. Scanometric Detection of Light Scattering

The unique light-scattering properties of GNPs have prompted interest in their potential application as labels for multiplexed analyte detection. Light-scattering particles can be used as fluorescent analog tracers in immune- and DNA sensing. It has been proved that GNPs show a light-scattering power equivalent to the signal generated from 10^6 fluorescein molecules.[77] Besides this, GNPs are advantageous in these applications because the signal from this particles is quench resistant and it is possible to detect the emission using a very simple and low-cost illumination system. Molecular fluorophores have been replaced in typical cDNA microarrays by resonance light-scattering nanoparticles due to their potential as labeling agents.

Scanometric detection of light scattering was used to detect single-base mismatches in oligonucleotides. The methodology for scanometric detection is based on a modified glass slide mounted on a microscope stage and was illuminated in the plane of the slide with white light. In this configuration, the slide served as a planar waveguide to prevent any light from reaching the micro scope objective by total internal reflectance. Wherever NP probes were attached to the surface, evanescently coupled light was scattered from the slide and was observed as bright, colored spots. This approach was used for the detection of GNP-labeled DNA molecules that were specifically bound to a DNA-functionalized surface. Enlargement of the GNPs by the catalytic reduction of silver ions and the deposition of silver metal on the GNPs resulted in a 100-fold amplification of the signal and thus increased the sensitivity, demonstrating the ability to detect genomic DNA in samples without PCR amplification.[78] This approach was used to detect target DNA at picomolar concentrations and allowed the detection of single-nucleotide polymorphism.

A novel method namely biobarcode assay, based on this detection system, has developed for detecting proteins down to the low attomolar level. This methodology utilizes antibody-coated magnetic beads to concentrate the target protein, and GNP probes that are conjugated with specific antibodies and DNA strands (barcodes) to label the analyte (Figure 10.3D). After the biorecognition event, GNP–DNA barcodes are released from the complex and detected via hybridization to a surface-immobilized DNA probe. The light scattered from the nanoparticles is detected after they were enlarged through silver deposition. The signal amplification was achieved because each GNP probe carried a large number of oligonucleotides per protein binding event.[79] Following this methodology, protein markers involved in disease such as prostate and breast cancer, Alzheimer's disease, and AIDS can be detected at levels unachievable with current techniques. The biobarcode assay was applied for detecting the PSA at attomolar level (six orders of magnitude higher than the ELISA). Recently, Georganopoulou *et al.* have been able to detect the Alzheimer's disease marker ADDL in the cerebral spinal fluid at picomolar levels, using this methodology.[80]

3.4. Surface-Enhanced Raman Scattering

In general, by the inelastic Raman scattering process, molecular vibrations and thereby molecular structure, composition, and interactions can be detected by the observation of scattered light with a shifted wavelength compared with the wavelength of an excitation source. Raman signals are quite weak and therefore a sufficient analyte concentration is needed in order to provide enough signal.

Spectroscopic effects can be affected strongly when they take place in the immediate vicinity of metal surfaces and nanostructures because of coupling to surface plasmons. Based on resonances with their surface plasmons, gold nanostructures also give rise to enhanced local optical fields. This results in a much higher scatter probability and thus in a gain of several orders of magnitude of the Raman-scattered light intensity that is detected. These local optical fields result in enhanced spectroscopic signals. Surface-enhanced Raman scattering (SERS) is one of the most impressive observations to demonstrate this effect. SERS is a technique very useful for detecting conformational changes and structural differences regarding preferred orientations of molecules on a metal surface.[81]

SERS can give detailed information on molecular structure and chemical composition in the nanometer-scaled environment of the nanoparticles. SERS on gold nanostructures can serve two different purposes: (i) they are used as SERS labels, a reporter molecule linked to a GNP is detected by its very specific Raman signature. (ii) GNPs can serve as nanosensors by delivering the enhanced Raman spectroscopic signatures of biological molecules and structures in their environment. Recent work indicates that SERS nanoparticle tags can be used for highly specific recognition and Raman detection of human cancer cells. Recent works reported the use of GNPs conjugated with a single-chain variable fragment ScFv antibody anti-EGFR (epidermal growth factor receptor) for active tumor targeting.[82,83] GNPs modified with Raman-active reporter molecules also have been used in the detection of DNA, proteins, or pathogen bacteria.[84,85]

3.5. Fluorescence Quenching

An alternative approach for the optical detection of biorecognition processes consists of the use of GNP as local quenchers of the fluorescence of dyes (Figure 10.3C). Several sensor strategies are based on this effect. For example, a molecule which changes its conformation upon binding of the analyte is used as a spacer to link fluorophores to GNPs. Without the presence of the analyte, the spacer molecule is extended and there is no quenching of the fluorescence of the fluorophore. On the other hand, when the analyte binds to the spacer, it changes its conformation in such a way that the attached fluorophore will be brought into close proximity to the nanoparticle surface which will result in the quenching of the fluorophore. The higher the concentration of analyte in solution, the lower the recorded fluorescence signal will be. The same principle can be used in the opposite way when binding of the analyte extends the spacer attached to the particles and the quenched fluorescence increases after binding.

Molecular beacons are commonly used for nucleic acid detection, and recent studies have improved the quenching efficiency using GNPs instead of the molecular quencher. In a DNA molecular beacon, which was terminated at its ends with a GNP and a dye, intramolecular quenching of the dye fluorescence persists. Opening of the DNA molecular beacon by hybridization with an analyte DNA probe led to the regeneration of the fluorescence of the dye because of the spatial separation of the nanoparticle and dye units.[86]

3.6. Lateral LFIAs

LFIA, also called lateral flow assay or the strip assay, is based on an immunochromatographic process that uses antigen–antibody interactions and enables rapid detection of the analyte. This technique has been under development for several years due to its multiple advantages with regard to other methodologies. Among this benefits, it could emphasize their simplicity, portability, affordable price, and, of course, the rapidity in the appearance of the signal (less than 10 min). Moreover, these kinds of test represent a well-established technology when applied to a wide variety of point-of-care or field use applications. However, these assays are basically designed for colorimetric detection by visual inspection given only qualitative or semiquantitative results. In addition, they show poor limit detection. Several strategies have been reported to increase its sensitivity such as labeling the antibody with enzymes, using magnetic nanoparticles, quantum dots, colloidal GNPs, or even mixes of them.[87–91]

Basically, LFIA consists of a nitrocellulose membrane with micrometric pores allowing the flow of liquid via capillarity, where a biological recognition agent has been positioned in the test line and traps the targeted analyte if present in the flowing liquid. More in detail, an LFIA is formed by three pads (sample, conjugate, and absorbent) and the nitrocellulose membrane. The sample pad is treated with a solution containing different reactives that allow the flow of the sample and avoid the unspecific interactions (BSA, Tween-20, sucrose...). The conjugated pad is saturated with a solution containing anti-analyte-functionalized NP (anti-analyte-NP) and NP funtionalized with other molecule (i.e., biotin). Near the top of the strip membrane, a solution of anti-analyte is applied (test line) and, close to this line, another solution with, in the case of biotin, streptavidin is deposited (control line). This way, if the analyte is present in the sample, two bands appears (control and test line) and the test is considered positive. On the other hand, the test is negative when only the control line is visible. Finally, if no signal is observed at the control line, the test is rejected.[92,93]

As it was described above, the low sensitivity of LFIA is one of the disadvantages that present the technique. The introduction of GNP has allowed improving both the sensitivity and the possibility of the quantification. The intensity of the red color from the GNP is proportional to the analyte concentration, and it could be determined by using a portable "strip reader" instrument.[94] Moreover, it has been possible to design competitive LFIA by using GNP (Figure 10.3E). This new approach is based on the appearance of the control band only when the sample contains a larger amount of analyte. If the concentration is too low, the control and the test bands are visible. Normally, this kind of test has immobilized an anti-antibody in the control line and a hapten-functionalized GNP in the test line. Hereby, after the reaction between the analyte and the anti-analyte-GNP, the conjugate migrates by capillarity and binds to the hapten placed in test line. The excess of conjugate is captured by the anti-antibody situated in the control line. Therefore, if the sample contains a large amount of analyte, the conjugate competes to immobilize hapten obtaining only the control band. However, if the sample does not contain the analyte, the anti-analyte-GNP reacts with the hapten and the excess with the anti-antibody visualizing, at the end, two bands. In this case, the intensity of the test line is inversely proportional to the amount of analyte present in the sample.[95,96]

Due to the diversity and selectivity of antibodies and the advantages of LFIA, this device has been used to determine a huge variety of analytes such as toxins, antibiotics in food, hormones in serum and urine, explosives, etc.[92–95] Conversely, other molecules can be attached on GNP (nucleic acids, enzymes, aptamers, etc.) increasing the versatility of this methodology.[23,24,97–101] Although GNP improves considerably the sensitivity, it is necessary to take into account some aspects such as its size and stability. Several works have been published showing that 40-nm GNPs are the best for strip assays. Smaller particles give low signals due to the scattering light while larger particles migrate slower.[102,103]

4. MASS-SENSITIVE SENSORS

Mass-sensitive detectors such as quartz crystal microbalance (QCM) devices operate on the basis of the piezoelectric effect, in this case, an oscillating crystal that resonates at a fundamental frequency. The resonant frequency shift of a QCM is based on the measurement of mass change by the adsorption of target molecules.[104]

The QCM technique has been used for a wide range of applications in the medical field to analyze biomolecule reactions such as antigen–antibody and DNA hybridization. However, for biosensing, in which the target detection molecules are always in trace quantity, the sensitivity of QCM remains to be further enhanced.[105] A novel strategy for the high-sensitive detection of target biomolecules on a QCM is based on GNPs as signal enhancement probes. The mechanisms of the sensitivity enhancement contribution of GNPs can be summarized as follows: surface enhancement effect and mass enhancement effect (Figure 10.4A). The high surface area-to-volume ratio and biocompatibility of GNPs, which provide a very large number of interaction sites, make GNPs potential candidates to amplify the sensing surface area and maintain the sensing bioactivity, thereby enhancing the quantity and activity of biological recognition elements. Moreover the relatively high density of GNP allows application as a mass enhancer.

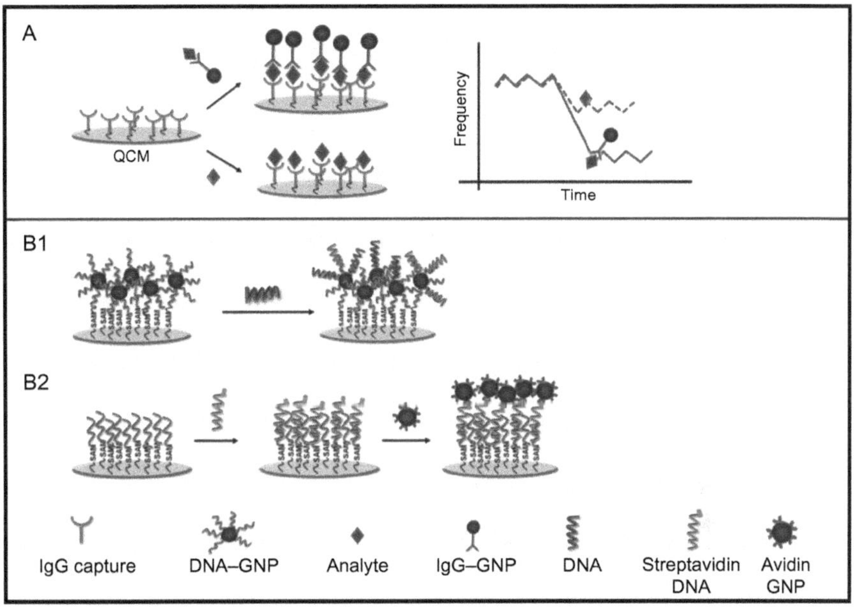

FIGURE 10.4 Schematic representation of QCM biosensors using gold nanoparticles. (A) Immune-gold nanoparticles as signal enhancement probes. (B1) GNPs amplify the sensing surface area and enhance the quantity of biological recognition elements. (B2) GNPs as mass enhancers. (For color version of this figure, the reader is referred to the online version of this chapter.)

Numerous research groups improved the analytical sensitivity by coupling GNPs with the QCM sensing process. DNA biosensor signal is amplified by immobilized GNPs onto the gold surface of QCM, followed by immobilization of oligonucleotide probes onto the GNPs (Figure 10.4B1). The target DNA was detected by hybridization reaction with the probes. The relatively large surface area of the GNPs was considered to increase the amount of probes onto the gold surface of the QCM, and thereby enhance the sensitivity. GNPs are also successfully used as amplification tags in some other DNA piezoelectric biosensors, such as a DNA point mutation detection method based on DNA ligase reaction and GNP amplification, single-base mismatches detection using GNP-catalyzed deposition of gold on GNP–avidin conjugate to amplify the signals (Figure 10.4B2), and *Escherichia coli* real-time detection using GNPs as a mass enhancer.[106,107]

REFERENCES

1. Faraday M. Experimental relations of gold (and other metals) to light. *Philos Trans R Soc Lond* 1857;**147**:145–81.
2. Chow G, Markowitz MA, Rayne R, Dunn DN, Singh A. Phospholipid mediated synthesis and characterization of gold nanoparticles. *J Colloid Interface Sci* 1996;**183**:135–42.

3. Chen S, Kimura K. Synthesis and characterization of carboxylate-modified gold nanoparticle powders dispersible in water. *Langmuir* 1999;**15**:1075–82.
4. Lin J, Zhou W, Kumbhar A, Wiemann J, Fang J, Carpenter EE, et al. Gold-coated iron (Fe@Au) nanoparticles: synthesis, characterization, and magnetic field-induced self-assembly. *J Solid State Chem* 2001;**159**:26–31.
5. Mukherjee P, Patra CR, Ghosh A, Kumar R, Sastry M. Characterization and catalytic activity of gold nanoparticles synthesized by autoreduction of aqueous chloroaurate ions with fumed silica. *Chem Mater* 2002;**14**:1678–84.
6. Chen MMY, Katz A. Synthesis and characterization of gold–silica nanoparticles incorporating a mercaptosilane core-shell interface. *Langmuir* 2002;**18**:8566–72.
7. Zhou JC, Wang X, Xue M, Xu Z, Hamasaki T, Yang Y, et al. Characterization of gold nanoparticle binding to microtubule filaments. *Mater Sci Eng C* 2010;**30**:20–6.
8. Faramarzi MA, Forootanfar H. Biosynthesis and characterization of gold nanoparticles produced by laccase from Paraconiothyrium variabile. *Colloids Surf B Biointerfaces* 2011;**87**:23–7.
9. Mahdihassan S. Cinnabar-gold as the best alchemical drug of longevity, called Makaradhwaja in India. *Am J Chin Med* 1985;**13**:93–108.
10. Shaw CF. Gold-based therapeutic agents. *Chem Rev* 1999;**99**:2589–600.
11. Panyala NR, Peña-Méndez EM, Havel J. Gold and nano-gold in medicine: overview, toxicology and perspectives. *J Appl Biomed* 2009;**7**:75–91.
12. Paciotti GF, Myer L, Weinreich D, Goia D, Pavel N, McLaughlin RE, et al. Colloidal gold: a novel nanoparticle vector for tumor directed drug delivery. *Drug Deliv* 2004;**11**:169–83.
13. Yang P-H, Sun X, Chiu J-F, Sun H, He Q-Y. Transferrin-mediated gold nanoparticle cellular uptake. *Bioconjug Chem* 2005;**16**:494–6.
14. Wieder ME, Hone DC, Cook MJ, Handsley MM, Gavrilovic J, Russell DA. Intracellular photodynamic therapy with photosensitizer-nanoparticle conjugates: cancer therapy using a "Trojan horse" *Photochem Photobiol Sci* 2006;**5**:727–34.
15. Saxena U, Chakraborty M, Goswami P. Covalent immobilization of cholesterol oxidase on self-assembled gold nanoparticles for highly sensitive amperometric detection of cholesterol in real samples. *Biosens Bioelectron* 2011;**26**:3037–43.
16. Haghighi B, Bozorgzadeh S, Gorton L. Fabrication of a novel electrochemiluminescence glucose biosensor using Au nanoparticles decorated multiwalled carbon nanotubes. *Sens Actuators B Chem* 2011;**155**:577–83.
17. de la Escosura-Muñiz A, Sánchez-Espinel C, Díaz-Freitas B, Gónzalez-Fernández A, Maltez-da Costa M, Merkoçi A. Rapid identification and quantification of tumor cells using an electrocatalytic method based on gold nanoparticles. *Anal Chem* 2009;**81**:10268–74.
18. Wang Q, Huang J, Yang X, Wang K, He L, Li X, et al. Surface plasmon resonance detection of small molecule using split aptamer fragments. *Sens Actuators B Chem* 2011;**156**:893–8.
19. Sepúlveda B, Angelomé PC, Lechuga LM, Liz-Marzán LM. LSPR-based nanobiosensors. *Nano Today* 2009;**4**:244–51.
20. Heurich M, Kadir MKA, Tothill IE. An electrochemical sensor based on carboxymethylated dextran modified gold surface for ochratoxin A analysis. *Sens Actuators B Chem* 2011;**156**:162–8.
21. Liu S, Yuan R, Chai Y, Su H. A label-free amperometric immunosensor based on horseradish peroxidase functionalized carbon nanotubes and bilayer gold nanoparticles. *Sens Actuators B Chem* 2011;**156**:388–94.
22. Rezaei B, Saghebdoust M, Sorkhe AM, Majidi N. Generation of a doxorubicin immunosensor based on a specific monoclonal antibody-nanogold-modified electrode. *Electrochim Acta* 2011;**56**:5702–6.

23. He Y, Zhang S, Zhang X, Baloda M, Gurung AS, Xu H, et al. Ultrasensitive nucleic acid biosensor based on enzyme-gold nanoparticle dual label and lateral flow strip biosensor. *Biosens Bioelectron* 2011;**26**:2018–24.
24. Girotti S, Eremin S, Montoya A, Moreno MJ, Caputo P, D'Elia M, et al. Development of a chemiluminescent ELISA and a colloidal gold-based LFIA for TNT detection. *Anal Bioanal Chem* 2010;**396**:687–95.
25. Urusov AE, Kostenko SN, Sveshnikov PG, Zherdev AV, Dzantiev BB. Ochratoxin A immunoassay with surface plasmon resonance registration: lowering limit of detection by the use of colloidal gold immunoconjugates. *Sens Actuators B Chem* 2011;**156**:343–9.
26. Wang X, Xu Y, Chen Y, Li L, Liu F, Li N. The gold-nanoparticle-based surface plasmon resonance light scattering and visual DNA aptasensor for lysozyme. *Anal Bioanal Chem* 2011;**400**:2085–91.
27. Hao R-Z, Song H-B, Zuo G-M, Yang R-F, Wei H-P, Wang D-B, et al. DNA probe functionalized QCM biosensor based on gold nanoparticle amplification for Bacillus anthracis detection. *Biosens Bioelectron* 2011;**26**:3398–404.
28. Chu X, Zhao Z-L, Shen G-L, Yu R-Q. Quartz crystal microbalance immunoassay with dendritic amplification using colloidal gold immunocomplex. *Sens Actuators B Chem* 2006;**114**:696–704.
29. Wei X-L, Mo Z-H, Li B, Wei J-M. Disruption of HepG2 cell adhesion by gold nanoparticle and Paclitaxel disclosed by in situ QCM measurement. *Colloids Surf B Biointerfaces* 2007;**59**:100–4.
30. Sener G, Ozgur E, Yılmaz E, Uzun L, Say R, Denizli A. Quartz crystal microbalance based nanosensor for lysozyme detection with lysozyme imprinted nanoparticles. *Biosens Bioelectron* 2010;**26**:815–21.
31. Ansari AA, Alhoshan M, Alsalhi MS, Aldwayyan AS. Prospects of nanotechnology in clinical immunodiagnostics. *Sensors* 2010;**10**:6535–81.
32. Rosi NL, Mirkin CA. Nanostructures in biodiagnostics. *Chem Rev* 2005;**105**:1547–62.
33. Pingarron JM, Yáñez-Sedeño P, González-Cortés A. Gold nanoparticle-based electrochemical biosensors. *Electrochim Acta* 2008;**53**:5848–66.
34. Siangproh W, Dungchai W, Rattanarat P, Chailapakul O. Nanoparticle-based electrochemical detection in conventional and miniaturized systems and their bioanalytical applications: a review. *Anal Chim Acta* 2011;**690**:10–25.
35. Willner I, Willner B. Biomolecule-based nanomaterials and nanostructures. *Nano Lett* 2010;**10**:3805–15.
36. Xiao Y, Patolsky F, Katz E, Hainfield JF, Willner I. Plugging intro enzymes: nanowiring of redox enzymes by a gold nanoparticle. *Science* 2003;**299**(5614):1877–81.
37. Ozsoz M, Erdem A, Kerman K, Ozkan D, Tugrul B, Topcuoglu N, et al. Electrochemical genosensor based on colloidal nanoparticles for the detection of Factor V Leiden mutation using disposable pencil graphite electrodes. *Anal Chem* 2003;**75**(9):2181–7.
38. Shulga O, Kirchhoff JR. An acetylcholinesterase enzyme electrode stabilized by an electrodeposited gold nanoparticle layer. *Electrochem Commun* 2007;**9**:935–40.
39. Zhu Y, Son JI, Shim Y-B. Amplification strategy base on gold nanoparticle-decorated carbon nanotubes for neomycin immunosensors. *Biosens Bioelectron* 2010;**26**:1002–8.
40. Carralero V, Mena ML, Gonzalez-Cortes A, Yañez-Sedeño P, Pingarron JM. Development of a high analytical performance-tyrosinase biosensor base on a composite graphite-Teflon electrode modified with gold nanoparticles. *Biosens Bioelectron* 2006;**22**(5):730–6.
41. Çubukçu M, Timur S, Anik U. Examination of performance of glassy carbon paste electrode modified with gold nanoparticle and xanthine oxidase for xanthine and hypoxanthine detection. *Talanta* 2007;**74**(3):434–9.

42. Liu S, Ju H. Electrocatalysis via direct electrochemistry of myoglobin immobilized on colloidal gold nanoparticles. *Electroanalysis* 2003;**15**(18):1488–93.
43. Zhang S, Wang N, Yu H, Niu Y, Sun C. Covalent attachment of glucose oxidase to an Au electrode modified with gold nanoparticles for use as glucose biosensor. *Bioelectrochemistry* 2005;**67**(1):15–22.
44. Mani V, Chikkaveeraiah BV, Patel V, Gutkind JS, Rusling JF. Ultrasensitive Immunosensor for cancer biomarker proteins using gold nanoparticle film electrodes and multienzyme-particle amplification. *ACS Nano* 2009;**3**(3):585–94.
45. Mena ML, Yáñez-Sedeño P, Pingarrón JM. A comparison of different strategies for the construction of amperometric enzyme biosensors using gold nanoparticle-modified electrode. *Anal Biochem* 2005;**336**(1):20–7.
46. Wang M, Sun C, Wang L, Ji X, Bai Y, Li T, et al. Electrochemical detection of DNA immobilized on gold colloid particles modified self-assembled monolayer electrode with silver nanoparticle label. *J Pharm Biomed Anal* 2003;**33**(5):1117–25.
47. Drummond TG, Hill MG, Barton JK. Electrochemical DNA sensors. *Nat Biotechnol* 2003;**21**(10):1192–9.
48. Pérez-López B, Merkoçi A. Nanoparticles for the development of improved (bio) sensing systems. *Anal Bioanal Chem* 2011;**399**:1577–90.
49. Chumbimuni-Torres KY, Dai Z, Rubinova N, Xiang Y, Pretsch E, Wang J, et al. Potentiometric biosensing for proteins with ultrasensitive ion-selective microelectrodes and nanoparticle labels. *J Am Chem Soc* 2006;**128**(42):13676–7.
50. Park S-J, Taton TA, Mirkin CA. Array-based electrical detection of DNA with nanoparticle probes. *Science* 2002;**295**(5559):1503–6.
51. Hrapovic S, Liu Y, Male KB, Luong JHT. Electrochemical biosensing platforms using platinum nanoparticles and carbon nanotubes. *Anal Chem* 2004;**76**:1083–8.
52. Willner I, Katz E. Integrated nanoparticle-biomolecule hybrid systems: synthesis, properties, and applications. *Angew Chem* 2004;**43**:6042–108.
53. Pérez-López B, Merkoçi A. Nanoparticles for the development of improved (bio) sensing systems. *Anal Bioanal Chem* 2011;**399**:1577–90.
54. Sperling RA, Gil PR, Zhang F, Zanella M, Parak WJ. Biological applications of gold nanoparticles. *Chem Soc Rev* 2008;**37**:1896–908.
55. Daniel MC, Astruc D. Gold nanoparticles: assembly, supramolecular chemistry, quantum-size-related properties, and applications toward biology, catalysis and nanotechnology. *Chem Rev* 2004;**104**:293–346.
56. Jain PK, Lee KS, El-Sayed I-H, El-Sayed MA. Calculated absorption and scattering properties of gold nanoparticles of different size, shape, and composition: applications in biological imaging and biomedicine. *J Phys Chem B* 2006;**110**:7238–48.
57. Elghanian R, Storhoff JJ, Mucic RC, Letsinger RL, Mirkin CA. Selective colorimetric detection of polynucleotides based on the distance-dependent optical properties of gold nanoparticles. *Science* 1997;**277**:1078–80.
58. Thaxton CS, Georganopoulou DG, Mirkin CA. Gold nanoparticle probes for the detection of nucleic acid targets. *Clin Chim Acta* 2006;**363**:120–6.
59. Kim Y, Johnson RC, Hupp JT. Gold nanoparticle based sensing of "spectroscopically silent" heavy metal ions. *Nano Lett* 2001;**1**(4):165–7.
60. Wang Z, Levy R, Fernin DG, Brust M. The peptide route to multifunctional gold nanoparticles. *Bioconjug Chem* 2005;**16**:497–500.
61. Thanh NTK, Rosenzweig Z. Development of an aggregation-base immunoassay for anti-protein A using gold nanoparticles. *Anal Chem* 2002;**74**:1624–8.

62. Hirsch LR, Jackson JB, Lee A, Halas NJ, West JL. A whole blood immunoassay using gold nanoshells. *Anal Chem* 2003;**75**:2377–81.
63. Hodnik V, Anderluh G. Toxin detection by surface plasmon resonance. *Sensors* 2009;**9**:1339–54.
64. Lu Y, Liu J. A colorimetric lead biosensor using DNAzyme-directed assembly of gold nanoparticles. *J Am Chem Soc* 2003;**125**(22):6642–3.
65. Kanaras AG, Wang Z, Brust M, Cosstick R, Bates AD. Enzymatic disassembly of DNA-gold nanostructures. *Small* 2007;**3**(4):590–4.
66. Stewart ME, Anderton CR, Thompson LB, Maria J, Gray SK, Rogers JA, et al. Nanostructured plasmonic sensors. *Chem Rev* 2008;**108**:494–521.
67. Matusi J, Akamatsu K, Hara N, Miyoshi D, Nawafune H, Tamaki K, et al. SPR sensor chip for detection of small molecules using molecularly imprinted polymer with embedded gold nanoparticles. *Anal Chem* 2005;**77**:2882–8.
68. He L, Musick MD, Nicewarner SR, Salinas FG, Benkovic SJ, Natan MJ, et al. Colloidal Au-enhanced surface plasmon resonance for ultrasensitive detection of DNA hybridization. *J Am Chem Soc* 2000;**122**(38):9071–7.
69. Lioubashevski O, Chegel VI, Patolsky F, Katz E, Willner I. Enzyme-catalyzed bio-pumping of electrons into Au-nanoparticles: a surface plasmon resonance and electrochemical study. *J Am Chem Soc* 2004;**126**:7133–43.
70. Haes AJ, Zou S, Schatz GC, Van Duyne RP. A nanoscale optical biosensor: the long range distance dependence of the localized surface plasmon resonance of nobel metal nanoparticles. *J Phys Chem B* 2004;**108**:109–16.
71. Raschke G, Kowarik S, Franzl T, Sönnichsen C, Klar TA, Feldmann J. Biomelecular recognition based on single gold nanoparticle light scattering. *Nano Lett* 2003;**3**(7):935–8.
72. Zhao J, Zhang X, Yonzan CR, Haes AJ, Van Duyne RP. Localized surface plasmon resonance biosensors. *Nanomedicine* 2006;**1**(2):219–28.
73. Lee KS, El-Sayed MA. Gold and silver nanoparticles in sensing and imaging: sensitivity of plasmon response to size, shape, and metal composition. *J Phys Chem B* 2006;**110**:19220–5.
74. Nehl CL, Liao HW, Hafner JH. Optical properties of star-shaped gold nanoparticles. *Nano Lett* 2006;**6**:683–8.
75. Dahlin A, Zach M, Rindzevicius T, Käll M, Sutherland DS, Höök F. Localized surface plasmon resonance sensing of lipid-membrane mediated biorecognition events. *J Am Chem Soc* 2005;**127**:5043–8.
76. Haes AJ, Hall WP, Chang L, Klein WL, Van Duyne RP. A localized surface plasmon resonance biosensor: first steps toward an assay for Alzheimer's disease. *Nano Lett* 2004;**4**(6): 1029–34.
77. Yguerabide J, Yguerabide EE. Light-scattering submicroscopic particles as highly fluorescent analogs and their use as traces labels in clinical and biological applications. *Anal Biochem* 1998;**262**:137–56.
78. Taton TA, Mirkin CA, Letsinger RL. Scanometric DNA array detection with nanoparticle probes. *Science* 2000;**289**:1757–60.
79. Nam JM, Thaxton CS, Mirkin CA. Nanoparticle-based bio-bar codes for the ultrasensitive detection of proteins. *Science* 2003;**301**:1884–6.
80. Georganopoulou DG, Chang L, Nam JM, Thaxton CS, Mufson EJ, Klein WL, et al. Nanoparticle-based detection in cerebral spinal fluid of a soluble pathogenic biomarker for Alzheimer's disease. *Proc Natl Acad Sci USA* 2005;**102**:2273–6.
81. Xiao X, Zhao HB, Ozaki Y. Surface-enhanced Raman scattering for protein detection. *Anal Bioanal Chem* 2009;**394**:1719–27.

82. Paez JG, Jänne PA, Lee JC, Tracy S, Greulich H, Gabriel S, et al. EGFR mutations in lung cancer: correlation with clinical response to gefitinib therapy. *Science* 2004;**304**:1497–500.
83. Lynch TJ, Bell DW, Sordella R, Gurubhagavatula S, Okimoto RA, Branningan BW, et al. Activating mutations in the epidermal growth factor receptor underlying responsiveness of non-small-cell lung cancer to gefitinib. *N Engl J Med* 2004;**350**:2129–39.
84. Kneipp J, Kneipp H, McLaughlin M, Brown D, Kneipp K. In vivo molecular probing of cellular compartments with gold nanoparticles and nanoaggregates. *Nano Lett* 2006;**6**(10):2225–31.
85. Kneipp K, Kneipp H, Kneipp J. Surface-enhanced Raman scattering in local optical fields of silver and gold nanoaggregates-from single-molecule Raman spectroscopy to ultrasensitive probing in live cells. *Acc Chem Res* 2006;**39**(7):443–50.
86. Dubertret B, Calame M, Libchaber AJ. Single-mismatch detection using gold-quenched fluorescent oligonucleotides. *Nat Biotechnol* 2001;**19**:365–70.
87. Choi DH, Lee SK, Oh YK, Bae BW, Lee SD, Kim S, et al. A dual gold nanoparticle conjugate based lateral flow assay (LFA) method for the analysis of troponin I. *Biosens Bioelectron* 2010;**25**:1999–2002.
88. Cao X, Ye Y, Liu S. Gold nanoparticle-based signal amplification for biosensing. *Anal Chem* 2011;**417**:1–16.
89. Zou Z, Du D, Wang J, Smith JN, Timchalk C, Li Y, et al. Quantum dot-based immunochromatographic fluorescent biosensor for biomonitoring trichloropyridinol, a biomarker of exposure to chlorpyrifos. *Anal Chem* 2010;**82**:5125–33.
90. Wang Y, Xu H, Wei M, Gu H, Xu Q, Zhu W. Study of superparamagnetic nanoparticles as labels in the quantitative lateral flow immunoassay. *Mater Sci Eng C* 2009;**29**:714–8.
91. Taton K, Johnson D, Guire P, Lange E, Tondra M. Lateral flow immunoassay using magnetoresistive sensors. *J Magn Magn Mater* 2009;**321**:1679–82.
92. Komano A, Maruko H, Sekiguchi H, Seto Y. Detection of saxitoxin in counterterrorism using a commercial lateral flow immunoassay kit. *Forensic Toxicol* 2010;**29**:38–43.
93. Noguera P, Posthuma-Trumpie GA, van Tuil M, van der Wal FJ, de Boer A, Moers APHA, et al. Carbon nanoparticles in lateral flow methods to detect genes encoding virulence factors of Shiga toxin-producing Escherichia coli. *Anal Bioanal Chem* 2011;**399**:831–8.
94. Xu H, Mao X, Zeng Q, Wang S, Kawde A-N, Liu G. Aptamer-functionalized gold nanoparticles as probes in a dry-reagent strip biosensor for protein analysis. *Anal Chem* 2009;**81**: 669–75.
95. Wang X, Li K, Shi D, Jin X, Xiong N, Peng F, et al. Development and validation of an immunochromatographic assay for rapid detection of sulfadiazine in eggs and chickens. *J Chromatogr B Analyt Technol Biomed Life Sci* 2007;**847**:289–95.
96. Shim W-B, Kim K-Y, Chung D-H. Development and validation of a gold nanoparticle immunochromatographic assay (ICG) for the detection of zearalenone. *J Agric Food Chem* 2009;**57**:4035–41.
97. Biagini RE, Sammons DL, Smith JP, MacKenzie BA, Striley CAF, Snawder JE, et al. Rapid, sensitive, and specific lateral-flow immunochromatographic device to measure anti-anthrax protective antigen immunoglobulin G in serum and whole blood. *Clin Vaccine Immunol* 2006;**13**:541–6.
98. Huang X, Xuan X, Verdida RA, Zhang S, Yokoyama N, Xu L, et al. Immunochromatographic test for simultaneous serodiagnosis of Babesia caballi and B. equi infections in horses. *Society* 2006;**13**:553–5.
99. Jiang T, Liang Z, Ren W, Chen J, Zhi X, Qi G, et al. Development and validation of a lateral flow immunoassay using colloidal gold for the identification of serotype-specific foot-and-mouth disease virus O, A and Asia 1. *J Virol Methods* 2011;**171**:74–80.

100. Mao X, Ma Y, Zhang A, Zhang L, Zeng L, Liu G. Disposable nucleic acid biosensors based on gold nanoparticle probes and lateral flow strip. *Anal Chem* 2009;**81**:1660–8.
101. Liu G, Mao X, Phillips JA, Xu H, Tan W, Zeng L. Aptamer-nanoparticle strip biosensor for sensitive detection of cancer cells. *Anal Chem* 2009;**81**:10013–8.
102. Zhou P, Lu Y, Zhu J, Hong J, Li B, Zhou J, et al. Nanocolloidal gold-based immunoassay for the detection of the N-methylcarbamate pesticide carbofuran. *J Agric Food Chem* 2004;**52**:4355–9.
103. Shim W-B, Yang Z-Y, Kim J-Y, Choi J-G, Je J-H, Kang S-J, et al. Immunochromatography using colloidal gold–antibody probe for the detection of atrazine in water samples. *J Agric Food Chem* 2006;**54**:9728–34.
104. Ferreira GNM, Da-Silva AC, Tomé B. Acoustic wave biosensors: physical models and biological applications of quartz crystal microbalance. *Trends Biotechnol* 2009;**27**(12):689–97.
105. Tang D-Q, Zhang D-J, Tang D-Y, Ai H. Amplification of the antigen-antibody interaction from quartz crystal microbalance immunosensors via back-filling immobilization of nano gold on biorecognition surface. *J Immunol Methods* 2006;**316**:144–52.
106. Zhou XC, O´Shea SJ, Li SFY. Amplified microgravimetric gene sensor using Au nanoparticle modified oligonucleotides. *Chem Commun* 2000;953–4.
107. Lin L, Zhao H, Li J, Tang J, Duan M, Jiang L. Study on colloidal Au-enhanced DNA sensing by quartz crystal microbalance. *Biochem Biophys Res Commun* 2000;**274**:817–20.

Chapter 11

Magnetic Nanoparticles for Application in Biomedical Sensing

David Alcantara and Lee Josephson

Center for Nuclear Medicine and Molecular Imaging, Massachusetts General Hospital-Harvard Medical School, Charlestown, Boston, Massachusetts, USA

1. INTRODUCTION

Nanoscale materials have found broad application in industrial processes, consumer products, therapeutics, and diagnostics. In each application, the driving force for nanoparticle use lies in the unique size-dependent chemical or physical properties, including catalytic, electrochemical, electron transport, magnetic, optical, and thermodynamic behaviors.[1] Magnetic nanomaterials are an important source of labels for biosensing due to their strong magnetic properties, which are not found in biological systems. Modulation of the composition, size, and magnetic properties of these materials permits their use in a variety of instruments and formats for biosensing.[2,3] New types of portable instrumentation are promising for the use of nanoscale magnetic materials in point-of-care (POC) sensors in a variety of applications. Magnetic biosensors are also under active development and may soon rival established biological detection methods that use surface-bond fluorescent tags.[4]

2. MAGNETIC RELAXATION PROPERTIES

Nanoparticles (generally materials with overall diameters between about 10 and 300 nm) have great potential for biomedical applications, mainly due to their similar size to biological molecules (e.g., ferritin 14 nm, IgG 11.9 nm, LDL 22 nm, viruses 50–250 nm) and because their properties may be fine tuned during chemical synthesis. When placed in a magnetic field with strength H, materials exhibit an induced magnetization M characterized by $M=\chi H$, where χ is the magnetic susceptibility (Figure 11.1). Materials with magnetic moments aligned parallel to H, and susceptibilities on the order of 10^{-10}–10^{-1} are described as paramagnetic. At very small crystal sizes, usually below about

Frontiers of Nanoscience, Vol. 4. DOI: 10.1016/B978-0-12-415769-9.00011-X

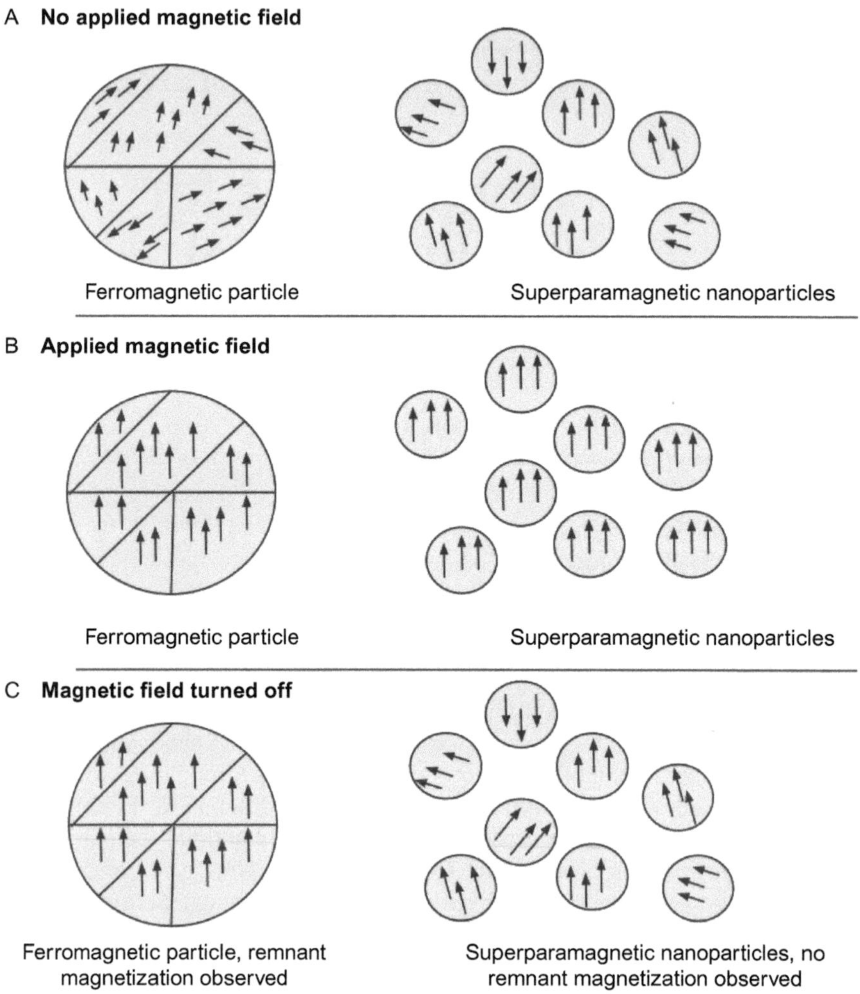

FIGURE 11.1 Superparamagnetic versus ferromagnetic particles in (A) the absence and (B) presence of an external magnetic field. The magnetic moment of superparamagnetic particles in the absence of an external field is time averaged to zero, similarly to the net magnetic moment of a superparamagnetic particle assembly. In the presence of external field, the magnetic moments of the particles align to it. When magnetic field is turned off (C), magnetic moments of superparamagnetic particles randomize and lack a remnant magnetization. (For color version of this figure, the reader is referred to the online version of this chapter.)

20 nm, iron oxide nanoparticles [magnetite (Fe_3O_4), maghemite (γ-Fe_2O_3), or nonstoichiometric magnetite], oxygen ions form a cubic close-packed (ccp) lattice, with smaller iron cations located at the interstices between the larger oxygen anions. Magnetic properties arise through *h* quantum tunneling (electron coupling), which gives rise to a single magnetic domain (single magnetization vector) in each crystal. As a result of the coupling and coalignment of electron

spins in the single domain, the individual magnetic dipole moments of superparamagnetic nanoparticles, per metal atom, are far larger than those of paramagnetic materials, termed high magnetic susceptibility. In the absence of a magnetic field, and at room temperature, nanoparticle magnetic moments jump easily between the crystallographic axis and an ensemble of superparamagnetic particles exhibits no net magnetic moment (Figure 11.1A). A relatively weak magnetic field (H) is sufficient to align the magnetic moments of superparamagnetic nanoparticles (high magnetic susceptibility, Figure 11.1B).

When the external magnetic field is turned off (Figure 11.1C), the magnetic moments of superparamagnetic nanoparticles randomize due to thermal agitation, termed a Neel relaxation, and, therefore, lack a remnant magnetization. Ferromagnetic particles have larger crystal sizes and larger magnetic domains than superparamagnetic particles. In common with superparamagnetic particles, they exhibit a high magnetic susceptibility. However, after their magnetic moments have been aligned by an applied magnetic field, and applied field then removed, their magnetic moments resist thermal agitation (no Neel relaxation). These properties (high magnetic susceptibility and loss of magnetic moment when the applied field is removed) enable superparamagnetic nanoparticles to avoid magnetic attractions between particles in solution, a feature that makes them suitable for many biomedical applications.

In MRI/NMR systems, magnetic particles (MPs) (typically superparamagnetic) shorten both the longitudinal (spin–lattice, T_1) and transverse (spin–spin, T_2) relaxation times of water protons. However, their effects on the transverse relaxivity (r_2) are significantly greater than longitudinal relaxivity (r_1), and their effects on T_2 are normally used for biosensing applications. (The relaxivity of particle r is defined as the change $1/T$ [usually in units of sec^{-1}] for a given change in particle concentration [usually in units of millimolar metal].)

3. MAGNETIC RELAXATION SWITCHES

Superparamagnetic iron oxide nanoparticles coated with biocompatible polymers are clinically proven magnetic resonance (MR) contrast agents and have been widely used in molecular imaging applications.[3,5] Recently, portable MR technology combined with surface-functionalized MNPs has produced a new approach for *in vitro* diagnostics. This technology, termed magnetic relaxation switch (MRSw), created a new class of MR diagnostic sensors that have the potential to provide sensitive, selective, immuno-, and molecular diagnostics in POC settings.[6–8] MRSw assays exploit the fact that for both MNPs and larger MPs transverse relaxation times (T_2) differ between dispersed and aggregated states. However, for Type I, NP-based systems, T_2 decreases with the aggregation, while with Type II, MP-based systems, T_2 increases with aggregation. The principle of Type I MRSw sensors is illustrated in Figure 11.2.

NPs switch between dispersed and aggregated states, and associated with the change in aggregation are changes in the spin–spin relaxation time (T_2),

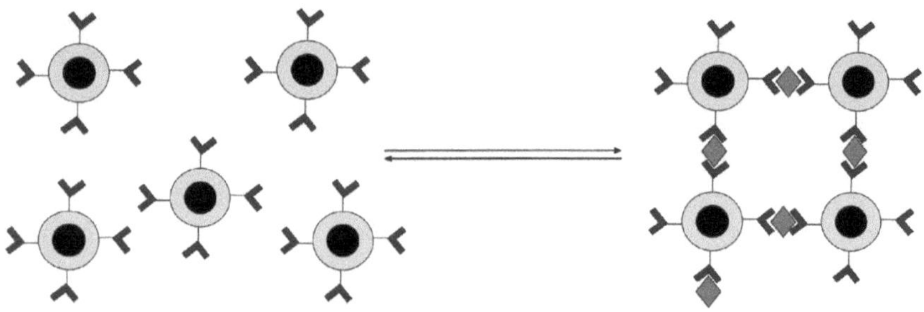

FIGURE 11.2 Principle of Type I MRSws sensors. Dispersed magnetic nanoparticles (NPs) form an aggregate upon binding with target analytes (red rhomboids). The aggregated form of the NPs dephases the spins of the surrounding protons of water molecules more efficiently than NPs present as the dispersed state. The effect is observed as a decrease in spin–spin relaxation time, T_2. (For interpretation of the references to color in this figure legend, the reader is referred to the online version of this chapter.)

the basis of which is discussed below.[9] The materials used in MRSw assays are either MNPs (diameter 5–300 nm) or micrometer-sized MPs (diameter 300–5000 nm). As shown in Figure 11.2, MRSws are homogeneous particle aggregation/disaggregation-based assays similar to aggregation assays using Latex particles, red blood cell hemagglutination, and antibody reactions with proteins (nephelometry). Unlike optically based assays, MRSws employ radio-frequency radiation which penetrates biological samples regardless of their optical properties.[10] Since the dispersed and aggregated states of NPs (or MPs) can be reversed by several factors such as temperature, pH, and a high concentration of competing analytes and hence are referred to as "relaxation switches." The aggregated and dispersed states of magnetic NPs or MPs have different transverse spin–spin relaxation times (values of T_2). NP aggregation and the size range of the resulting aggregates depend on the type of analyte and analyte concentration.[11] On the contrary, longitudinal relaxivity is not affected by clustering, and, therefore, r_1 can be used as a measure of total nanoparticle concentration independently of aggregation status.

3.1. Mechanism of MRSws

The mechanistic basis of Type I and Type II MRSw is as follows.[11–13] The general theory of how magnetic spheres alter T_2 is termed outer sphere relaxation theory. This theory uses two parameters of D_w and t_D. D_w is the difference in angular frequencies between the local field experienced by a proton at the equatorial line of the sphere's surface and the bulk ($D_w = m_O Mg/3$, where m_O is the vacuum magnetic permeability, M is the particle magnetization, and g is the proton gyromagnetic ratio). Then t_D is the translational diffusion time of water around the sphere ($t_D = R_a 2/D$, where R_a is the sphere radius and D is the water diffusion coefficient). The outer sphere diffusion theory is applied when the motional average condition is fulfilled as $D_w t_D < 1$.[12,13] In this condition, the relaxation rate $R_2(=1/T_2)$ increases as the sphere's size is increased. As the definitions of D_w and t_D imply, the motional average condition is not fulfilled with increased size of the particles such as MPs ($D_w t_D > 1$) and the relaxation rate of $1/T_2$ decreases with the formation of MP aggregates. See the detailed discussion of this phenomenon in a review.[13]

Thus, when present in solution, magnetic NPs (or MPs) induce local magnetic field inhomogeneities, which cause a dephasing (loss of phase coherence) of the proton spin precession, and these inhomogeneities lead to a reduction of the T_2 relaxation time. When NPs aggregate (Type I MRSw), a smaller number of larger magnetic field inhomogeneities result. These larger inhomogeneities are more effective dephasers of proton relaxation and T_2 drops. Here, $D_w t_D < 1$. When MPs aggregate (Type II MRSw), a smaller number of larger magnetic field inhomogeneities again result. However, there are now so few aggregates, and spaces between them so great, that many water protons fail to diffuse in and out of these homogeneities during the time course of the measurement. This is termed the "diffusion limited case" for the enhancement of proton relaxation by magnetic microspheres. Here, $D_w t_D > 1$.

3.2. Type I MRSw

Small cross-linked iron oxide (CLIO) nanoparticles constitute a good material for MRSw applications. CLIO nanoparticles contain a superparamagnetic core (3–6 nm monocrystalline iron oxide) composed of magnetite (Fe_3O_4) and/or maghemite (γ-Fe_2O_3). The iron oxide core is coated with biocompatible dextran, cross-linked with epichlorohydrin, and activated with ammonia to provide an amine group. Further surface functionalization permits surfaces which react with a variety of targets reviewed in Table 11.1.

To address the need for fast, simple sensors that can be used in different environments, miniaturized MR-based sensors have been developed. One such system is called diagnostic magnetic resonance (DRM) and can measure Type I MRSw T_2 changes. The system achieves a high assay sensitivity by reducing the sample volume to less than 5 μl (Figure 11.3).[15] Newer DRM instruments have been integrated onto a CMOS chip, and the sample volume reduced to 1 μl. This configuration resulted in more homogeneous magnetic fields, less electrical resistance, and permitted *Staphylococcus aureus* detection with as few as 10 colony-forming units (CFUs).[15]

3.3. Type II MRSw

Type II MRSws exhibit an increased T_2 when binding to a target analyte. In this case, biomolecules are attached to microparticles that, with a greater magnetic moment per particle, can be used at concentrations far below than that of MNPs in MRSw assays. Lower concentration of particles entails lower

TABLE 11.1 Classes of Analyte Detected with MRSw Assay Technology

Analyte	Target	Reference(s)
DNA	Telomeres	14
RNA	GFP	10
Proteins	Avidin	15,16
	GFP	10
	B-hCG	17
	Telomerase	18
	CA-125	15
	VEGF	15
	α-Fetoprotein	15
	HSA	19
Enzymes	BamH1	20
	Methylase, MboI, DpnI	20
	Caspase-3	10
	Renin	21
	Trypsin	21
	MMP-2	21
	Telomerase	14,18
	Peroxidases	22
Small molecules	Enantiomers	23
	Folate	24
	Glucose	24
	HA peptide	24
	Calcium	25,26
	Influenza Tag peptide	11,27
Pathogens	Herpes simplex virus	28
	Adenovirus-5	28
	S. aureus	15
	MTB/BCG	29
Cells	Tumor cell lines	15
	FNA (mouse xenograft)	30
	Tumor cell lines (BOND amplification)	31

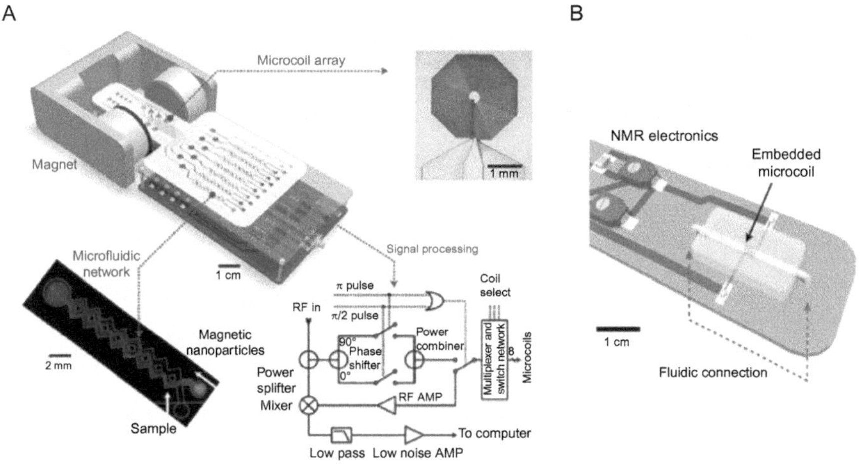

FIGURE 11.3 Schematic diagram of the micro-NMR device system (DRM). (A) The original system is composed of an array of planar microcoils for NMR measurements, minituarized NMR electronics, microfluidic networks for sample mixing, and a small magnet for polarizing magnetic field generation. (B) The second-generation consists of a solenoidal coil embedded in a microfluidic device. This improved device has a higher filling factor, better signal-to-noise ratio, and reduced sample volume requirement to 1 μl. (Reproduced with permissions from Ref. 15. Copyright 2008 Nature Publishing Group. Reproduced with permission from Ref. 30. Copyright 2009 National Academy of Sciences, USA.) (For color version of this figure, the reader is referred to the online version of this chapter.)

concentration of analyte needed to induce agglomeration, improving therefore the sensitivity of the switch.[11,27]

When inserted in a homogeneous magnetic field, large MPs with large magnetic moments will aggregate due to magnetic attractions between them, while smaller NPs will not respond in this manner (Figure 11.4A).[11,12,27,32–35] The magnetic field-induced MP aggregation is lost when the magnetic field was removed and Brownian effects break down aggregates. The speed of self-assembly formation of MPs in a magnetic field is a function of viscosity and can be used to make a T_2-based viscometer.[12] Magnetic field-induced MP aggregation has also been used to step up analyte-mediated formation of MP aggregates.[27,33,36] The applied magnetic field enhanced the molecular interactions between multivalent analytes (e.g., a monoclonal Tag antibody) and multivalent MPs displaying the Tag peptide. This approach is referred to as magnetic-field-enhanced target aggregation and shown in Figure 11.4B.

3.4. Aggregation-Based Implantable MR Devices

Surface-functionalized MNPs have also been used as components of implantable MRI-based sensors. Here, MNPs are placed inside devices consisting of semipermeable membranes with size cutoffs that allow small analytes to enter and leave the device, while retaining the larger MNPs. The principle

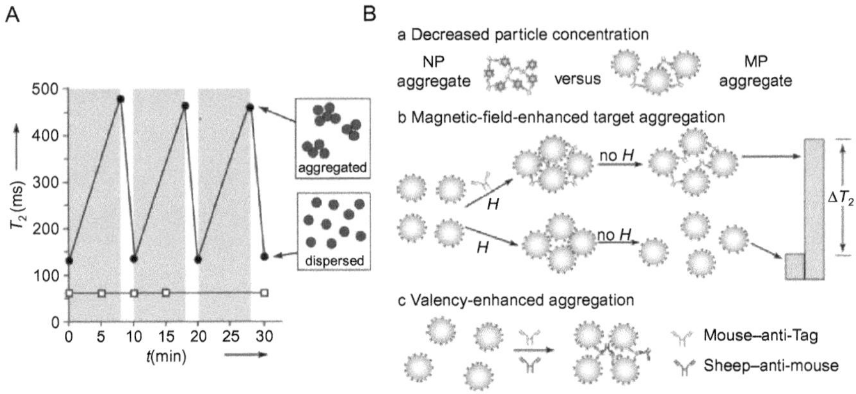

FIGURE 11.4 Methods for the improvement in MRSw assay sensitivities. (A) MPs (black dots) aggregate in a homogeneous magnetic field, whereas NPs (white squares) do not respond. A T_2 increase in time is observed in a 0.47-T field (gray) in an MP solution, but not in an NP solution. The T_2 value of the MP solution decreases as the MPs are dispersed with the field turn-off (white). (B) Three strategies for enhancing the sensitivities with a Type II MRSw assay. (a) A decreased concentration of MPs formed aggregates at a lower concentration of analyte (anti-Tag antibody) than that of NPs. MPs are larger than NPs and used at a lower concentration. (b) Application of a magnetic field (0.47 T) induced aggregation of MPs as in (A) and accelerated the interaction between MPs and analytes. (c) Target valency enhancement by addition of a secondary antibody (sheep–anti-mouse). The valency increase of targets from two (anti-Tag) to four (anti-Tag:anti-mouse) enhanced MRSw sensitivities. Reproduced with permission from Ref. 27. (For color version of this figure, the reader is referred to the online version of this chapter.)

of MRI sensors was first demonstrated using MNPs functionalized with the glucose binding protein Concanavalin A, when ConA-MNPs were placed in a device, with a pore size that allowed glucose to enter or leave. T_2 readings of the device then reflected cycling glucose concentrations in the normal range outside the device (Figure 11.5A and B).[24]

The first implantable MR relaxation sensor used MNP binding the cancer marker human chorionic gonadotropin (hCG).[37,38] The device had a reservoir covered with a semipermeable membrane and was filled with CLIO particles functionalized with antibodies to the hCG cancer biomarker (Figure 11.5C). Polydimethylsiloxane was chosen as the substrate material for the first-generation sensing devices because it is magnetically inert and biocompatible. Implanted in a mouse tumor model, the sensor showed significant decreases of MR signal in 1–4 days due to diffusion of the cancer biomarker hCG into the reservoir producing the nanoparticles aggregation. Recently, Cima *et al.* used this device in the detection of pathologically elevated level of myoglobin, cTnI, and CK-MB, functional proteins released by the ischemic myocardium after an acute infarct.[39] The sensor, designed to sample biomarkers in a subcutaneous space within the animal's flank, exhibited a sensitivity in the ng/ml range that could be improved to the pg/ml range and was stable for several months.

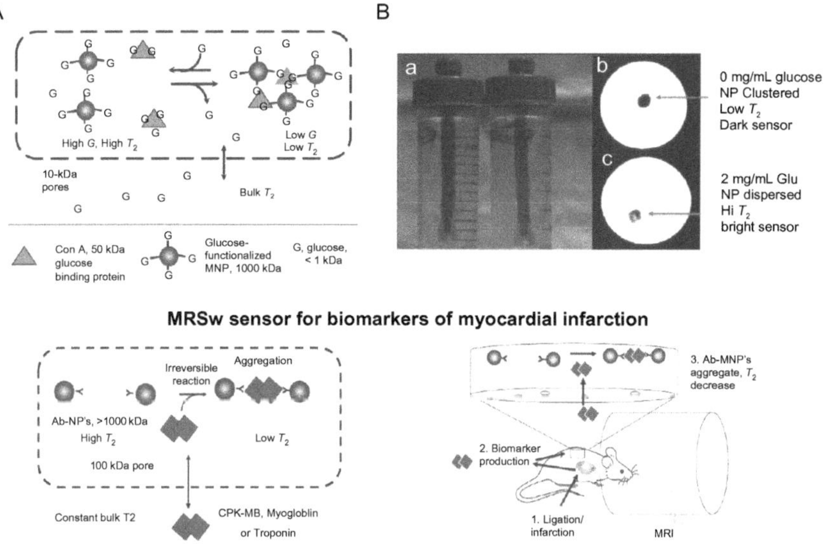

FIGURE 11.5 Membrane MRSw concept. (A) Glucose molecule can freely diffuse inside or outside the membrane and make the sensor aggregates or disaggregates depending on glucose concentration. (B) MRI signals of the membrane sensor (a) [(b) and (c) insets showed a dark or bright MRI images corresponding to a low or high glucose concentration, respectively]. (C) Implantable sensor registered CK-MB, myoglobin, or cTnI biomarkers *in vivo* by using MRI. (For color version of this figure, the reader is referred to the online version of this chapter.)

3.5. Post Filtration MR Sensors

Some versions of DRM devices determine analyte levels after separations, such as by filtration, concentrate MPs, and their target. This configuration allowed the detection of tuberculosis (TB) bacterium via the cell tagging method. Without filtration, NMR measurements were as sensitive as acid-fast bacilli smear microscopy, with a detection threshold of about 10^3 CFU ml^{-1}. Filters can concentrate targets larger sample volumes, as was demonstrated when 20 CFUs could be detected in 1 μl of sputum.[29]

4. APPLICATIONS OF MRSw

A major advantage of MRSw sensors is their ability to detect analytes (without separation step) in opaque samples, a feature which results because MR uses penetrating radiofrequency radiation. Therefore, the transmission of light through the sample is not required, and MRSw assays are indifferent to light-based interferences. A second advantage of MRSw sensors is their use of multivalent interactions between nanoparticle surfaces and targets in solution. The ability to place high numbers of ligand-binding targets on the surface of nanoparticles can generate multivalent effects between the nanoparticle and its target; this results increases in affinity and assay sensitivity.[40] A third major advantage of MRSw is the ability to detect chemically diverse types of target analytes. Nanoparticles can be surface modified for the MR-based detection of proteins, enzymes, small molecules, nucleic acids, and cells (see Table 11.1 and below sections).

4.1. Detection of Proteins

Early MRSw utilized Type I effects with protein molecule targets. MNPs were functionalized with biotin and reacted with avidin, resulting in clustering of MNPs and an avidin concentration-dependent T_2 change.[10] Four orders of dynamic ranges were achieved adjusting MNPs concentration, indicating the robust working range of the sensor. In other experiment, green fluorescent protein (GFP)-sensitive nanoparticles were synthesized by conjugation of CLIO MNPs with anti-GFP polyclonal antibodies. When GFP was introduced to a solution of anti-GFP decorated nanoparticles (45 nm diameter), there was a time- and dose-dependent response in the measured T_2 values. After about 30 min, the T_2 signal stabilized, indicating that the clustering reaction had reached completion. Concentrations of GFP as low as single nanomolar were detected with T_2 changes on the order of 10–20 ms.[10]

The protein hormone hCG, a biomarker implicated in prostate and ovarian cancer, was detected by antibody-decorated nanoparticles by Michael Cima's team and using the MRSw system.[17] Two monoclonal antibodies were attached to separate particles population to generate a sandwich assay MRSw. hCG was successfully detected using this system with a limit of detection of 3.6 nM. Other tumor biomarkers like CA-125, VEGF, and α-fetoprotein have been detected in blood samples using MNPs coupled with μNMR device.[15]

Very recently, MRSw strategy has been used to identify different types of lectins, proteins that specifically recognize and binds to carbohydrates. Here, MNP functionalized with specific carbohydrates was reacted with the corresponding lectins. Differences in the structure, number of binding sites, and topological arrangement of the binding sites in proteins lead to a different initial aggregation rate with the carbohydrate-coated beads.[41]

4.2. Detection of Enzyme Activities

Enzymatic activity detection by MRSws has employed Type I MRSws (decreased T_2 when dispersed NPs form aggregates) in two configurations:

- Analyte-induced aggregation of dispersed magnetic NPs (T_2 drops)
- Analyte-induced dispersion (disassembly) of preformed aggregates of magnetic NPs (T_2 increases).

Aggregation MRSw: MRSw has been designed to assess human telomerase (hTERT) activity by hybridizing MNPs with the 30-base pair telomeric repeat complimentary sequences produced by hTERT activity.[14] More recently, a dual sensor capable of determining telomerase activity and the amount of hTERT protein has been developed and used to demonstrate a relationship between hTERT phosphorylation and enzymatic activity.[18] The activity of telomerase can be elevated in tumor malignancies, so this technique may have important applications in anti-telomerase therapies.

Enzyme activity can also be sensed by means of activating surface groups on nanoparticles to facilitate particle agglomeration. Myeloperoxidase (MPO) activity, which plays a role in atherosclerosis and inflammation, was sensed by attaching serotonin molecules to CLIO nanoparticles.[22] The peroxidase-catalyzed reduction of H_2O_2 converted the phenol group on serotonin to a radical, which led to radical-based cross-linking of the nanoparticles and the formation of nanoassemblies. Using a similar approach, leukocyte-derived MPO has been shown to play a critical role in the pathogenesis of atherosclerotic plaques.[42]

Dispersion MRSw: Dispersion-based MRSw sensors have been designed to quantify and detect proteases,[43,44] methylases, and endonucleases.[20] In this configuration, the enzyme activity disaggregates previously formed nanoparticle clusters, transducing the enzymatic activity into a detectable T_2 signal. In the first case, MNP aggregates were formed using the peptide sequence biotin-GDEVDGC, which was attached to avidin-conjugated CLIO particles (via the thiol provided by the terminal cysteine in the peptide). Addition of caspase-3 disassembled the clusters by cleaving at DEVD site, with the corresponding increase in T_2 relaxation time.[10] A similar reverse switching strategy has been used to detect renin, trypsin, and matrix metalloproteinase 2 activities.[21]

4.3. Detection of Small Molecules

The detection of small molecules with MRSw often involves modifying the nanoparticle surface with low-molecular-weight groups similar to the analyte and allowing aggregation to occur with the use of a dimeric antibody, a tetrameric ConA, or tetrameric avidin. This configuration was used to measure glucose, the influenza hemagglutinin peptide (HA), and folic acid.[24] Here, the small molecules were attached to the surface of CLIO nanoparticles following aggregation with ConA (for glucose particles) or an antibody (for folic acid and HA). The addition of the free analyte as a competing agent showed an increase in T_2 time achieving detection limits of 50 and 3 nM for HA and folic acid, respectively. The most important discovery here was the reversible T_2 change when the analyte was removed from the solution and aggregates were again reassembled. This led to the construction of a device with a semipermeable membrane capable of sensing analytes in real time (as described in Section 3.4). Initially used to sensing glucose[24] his capabilities have been expanded to other molecules as hCG,[38] CA-125, VEGF, and α-fetoprotein.[39] The ability to attach large numbers of small molecule to the nanoparticle surface increases possible multivalent effects between the particle and its target.[40]

The versatility of the MRSw has been used to differentiate low-molecular-weight enantiomeric drugs.[23] MRSw designs have also been used to detect analytes as small calcium ions, by making use of selective chelators[25] or the calcium-dependent interaction between calmodulin and the kinase M13 peptide-coupled particles.[26]

4.4. Detection of Nucleic Acids

In the first published paper on MRSw assays, Josephson *et al.* presented a biosensor that was sensitive to the presence of a 24-base pair synthetic oligonucleotide sequence. Sensors were made by attaching different complimentary 12-base pair oligonucleotides to CLIO particles surface. Addition of the target oligonucleotide sequence resulted in fast nanoparticle aggregation and a simultaneous decrease in T_2 relaxation time.[9] A follow-up study by Perez *et al.* demonstrated the potential for nucleic acid sensing and introduced the capability of detecting single-base pair mismatches.[10] Perez *et al.* also demonstrated the detection of a target mRNA from a transfected GFP gene in various eukaryotic cell lines.

4.5. Detection of Viruses

Initial experiments in pathogen detection were performed using antibody-conjugated MNPs targeting adenovirus-5 and herpes simplex virus-1 and showing a lower limit of five viral particles in 10 μl.[28]

4.6. Detection of Cancer Cells

Detecting multiple biomarkers and circulating cells in human body fluids is an especially crucial task for diagnosis and prognosis of complex diseases such as cancer and metabolic disorders. Using their first-generation μNMR device, Lee *et al.* made a fast and sensitive detection of tumor cell surface markers Her2/neu, EGFR, and CD326 (EpCAM) on mammalian cells.[15] Magnetic nanoparticles were functionalized with the corresponding monoclonal antibodies and incubated for 30 min with cancer cell lines and fibroblast as a control. After washing steps, selective detection of tumor markers was achieved by means of significant differences in T_2 relaxation times. More recently, the use of BOND-2 (biorthogonal nanoparticle detection) strategy has further advanced the profiling abilities of the μNMR device.[31] At low cell count (~ 1000 cells/sample), parallel measurements could be performed.

Detection sensitivity was improved with the second-generation μNMR device. Here, the use of a solenoidal coil and new highly magnetic nanoparticles ($MnFe_2O_4$) improved remarkably the sensitivity for cell sensing (Figure 11.6A).[30] This new device enabled detection of Her2/neu tumoral marker from as few as 200 breast cancer cells in 1 μl sample volume using CLIO MNPs. As shown in Figure 11.6B, the detection threshold was reduced to a single-cell level beating other conventional clinical methods. Close correlation between μNMR measurements, flow cytometry, and Western blot analysis were found (Figure 11.6C).

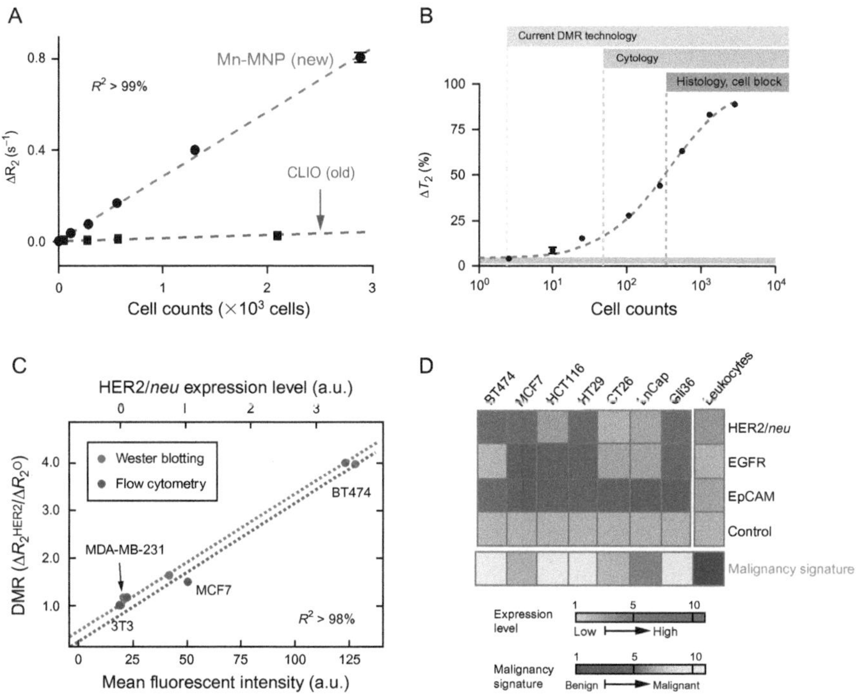

FIGURE 11.6 Tumor detection and profiling with a miniaturized NMR device. (A) Human breast cancer cells (BT474) were labeled with CLIO and Mn-MNP–HER2 and introduced in a microfluidic chip for measurements. Excellent linear relationship ($R^2 > 99\%$) between the cell number and the transverse relaxation rate (R^2) and 10 times higher sensitivity was observed when using Mn-MNPs (higher magnetic moment). (B) The detection limit was nearly at the single-cell level (≈2 cells), superior to the current clinical methods (cytology and histology). (C) Accuracy benchmarking of the assay against gold standards in molecular analysis, flow cytometry, and Western blot analysis, by measuring HER2/neu expression on breast cancer cells. The results showed good correlation ($R_2 > 98\%$) with the standard methods but requiring fewer cells. (D) Molecular profiling of mice tumors xenograft. Three cancer biomakers (HER2/neu, EGFR, and EpCAM) were profiled to increase diagnosis accuracy. (Reproduced with permission from Ref. 30 Copyright 2009 National Academy of Sciences, USA.) (For color version of this figure, the reader is referred to the online version of this chapter.)

Profiling of fine-needle aspirates from a panel of mouse xenograft tumors was analyzed for Her2/neu, EGFR, and EpCAM expression, and improvement of cancer diagnosis was also shown (Figure 11.6D).

Finally, a very recent work of Weissleder *et al.* with their third-generation portable micro-NMR (or DMR) device showed accurate diagnosis of malignant tumors from the scarce cancer cells present in fine-needle aspirate biopsies in nearly 60 min.[45] The system use MNPs functionalized with affinity ligands to detect specific cancer protein markers. Using nine-marker signature detection by NMR, they correctly identified 44 patients as having malignant tumors which were verified by standard techniques like biopsy. The accuracy of the micro-NMR system was further increased to 96% by using a four-protein marker signature (MUC-1, HER2, epidermal growth factor receptor, and EpCAM), surpassing the 84% accuracy of immunohistochemistry, the gold standard on cancer diagnosis. The new DMR-3 has the advantages of speed and accuracy and is easy to operate with even a smart phone or tablet pc technologies.

4.7. Detection of Bacteria

The detection of pathogenic bacteria is key to many problems related to health and safety. Many detection methods involve culturing bacteria and may take up to 7 or 8 days to yield an answer which is clearly inadequate in many healthcare and food preparation settings. Consequently, many efforts have been made toward the development of rapid detection methods for bacteria.[46]

Detection of *S. aureus* was reported with the first-generation μNMR device. *S. aureus* was detected using MNPs decorated with vancomycin, a tricyclic glycopeptide that binds to D-alanyl-D-alanine moieties in the bacterial cell wall to block the formation of cross-linked peptidoglycanes.[15] As low as few CFUs per microliter of sample were detected pointing out the great sensitivity of the technique. Furthermore, the bacterium *Mycobacterium avium* subspecies *paratuberculosis*, a difficult microorganism to culture and identify, was satisfactorily sensed using MRSw in complex media like blood and milk.[47] Assessment of metabolic activity and determination of antimicrobial susceptibility (*E. Coli* and *S. marcescens*) in complex matrices, such as blood, through magnetic relaxation were also studied by Kaittanis *et al.*[48] In this case, MNPs were functionalized with the mannose binding lectin ConA and their clustering translated into the polysaccharide levels associated with bacterial metabolism and growth through a change in T_2 relaxation time.

In a recent study, TB bacteria have been detected by combining new highly magnetic nanoparticles (Fe-core/ferrite shell, "Cannonballs") with the second generation μMNR device.[29] Bacillus Calmette-Guerin (BCG), used as a surrogate for *Mycobacterium tuberculosis*, was spiked into sputum samples, and after liquefaction, they were incubated with cannonballs functionalized with an anti-BCG monoclonal antibody. The bacteria were concentrated using a membrane filter embedded within the device, allowing remarkable detection sensitivity and speed. As few as 20 CFUs could be detected in 1 μl of sputum sample, in less than 30 min.

5. OTHER TYPES OF MP-BASED SENSORS

5.1. Magnetoresistive Sensors

There are a number of sensors where the magnetism of MPs is detected, rather than their effects on MR proton relaxation times as occurs with MRSw sensors. Giant magnetoresistance (GMR), spin valve (SV), or magnetic tunnel junction sensors have been successfully used to sense MPs and have been given attention because of the possibility of fabricating sensor arrays. Magnetoresistive sensors are based on the binding of MPs to a sensor surface, and the magnetic fields of the particles alter the resistance and electrical current within the sensor. Sensors are composed of multiple layers of ferromagnetic materials. A biologically active molecule can be deposited on an Au layer or SiO_2 layer to obtain a surface for the attachment of biomolecules. For a review of the structure of magnetoresistive sensors, see reference.[49]

The possibility of using the GMR effect for detecting magnetic labels was first described by Baselt *et al.* in 1998,[50] and the concept has been subsequently developed by other groups, where arrays of GMR sensors have been fabricated and used for multiplexed immunoassays and nucleic acid assays.[51,52] Superparamagnetic particles with different sizes have been used in magnetoresistive biosensing. Earlier applications used relatively large MPs, with diameters between 0.1 and 3 μm.[53] Micrometer-sized particles normally have higher particle-based magnetic moment that permits detection of very small numbers of particles. However, recently, magnetic NPs have replaced the larger particles because the NPs are stable in suspension and are less prone to particle clustering in an applied magnetic field.[49,54–56]

5.2. Spin Valve Sensors

SV sensors are also based on the GMR effect, but in this case, each sensing resistor is composed of only two ferromagnetic layers, separated by a nonmagnetic metallic layer. The direction of the magnetization in the magnetic layer furthest from the sensing surface is pinned by interfacing it with a strong antiferromagnetic material. When magnetic labels bind to the sensing surface, they cause the direction magnetization in the unpinned magnetic layer to rotate, which induces a decrease in electrical resistance.

The combination of SV sensor arrays and MNPs constitutes a promising architecture for a sensitive, quantitative, nonoptical detection system for microarrays, a universal platform for many different biological assays. By using 50-nm MACS magnetic nanoparticles, Wang and collaborators demonstrated cancer marker detection in 50% serum at subpicomolar concentrations on a chip configuration.[55] When prespotted with hCG (a pregnancy marker) antibody probes, the magneto-nano chip showed detection of hCG protein levels as low as 5 pM (0.2 ng/ml) in spiked blood samples, further confirming the viability and sensitivity of magneto-nano chips for use with real-world samples.[49] Another effort to synthesize magnetic nanoparticles with high magnetic moment utilized cubic-shaped FeCo nanoparticles of 12.8 nm in a GMR-based sensor.[56] The cubic nanoparticles were surface functionalized with silane chemistry for attachment of streptavidin or antibodies. Direct labeling of biotinylated surface with streptavidin-coated nanoparticles allowed detection of 600 nanoparticle binding. Indirect labeling in ELISA type assay produced signals as low as 2×10^6 molecules of a biomarker protein.

Improvement of SV sensors was achieved by reducing the passivation layer to 30 nm and led to an enhanced sensitivity. A signal amplification strategy that had multiple layers of streptavidin-coated NPs and biotinylated antibodies in the sandwich type immunoassay also showed enhanced signals. Multiplex sensing of different protein markers in serum was demonstrated on a single chip by carefully selecting antibodies and by employing the signal enhancing strategy with multiple layers of NPs. Wang and his group at Stanford University used nanoimprint lithography to synthesize antiferromagnetic nanoparticles of 100 nm size with high magnetic moment and zero remanence.[57] The antiferromagnetic nanoparticles, that have a disk shape, were composed of multiple layers of ferromagnetic material separated by a nonmagnetic interlayer. NPs with high magnetic moments were functionalized with streptavidin and permitted the detection of DNA at concentrations as low as 10 pM.[54]

5.3. Magnetophoretic Sensors

Magnetophoresis describes the migration that occurs when a magnetic field and gradient are applied to an MNP and the particle moves to regions and of higher field and gradient. If the size of MNP changes due to the presence of an analyte, its magnetic field-induced mobility will also change, providing a magnetophoretic detection system. See Table 11.2 for a summary of magnetophoretic biosensors.

Park and coworkers developed magnetophoretic assays principles, describing a microfluidic immunoassay which utilizes the binding of superparamagnetic NPs

TABLE 11.2 Magnetophoresis Applications in Biosensing

Nanoparticles	Technique	Target (Application)	References
50-nm polystyrene beads	Microfluidic immunoassay	Rabbit and mouse IgG	58–62
Polystyrene beads	Microfluidic immunoassay	IgE (allergy diagnosis)	59
Colored microspheres/ neutravidin SMNPs	Multiplexed microfluidic immunoassay	Goat IgG, rabbit IgG, mouse IgG	63
2.8-μm dynabeads particles	DNA hybridization	DNA	64
1-μm streptavidin-coated beads	Microfluidic immunoassay	Protein C3 (inflammation in CPB)	65

to microbeads and the subsequent deflection of microbeads in a magnetic field as the signal for measuring analytes.[58] Furthermore, Park *et al.*, using a sandwich methodology, demonstrated that magnetophoretic detection can improve the sensitivity of immunoassay.[59–61] In a microchannel, polystyrene particle, onto which antigen was bioconjugated, reacted with analyte antibody, and then the particle with analyte was immunologically labeled with magnetic nanoparticles. Since the magnetic susceptibility of the particle depended on the number of label, it was possible to determine the analyte concentration from the magnetophoretic velocity or deflection in the microchannel. They demonstrated the determination of allergen-specific human immunoglobulin E from the magnetophoretic velocity and achieved one-order enhancement of sensitivity compared with a conventional immunoassay.[59] The detection limits for *D. farinae* and *D. pteronyssinus* (the two common types of dust mite analyzed in the experiment) were estimated as 565 fM (0.045 IU/ml) and 268 fM (0.021 IU/ml) respectively.

A multiplexed assay can provide quantitative information on target analytes of major physiological significance in a high-throughput manner, accelerating both disease diagnosis and biomedical studies.[63] Park *et al.* proposed a multiplexed immunoassay by using colored microspheres and succeeded in simultaneously determining the three types of antibodies (goat IgG, rabbit IgG, and mouse IgG) from the magnetophoretic deflection.[61] The detection limit for goat IgG, rabbit IgG, and mouse IgG were 10.9, 30.6, and 12.1 fM, respectively.

Magnetophoretic assay systems are adaptable to microfluidic configurations. Utilization of a microfluidic device makes possible a reduction of measurement time and sample amount, because several steps are integrated within the device. Pamme *et al.* used magnetic microparticles as a solid support and a different microfluidic configuration.[64,66,67] Here, multiple reagent streams were generated in their microfluidic device, and the magnetic microparticles migrated across these streams by magnetophoresis. Therefore, the multistep operation required in bioassay was confined to one channel and was extremely rapid, since reaction, washing, particle isolation, and detection were performed by continuous flow on a particle by particle basis. For demonstration, they employed streptavidin-conjugated magnetic microparticles and fluorescently labeled biotin streamline, which was flowed between two washing buffer streams. After passing the biotin stream, the fluorescence from the magnetic microspheres was measured to detect the streptavidin and biotin complex. This method reduced the processing time greatly from 60 to 1 min. Their technique was applicable to immunoassay[67] and DNA hybridization.[64] Recently, Sasso *et al.* proposed a novel bead manipulation scheme and fabricated a microfluidic device for a magnetophoresis-based bioassay.[65] In this case, a sandwich immunoassay was able to monitor complement C3a protein concentration up to 625 ng/ml; authors stated the device might be used to check inflammation during cardiac surgeries involving CPB (cardiopulmonary bypass). Further applications of magnetophoresis are given below and have been recently reviewed.[68]

6. CONCLUSIONS

MRSw assays and sensors have demonstrated key advantages such as high sensitivity, rapid target measurement, low sample volumes, indifference to light-based interferences, and the ability to measure chemically diverse types of analytes. MRSw technology has the potential to be practical technology in many settings such as the operating room, the clinical laboratory, with patient POC tests, and in the field settings around the world.

REFERENCES

1. Wittenberg NJ, Haynes CL. Using nanoparticles to push the limits of detection. *Wiley Interdiscip Rev Nanomed Nanobiotechnol* 2009;**1**(2):237–54.
2. Lee JH, Huh YM, Jun YW, Seo JW, Jang JT, Song HT, et al. Artificially engineered magnetic nanoparticles for ultra-sensitive molecular imaging. *Nat Med* 2007;**13**(1):95–9.
3. Weissleder R, Pittet MJ. Imaging in the era of molecular oncology. *Nature* 2008;**452**(7187):580–9.
4. Graham DL, Ferreira HA, Freitas PP, Cabral JM. High sensitivity detection of molecular recognition using magnetically labelled biomolecules and magnetoresistive sensors. *Biosens Bioelectron* 2003;**18**(4):483–8.
5. Weissleder R. Molecular imaging in cancer. *Science* 2006;**312**(5777):1168–71.
6. Koh I, Josephson L. Magnetic nanoparticle sensors. *Sensors* 2009;**9**(10):8130–45.
7. Shao H, Yoon T-J, Liong M, Weissleder R, Lee H. Magnetic nanoparticles for biomedical NMR-based diagnostics. *Beilstein J Nanotechnol* 2010;**1**:142–54.
8. Demas V, Lowery TJ. Magnetic resonance for in vitro medical diagnostics. *New J Phys* 2011;**13**:025005.
9. Josephson L, Perez JM, Weissleder R. Magnetic nanosensors for the detection of oligonucleotide sequences. *Angew Chem Int Ed* 2001;**40**(17):3204.
10. Perez JM, Josephson L, O'Loughlin T, Hogemann D, Weissleder R. Magnetic relaxation switches capable of sensing molecular interactions. *Nat Biotechnol* 2002;**20**(8):816–20.
11. Koh I, Hong R, Weissleder R, Josephson L. Nanoparticle-target interactions parallel antibody-protein interactions. *Anal Chem* 2009;**81**(9):3618–22.
12. Hong R, Cima MJ, Weissleder R, Josephson L. Magnetic microparticle aggregation for viscosity determination by MR. *Magn Reson Med* 2008;**59**(3):515–20.
13. Laurent S, Forge D, Port M, Roch A, Robic C, Vander Elst L, et al. Magnetic iron oxide nanoparticles: synthesis, stabilization, vectorization, physicochemical characterizations, and biological applications. *Chem Rev* 2008;**108**(6):2064–110.
14. Grimm J, Perez JM, Josephson L, Weissleder R. Novel nanosensors for rapid analysis of telomerase activity. *Cancer Res* 2004;**64**(2):639–43.
15. Lee H, Sun E, Ham D, Weissleder R. Chip-NMR biosensor for detection and molecular analysis of cells. *Nat Med* 2008;**14**(8):869–74.
16. Taktak S, Sosnovik D, Cima MJ, Weissleder R, Josephson L. Multiparameter magnetic relaxation switch assays. *Anal Chem* 2007;**79**(23):8863–9.
17. Kim GY, Josephson L, Langer R, Cima MJ. Magnetic relaxation switch detection of human chorionic gonadotrophin. *Bioconjug Chem* 2007;**18**(6):2024–8.
18. Perez JM, Grimm J, Josephson L, Weissleder R. Integrated nanosensors to determine levels and functional activity of human telomerase. *Neoplasia* 2008;**10**(10):1066–72.

19. Hogemann D, Josephson L, Weissleder R, Basilion JP. Improvement of MRI probes to allow efficient detection of gene expression. *Bioconjug Chem* 2000;**11**(6):941–6.
20. Perez JM, O'Loughin T, Simeone FJ, Weissleder R, Josephson L. DNA-based magnetic nanoparticle assembly acts as a magnetic relaxation nanoswitch allowing screening of DNA-cleaving agents. *J Am Chem Soc* 2002;**124**(12):2856–7.
21. Zhao M, Josephson L, Tang Y, Weissleder R. Magnetic sensors for protease assays. *Angew Chem Int Ed Engl* 2003;**42**(12):1375–8.
22. Perez JM, Simeone FJ, Tsourkas A, Josephson L, Weissleder R. Peroxidase substrate nanosensors for MR imaging. *Nano Lett* 2004;**4**(1):119–22.
23. Tsourkas A, Hofstetter O, Hofstetter H, Weissleder R, Josephson L. Magnetic relaxation switch immunosensors detect enantiomeric impurities. *Angew Chem Int Ed Engl* 2004;**43**(18):2395–9.
24. Sun EY, Weissleder R, Josephson L. Continuous analyte sensing with magnetic nanoswitches. *Small* 2006;**2**(10):1144–7.
25. Taktak S, Weissleder R, Josephson L. Electrode chemistry yields a nanoparticle-based NMR sensor for calcium. *Langmuir* 2008;**24**(14):7596–8.
26. Atanasijevic T, Shusteff M, Fam P, Jasanoff A. Calcium-sensitive MRI contrast agents based on superparamagnetic iron oxide nanoparticles and calmodulin. *Proc Natl Acad Sci USA* 2006;**103**(40):14707–12.
27. Koh I, Hong R, Weissleder R, Josephson L. Sensitive NMR sensors detect antibodies to influenza. *Angew Chem Int Ed Engl* 2008;**47**(22):4119–21.
28. Perez JM, Simeone FJ, Saeki Y, Josephson L, Weissleder R. Viral-induced self-assembly of magnetic nanoparticles allows the detection of viral particles in biological media. *J Am Chem Soc* 2003;**125**(34):10192–3.
29. Lee H, Yoon TJ, Weissleder R. Ultrasensitive detection of bacteria using core-shell nanoparticles and an NMR-filter system. *Angew Chem Int Ed Engl* 2009;**48**(31):5657–60.
30. Lee H, Yoon T-J, Figueiredo J-L, Swirski FK, Weissleder R. Rapid detection and profiling of cancer cells in fine-needle aspirates. *Proc Natl Acad Sci USA* 2009;**106**(30):12459–64.
31. Haun JB, Devaraj NK, Hilderbrand SA, Lee H, Weissleder R. Bioorthogonal chemistry amplifies nanoparticle binding and enhances the sensitivity of cell detection. *Nat Nano* 2010;**5**(9):660–5.
32. Doyle PS, Bibette J, Bancaud A, Viovy JL. Self-assembled magnetic matrices for DNA separation chips. *Science* 2002;**295**(5563):2237.
33. Baudry J, Rouzeau C, Goubault C, Robic C, Cohen-Tannoudji L, Koenig A, et al. Acceleration of the recognition rate between grafted ligands and receptors with magnetic forces. *Proc Natl Acad Sci USA* 2006;**103**(44):16076–8.
34. Singh H, Laibinis PE, Hatton TA. Rigid, superparamagnetic chains of permanently linked beads coated with magnetic nanoparticles. Synthesis and rotational dynamics under applied magnetic fields. *Langmuir* 2005;**21**(24):11500–9.
35. Zerrouki D, Baudry J, Pine D, Chaikin P, Bibette J. Chiral colloidal clusters. *Nature* 2008;**455**(7211):380–2.
36. Cohen-Tannoudji L, Bertrand E, Baudry J, Robic C, Goubault C, Pellissier M, et al. Measuring the kinetics of biomolecular recognition with magnetic colloids. *Phys Rev Lett* 2008;**100**(10):108301.
37. Daniel KD, Kim GY, Vassiliou CC, Galindo M, Guimaraes AR, Weissleder R, et al. Implantable diagnostic device for cancer monitoring. *Biosens Bioelectron* 2009;**24**(11):3252–7.
38. Daniel KD, Kim GY, Vassiliou CC, Jalali-Yazdi F, Langer R, Cima MJ. Multi-reservoir device for detecting a soluble cancer biomarker. *Lab Chip* 2007;**7**(10):1288–93.

39. Ling Y, Pong T, Vassiliou CC, Huang PL, Cima MJ. Implantable magnetic relaxation sensors measure cumulative exposure to cardiac biomarkers. *Nat Biotechnol* 2011;**29**(3):273–7.
40. McCarthy JR, Weissleder R. Multifunctional magnetic nanoparticles for targeted imaging and therapy. *Adv Drug Deliv Rev* 2008;**60**(11):1241–51.
41. Kulkarni AA, Weiss AA, Iyer SS. Detection of carbohydrate binding proteins using magnetic relaxation switches. *Anal Chem* 2010;**82**(17):7430–5.
42. Jaffer FA, Weissleder R. Seeing within: molecular imaging of the cardiovascular system. *Circ Res* 2004;**94**(4):433–45.
43. Perez JM, Josephson L, Weissleder R. Use of magnetic nanoparticles as nanosensors to probe for molecular interactions. *Chembiochem* 2004;**5**(3):261–4.
44. Schellenberger E. Bioresponsive nanosensors in medical imaging. *J R Soc Interface* 2010;**7** (Suppl. 1):S83–S91.
45. Haun JB, Castro CM, Wang R, Peterson VM, Marinelli BS, Lee H, et al. Micro-NMR for rapid molecular analysis of human tumor samples. *Sci Transl Med* 2011;**3**(71):71ra16.
46. Lazcka O, Campo FJD, MuÒoz FX. Pathogen detection: a perspective of traditional methods and biosensors. *Biosens Bioelectron* 2007;**22**(7):1205–17.
47. Kaittanis C, Naser SA, Perez JM. One-step, nanoparticle-mediated bacterial detection with magnetic relaxation. *Nano Lett* 2007;**7**(2):380–3.
48. Kaittanis C, Nath S, Perez JM. Rapid nanoparticle-mediated monitoring of bacterial metabolic activity and assessment of antimicrobial susceptibility in blood with magnetic relaxation. *PLoS One* 2008;**3**(9):e3253.
49. Wang SX, Li G. Advances in giant magnetoresistance biosensors with magnetic nanoparticle tags: review and outlook. *IEEE Trans Magn* 2008;**44**:1687–702.
50. Baselt DR, Lee GU, Natesan M, Metzger SW, Sheehan PE, Colton RJ. A biosensor based on magnetoresistance technology. *Biosens Bioelectron* 1998;**13**(7–8):731–9.
51. Mulvaney SP, Cole CL, Kniller MD, Malito M, Tamanaha CR, Rife JC, et al. Rapid, femtomolar bioassays in complex matrices combining microfluidics and magnetoelectronics. *Biosens Bioelectron* 2007;**23**(2):191–200.
52. Rife JC, Miller MM, Sheehan PE, Tamanaha CR, Tondra M, Whitman LJ. Design and performance of GMR sensors for the detection of magnetic microbeads in biosensors. *Sens Actuators A Phys* 2003;**107**(3):209–18.
53. Graham DL, Ferreira HA, Freitas PP. Magnetoresistive-based biosensors and biochips. *Trends Biotechnol* 2004;**22**(9):455–62.
54. Fu A, Hu W, Xu L, Wilson RJ, Yu H, Osterfeld SJ, et al. Protein-functionalized synthetic antiferromagnetic nanoparticles for biomolecule detection and magnetic manipulation. *Angew Chem Int Ed Engl* 2009;**48**(9):1620–4.
55. Osterfeld SJ, Yu H, Gaster RS, Caramuta S, Xu L, Han SJ, et al. Multiplex protein assays based on real-time magnetic nanotag sensing. *Proc Natl Acad Sci USA* 2008;**105**(52): 20637–40.
56. Srinivasan B, Li Y, Jing Y, Xu Y, Yao X, Xing C, et al. A detection system based on giant magnetoresistive sensors and high-moment magnetic nanoparticles demonstrates zeptomole sensitivity: potential for personalized medicine. *Angew Chem Int Ed Engl* 2009;**48**(15): 2764–7.
57. Hu W, Wilson RJ, Koh A, Fu A, Faranesh AZ, Earhart CM, et al. High-moment antiferromagnetic nanoparticles with tunable magnetic properties. *Adv Mater* 2008;**20**(8):1479–83.
58. Kim KS, Park JK. Magnetic force-based multiplexed immunoassay using superparamagnetic nanoparticles in microfluidic channel. *Lab Chip* 2005;**5**(6):657–64.

59. Hahn YK, Jin Z, Kang JH, Oh E, Han M-K, Kim H-S, et al. Magnetophoretic immunoassay of allergen-specific IgE in an enhanced magnetic field gradient. *Anal Chem* 2007;**79**(6):2214–20.
60. Jin Z, Hahn YK, Oh E, Kim Y-P, Park J-K, Moon SH, et al. Magnetic nanoclusters for ultra-sensitive magnetophoretic assays. *Small* 2009;**5**(20):2243–6.
61. Hahn YK, Chang J-B, Jin Z, Kim H-S, Park J-K. Magnetophoretic position detection for multiplexed immunoassay using colored microspheres in a microchannel. *Biosens Bioelectron* 2009;**24**(7):1870–6.
62. Zborowski M, Fuh CB, Green R, Sun L, Chalmers JJ. Analytical magnetapheresis of ferritin-labeled lymphocytes. *Anal Chem* 1995;**67**(20):3702–12.
63. Trau M, Battersby BJ. Novel colloidal materials for high-throughput screening applications in drug discovery and genomics. *Adv Mater* 2001;**13**(12–13):975–9.
64. Vojtìsek M, Iles A, Pamme N. Rapid, multistep on-chip DNA hybridisation in continuous flow on magnetic particles. *Biosens Bioelectron* 2010;**25**(9):2172–6.
65. Sasso L, Ündar A, Zahn J. Autonomous magnetically actuated continuous flow microimmunofluorocytometry assay. *Microfluid Nanofluid* 2010;**9**(2):253–65.
66. Peyman SA, Iles A, Pamme N. Rapid on-chip multi-step (bio)chemical procedures in continuous flow-manoeuvring particles through co-laminar reagent streams. *Chem Commun* 2008;**10**:1220–2.
67. Peyman SA, Iles A, Pamme N. Mobile magnetic particles as solid-supports for rapid surface-based bioanalysis in continuous flow. *Lab Chip* 2009;**9**(21):3110–7.
68. Suwa M, Watarai H. Magnetoanalysis of micro/nanoparticles: a review. *Anal Chim Acta* 2011;**690**(2):137–47.

Chapter 12

Quantum Dot Nanoparticles for *In Vitro* Sensing

Zongwen Jin and Niko Hildebrandt
Université Paris-Sud, Institut d'Electronique Fondamentale, Orsay, France

1. INTRODUCTION

Biological systems are complex entities consisting of various components such as proteins, DNAs, RNAs, lipids, carbohydrates, etc. These building blocks usually have multiple interactions with other components, forming complicated networks to sustain life phenomena.

A biosensor is an analytical device that detects the presence and/or function of one or multiple parameters in such a system (e.g., concentration or activity) by adopting biological components (such as antibodies, enzymes, DNAs, RNAs, lipids, whole cells, etc). The transducer of such a device is usually conjugated with the biological component to endow biological specificity and sensitivity. When biological binding or activity takes place at the present of a target analyte, the transducer transforms the event into a measurable signal, such as electric, magnetic, or optical signals. This allows a comfortable analysis and quantification of the target analyte.

Currently, many research applications focus on the integration of nanotechnology into biosensor design, by introducing nanostructures, such as nanoparticles (NPs). Gold, silver and magnetic nanoparticles as well as semiconductor quantum dots (QDs) are among the most often used NPs for biosensor development. These NPs provide the sensors with unique physicochemical properties, for example, localized surface plasmon resonance or colorimetric change (upon aggregation) with gold NPs[1–4]; magnetic relaxation switching or giant magneto resistance with magnetic NPs[5,6] or Förster resonance energy transfer (FRET)[7–23]; and fluorescence intermittency or blinking,[24–27] in the case of QDs.

Despite many advantages, application of NPs in biomedical sciences has been hampered by the cytotoxical uncertainty (especially on the long term), which needs to be thoroughly studied. Currently, only few NPs have passed safety regulation for clinical trials[28] mainly caused by the enormous efforts in research and financial investment necessary for acquiring approval.

Frontiers of Nanoscience, Vol. 4. DOI: 10.1016/B978-0-12-415769-9.00012-1

On the contrary, application of NPs for developing *in vitro* diagnostic devices is subjected to less stringent regulation and has relatively short developing time compared to applications for *in vivo* biosensor developments.

In this chapter, we will give a short overview of NPs for *in vitro* biosensing with a focus on QDs. We will introduce some important biological targets, define the most important requirements for a biosensor, present some concepts of NP functionalization, and highlight some recent applications using QD-based biosensors.

2. TARGETS OF BIOSENSOR

There are many biological targets, which can be analyzed by biosensors. These include pathogens, toxins, and metal ions for food and environmental safety, small molecules and metabolites for drug detection and forensics, single nucleotide polymorphism and DNA cleavages, serum biomarkers (such as immunoglobulin, circulating micro DNAs and circulating tumor cells) for diagnosis of diseases.[29–35] Considering the large variety of available targets, we limit our discussion to disease diagnosis, where blood or serum level biomarkers are the most important source of analysis. While albumins and globulins are the main constituents of serum proteins to carry out their native functions in the circulating blood system, there are also numerous proteins, which exist in very low amounts (compared to albumin) in human serum, apart from their tissue of origin. The biological functions of these low level proteins often remain unknown. However, the number of serum proteins continues to increase, mainly driven by the advances in ultrasensitive proteomics based on multidimensional liquid chromatography followed by tandem mass spectrometry, which can detect and identify as low as a few dozens of proteins. Such low level proteins have the potential for early diagnostics (e.g., for Alzheimer's disease[36]) or for reflecting the state of a specific disease.[37] The abundance of different types of proteins in blood is depicted in Figure 12.1.[38] Taking into account the extreme variations in their concentration, a biosensor targeting different biomarkers should be designed and optimized to meet the different specific needs.

3. BIOSENSOR REQUIREMENTS

Biosensor performance can be assessed by many different aspects. In general, sensitivity, selectivity, and dynamic range are the three most important factors to be considered. However, in more specific applications, such as real-life clinical diagnostics, invasiveness, sample volume, time, multiplexing capability, cost, and reimbursement are additional factors for a successful biosensor application, especially for sensors in point-of-care testing (POCT). Biosensor requirements and general technical approaches to meet these conditions are presented in Table 12.1.

The sensitivity of a biosensor is largely dependant on the signal to noise ratio of the sensing probe. This is, for example, determined by an efficient

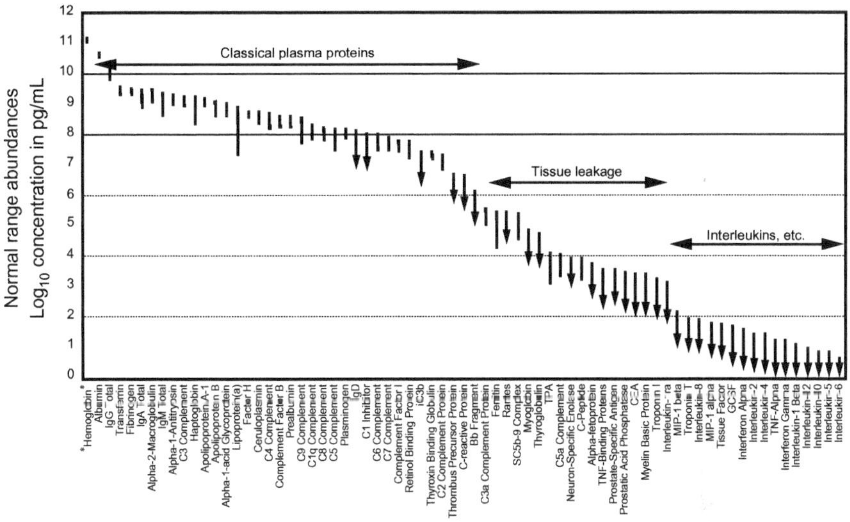

FIGURE 12.1 Concentration intervals for 70 protein analytes in plasma, where their abundance is covering a concentration range over 12 orders of magnitude (where only the upper limit is quoted, the lower end of the interval line shows an arrowhead). The classical plasma proteins are clustered to the left (high abundance). Hemoglobin is included (far left) for comparison. Figure reprinted with permission from Ref. 38. Copyright by the American Society for Biochemistry and Molecular Biology.

TABLE 12.1 Biosensor Requirements

Biosensor Requirement	Technical Approach
High sensitivity	Generation of high signals and low background noise
High selectivity	Selection of highly specific capture molecules
Wide dynamic range	Adoption of chemical or enzymatic amplification strategies
Reproducibility	Reduction of background noise and signal deviation
Multiplexing	Generation and separation of different specific signals
Economy	Small and fast devices with low sample volumes

coating of the sensor surface with active capture molecules in order to generate a high amount of positive signal, and to suppress any nonspecific signals (e.g., caused by nonspecific binding) from competing surrounding molecules. In the case of an immune sensor, this means that one should select a high-purity and high-affinity immunoglobulin (Ig) which is specific only for the analyte molecules, and an efficient homogenous blocking layer to prevent adsorption of abundant serum proteins, such as albumin, to the sensor surface. For signal amplification, enzymatic strategies are usually adopted, for example, densely loaded DNA oligomers (for later usage in PCR) on gold NPs for improved signals within nanoparticle-based barcode sensor systems,[39] or substrate recycling methods for electrochemical signal amplification, where tyrosine catalyzes substrate recycling between catechol and an electrochemically measurable *o*-quinone.[40] Nanobiosensors with amplification systems often have a wider dynamic range compared to those without amplification systems. The dynamic range should be considered depending on the actual analyte of interest, because the concentrations of analytes in a diseased state can be very different from normal conditions.

For optical sensors, the generation of specific signals and the suppression of background noise are usually realized by using wavelengths (colors), which do not interfere with the biological sample (e.g., exciting a fluorophore at a wavelength that does not generate sample autofluorescence). Often, the background is further reduced by using washing steps in between the incubation steps of the bioassays in order to remove any components that might cause nonspecific signals. Since these steps are time consuming, so-called homogeneous assays, in which the measurement directly follows incubation, are preferred. The combination of lanthanide complexes and FRET is an ideal solution for efficient background suppression (using time-resolved or time-gated measurements of the long-lived lanthanide luminescence) in combination with a homogeneous assay format (only the FRET signal is specific for the biological recognition and washing or separation steps are not necessary).[41] The combination of lanthanides with QDs can add a multiplexing option to the sensor and push the sensitivity down to subpicomolar detection limits.[16,18,19,42] Creating luminescence without any external excitation source, for example, by bioluminescence or chemiluminescence resonance energy transfer (BRET, CRET) to QDs is another elegant way to circumvent direct excitation of any nonspecific components of the sample.[43,44]

In real-life testing, such as POCT, other requirements become crucial for a successful biosensor application. These biosensors should be easy to handle because the devices need to be operated by untrained individuals (the patients themselves). This means that collection of body fluid must be simple and minimal-invasive, for example, by using a painless needle, often applied in finger prick blood tests. Such miniaturized needles are designed to penetrate skin without affecting the layer of pain receptor under the skin.[45] The cost of a single test using a biosensor is very much dependent on the application. High-throughput screening tests are usually inexpensive per single test because many are performed at once. A single POCT can be much more expensive but still needs to be affordable by individuals. Assays for new markers or with higher sensitivity than the tests from the competition can also be more expensive than the already available tests if the advantage of the new assay is obvious. However, in all of these applications (and as with all commercial products), the price needs to be as low as possible. Technologically, this can be realized by reduction of the sensor size, minimization of sample or reagent consumption, using cheap disposables or (on the contrary) reusable devices, or multiplexing for the simultaneous measurement of several different biomarkers. For some applications, biosensors also need to be very robust to withstand coarse storage or testing environment, especially in cases where a cold chain (uninterrupted series of storage and distribution) cannot be guaranteed. In summary, a good biosensor does not only need to fulfill the technical requirements to achieve an overall high performance, (cf. Table 12.1) but it also needs to be designed to meet the daily clinical use requirements which are often much different from optimized laboratory conditions.

4. FUNCTIONALIZATION OF NPs

An appropriate surface functionalization of NP is inevitable for their application in biosensing. As a basis for modification, chemical groups such as carboxylates, amines, or thiols are introduced to the NP surface during ligand exchange steps after organic synthesis or directly during aqueous synthesis. These groups are targeted by chemical cross-linkers consisting of reactive chemical groups toward them (Figure 12.2). One or two groups (homobifunctional, heterobifunctional, or multifunctional) are usually connected by spacer arms with various lengths and functions, imparting flexibility, solubility, or minimized biofouling (e.g., alkane, polyethylene glycol, cleavable S—S spacer, etc.). Many standard techniques for conjugating biomolecules with fluorophores or other labels can also be applied for NPs. These conjugation methods are extensively described in the following books.[47,48]

Spacer lengths can be modified to avoid problems for conjugated biomolecules, which are prone to spatial hindrances, for example, in preservation of protein–protein interaction, DNA–DNA hybridization, and substrate access to active sites of an enzyme. These commercially available conjugation methods are generally developed for random immobilization.

Recently, thanks to advancements in synthetic biology, more and more options become available for delicate immobilization of biomolecules. Incorporation of chemical groups which are less abundant (such as thiol modification) or even orthogonal (such as site-specific unnatural amino acid incorporation into proteins[49–51]) are novel methods for advanced biosensor design to achieve improved detection sensitivities. The latter method, although it is relatively complex to perform and structural tolerance of modification will vary between the modifications, enables a way of interfacing biomolecules and NPs in a three dimensionally controlled manner, which would not be possible by random immobilization or very limited by introduction of N- or C-terminus tags to a protein. In bioorthogonal chemistry, for example, an artificially modified pair of orthogonal aminoacyl tRNA–tRNA synthetase was introduced to a host expression system. These are designed to recognize an amber codon, which the host system does not recognize. Thus, one can introduce an unnatural amino acid

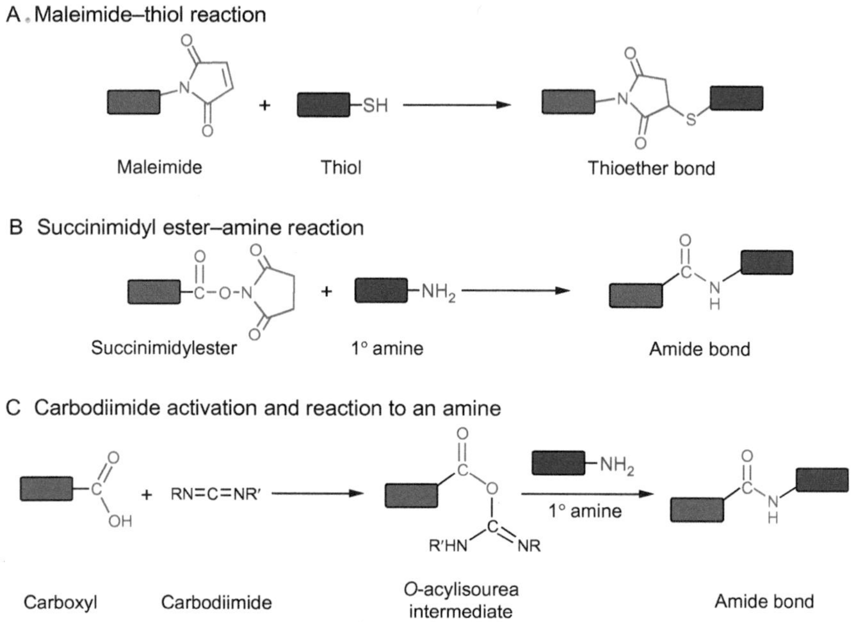

FIGURE 12.2 Standard bioconjugation reactions, including (A) maleimide–thiol, (B) succinimidyl ester–amine, and (C) carbodiimide-mediated coupling between carboxyls and amines. Figure reprinted with permission from Ref. 46. Copyright by the American Chemical Society. (For color version of this figure, the reader is referred to the online version of this chapter.)

(e.g., acetylene and azido amino acids for downstream conjugation by click chemistry) into specific sites of a protein, by substituting a particular site of the protein coding codon sequence with the amber codon. These techniques are especially powerful, given the increasing availability of 3D structures of target proteins in public domains. Although the incompatibility of QDs to copper ions used as a catalyst in click chemistry posed fundamental limitations to such methods, solutions to circumvent such a problem by copper-free click chemistry have been developed.[52] An application of a copper-free chemistry for QDs has been recently demonstrated by Bernardin *et al.*[53] For a more detailed discussion on bioconjugation, we recommend the interested reader to consult a recent review by R. Algar and I.L. Medintz,[46] in which standard biofunctionalization of various NPs as well as highly controlled immobilization strategies are elaborated.

5. RECENT APPLICATIONS

As applications of magnetic, gold, and silver NPs are already covered within the other chapters of this book, we focus our chapter on the use of semiconductor QDs. In particular, we will present some recent work on multiplexing approaches with the combination of QDs with gold nanoparticles (AuNPs) in energy transfer, the application of QDs in charge transfer (CT) with ruthenium complexes, and the use of chemiluminescence and luminescent lanthanide complexes (LLCs) for transferring energy to QD FRET acceptors.

QDs are semiconductor nanoparticles with sizes (diameters) ranging from ca. 1 to 10 nm. After first discovery in the 1980s by Alexei Ekimov,[54] QDs have drawn enormous attention in the past decade, especially in biosensor development and biomedical applications.[55] The dimensions of a QD are comparable to a small protein, and they exhibit high quantum yields, good separation between excitation and emission wavelengths (large "stokes shift"), excellent photostability, broad absorption spectrum, and narrow and symmetric emission bands. One of the most important properties of QD is their size tunability. Changing the diameter of a QD from small to large leads to a shift of absorption and emission from the blue to the NIR wavelength range. This unique photophysical property makes QDs so versatile fluorophores for applications in biosensing. QDs can be simply used as a fluorescent label to complement organic dyes and/or to overcome their photophysical deficiencies,[56–58] or as a transducing core component in biosensor design, in which the QDs luminescence properties are modulated upon the presence of the target analyte. These kinds of modulations usually include FRET, BRET or CRET, CT, and electrochemiluminescence. For details concerning these transfer mechanisms of QDs, please refer to the recent literature (e.g., Ref. 59).

5.1. QDs and AuNPs

In FRET applications, QDs are mainly used as energy donors. Their large luminescence quantum yields allow long Förster distances (donor–acceptor distance for which energy transfer is 50% efficient), and due to the size tunability, their absorption as well as emission spectra can be perfectly matched to many different FRET acceptors. Several groups of researchers have introduced gold nanoparticles (AuNPs) as energy transfer acceptors for QDs in biosensing (e.g., Refs. 60,61). Although nanosurface energy transfer (NSET) with an r^{-4} distance dependence (due to "FRET" to the surface of the AuNPs) provides a much better description for energy transfer to AuNPs[62,63], many authors wrongly use the expression FRET. In a proof of concept study, Oh *et al.*[61] realized an inhibition assay based on NSET between the two nanoparticles. Biotin–AuNPs were first incubated with avidin followed by the addition of streptavidin-conjugated QDs. The more avidin is present in the assay, the less AuNP–biotin–streptavidin–QD complexes occur and thus photoluminescence quenching is decreased. Therefore, the luminescence intensity is directly proportional to the avidin concentration. The authors were able to detect avidin down to 10 nM with a dynamic range up to 2 μM. Their assay system was then extended to detect protein glycosylation based on NSET between carbohydrate-conjugated QDs and lectin-conjugated AuNPs.[64] The two functionalized nanoparticles were preassembled, followed by addition of the glycoprotein, which can efficiently compete with the carbohydrate-conjugated

QDs for binding to the lectin-conjugated AuNPs. As a result, QD photoluminescence is recovered dependent on the amount of added glycoprotein. The assay system was also shown to be effective in distinguishing avidin (a natural glycoprotein) and neutravidin (deglycosylated form of avidin). The QD/AuNP couple can be easily modified by changing the biological interaction that mediates the assembly of the nanoparticles. Thus, it is a very versatile tool for constructing simple and rapid assay systems which are even compatible with microarray platforms. The NP assembly can be immobilized on substrates for spatial registration toward high-throughput assay formats.[65,66] As an example, a protease inhibitor screening assay was developed based on protease susceptible peptide-linked QD/AuNP conjugates.[65] The linker peptide was designed to be recognized by a specific protease. The photoluminescence is recovered if the linker peptide is cleaved to release the AuNPs that quench the QDs (cf. Figure 12.3). For multiplexing purposes, three differently colored QDs (QD525, QD605, and QD655) were conjugated to three different linker peptides with conserved cleavage sequences for MMP7, caspase-3, and thrombin, respectively.

The chip-based QD/AuNP FRET system can be extended to higher fold multiplexing by spatial registration of QDs, for example, on microarrayed glass surfaces. Taking consideration that high-density and high-throughput assay systems are important requirements for pharmaceutical screening application, the chip-based QD FRET system is very useful for activity-based inhibitor screening of pharmaceutically important target enzymes (such as proteases or kinases).

5.2. QDs in Charge Transfer

An alternative to use AuNPs as energy transfer quenchers is the application of CT. In the case of CT, redox-active molecules in close proximity to the QDs can transfer electrons to the surface states of the QDs, which will lead to a partial quenching of QD luminescence, whereas transfer to the QD core will lead to complete quenching.[67] In a recent publication, Medintz *et al.* have used polyhistidine-tagged peptides labeled with redox-active ruthenium phenanthroline complexes which self-assembled to the Zn-rich QD surfaces.[68] The main advantage of this approach is the "universal" quenching, because CT with the same ruthenium complex can be performed to many differently colored QDs (Figure 12.4). This allows the realization of high orders of multiplexing, and the authors demonstrated the deconvolution of eight different QD emission spectra. Although the overlap of the different spectra will limit the sensitivity of an eightfold multiplexed bioassay, the CT system showed the flexibility of QDs for color-coded multiplexing and the advantage of using one "universal" quencher for many different QDs.

5.3. QD as CRET Acceptors

Although QDs are theoretically very good FRET acceptors, because their high extinction coefficients over a broad wavelength range allow large overlap

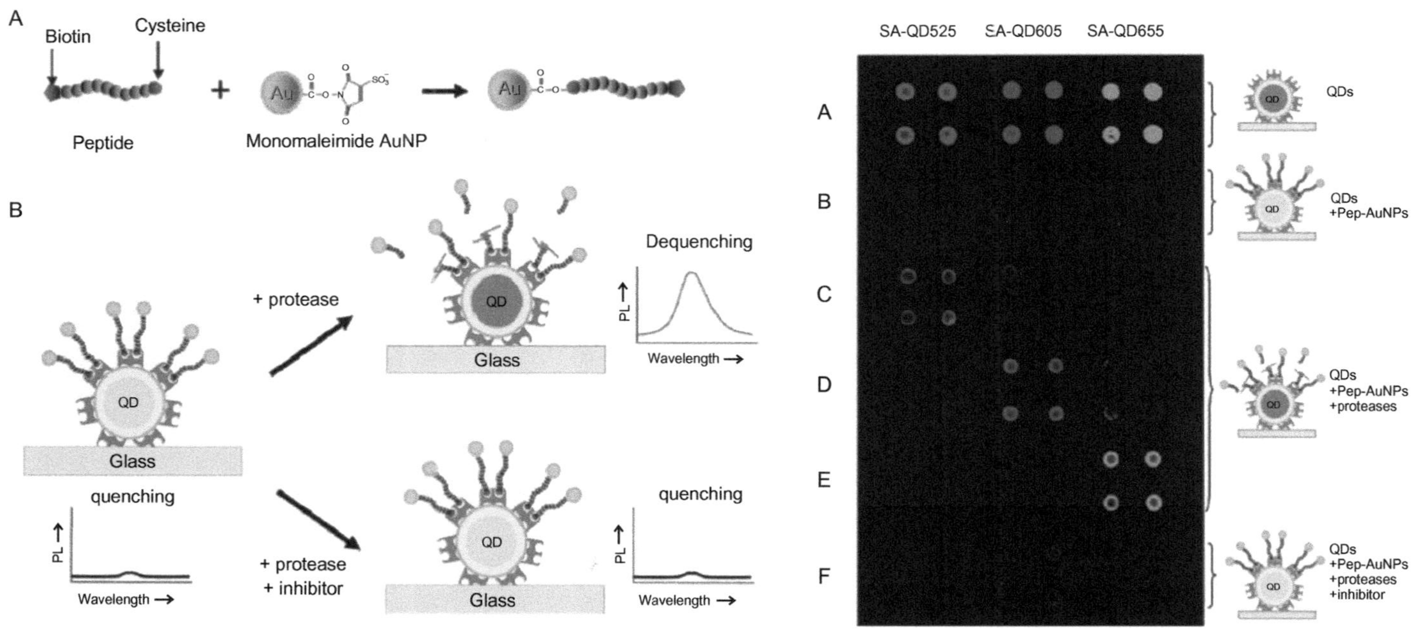

FIGURE 12.3 Left panel: Schematic diagram of QD-AuNP NSET sensor chip for protease inhibitor. (A) Construction of the peptide-conjugated AuNP. (B) Changes of PL intensities of nanoprobes by the protease action in the absence and the presence of its inhibitor. Right panel: Multiplexed assay of proteases by using streptavidin (SA)-conjugated QDs with different colors (QD525, QD606, and QD655 from left to right) on a glass slide. Biotinylated peptide substrates for MMP-7, caspase-3, and thrombin were conjugated to AuNPs, and the resulting Pep-AuNPs were associated with the different SA-QDs. (A) SA-QDs only, (B) SA-QDs + respective Pep-AuNPs, (C) SA-QDs + Pep-AuNPs + MMP-7, (D) SA-QDs + Pep-AuNPs + caspase-3, (E) SA-QDs + Pep-AuNPs + thrombin, (F) QDs + Pep-AuNPs + mixture of the respective protease and its inhibitor. Reproduced with permission from Ref. 65. Copyright by the American Chemical Society. (For interpretation of the references to color in this figure legend, the reader is referred to the online version of this chapter.)

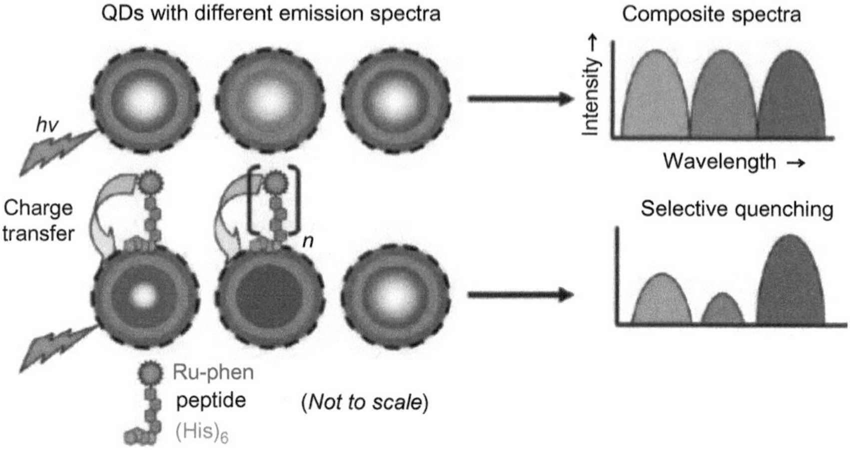

FIGURE 12.4 Schematic diagram of selective quenching by charge transfer between QD and surface-assembled ruthenium complex and the resulting luminescence spectra. Reproduced with permission from Ref. 68. Copyright by the American Chemical Society. (For color version of this figure, the reader is referred to the online version of this chapter.)

integrals (and thus long Förster distances) with many different donors, QDs are mainly applied as FRET donors. The reason for this is the direct excitation of the QDs at almost every wavelength that will excite a possible donor candidate. If both FRET partners have similar excited state lifetimes, the simultaneous excitation of donor and acceptor will render FRET very inefficient (because the acceptor needs to be in the ground state in order to receive the donor energy). The inefficiency of QDs as FRET acceptors with organic dye donors has already been shown.[69] One possibility to circumvent direct excitation of QD acceptors is the abstinence of an external excitation source. This can be realized by BRET or CRET.[70,71] Another advantage of the lacking external excitation is the possibility of an efficient use of this technique in tissues, where penetration of light in deep regions is usually problematic. BRET has been elegantly realized by Rao and coworkers for *in vivo* imaging in mice.[44] More recently, aptamer-based Hg^{2+}, adenosine-5′-triphosphate (ATP), and DNA sensors, in which CRET to QDs was generated by the formation of active hemin–G-quadruplex DNAzyme structures, have been developed.[72] In the study, the authors were able to demonstrate chemiluminescence generated during luminol oxidization, which was sufficient to trigger QD luminescence by CRET to the QDs. Hg^{2+}- or ATP-specific aptamers (from which one was labeled to the QD) or hairpin nucleic acids (labeled to the QD) opened by specific target DNA were applied for different detection schemes (cf. Figure 12.5 for DNA and ATP detection). Multiplexing was demonstrated by using three different QDs for the detection of three different DNA targets by CRET.

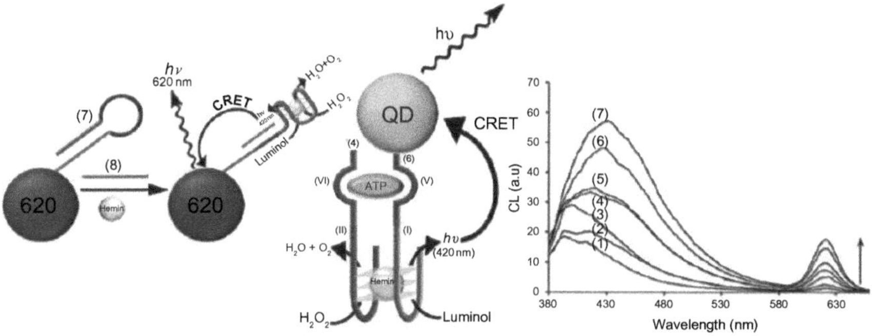

FIGURE 12.5 Chemiluminescent analysis of ATP by two subunits consisting of the conjugated anti-ATP and HRP-DNAzyme subunits. Chemiluminescence generated during hemin–G-quadruplex (assembled to the QD) catalysis of luminol oxidization by H_2O_2 is sufficient to trigger QD luminescence. Reproduced with permission from Ref. 72. Copyright by the American Chemical Society. (For color version of this figure, the reader is referred to the online version of this chapter.)

5.4. QDs as FRET Acceptors

Another possibility to circumvent inefficient FRET to QD acceptors due to direct excitation of QDs is the use of FRET donors with long excited state lifetimes such as LLCs (e.g. Ref. 13). This approach requires an external light source, but the excited state lifetimes of the LLCs are so much longer than the ones of the QDs (ca. 1 ms compared to ca. 10 ns for QDs) that one can simply wait until all the directly excited QDs have decayed back to their ground states. Then there are still many excited LLCs available for FRET to QDs, and the energy transfer becomes very efficient. LLCs are small (compared to antibodies) so that their valance can be easily controlled, which is an important aspect for bioapplications. Moreover, several LLCs can be placed around one single QD, which allows sequential FRET and thus high brightness and sensitivity for such FRET pairs. Another big advantage of the long LLC luminescence lifetimes is the possibility of time gating for nearly background-free measurements (the nano- to microsecond background emission from sample autofluorescence is simply cut off). In a recent publication, we could demonstrate a fivefold multiplexed assay with subpicomolar detection limits for all five FRET systems (one luminescent terbium complex and five different QDs) within one sample.[16] This highly sensitive multiplexed detection was demonstrated on a proof-of-principle basis with biotin–streptavidin (cf. Figure 12.6). The system was further exploited for the use as a multiplexed molecular ruler for measuring distances between ca. 5 and 11 nm.[22] Further studies for the integration of this highly sensitive multiplexed tool for real-life bioanalysis are ongoing.

6. CONCLUSION

In conclusion, biological systems can become quite complex, especially if conjugated with nanoparticles such as QDs. A good biosensor needs to reflect

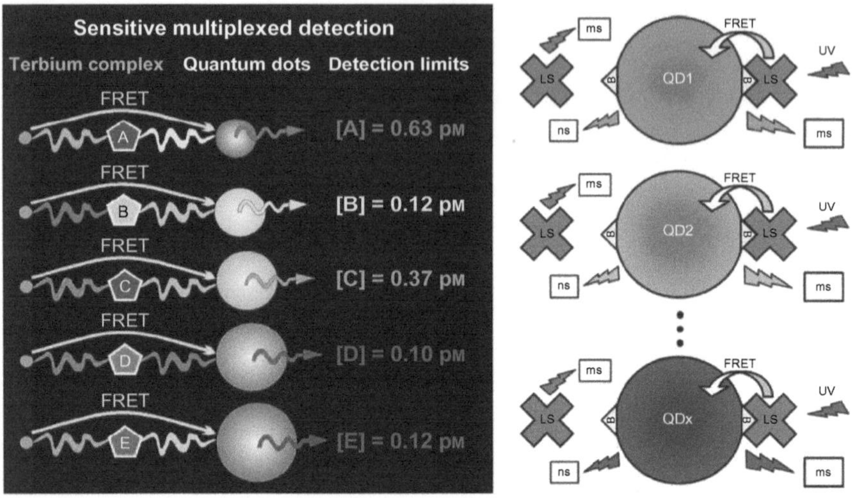

FIGURE 12.6 Fivefold multiplexed detection from one luminescent terbium complex to five different QDs with subpicomolar detection limits (left), realized for a biotin–streptavidin binding system, where long-lived (ms) emission from the different biotinylated QDs is generated by FRET from terbium-labeled streptavidin (LS) due to streptavidin–biotin binding (right). Reproduced with permission from Ref. 16. Copyright by Wiley-VCH Verlag GmbH & Co. KGaA. (For color version of this figure, the reader is referred to the online version of this chapter.)

subtle changes in the biosystem of interest and thus requires some extent of sophisticated development. Due to the large versatility of biological systems, there is no "general" biosensor. One must carefully observe the biosensor requirements for each specific analyte and chose the appropriate commercial or noncommercial solution, which might require some additional development. Nevertheless, there are manifold possibilities available (commercial and noncommercial solutions) to design an efficient biosensor for almost every diagnostic application. Nanoparticles such as semiconductor QDs have unique photophysical properties, which make them extremely useful for optical sensing. Especially in combination with energy or CT processes, the QD properties can be efficiently exploited for the design of very sensitive multiplexed biosensors. Possibly, the still most problematic issue of semiconductor QDs is their integration into real-life clinical applications, which requires stable, easily storable, and reproducible nanoparticle bioconjugates in order to satisfy the stringent regulations for such products. Research with QDs for biosensing has brought up many very interesting and useful applications and our chapter only covered a very small selection. All these developments and the ongoing interest of researchers and the industry in highly sensitive multiplexed biosensors point toward a successful integration of QDs into real-life diagnostics in the near future and the appearance of novel innovative solutions for biosensing applications.

ACKNOWLEDGEMENT

We acknowledge the European Commission (FP7 project NANOGNOSTICS) and the Innovative Medicines Initiative (project OncoTrack) for financial support.

REFERENCES

1. Baptista P, Pereire E, Eaton P, Doria G, Miranda A, Gomes I, et al. Gold nanoparticles for the development of clinical diagnosis methods. *Anal Bioanal Chem* 2008;**391**:943–50.
2. Liu J, Lu Y. Colorimetric biosensors based on DNAzyme-assembled gold nanoparticles. *J Fluoresc* 2004;**14**:343–54.
3. Murray WA, Barnes WL. Plasmonic materials. *Adv Mater* 2007;**19**:3771–82.
4. Zhao W, Brook MA, Li Y. Design of gold nanoparticle-based colorimetric biosensing assays. *Chembiochem* 2008;**9**:2363–71.
5. Gaster RS, Xu L, Han SJ, Wilson RJ, Hall DA, Osterfeld SJ, et al. Quantification of protein interactions and solution transport using high-density GMR sensor arrays. *Nat Nanotechnol* 2011;**6**:314–20.
6. Haun JB, Yoon TJ, Lee HH, Weissleder R. Magnetic nanoparticle biosensors. *Wiley Interdiscip Rev Nanomed Nanobiotechnol* 2010;**2**:291–304.
7. Algar WR, Krull UJ. New opportunities in multiplexed optical bioanalyses using quantum dots and donor-acceptor interactions. *Anal Bioanal Chem* 2010;**398**:2439–49.
8. Boeneman K, Mei BC, Dennis AM, Bao G, Deschamps JR, Mattoussi H, et al. Sensing caspase 3 activity with quantum dot-fluorescent protein assemblies. *J Am Chem Soc* 2009;**131**:3828–9.
9. Chen Z, Li G, Zhang L, Jiang J, Li Z, Peng Z, et al. A new method for the detection of ATP using a quantum-dot-tagged aptamer. *Anal Bioanal Chem* 2008;**392**:1185–8.
10. Clapp AR, Medintz IL, Mauro JM, Fisher BR, Bawendi MG, Mattoussi H. Fluorescence resonance energy transfer between quantum dot donors and dye-labeled protein acceptors. *J Am Chem Soc* 2004;**126**:301–10.
11. Clapp AR, Medintz IL, Uyeda HT, Fisher BR, Goldman ER, Bawendi MG, et al. Quantum dot-based multiplexed fluorescence resonance energy transfer. *J Am Chem Soc* 2005;**127**:18212–21.
12. Clapp AR, Pons T, Medintz IL, Delehanty JB, Melinger JS, Fisher BR, et al. Two-photon excitation of quantum-dot-based fluorescence resonance energy transfer and its applications. *Adv Mater* 2007;**19**:1921–6.
13. Charbonnière LJ, Hildebrandt N. Lanthanide complexes and quantum dots: a bright wedding for resonance energy transfer. *Eur J Inorg Chem* 2008;**21**:3241–51.
14. Charbonnière LJ, Hildebrandt N, Ziessel RF, Löhmannsröben H-G. Lanthanides to quantum dots resonance energy transfer in time-resolved fluoro-immunoassays and luminescence microscopy. *J Am Chem Soc* 2006;**128**:12800–9.
15. Curutchet C, Franceschetti A, Zunger A, Scholes GD. Examining Förster energy transfer for semiconductor nanocrystalline quantum dot donors and acceptors. *J Phys Chem C* 2008;**112**:13336–41.
16. Geißler D, Charbonnière LJ, Ziessel RF, Butlin NG, Löhmannsröben H-G, Hildebrandt N. Quantum dot biosensors for ultrasensitive multiplexed diagnostics. *Angew Chem Int Ed* 2010;**49**:1396–401.

17. Härmä H, Soukka T, Shavel A, Gaponik N, Weller H. Luminescent energy transfer between cadmium telluride nanoparticle and lanthanide(III) chelate in competitive bioaffinity assays of biotin and estradiol. *Anal Chim Acta* 2007;**604**:177–83.
18. Hildebrandt N, Charbonnière LJ, Beck M, Ziessel RF, Ziessel RF, Löhmannsröben H-G. Quantum dots as efficient energy acceptors in a time-resolved fluoroimmunoassay. *Angew Chem Int Ed* 2005;**44**:7612–5.
19. Hildebrandt N, Charbonnière LJ, Löhmannsröben H-G. Time-resolved analysis of a highly sensitive Förster resonance energy transfer immunoassay using terbium complexes as donors and quantum dots as acceptors. *J Biomed Biotechnol* 2007;6. doi:10.1155/2007/79169 Article ID 79169.
20. Hildebrandt N, Löhmannsröben H-G. Quantum dot nanocrystals and supramolecular lanthanide complexes—energy transfer systems for sensitive in vitro diagnostics and high throughput screening in chemical biology. *Curr Chem Biol* 2007;**1**:167–86.
21. Medintz IL, Clapp AR, Mattoussi H, Goldman ER, Fisher B, Mauro JM. Self-assembled nanoscale biosensors based on quantum dot FRET donors. *Nat Mater* 2003;**2**:630–8.
22. Morgner F, Geißler D, Stufler S, Butlin NG, Löhmannsröben H-G, Hildebrandt N. A quantum-dot-based molecular ruler for multiplexed optical analysis. *Angew Chem Int Ed* 2010;**49**:7570–4.
23. Roberti MJ, Giordano L, Jovin TM, Jares-Erijman EA. FRET imaging by k(t)/k(f). *ChemPhysChem* 2011;**12**(3):563–6.
24. Bachir AI, Kolin DL, Heinze KG, Hebert B, Wiseman PW. A guide to accurate measurement of diffusion using fluorescence correlation techniques with blinking quantum dot nanoparticle labels. *J Chem Phys* 2008;**128**:225105.
25. Bharadwaj P, Novotny L. Robustness of quantum dot power-law blinking. *Nano Lett* 2011;**11**:2137–41.
26. Lee SF, Osborne MA. Brightening, blinking, bluing and bleaching in the life of a quantum dot: friend or foe? *ChemPhysChem* 2009;**10**:2174–91.
27. Yu M, Van OA. Enhanced fluorescence intermittency of CdSe-ZnS quantum-dot clusters. *Phys Rev Lett* 2006;**97**:237402.
28. Raj B. NanoBiotech 2008: exploring global advances in nanomedicine. *Nanomedicine* 2009;**5**:5–7.
29. Gutiérrez JC, Amaro F, Martín-González A. From heavy metal-binders to biosensors: ciliate metallothioneins discussed. *Bioassays* 2009;**31**:805–16.
30. Kantiani L, Llorca M, Sachís J, Farré M, Barceló D. Emerging food contaminants: a review. *Anal Bioanal Chem* 2010;**98**:2413–27.
31. Lazka O, Campo JD, Munõz FX. Pathogen detection: a perspective of traditional methods and biosensors. *Biosens Bioelectron* 2007;**22**:1205–17.
32. Rusling JF, Kumar CV, Gutkind JS, Patel V. Measurement of biomarker proteins for point-of-care early detection and monitoring of cancer. *Analyst* 2010;**135**:2496–511.
33. Vaddiraju S, Tomazos I, Burgess DJ, Jain FC, Papadimitrakopoulos F. Emerging synergy between nanotechnology and implantable biosensors: a review. *Biosens Bioelectron* 2010;**25**:1553–65.
34. Wei F, Lillehoj PB, Ho CM. DNA diagnostics: nanotechnology-enhanced electrochemical detection of nucleic acids. *Pediatr Res* 2010;**67**:458–68.
35. Yu M, Stott S, Toner M, Maheswaran S, Haber DA. Circulating tumor cells: approaches to isolation and characterization. *J Cell Biol* 2011;**192**:373–82.
36. Ray S, Britschgi M, Herbert C, Takeda-Uchimura Y, Boxer A, Blennow K, et al. Quantum dots versus organic dyes as fluorescent labels. *Nat Methods* 2008;**5**:763–75.

37. Jacobs JM, Adkins JN, Qian WJ, Liu T, Shen Y, Camp II DG, et al. Utilizing human blood plasma for proteomic biomarker discovery. *J Proteome Res* 2005;**4**:1073–85.
38. Anderson NL, Anderson NG. The human plasma proteome: history, character, and diagnostic prospects. *Mol Cell Proteomics* 2002;**1**(11):845–67.
39. Nam JM, Thaxton CS, Mirkin CA. Nanoparticle-based bio-bar codes for the ultrasensitive detection of proteins. *Science* 2006;**301**:1884–6.
40. Piao Y, Jin Z, Lee D, Lee HJ, Na HB, Hyeon T, et al. Sensitive and high-fidelity electrochemical immunoassay using carbon nanotubes coated with enzymes and magnetic nanoparticles. *Biosens Bioelectron* 2011;**26**:3192–9.
41. Geißler D, Hildebrandt N. Lanthanide complexes in FRET applications. *Curr Inorg Chem* 2011;**1**:17–35.
42. Sellrie F, Beck M, Hildebrandt N, Micheel B. A homogeneous time-resolved fluoroimmunoassay (TR-FIA) using antibody mediated luminescence quenching. *Anal Meth* 2010;**2**:1298–301.
43. Li Z, Wang Y, Zhang G, Xu W, Han YJ. Chemiluminescence resonance energy transfer in the luminol CdTe quantum dots conjugates. *Luminescence* 2010;**30**(6):995–9.
44. So MK, Xu CJ, Loening AM, Gambhir SS, Rao JH. Self-illuminating quantum dot conjugates for in vivo imaging. *Nat Biotechnol* 2006;**24**:339.
45. Moon SJ, Lee SS, Kwon TH. Fabrication of microneedle array using LIGA and hot embossing process. *Microsyst Technol* 2005;**11**:311–8.
46. Algar WR, Prasuhn DE, Steward MH, Jennings TL, Blanco-Canosa JB, Dawson PE, et al. The controlled display of biomolecules on nanoparticles: a challenge suited to bioorthogonal chemistry. *Bioconjug Chem* 2011;**22**:825–58.
47. Haugland RP. *The handbook: a guide to fluorescent probes and labeling technologies.* 10th ed. San Diego, USA: Invitrogen; 2005.
48. Hermanson GT. *Bioconjugate techniques.* 2nd ed. San Diego, USA: Academic Press; 2008.
49. Lim RK, Lin Q. Bioorthogonal chemistry: recent progress and future directions. *Chem Commun* 2010;**46**(10):1589–600.
50. Prescher JA, Bertozzi CR. Chemistry in living systems. *Nat Chem Biol* 2005;**1**:13–21.
51. Sletten EM, Bertozzi CR. Bioorthogonal chemistry: fishing for selectivity in a sea of functionality. *Angew Chem Int Ed Engl* 2009;**48**(38):6974–98. doi:10.1002/anie.200900942.
52. Baskin JM, Prescher JA, Laughlin ST, Agard NJ, Chang PV, Miller IA, et al. Copper-free click chemistry for dynamic in vivo imaging. *Proc Natl Acad Sci USA* 2007;**104**:16793–7.
53. Bernardin A, Cazet A, Guyon L, Delannoy P, Vinet F, Bonnaffé D, et al. Copper-free click chemistry for highly luminescent quantum dot conjugates: application to in vivo metabolic imaging. *Bioconjug Chem* 2010;**21**(4):583–8.
54. Ekimov AI, Onushchenko AA. Quantum size effect in three-dimensional microscopic semiconductor crystals. *JETP Lett* 1981;**34**:345–9.
55. Hildebrandt N. Biofunctional quantum dots: controlled conjugation for multiplexed biosensors. *ACS Nano* 2011;**5**(7):5286–90.
56. Long F, He M, Zhu A, Song B, Sheng J, Shi H. Compact quantitative optic fiber-based immunoarray biosensor for rapid detection of small analytes. *Biosens Bioelectron* 2010;**26**:16–22.
57. Morarka A, Agrawal S, Kale S, Kale A, Ogale S, Paknikar K, et al. Quantum dot based immunosensor using 3D circular microchannels fabricated in PDMS. *Biosens Bioelectron* 2011;**26**:3050–3.
58. Zhang Y, Zeng Q, Sun Y, Liu X, Tu L, Kong X, et al. Multi-targeting single fiber-optic biosensor based on evanescent wave and quantum dots. *Biosens Bioelectron* 2010;**26**:149–54.

59. Algar WR, Tavares AJ, Krull UJ. Beyond labels: a review of the application of quantum dots as integrated components of assays, bioprobes, and biosensors utilizing optical transduction. *Anal Chim Acta* 2010;**673**:1–25.
60. Gueroui Z, Libchaber A. Single-molecule measurements of gold-quenched quantum dots. *Phys Rev Lett* 2004;**93**:166108.
61. Oh E, Hong MY, Lee D, Nam SH, Yoon HC, Kim HS. Inhibition assay of biomolecules based on fluorescence resonance energy transfer (FRET) between quantum dots and gold nanoparticles. *J Am Chem Soc* 2005;**127**:3270–1.
62. Jennings TL, Singh MP, Strouse GF. Fluorescent lifetime quenching near d = 1.5 nm gold nanoparticles: probing NSET validity. *J Am Chem Soc* 2006;**128**(16):5462–7.
63. Pons T, Medintz IL, Sapsford KE, Higashiya S, Grimes AF, English DS, et al. On the quenching of semiconductor quantum dot photoluminescence by proximal gold nanoparticles. *Nano Lett* 2007;**7**(10):3157–64.
64. Oh E, Lee D, Kim YP, Cha SY, Oh DB, Kang HA, et al. Nanoparticle-based energy transfer for rapid and simple detection of protein glycosylation. *Angew Chem Int Ed* 2006;**45**:7959–63.
65. Kim YP, Oh YH, Oh E, Ko S, Han MK, Kim HS. Energy transfer-based multiplexed assay of proteases by using gold nanoparticle and quantum dot conjugates on a surface. *Anal Chem* 2008;**80**:4634–41.
66. Kim YP, Park S, Oh E, Oh YH, Kim HS. On-chip detection of protein glycosylation based on energy transfer between nanoparticles. *Biosens Bioelectron* 2009;**24**:1189–94.
67. Medintz IL, Pons T, Trammell SA, Grimes AF, English DS, Blanco-Canosa JB, et al. Interactions between redox complexes and semiconductor quantum dots coupled via a peptide bridge. *J Am Chem Soc* 2008;**130**:16745–56.
68. Medintz IL, Farrell D, Susumu K, Trammell SA, Deschamps JR, Brunel FM, et al. Multiplex charge-transfer interactions between quantum dots and peptide-bridged ruthenium complexes. *Anal Chem* 2009;**81**:4831–9.
69. Clapp AR, Medintz IL, Fisher BR, Anderson GP, Mattoussi H. Can luminescent quantum dots be efficient energy acceptors with organic dye donors? *J Am Chem Soc* 2005;**127**:1242.
70. Huang XY, Li L, Qian HF, Dong CQ, Ren JC. A resonance energy transfer between chemiluminescent donors and luminescent quantum-dots as acceptors (CRET). *Angew Chem* 2006;**118**:5264 *Angew Chem Int Ed* 2006;**45**:5140–3.
71. Yao HQ, Zhang Y, Xiao F, Xia ZY, Rao JH. Quantum dot/bioluminescence resonance energy transfer based highly sensitive detection of proteases. *Angew Chem* 2007;**119**:4424 *Angew Chem Int Ed* 2007;**46**:4346.
72. Freeman R, Liu X, Willner I. Chemiluminescent and chemiluminescence resonance energy transfer (CRET) detection of DNA, metal ions, and aptamer-substrate complexes using hemin/G-quadruplexes and CdSe/ZnS quantum dots. *J Am Chem Soc* 2011;**133**:11597–604.

Part IV

Nanoparticles for Therapy

Chapter 13

Hyperthermia Using Inorganic Nanoparticles

Pablo del Pino and Beatriz Pelaz

Nanotherapy and Nanodiagnostics Group (GN^2), Instituto de Nanociencia de Aragón, Universidad de Zaragoza, Edificio I+D, Mariano Esquillor s/n, Campus Rio Ebro, Zaragoza, Spain

1. INTRODUCTION

The suitability of increased temperatures to achieve therapeutic gain has been known for centuries; where the earliest report is likely to have been in 3000 BC, known as the Egyptian Edwin Smith surgical papyrus.[1] Hippocrates (460–360 BC) already describes in one of his aphorisms "hot-irons, concerns higher temperatures, such as those used in cauterization." In the nineteenth and twentieth centuries, researchers started to use artificial fever (generated by infection and toxins) and novel radiofrequency approaches as treatments based on increased temperature. Although at the end of the nineteenth century, technology was able to match the need of researchers regarding the use of radiation to heat tissue, a wider interest in hyperthermia had to wait until 1975, during the first international congress on hyperthermia in oncology (Washington, 1975). In recent years, hyperthermia appears to have gained renewed interest from the scientific community. This might well have been motivated by the latest progress in the generation and control of nonionizing radiation and the capability of nanotechnology to deliver materials which can act as heat transducers (nanoheaters). The general consensus now is that hyperthermia is a very promising treatment which has remarkable biological effect in tumor cells *in vivo*; however, suitable heat delivery and control represents a challenge, both from the technical and physiological point of view.[2]

What do we understand by oncological hyperthermia? It is the use of heat to kill tumor cells or at least make them more susceptible to be killed by other methods such as chemo- and/or radiotherapy. A more general definition of hyperthermia is "the elevation of temperature above the physiological level with the objective to achieve therapeutic gain."[2] Hyperthermia can kill cells directly; however, clinical results have shown that the combination of radiotherapy and hyperthermia and chemotherapy and hyperthermia (thermochemotherapy) translates into a substantial (not only additive) therapeutic improvement.[3–5] There are currently two main hyperthermia approaches for fighting cancer:

Frontiers of Nanoscience, Vol. 4. DOI: 10.1016/B978-0-12-415769-9.00013-3

i. Mild hyperthermia where the temperature (of the region to treat) is raised to 40–44 °C. Given that *in vivo* tumor cells are more vulnerable to heat than healthy cells, mild hyperthermia can be used to weaken tumor cells prior to radio- and/or chemotherapy. In addition, heat can be used to stimulate the immune response for nonspecific immunotherapies.[6]

ii. Thermal ablation (thermoablation) where temperature is raised above 46 °C and leads to tumor elimination via necrosis, coagulation, or carbonization.[7]

Cancerous cells are not inherently more susceptible to damage by heat than healthy cells. With the exception of tumor cells of cancers that affect blood, bone marrow, and lymph nodes (hematological malignancies),[8] normal and tumor cells are equally susceptible to increased temperatures *in vitro*. In general, the thermal energy dose required for induction of cell death in cell cultures appears to be related to the amount of energy to produce protein denaturation; it ultimately affects the cell by damaging structures like cytoskeleton and membrane, and enzyme complexes for DNA synthesis and repair.[9] Therefore, the energy dose required should carry the cell above a threshold temperature; this temperature might slightly vary between cell lines and cell in different stages of the cell cycle. Another scenario applies, however, *in vivo* where tumor cells, in contrast to normal tissue, can be irreversibly damaged when exposed to temperatures between 40 and 44 °C; this is motivated by the characteristic physiology which most types of tumors share, that is, the vasculature of solid tumors is chaotic and the blood supply is therefore deficient, resulting in regions with hypoxia and low pH which ultimately makes cells more susceptible to hyperthermia treatment. This is normally not the case in nontumor tissues which remain unaffected even after an exposure time of 1 h to an environment temperature of 44 °C.[10] Only nervous tissues seem to be more temperature sensitive.[11]

It has been demonstrated that after hyperthermia treatment, a class of proteins called heat-shock proteins (HSPs) are expressed on the surface of tumor cells but not on the corresponding normal tissues.[12,13] HSPs mediate the stability of tumor cells following environmental stress, and they are commonly regarded as chaperones assisting protein folding and translocation; however, HSPs can also serve as cytokines that can stimulate dendritic cells and macrophages to produce proinflammatory cytokines and chemokines. Tumor cells expressing HSPs on their surface are more susceptible to lysis by natural killer cells. Moreover, the release of HSPs following necrotic cell death, but not apoptotic death, stimulates macrophages and dendritic cells to produce cytokines and, therefore, activate antigen-presenting cells.[14] There is direct evidence for the important role of releasable HSPs in the initiation of antitumor immunity.[15]

As previously mentioned, hyperthermia works very well as an adjuvant for the more classical radio- and chemotherapy. Hyperthermia has been proven to drastically enhance the suitability of radiotherapy.[3] Two main mechanisms are responsible for this more than additive enhancement. First, hyperthermia has been reported to increase blood flow into the tumor; thus, an improved tissue oxygenation can prevent the hypoxic problem and radiosensitize the hypoxic area.[16] How persistent is the improved tissue oxygenation is a controversial matter.[17] Second, it has been demonstrated that also *in vitro*, where the hypoxic problem does not apply, hyperthermia fosters radiation effects as well. This may well be result of destructive interference of hyperthermia with DNA repair after radiotherapy, probably by damaging cellular proteins.[18] Overall, hyperthermia seems to palliate radioresistance of tumors and to greatly improve the killing efficiency of ionizing radiation (without entering in "forbidden" radiation doses) by affecting the DNA repair mechanism.[19]

On the other hand, and despite the lack of interaction of some typical antitumor drugs and heat, the suitability of hyperthermia to enhance the cytotoxicity of many chemotherapeutic agents has been widely confirmed *in vivo*.[5] Mechanisms for the thermal enhancement include increased rate constants of alkylation, increased drug uptake, and inhibition of repair of drug-induced lethal or sublethal damage. Studies on drug-heat sequence show, in general, that drugs administered immediately before hyperthermia are most effective.[20]

Modalities of classical oncological hyperthermia may be classified according to the nature of the applicator, from the whole body to the tumor cell level (Table 13.1). There are three main heat sources: contact with externally heated liquid, contactless applicator (e.g., ultrasound, microwave radiofrequency, and infrared devices), and inserted heat sources (e.g., antennas, laser fibers, probes, and mediators). Nevertheless, none of these devices is able to destroy selectively deeply situated cancer, without destroying the surrounding normal tissues. This fact has led to the development of technologies based on inserted heat sources.[21]

Nanotechnology can offer a wide variety of nanomaterials with great potential as "ideal" nanoheaters. As the degree of control to fabricate "smart" nanomaterials evolves, the application of nanotechnology to medicine (nanomedicine) can find more applications where it can have a great impact. This is the case of hyperthermia field where superparamagnetic nanoparticles (SPM-NPs) as well as plasmonic NPs are currently investigated as nanoheaters or nanotransducers which can be remotely activated by radiation which does not (or minimally) interact with physiological tissue and fluids. In addition, the design of complex functional nanomaterials has undergone an extraordinary development in recent years; this has provided us with complex nanomaterials not only with extraordinary physical properties but also with a vast array of molecules with biological relevance such as antibodies, genetic material, or tumoral markers of the most diverse nature; these molecules can also be thought as functional surfactants.

Summarizing, nanotechnology provides us with the opportunity of achieving smart nanostructures with complex functionalities including local heating, targeting (passive or active), improved uptake, delivery, biocompatibility, suitable biodistribution, or nonimmunogenicity, to name a few. The use of local

TABLE 13.1 Typical Hyperthermia Strategies in Oncology

		Overheated Region			
Hyperthermia Strategy		***Whole-Body***	***Organ***	***Tumor***	***Cell***
Hot source		Hot bath, air, wax, blanket, suits, etc.	Isolated organ perfusion (e.g., liver)	Direct injection of hot water (96 °C)	
Ultrasound applicator				Scanned focused ultrasound monitored by MRI	
Electromagnetic source	Radiative applicator	Water-filtrated infrared exposure		Interstitial laser photocoagulation (direct insertion of laser fiber)	
	Capacitive applicator	Radiofrequency capacitance hyperthermia through two electrodes coupled at the body surface	Focalized microwave beam through one single element applicator coupled at the body surface	RF-antennas based on unipolar or bipolar interstitial electrodes or interstitial microwave antennas	
	Inductive applicator (magnetic hyperthermia)			Magnetic interstitial implants hyperthermia Arterial embolization hyperthermia or direct injection hyperthermia of magnetic nanoparticles	Intracellular hyperthermia through ligand mediates magnetic particle (i.v. administration)
	NIR light source (plasmonic heating)			Gold nanoparticles[22,23]	

Adapted from *Mornet et al.*[21]

heating by bioactive NPs can drastically reduce the side effects (cell toxicity and/or tissue radiation damage) of traditional treatments when used in combined therapies. There is also another approach to destroy tumors by NP-based heating, that is, increasing the temperature above 46 °C and, therefore, causing cell death by necrosis; in contrast to apoptosis (the other known cellular death), necrosis occurs when cells suffer an irreversible damage.

2. HYPERTHERMIA BASED ON SPM-NPs: MAGNETIC FLUID HYPERTHERMIA

The original idea of using magnetic NPs (MNPs) to heat and destroy tumors was first introduced by Gilchrist[24] back in 1957 and further explored by the groups of Gordon, Hines, and Nielsen.[25] However, subsequent studies lacked suitable animal models, temperature control, or the adequate technology (frequency and amplitude of alternating magnetic field, AMF) to push this new technique to any reasonable clinical study. Hyperthermia based on magnetic materials had to wait until the 1990s when the group of Jordan introduced the concept of magnetic fluid hyperthermia (MFH).[26,27] They found that subdomain magnetic particles (nanometer in size) absorb much more power at tolerable AMF than is obtained by well-known hysteresis heating of multidomain (microns in size) magnetic particles. The development of MFH has gained since increasing attention of both materials and oncology researchers. At present, MFH has obtained successful results in clinical trials in humans.[1,28]

In general, magnetic materials present magnetic dipoles generated by the spinning of some of their electrons; these polarized electrons can be aligned parallel or antiparallel with respect to the neighboring ones and respond very differently to an applied magnetic field. Depending on how these polarized electrons interact with each other, magnetic materials fall into one of the following categories: paramagnets, ferromagnets, ferrimagnets, or antiferromagnets. However, it is important to highlight here that falling into one of these categories strongly depends on the size of the material; therefore, the magnetic behavior of particular materials can be tuned by adjusting its size.[29]

When the magnetic dipoles of a material are randomly oriented and can be only oriented by the presence of an external magnetic field, this material is considered to be in a paramagnetic phase. In this phase, magnetic dipoles will randomly relax to their equilibrium state (initial zero net magnetic moment), as the magnetic field is switched off, with no extra energy required.

Alternatively, a ferromagnetic phase is characterized by an enhanced collective parallel alignment of magnetic dipoles, even in the absence of a magnetic field. Bulk metals such as Fe, Pt, or Co, as well as some of their alloys (FePt, FeCo) are ferromagnetic materials. In contrast to the ferromagnetic phase, neighboring magnetic dipoles can collectively align in an antiparallel fashion. This type of magnetic exchange coupling gives rise to antiferromagnetic and ferrimagnetic phases. The former one applies when the material shows a net zero magnetization which results from the destructive coupling of magnetic dipoles with the same magnetic moment; typical examples of antiferromagnetic bulk materials include hematite (one of the many mineral forms of Fe_2O_3), metals such as Cr, alloys such as FeMn, and oxides such as NiO. In contrast, if the neighboring antiparallel magnetic moments have different values, their destructive coupling gives rise to a net magnetization in the so-called ferrimagnetic phase. The principal examples of ferrimagnetic materials are magnetite (iron (II, III) oxide; Fe_3O_4) and a variety of ferrites such as Co ferrite ($CoFe_2O_4$), MnZn ferrite ($Mn_aZn_{(1-a)}Fe_2O_4$), and NiZn ferrite ($Ni_aZn_{(1-a)}Fe_2O_4$).

The main parameters required to define the behavior of magnetic materials are

i. Coercive field (H_C), the external field (of opposite sign) required to reduce the net magnetization back to zero; this parameter defines the minimum energy required for the reversal of the magnetization of the material. H_C is strictly related to the magnetic anisotropic constant (K_a); this determines the energy barrier to be overcome in order to invert the direction of the magnetic dipoles of the material. In the case of NPs, K_a is determined by the symmetry of the crystal lattice, size, and shape of the NP.

ii. Saturation magnetization (M_S), the maximum value of net magnetization of a material under the influence of sufficiently high magnetic fields.

iii. Remanent magnetization (M_R), the residual magnetization at zero applied field. These three parameters define the hysteresis loop generated in field-dependent magnetization measurements (Figure 13.1).

Magnetic parameters like the coercivity can be finely tuned by adjusting the size, shape, and structure of materials in the nanometer range (also in

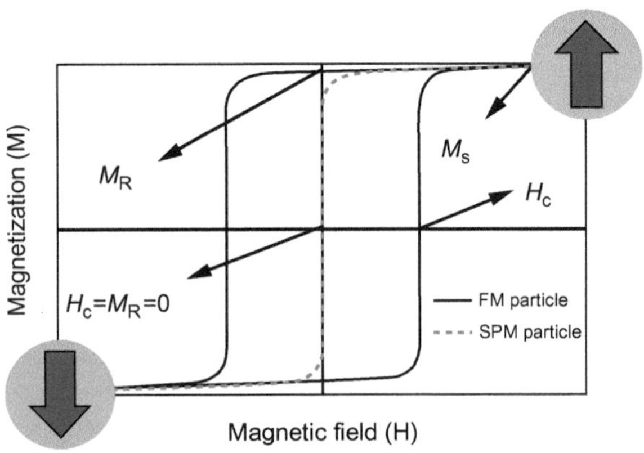

FIGURE 13.1 Magnetization versus magnetic field curve for superparamagnetic particles (SPM) and ferro- and ferrimagnetic particles (FM). Coercive field (H_C), saturation magnetization (M_S), and remanence magnetization (M_R) are indicated (adapted from Figuerola *et al.*).[30]

the micrometer range); this represents an important advantage of magnetic nanoparticles compared to their bulk counterparts since, for instance, they can be custom made for specific bioapplications with particular practical requirements such as magnetic hyperthermia. In fact, for spherical magnetic NPs with sizes bellow the so-called superparamagnetic radius (r_{SP}), ferro- and ferrimagnetic NPs become superparamagnetic; in the superparamagnetic regime, the application of an external magnetic field induces high magnetic moments in NPs, whereas switching-off the field leaves zero remanence (M_R). Reducing the size of magnetic NPs impairs the presence of domain walls in the particle (multidomain, MD) and gives rise to single domain particles which are energetically favorable for sizes bellow the r_{SP}; in the superparamagnetic regime, the thermal energy is sufficiently high to overcome the energy of magnetic stabilization of the particle (Figure 13.2).

The concept of superparamagnetism is intrinsically linked to the nanometer range. SPM-NPs are very suitable candidates for *in vivo* bioapplications since the absence of coercivity, among other reasons, prevents potential NP aggregation; in the context of *in vivo* experimentation, NPs aggregation represents a potential danger that can lead to embolisms.[30]

In principle, ferro- and ferrimagnetic particles (FM particles, in the micrometric size range) release heat by the hysteresis loss mechanism; in these regimes, the heating efficiency comes determined here by the area of the hysteresis loop, which, in turn, depends on three parameters given by the sample, that is, coercivity, remanence, and saturation at a given AMF (defined by values of magnetic field B_0 and frequency f); thus, the most efficient heater is the one that exhibits the greatest values of these three parameters.

Large FM particles contain several domains characterized by a variety of magnetization vectors. When FM particles are exposed to an applied magnetic field, those domains "parallel" to the external field will "grow" whereas the remaining domains will "shrink"; this phenomenon gives rise to irreversible domain wall displacements, that is, magnetization curves for increasing and

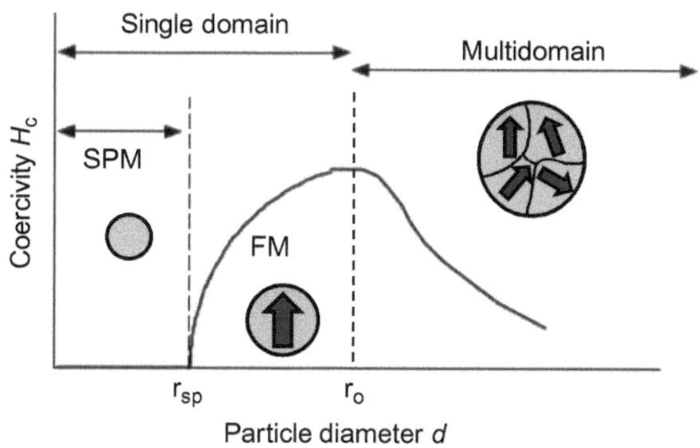

FIGURE 13.2 Size-dependence of the coercivity showing the different regimes as the size of particles increases from the nano- to the micro-range. Adapted from Figuerola *et al.*[30]

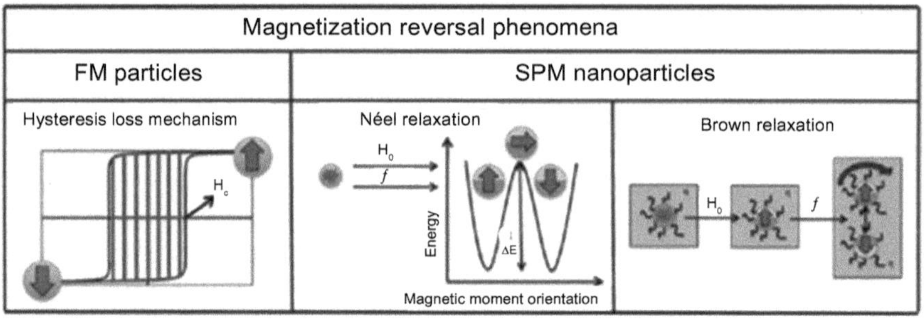

FIGURE 13.3 Relaxation processes involved in heating mechanism driven by AMF. Left panel, hysteresis losses in ferro- and ferromagnetic particles (FM); middle and right panels correspond to Néel and Brown relaxation processes in SPM-NPs, respectively. Adapted from Figuerola *et al.*[30]

decreasing magnetic field amplitudes are different (left panel in Figure 13.3). The heat produced per unit of volume is proportional to the hysteresis loop as shown in the following expression (Equation 13.1):

$$P_{FM} = \mu_0 f \oint H \delta M \qquad (13.1)$$

here μ_0 and f are the magnetic permeability in vacuum and the applied frequency, respectively. According to Equation (13.1), the value of P_{FM} can be experimentally determined with magnetometers of the type VSM (vibrating sample magnetometer) or SQUID (superconducting quantum interference device).

In single domain particles (superparamagnetic regime, in the nanosize range), the scenario changes drastically; heating due to hysteresis losses cannot occur due to the absence of domain walls. However, SPM-NPs can produce heat by the Néel and/or Brown relaxations mechanisms; although both mechanisms can be present simultaneously, the fastest mechanism for a particular superparamagnetic material under AMF (defined by B_0 and f) will predominate in the heating process. Briefly, the Néel relaxation time depends on the anisotropic energy barrier (ΔE) between the parallel and antiparallel states (middle panel in Figure 13.3). Ultimately, ΔE is dictated by the anisotropy (material, surface, and shape) and volume of the SPM-NPs. On the other hand, the Brown relaxation time depends on the probability of SPM-NPs to invert its magnetization by physically rotating in the medium (right panel in Figure 13.3); the ability to rotate depends on the viscosity of the immersion medium and the hydrodynamic radius of the SPM-NPs, that is, inorganic core plus organic and hydration layers. In theory, there is a critical diameter at which both mechanisms contribute equally (same relaxation time). Above this critical size, Brownian heating will predominate, whereas bellow it, Néel heating will apply.

The heating ability of MNPs under AMF is defined by the parameter called specific absorption rate (SAR) which can be calculated with the following expression (Equation 13.2):

$$\mathrm{SAR} = C \times \left(\frac{\Delta T}{\Delta t}\right) \mathrm{m}^{-1} \tag{13.2}$$

here m is the amount of the magnetic material used in the assay, C is the specific heat capacity of the sample (both the nanoparticle material and the dispersion medium are considered), and $\Delta T/\Delta t$ is the initial slope of the temperature increment versus time.

The heat produced per unit of volume (P_{FM} or P_{SPM}) can be experimentally obtained from the product of the SAR value and the concentration of NPs. It is very challenging to give a general theoretical expression of SAR due to the vast number of parameters which should be taken into account including size, size distribution, shape, and chemical composition of the particles as well as the experimental conditions of AMF (B_0 and f).

Table 13.2 lists the SAR values corresponding to different iron oxide-based NPs under different experimental conditions. Given the vast amount of parameters and AMF conditions to take into account regarding SAR values (Table 13.2), the answer is far from simple. In principle, SAR values of iron oxide particles vary following the same trend as the corresponding coercivity values (Figure 13.4).[31] In order to reach the maximum SAR values, the sample should be brought to saturation, which in the case of FM particles usually requires very high magnetic fields (B_0), normally above 100 kA m^{-1}. Besides the technical limitations to reach such AMF working conditions, there are also physiological limitations to the application of very high B_0 and f conditions since they can lead to damage of untargeted (without particles) healthy tissue. For *in vivo* experimentation, the general consensus is that the upper limit for B_0 and f are 15 kA m^{-1} and 1.2 MHz, respectively; in addition, the product $B_0 \times f$ should be below 5×10^9 $\mathrm{A(m^{-1}s^{-1})}$.[32] For FM NPs, this limitation adds to the aforementioned possible problems associated with aggregation due to remanence after treatment and, thus, limits the use of FM particles as hyperthermia agents.

In contrast, at "physiological AMF regimes," SPM-NPs in the range of 10–20 nm exhibit higher SAR values than their equivalent micrometric counterparts (FM particles), although larger NPs can achieve greater coercivity values.[31] Figure 13.5 shows theoretical (panel A) and experimental (panel B) SAR values for iron oxide NPs. In summary, SPM-NPs are the most suitable candidates for applications where a large value of SAR is required such as hyperthermia treatment or drug release mediated by heat.

TABLE 13.2 SAR Values for Different Iron Oxide Particles

Magnetic NP			Experimental Conditions			
Magnetic Compound	*Core Diameter (nm)*	*Corona*	H *(kAm^{-1})*	f *(kHz)*	*Dispersion Medium*	*SAR (W g^{-1})*
SD ferrite[33]	10–12	Dextran	7.2	880	Physiological solution	210 ± 8
SD ferrite	6–12	Carboxymethyl dextran	7.2	880	Physiological solution	90 ± 4
MD Fe_3O_4	100–150	None	7.2	880	Physiological solution	90 ± 4
MD γ-Fe_2O_3	100–150	None	7.2	880	Physiological solution	42 ± 3
SD Fe_3O_4[34]	8	None	6.5	300	Water	21
Fe_3O_4 (Endorem®)	6	Dextran	6.5	300	Water	<0.1
SD γ-Fe_2O_3[35]	3	Dextran	12.5	500	Water	106
SD γ-Fe_2O_3	5	Dextran	12.5	500	Water	524
SD γ-Fe_2O_3	7	Dextran	12.5	500	Water	626
Fe_3O_4	7.5	None	32.5	80	Water	15.6
Fe_3O_4	13	None	32.5	80	Water	39.4
Fe_3O_4	46	None	32.5	80	Water	75.6
Fe_3O_4	81	None	32.5	80	Water	63.7
Fe_3O_4	282	None	32.5	80	Water	32.5
Fe_3O_4	416	None	32.5	80	Water	28.9

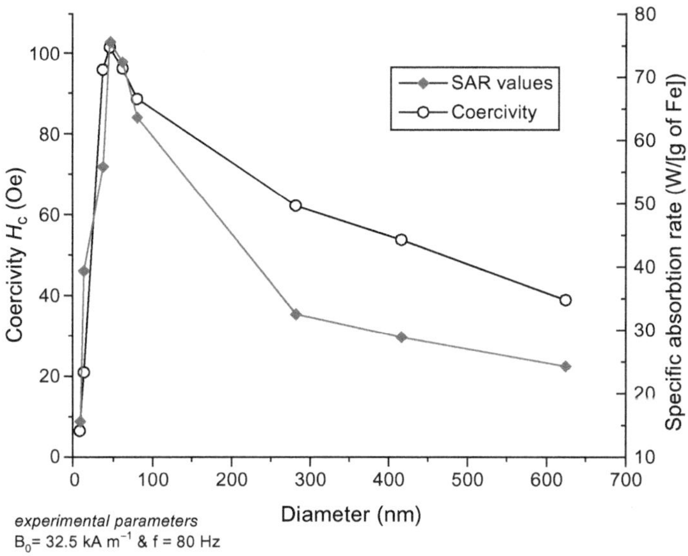

FIGURE 13.4 Coercivity (H_C) and the SAR values as functions of the average particle diameters of magnetite (reproduced from Ma *et al.*).[31]

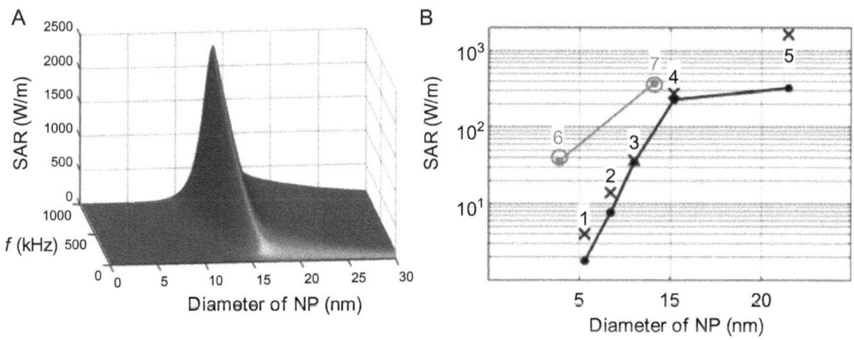

FIGURE 13.5 (A) Theoretical SAR values of monodispersed iron oxide (maghemite) nanoparticles as a function of particle diameter and magnetic field frequency at a fixed field amplitude $B_0 = 24.8$ kA m^{-1}. (B) Experimental SAR values for maghemite (1–5 black cross) and cobalt ferrite (6 and 7 gray circle) compared to the corresponding theoretical lines (black and gray solid lines). Reproduced from Fortin *et al.*[36]

3. HYPERTHERMIA BASED ON GOLD ANISOTROPIC NPs: PHOTOTHERMAL THERAPY

Optical properties of gold NPs (GNPs) have been studied over the past centuries intensively; recently, the ability of plasmonic NPs to drive photon-to-heat conversion has attracted much attention, in bioapplications specially. In this context, the therapeutic application of heat-generated following light absorption of NPs

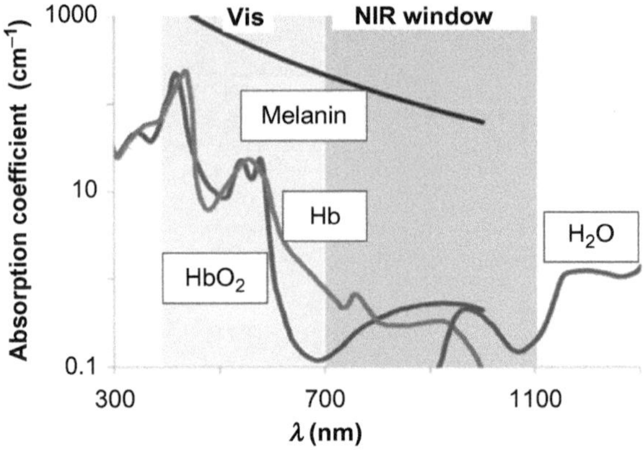

FIGURE 13.6 Range of the biological window or NIR window remarked over the absorption spectra of components of tissues (adapted from Qin *et al.*).[38]

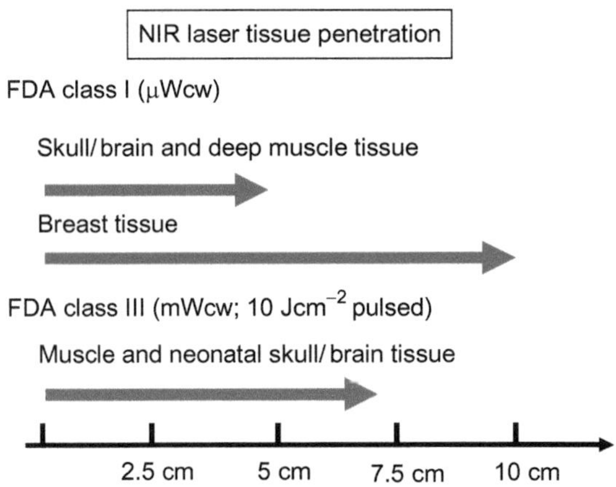

FIGURE 13.7 NIR-laser irradiation depth penetrability for medical uses in different tissues (adapted from Dreaden *et al.*).[23]

(typically anisotropic GNPs) is usually referred to as optical hyperthermia; the most relevant application is photothermal therapy (PTT) for cancer treatment.[37]

Importantly in the context of light-based therapies, the light penetration depends on the different absorption coefficients of the major components in tissue such as blood, fat, or water. This range is limited by blood, hemoglobin (Hb), and oxyhemoglobin (HbO_2), at low wavelengths and by water absorption at high wavelengths (Figure 13.6).

The potential of PTT based on anisotropic GNPs resides in the transparency of biological tissues in the so-called NIR-biological window where NIR light exhibits suitable penetration depths (Figure 13.7).

The optical properties (scattering and absorption of light) of GNPs are determined by their size, shape, and structure (solid or hollow) as well as the nature of the surrounding medium. For anisotropic GNPs, such as nanorods, nanoprisms,

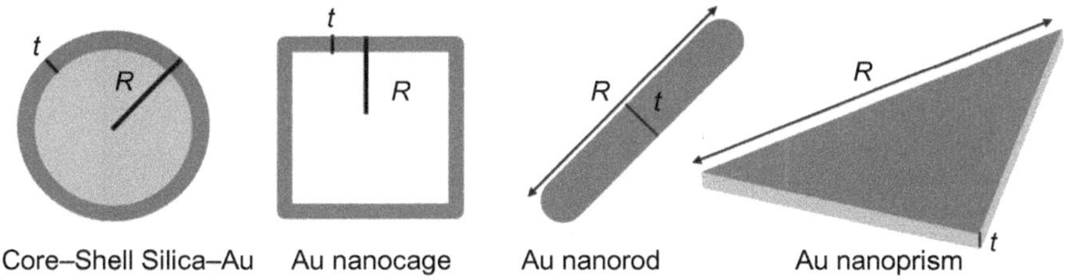

FIGURE 13.8 NIR-absorbing GNPs structures.

TABLE 13.3 Table of Typical Reported Aspect Ratios for NIR Absorbing GNPs

Nanoparticle	AR = *R*/*t* (nm/nm)	Typical AR Values
Nanorods (GNRs)	(40; 50; 60; 70; 80; 90)/20	2; 2.5; 3; 3.5; 4; 4.5[39,40]; 1.61; 1.89; 2.5; 3.38; 4.23[41]
Nanocages	(36.7/3.3)[42]; (50/10)[43]	11.12[42]; 10[43]
Nanoprisms	144/7.8[44]; (140–30)/8[45]; (310/28); (158/18); (94/14)[46]	18.5[44]; (17.5–3.75)[45]; (11, 8.7, 6.7)[46]
Core–shell silica/Au (GNSs)	60/(2–15)[8,9]	30–4[47,48]

or nanodisks, the absorption of light, that is, LSPR, is largely affected by the symmetry and aspect ratio (AR). The AR of a shape is typically defined as the length of the major axis divided by the width of the minor axis. Thus, spheres have an AR of 1. For core–shell GNPs such as silica–Au or hollow GNPs, the optical properties depend mainly on the thickness of the Au shell (t) relative to the radius (R) of the NPs; here, the aspect ratio, $AR = R/t$, deviates from the aforementioned definition. Figure 13.8 displays the geometry of the most common Au-based nanostructures in which LSPR bands lay within the NIR window.

The following table lists the parameters that define the AR for a number of NIR-absorbing NPs (Table 13.3).

For nanorods and nanoprisms, due to their anisotropy, LSPR bands appear depending on the orientation of the NP with respect to the incoming radiation. In the case of nanorods, transverse and longitudinal modes are present. As for triangular nanoprisms, four LSPR modes have been confirmed both theoretically and experimentally, namely, in-plane and out-of-plane modes, both which, in turn, can be dipolar and quadrupolar modes (Figure 13.9); the major contribution to the UV–vis–NIR spectrum corresponds to the in-plane dipolar mode laying in the NIR range.

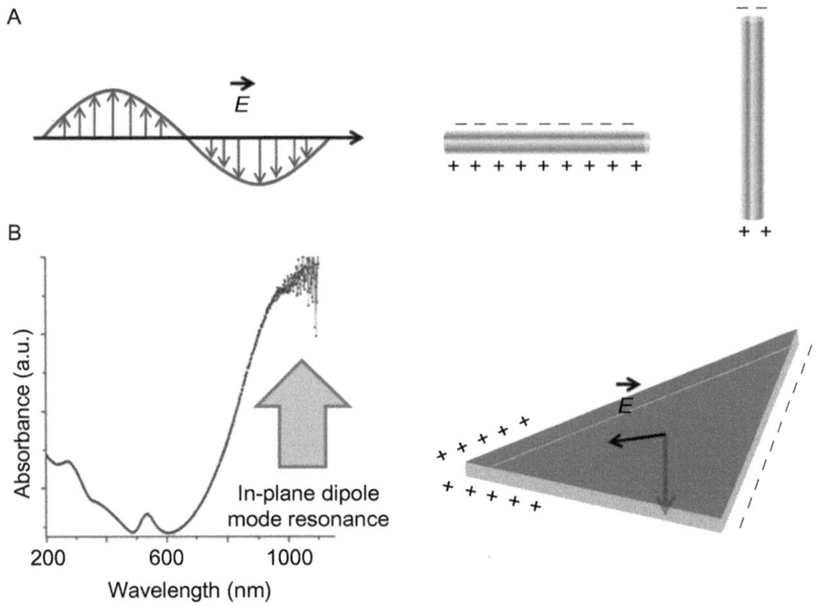

FIGURE 13.9 (A) GNRs longitudinal and transversal modes and (B) left, dipolar and quadrupolar modes representation in gold nanoprisms; right, absorption spectra of gold nanoprisms show in in-plane mode resonance contribution.

Since the pioneering work of the group of Halas,[47] the popularity of NIR-absorbing GNPs in the field of nanomedicine has dramatically increased.[49,50] This has been motivated by two complementary facts:

i. GNPs can produce heat when they are exposed to light matching their LSPR band.
ii. NIR light has the most suitable penetration depth in physiological tissue.

Importantly, the NIR band of GNPs can be excited to produce heat very efficiently.[51] NIR excitation of NPs is the most favorable scenario for biomedical applications.[52,23]

Light-to-heat conversion can be used to design applications, such as hyperthermia[53,54] and drug release[55–57] where anisotropic GNPs act as nanoheaters which can be remotely triggered by NIR radiation. In fact, Au nanorods[22] and nanocages[58] (i.e., hollow Au nanocubes) have been successfully used for remote ablation of tumors in mice. Nevertheless, many of the methods to produce NIR-absorbing GNPs rely on challenging syntheses with several steps[42,59] and/or seed-mediated approaches involving the cationic surfactant cetyltrimethylammonium bromide (CTAB).[60–64] Table 13.4 summarizes the most relevant methods to produce NIR-absorbing GNPs in high yield with control over their architectural parameters.

TABLE 13.4 Most Widely Used Method to Obtain Asymmetric GNPs

Nanoparticle	Synthetic Method/Coating	Group
Nanorods	Seed-mediated/CTAB	Murphy[65]
Nanocages	Galvanic replacement/PVP	Xia[59]
Nanoprisms	Seed-mediated/CTAB	Mirkin[44]
Core–shell silica/Au	Silica NP (Stöber method) and gold coating	Halas[47]

In general, metallic nanoparticles exhibit very low optical quantum yield and high absorption coefficients.[66] Parameters such as fluence of the light source, GNPs type, and the nature of the surrounding media can greatly modulate the thermophysical response of GNPs under illumination. During GNPs laser irradiation, energy is transferred from the light source (photons) to GNPs; this energy (photons) can be absorbed or scattered. In general, light can be (i) absorbed by GNPs and/or (ii) reemitted by the particle at same frequency (Rayleigh scattering, an elastic process) or at shifted frequency (Raman scattering, inelastic process).[38] When light interacts with metallic particles, the incoming electromagnetic field induces oscillations in the electronic cloud (mobile electrons) of the particle. If the frequency of the incident light is close to that of the collective resonance (LSPR) of the mobile electrons of the NP, this interaction will ultimately lead to heating. At the nanoscale, heat generation involves a cascade of processes as shown in Figure 13.10.[67,68] When photons from the light source excite mobile electrons from GNPs, these electrons jump to an excited energy state. As the energy of excited electrons is higher than that of the lattice energy (or temperature), electron–phonon interaction process drives an energy transfer. This process is known as electron–phonon relaxation which is in the time scale of picoseconds. Ultimately, the energy of phonons is dissipated with the surrounding media by phonon–phonon scattering. This process strongly depends on the size and shape of GNPs.[69]

Analogously as for magnetic nanoparticles but in the context of optical hyperthermia, the evaluation of the heating ability of GNPs can also be defined by a parameter typically referred to as SAR. The experimental expression of SAR is given by Equation (13.3):

$$\mathrm{SAR} = C\rho_{\mathrm{medium}}\left(\frac{\Delta T}{\Delta t}\right)_{\mathrm{initial}} \tag{13.3}$$

here, the experimental value of $(\Delta T/\Delta t)_{\mathrm{initial}}$ is calculated by performing a linear fit of the hyperthermia data (temperature vs. time) in the initial time interval, $t=[1\text{–}10]$ s; C and ρ_{medium} are the NPs concentration and the density of the surrounding media, respectively.[38]

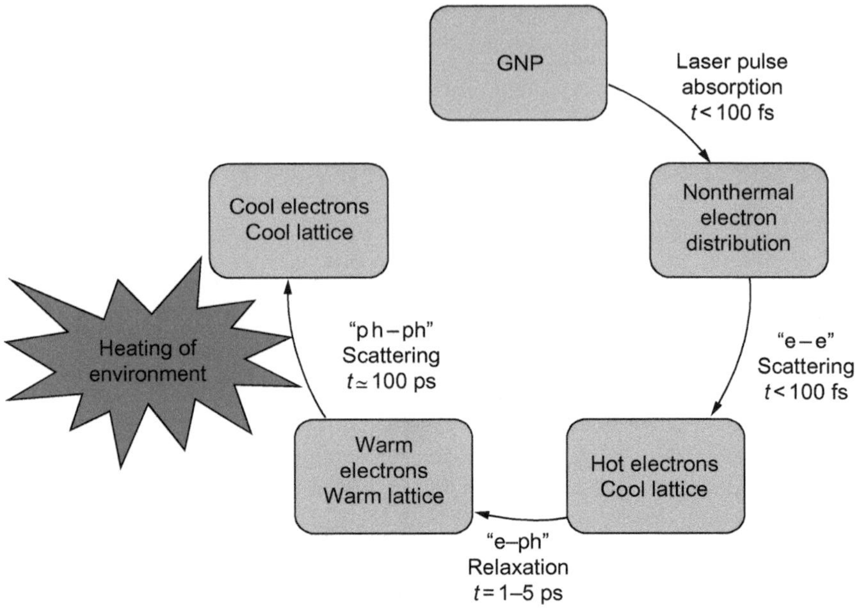

FIGURE 13.10 Scheme of process and time of each produce when a laser interact with the LSRP of a GNP to produce heating; e and ph stand for electron and phonon, respectively (adapted from Pelton).[70]

On the other hand, the theoretically SAR value can be defined as given in the following expression (Equation 13.4):

$$\mathrm{SAR} = NQ_{\mathrm{nano}} = NC_{\mathrm{abs}}I = \mu_{\mathrm{a}}I \tag{13.4}$$

here, N is defined as the number of nanoparticles per cubic meter and Q_{nano} is the heat generated by each nanoparticle under laser light; Q_{nano} can be expressed as the product of the absorption cross-section area of each NP (C_{abs}) and the power density (I given in W m^{-2}); μ_{a} represents the simultaneous absorption of light of a group N of NPs under laser irradiation.

In this Equation (13.4), there are factors which are difficult to estimate experimentally; first, the actual absorption and scattering of light by each NP depends on the shape and size of the particle; second, the local laser fluence is a complex function of the laser and the composition of the media where NPs are embedded. Additionally, this can be greatly complicated when working media is a cell or tissue.[38]

The heating profile of one particle and its temperature decay with distance have been evaluated using different models.[71–73] The temperature of a NP decays quickly from the NP surface (Figure 13.11).[51]

Some theoretical studies have shown that heat generation depends strongly on the NP shape which is ultimately related to the magnitude of the surface

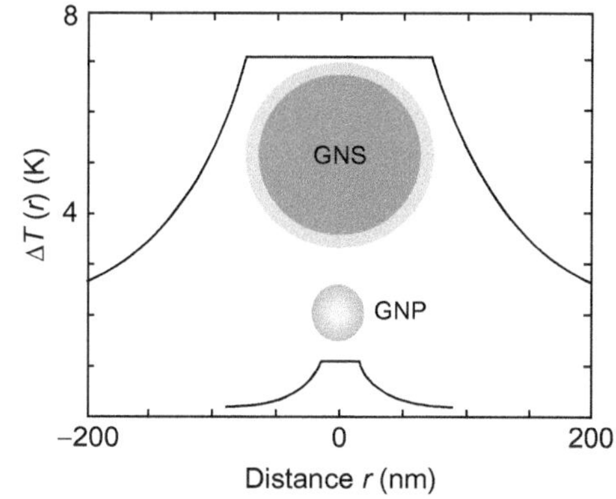

FIGURE 13.11 Calculation of temperature profile of two different types of NPs, that is, gold nanoshells of 150 nm (GNS) and a GNP of 30 nm (adapted from Bischof).[38,51]

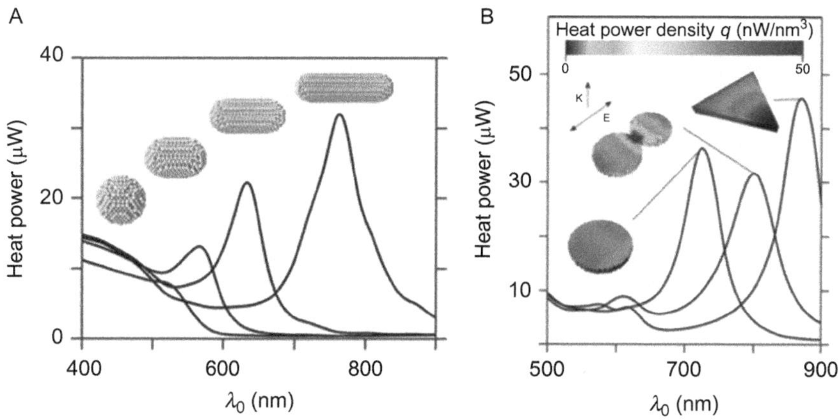

FIGURE 13.12 Theoretical calculation of heating power from different structures with the same volume; (A) comparison of a sphere and different-length rods; (B) comparison of a disk, two adjacent disks and a tabular nanoprisms (adapted from Baffou).[69]

plasmon band of NPs. For instance, for systems (of the same material) occupying the same volume, the sphere was founded to be the less effective heating shape.[69,74] Shapes exhibiting red-shifted plasmon bands are more efficient to produce heat; for instance, nanorods are 60% more efficient nanoheaters than spherical NPs (Figure 13.12A).

Another theoretical work compares different planar structures with same volume (Figure 13.12B); they found that the presence of sharp edges promote heat generation, although nanoparticle temperature is equal in all its surface due to the good gold conductivity.[69] This profile of heat generation indicates that tabular nanoprisms are very suitable candidates to PTT and reveal that, for instance, two small adjacent nanodisks instead one (of the same volume) is not an advantage in terms of heat production although their plasmon band is red-shifted relative to that of the two adjacent nanodisks.

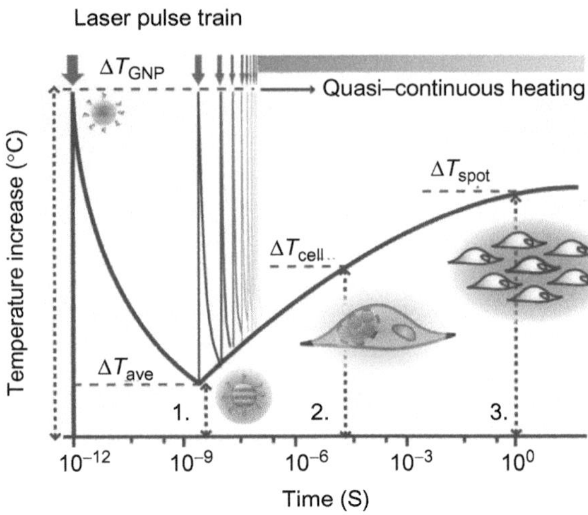

FIGURE 13.13 Representation of time scales in nanoparticle heating (1), single-cell heating (2), and an area of illumination containing several cells (3) (adapted from Bartczak).[75]

Recently, Bartczak *et al.* reported on a theoretical/experimental work where they evaluated the effect of laser irradiation on a single NP and compared it to the cumulative heating of a group of cells containing different kinds of GNPs (Figure 13.13)[75]; they studied the heating ability of a gold sphere of 15 nm (SP), a core–shell silica–gold NP (CR) of 43 nm (core) and 7 nm (shell), a nanorod (NR) of 47 nm (length) $\times$ 17 nm (thickness), and a hollow sphere of 91 nm and a shell of 9 nm (HG). The laser power density selected in this chapter was 30 W cm^{-2}, which correspond with mild hyperthermia conditions typically used in this kind of experiment.[76]

Bartczak *et al.* argued that the temperature in a single-particle irradiated by a laser pulse train decays rapidly, only a low average temperature remains (ΔT_{ave}); the average temperature is several orders of magnitude smaller than the peak of heat generated by only one laser pulse train (Figure 13.14A and B). Their results showed that the heating ability, as previously explained, depends on the NP absorption cross-section; in the case of cells, the laser pulse train excitation is replaced by a quasi-continue irradiation as it happens in a larger time–length scale. Based on their theoretical predictions, the authors claim that temperature generated in one cell containing NPs could only rises up to a small fraction of degree which is not enough to explain their experimental results. However, they predict increments of few degrees when the irradiated area includes a group of NP-containing cells; this increment would be enough to damage cells (Figure 13.14C and D).[75]

Other works have studied the behavior of GNPs solutions under laser irradiation. This experiments used different approaches such as drops[51] or cuvettes with magnetic stirring in a vacuum chamber,[77] or not.[78] In general, an energy balance is proposed in these studies; the system has to reach the steady state

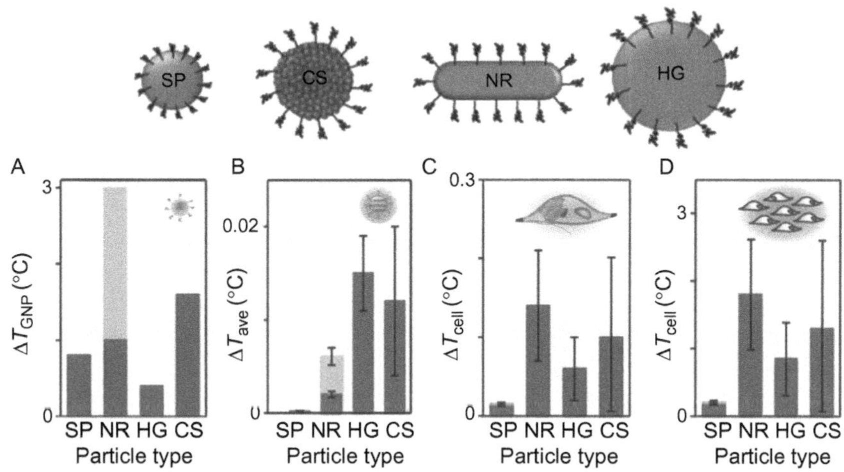

FIGURE 13.14 Representation of temperature increment in different types of GNPs in (A) single NP irradiated by a laser pulse train, (B) residual temperature after single NP laser pulse irradiation, (C) temperature increment under a quasi-continuous laser irradiation of a cell filled with GNPs, and (D) temperature increment in a group of cells containing NPs after quasi-continuous irradiation. For the gold NRs, two temperatures are shown corresponding to random polarization of the incident light (dark gray bar) and polarization parallel to the NR long-axis (light gray bar) (adapted from Bartczak).[75]

under laser illumination in order to evaluate the photothermal conversion process and determine the photothermal conversion efficiency.[79] This energy balance is expressed in Equation (13.5) as follows,

$$\left(m_s c_{p,s} + m_c c_{p,c}\right)\frac{d\Delta T}{dt} = Q_{laser} - Q_{loss} \qquad (13.5)$$

here, m_s is the mass of the solution; $c_{p,s}$ is the constant-pressure heat capacity of the solution (of value 4.187 J g^{-1} K^{-1} for aqueous solutions); m_c is the mass of the recipient; $c_{p,c}$ corresponds to the heat capacity of the recipient material (in the case of a quartz cuvette, it is 0.839 Jg^{-1} K^{-1}); ΔT is the increment of temperature obtained in a time t; Q_{laser} is the energy from the laser, and Q_{loss} is the energy dissipated to the surrounding environment. Chen *et al.* used this expression to predict temperature increments of solutions of GNRs under illumination at their longitudinal plasmon band. Experimentally, they found that a solution of GNRs under illumination can reach temperatures up to ca. 75 °C in 30 min at the selected concentration where the solution exhibited an absorbance peak of 2 OD at their plasmon band. Their theoretical calculations confirmed this result.[78] They also calculated theoretical curves of heating over time and compared them with experimental results using GNRs with different longitudinal plasmon bands. Equilibrium temperatures reached by solutions of GNRs with different longitudinal plasmon bands in the range of 760–860 nm differed less than 5 °C when illuminated by a 809-nm laser.[78]

In this very same work, the aggregation state of GNPs was also proven to influence in the process of heat generation; this can be easily correlated with the plasmon band change expected when aggregation occurs. Spherical rods were aggregated selectively and irradiated by a 809-nm laser; spherical GNPs aggregation leaded to an enhancement on heat generation due to the plasmon band shift toward the wavelength of the laser. On the other hand, GNRs aggregation translated into a decrease on the absorption at 809 nm and consequently, a decrease of heat generation was observed.[78]

Works addressing the heating ability of GNPs agree on the exponential tendency of temperature variation over time when solutions are irradiated; the cooling behavior has also been found to present a tendency typical of an exponential decay.[77–79]

Due to the different length-scale of different systems (single NP, group of NPs, NPs contained in a cell, etc.), it is possible to make a classification of the bioapplications of GNPs concerning their heating ability (Table 13.5).

Many theoretical studies have addressed the topic of light-to-heat conversion driven by GNPs as it was discussed previously. However, many of these approaches are not suitable when GNPs are irradiated inside biological

TABLE 13.5 Bioapplications of Laser-GNPs Interaction

Scale	Interacting biosystem	Example applications
Nano	Biomolecules	Nanoscale membrane melting[80–82]
		Ultrafast DNA melting assay[83]
		Drug release from GNP[83–85]
		Selective protein denaturation[86]
		Sub-wave-length molecular surgery[87]
Micro	Cells	Transmembrane drug delivery[88]
		Liposome/endosomal release[89]
		Single-cell ablation[90,91]
Macro	Organs/tissues	Photothermal therapy (PTT)[92]
		Photothermal enhancement of drug targeting[93]
		Photothermal enhancement of drug release[94,95]
		Photothermal enhancement of radiation therapy[96]
		GNP preconditioning for enhanced thermal therapy

Adapted from Qin *et al.*[38]

environments. It should be highlighted that, in general, cells challenged with NPs will store them in cytosolic vesicles where both the local concentration of NPs and the interparticle couplings will be very high; plasmonic coupling of NPs can give rise to greatly enhanced heating responses following irradiation.[97] Several works have reported on the ability of NIR-laser irradiation to damage cells loaded with GNPs, both *in vitro* and *in vivo*.[22,53,54,76,92,98,99,7,100] Upon irradiation, the mechanism responsible for damaging the cells loaded with GNPs is typically related to the thermophysical response of GNPs and the surrounding media (for detail discussion in this matter, see the recent and excellent review by Quin and Bischof)[38]; the production of reactive oxidative species (ROS) has also been recently pointed as responsible for cell injury.[54,101] In any case, the thermophysical response following irradiation of GNPs inside cells is a complex process that depends on several parameters including the local distribution, con centration and type of GNPs inside cells, and the wavelength and power density of the light source. Therefore, it is very difficult to compare the results of different works addressing the heating ability of GNPs inside cells.

Recently, Qin and Bischof proposed a combination of equations to predict and compare the nano-, micro-, and macroscale plasmonic heating; based on these, they predicted the laser power density required to achieve a 10-°C temperature rise for different biological systems with increased complexity, from the single cell to 10^9 cells which typically correspond to 10-mm tumor (Table 13.6). It should be noticed here that they had to assume a uniform GNP concentration of 5 μg Au g^{-1} of the biological system (a common value that is typically achieved by systemic delivery). Although they had to take many other important assumptions to compare these systems, assuming a uniform concentration of GNPs and neglecting collective coupling of GNPs inside cells[97,102] are a too restricting boundary conditions.[97,102]

TABLE 13.6 Predicted Power Density Required for Achieving 10 °C of Heating for Different Scaled Systems at Steady State

BioSystem	Size (Diameter)	Power Density (W cm^{-2})
Single GNP	30 nm	8.8×10^4
Single cell	10 μm	2.1×10^6
10 cells	22 μm	4.4×10^5
100 cells	46 μm	9.5×10^4
10^3 cells	100 μm	2.5×10^4
10^6 cells	1 mm	2.1×10^2
10^9 cells	10 mm	2.1

Adapted from Qin *et al.*[38]

REFERENCES

1. Nielsen OS, Horsman M, Overgaard J. A future for hyperthermia in cancer treatment? *Eur J Cancer* 2001;**37**:1587–9.
2. van der Zee J, Vujaskovic Z, Kondo M, Sugahara T. The Kadota Fund International Forum 2004–Clinical group consensus*. *Int J Hyperth* 2008;**24**:111–22.
3. van der Zee J, González D, van Rhoon GC, van Dijk JDP, van Putten WLJ, Hart AAM. Comparison of radiotherapy alone with radiotherapy plus hyperthermia in locally advanced pelvic tumours: a prospective, randomised, multicentre trial. *Lancet* 2000;**355**:1119–25.
4. Rolf DI. Hyperthermia adds to chemotherapy. *Eur J Cancer* 2008;**44**:2546–54.
5. Urano M. Invited review: for the clinical application of thermochemotherapy given at mild temperatures. *Int J Hyperth* 1999;**15**:79–107.
6. Sheng W-Y, Huang L. Cancer immunotherapy and nanomedicine. *Pharm Res* 2011;**28**:200–14.
7. El-Sayed IH, Huang XH, El-Sayed MA. Selective laser photo-thermal therapy of epithelial carcinoma using anti-EGFR antibody conjugated gold nanoparticles. *Cancer Lett* 2006; **239**:129–35.
8. Moriyama Y, Narita M, Sato K, Urushiyama M, Koyama S, Hirosawa H, et al. Application of hyperthermia to the treatment of human acute-leukemia—purging human-leukemic progenitor cells by heat. *Blood* 1986;**67**:802–4.
9. Hildebrandt B, Wust P, Ahlers O, Dieing A, Sreenivasa G, Kerner T, et al. The cellular and molecular basis of hyperthermia. *Crit Rev Oncol Hematol* 2002;**43**:33–56.
10. Fajardo LF. Pathological effects of hyperthermia in normal-tissues. *Cancer Res* 1984; **44**:4826–35.
11. Wondergem J, Haveman J, Rusman V, Sminia P, Vandijk JDP. Effects of local hyperthermia on the motor function of the rat sciatic-nerve. *Int J Radiat Biol* 1988;**53**:429–38.
12. van der Zee J. Heating the patient: a promising approach? *Ann Oncol* 2002;**13**:1173–84.
13. Multhoff G, Botzler C, Wiesnet M, Müller E, Meier T, Wilmanns W, et al. A stress-inducible 72-kDa heat-shock protein (HSP72) is expressed on the surface of human tumor cells, but not on normal cells. *Int J Cancer* 1995;**61**:272–9.
14. Basu S, Binder RJ, Suto R, Anderson KM, Srivastava PK. Necrotic but not apoptotic cell death releases heat shock proteins, which deliver a partial maturation signal to dendritic cells and activate the NF-κB pathway. *Int Immunol* 2000;**12**:1539–46.
15. Chen T, Guo J, Han C, Yang M, Cao X. Heat shock protein 70, released from heat-stressed tumor cells, initiates antitumor immunity by inducing tumor cell chemokine production and activating dendritic cells via TLR4 pathway. *J Immunol* 2009;**182**:1449–59.
16. Sun X, Li X-F, Russell J, Xing L, Urano M, Li GC, et al. Changes in tumor hypoxia induced by mild temperature hyperthermia as assessed by dual-tracer immunohistochemistry. *Radiother Oncol* 2008;**88**:269–76.
17. Kelleher DK, Vaupel P. No sustained improvement in tumor oxygenation after localized mild hyperthermia oxygen transport to tissue XXXI. In: Takahashi E, Bruley DF, editors. USA: Springer; 2010. p. 393–8, vol. 662.
18. Dikomey HHKE. Hyperthermic radiosensitization: mode of action and clinical relevance. *Int J Radiat Biol* 2001;**77**:399–408.
19. Franckena M, Fatehi D, Bruijne Md, Canters RAM, Norden Yv, Mens JW, et al. Hyperthermia dose-effect relationship in 420 patients with cervical cancer treated with combined radiotherapy and hyperthermia. *Eur J Cancer* 2009;**45**:1969–78.
20. Hohenberger P, Wysocki WM. Neoadjuvant treatment of locally advanced soft tissue sarcoma of the limbs: which treatment to choose? *Oncologist* 2008;**13**:175–86.

21. Mornet S, Vasseur S, Grasset F, Duguet E. Magnetic nanoparticle design for medical diagnosis and therapy. *J Mater Chem* 2004;**14**:2161–75.
22. Dickerson EB, Dreaden EC, Huang X, El-Sayed IH, Chu H, Pushpanketh S, et al. Gold nanorod assisted near-infrared plasmonic photothermal therapy (PPTT) of squamous cell carcinoma in mice. *Cancer Lett* 2008;**269**:57–66.
23. Dreaden EC, Mackey MA, Huang X, Kang B, El-Sayed MA. Beating cancer in multiple ways using nanogold. *Chem Soc Rev* 2011;**40**:3391.
24. Gilchrist RK, Medal R, Shorey WD, Hanselman RC, Parrott JC, Taylor CB. Selective inductive heating of lymph nodes. *Ann Surg* 1957;**146**:596–606.
25. Gordon RT, Hines JR, Gordon D. Intracellular hyperthermia a biophysical approach to cancer treatment via intracellular temperature and biophysical alterations. *Med Hypotheses* 1979;**5**:83–102.
26. Jordan A, Scholz R, Maier-Hauff K, Johannsen M, Wust P, Nadobny J, et al. Presentation of a new magnetic field therapy system for the treatment of human solid tumors with magnetic fluid hyperthermia. *J Magn Magn Mater* 2001;**225**:118–26.
27. Jordan A, Scholz R, Wust P, Fahling H, Felix R. Magnetic fluid hyperthermia (MFH): cancer treatment with AC magnetic field induced excitation of biocompatible superparamagnetic nanoparticles. *J Magn Magn Mater* 1999;**201**:413–9.
28. Maier-Hauff K, Ulrich F, Nestler D, Niehoff H, Wust P, Thiesen B, et al. Efficacy and safety of intratumoral thermotherapy using magnetic iron-oxide nanoparticles combined with external beam radiotherapy on patients with recurrent glioblastoma multiforme. *J Neurooncol* 2011;**103**:317–24.
29. Lu A-H, Salabas EL, Schüth F. Magnetic nanoparticles: synthesis, protection, functionalization, and application. *Angew Chem Int Ed* 2007;**46**:1222–44.
30. Figuerola A, Di Corato R, Manna L, Pellegrino T. From iron oxide nanoparticles towards advanced iron-based inorganic materials designed for biomedical applications. *Pharmacol Res* 2010;**62**:126–43.
31. Ma M, Wu Y, Zhou J, Sun Y, Zhang Y, Gu N. Size dependence of specific power absorption of Fe3O4 particles in AC magnetic field. *J Magn Magn Mater* 2004;**268**:33–9.
32. Hergt R, Dutz S. Magnetic particle hyperthermia-biophysical limitations of a visionary tumour therapy. *J Magn Magn Mater* 2007;**311**:187–92.
33. Bogdanov AA, Martin C, Weissleder R, Brady TJ. Trapping of dextran-coated colloids in liposomes by transient binding to aminophospholipid—preparation of ferrosomes. *Biochim Biophys Acta Biomembr* 1994;**1193**:212–8.
34. Hergt R, Andra W, d'Ambly CG, Hilger I, Kaiser WA, Richter U, et al. Physical limits of hyperthermia using magnetite fine particles. *IEEE Trans Magn* 1998;**34**:3745–54.
35. Batich C. BOOK REVIEW: Scientific and Clinical Applications of Magnetic Carriers, edited by U. Häfeli, W. Schütt, J. Teller, and M. Zborowski. *Annals of Biomedical Engineering 1998*;**26**: 904–904.
36. Fortin JP, Wilhelm C, Servais J, Menager C, Bacri JC, Gazeau F. Size-sorted anionic iron oxide nanomagnets as colloidal mediators for magnetic hyperthermia. *J Am Chem Soc* 2007;**129**:2628–35.
37. Huang X, Jain PK, El-Sayed IH, El-Sayed MA. Plasmonic photothermal therapy (PPTT) using gold nanoparticles. *Lasers Med Sci* 2007;**23**:217–28.
38. Qin Z, Bischof JC. Thermophysical and biological responses of gold nanoparticle laser heating. *Chem Soc Rev* 2012;**41**:1191–217.
39. Hu M, Chen JY, Li ZY, Au L, Hartland GV, Li XD, et al. Gold nanostructures: engineering their plasmonic properties for biomedical applications. *Chem Soc Rev* 2006;**35**:1084–94.

40. Perez-Juste J, Liz-Marzan LM, Carnie S, Chan DYC, Mulvaney P. Electric-field-directed growth of gold nanorods in aqueous surfactant solutions. *Adv Funct Mater* 2004;**14**:571–9.
41. Harris N, Ford M, Mulvaney P, Cortie M. Tunable infrared absorption by metal nanoparticles: the case for gold rods and shells. *Gold Bull* 2008;**41**:5–14.
42. Chen J, Saeki F, Wiley BJ, Cang H, Cobb MJ, Li Z-Y, et al. Gold nanocages: bioconjugation and their potential use as optical imaging contrast agents. *Nano Lett* 2005;**5**:473–7.
43. Lu X, Au L, McLellan J, Li Z-Y, Marquez M, Xia Y. Fabrication of cubic nanocages and nanoframes by dealloying Au/Ag alloy nanoboxes with an aqueous etchant based on Fe (NO3)3 or NH4OH. *Nano Lett* 2007;**7**:1764–9.
44. Millstone JE, Park S, Shuford KL, Qin L, Schatz GC, Mirkin CA. Observation of a quadrupole plasmon mode for a colloidal solution of gold nanoprisms. *J Am Chem Soc* 2005;**127**: 5312–3.
45. Fan X, Guo ZR, Hong JM, Zhang Y, Zhang JN, Gu N. Size-controlled growth of colloidal gold nanoplates and their high-purity acquisition. *Nanotechnology* 2010;**21**:105602.
46. Ah CS, Yun YJ, Park HJ, Kim W-J, Ha DH, Yun WS. Size-controlled synthesis of machinable single crystalline gold nanoplates. *Chem Mater* 2005;**17**:5558–61.
47. Oldenburg SJ, Averitt RD, Westcott SL, Halas NJ. Nanoengineering of optical resonances. *Chem Phys Lett* 1998;**288**:243–7.
48. Averitt RD, Westcott SL, Halas NJ. Ultrafast optical properties of gold nanoshells. *J Opt Soc Am B Opt Phys* 1999;**16**:1814–23.
49. Dreaden EC, Alkilany AM, Huang X, Murphy CJ, El-Sayed MA. The golden age: gold nanoparticles for biomedicine. *Chem Soc Rev* 2012;**41**:2740–79.
50. Huang X, El-Sayed MA. Gold nanoparticles: optical properties and implementations in cancer diagnosis and photothermal therapy. *J Adv Res* 2010;**1**:13–28.
51. Govorov AO, Richardson HH. Generating heat with metal nanoparticles. *Nano Today* 2007;**2**:30–8.
52. Bardhan R, Lal S, Joshi A, Halas NJ. Theranostic nanoshells: from probe design to imaging and treatment of cancer. *Acc Chem Res* 2011;**44**:936–46.
53. Huang X, El-Sayed IH, Qian W, El-Sayed MA. Cancer cell imaging and photothermal therapy in the near-infrared region by using gold nanorods. *J Am Chem Soc* 2006; **128**:2115–20.
54. Krpetić Z, Nativo P, Sée V, Prior IA, Brust M, Volk M. Inflicting controlled nonthermal damage to subcellular structures by laser-activated gold nanoparticles. *Nano Lett* 2010;**10**: 4549–54.
55. Huschka R, Neumann O, Barhoumi A, Halas NJ. Visualizing light-triggered release of molecules inside living cells. *Nano Lett* 2010;**10**:4117–22.
56. Huschka R, Zuloaga J, Knight MW, Brown LV, Nordlander P, Halas NJ. Light-induced release of DNA from gold nanoparticles: nanoshells and nanorods. *J Am Chem Soc* 2011; **133**:12247–55.
57. Munoz Javier A, del Pino P, Bedard MF, Ho D, Skirtach AG, Sukhorukov GB, et al. Photoactivated release of cargo from the cavity of polyelectrolyte capsules to the cytosol of cells. *Langmuir* 2008;**24**:12517–20.
58. Chen J, Glaus C, Laforest R, Zhang Q, Yang M, Gidding M, et al. Gold nanocages as photothermal transducers for cancer treatment. *Small* 2010;**6**:811–7.
59. Skrabalak SE, Chen J, Sun Y, Lu X, Au L, Cobley CM, et al. Gold nanocages: synthesis, properties, and applications. *Acc Chem Res* 2008;**41**:1587–95.

60. Jana NR, Gearheart L, Murphy CJ. Seed-mediated growth approach for shape-controlled synthesis of spheroidal and rod-like gold nanoparticles using a surfactant template. *Adv Mater* 2001;**13**:1389–93.
61. Nikoobakht B, El-Sayed MA. Preparation and growth mechanism of gold nanorods (NRs) using seed-mediated growth method. *Chem Mater* 2003;**15**:1957–62.
62. Murphy CJ, Sau TK, Gole AM, Orendorff CJ, Gao J, Gou L, et al. Anisotropic metal manoparticles: synthesis, assembly, and optical applications. *J Phys Chem B* 2005;**109**: 13857–70.
63. Sau TK, Murphy CJ. Room temperature, high-yield synthesis of multiple shapes of gold nanoparticles in aqueous solution. *J Am Chem Soc* 2004;**126**:8648–9.
64. Sau TK, Rogach AL. Nonspherical noble metal nanoparticles: colloid-chemical synthesis and morphology control. *Adv Mater* 2010;**22**:1781–804.
65. Jana NR, Gearheart L, Murphy CJ. Wet chemical synthesis of high aspect ratio cylindrical gold nanorods. *J Phys Chem B* 2001;**105**:4065–7.
66. Richardson HH, Carlson MT, Tandler PJ, Hernandez P, Govorov AO. Experimental and theoretical studies of light-to-heat conversion and collective heating effects in metal nanoparticle solutions. *Nano Lett* 2009;**9**:1139–46.
67. Burda C, Chen XB, Narayanan R, El-Sayed MA. Chemistry and properties of nanocrystals of different shapes. *Chem Rev* 2005;**105**:1025–102.
68. Link S, El-Sayed MA. Shape and size dependence of radiative, non-radiative and photothermal properties of gold nanocrystals. *Int Rev Phys Chem* 2000;**19**:409–53.
69. Baffou G, Quidant R, Girard C. Heat generation in plasmonic nanostructures: influence of morphology. *Appl Phys Lett* 2009;**94**: 153109.
70. Pelton M, Aizpurua J, Bryant G. Metal-nanoparticle plasmonics. *Laser Photonics Rev* 2008;**2**:136–59.
71. Ekici O, Harrison RK, Durr NJ, Eversole DS, Lee M, Ben-Yakar A. Thermal analysis of gold nanorods heated with femtosecond laser pulses. *J Phy D Appl Phys* 2008;**41**:185501.
72. Letfullin RR, George TF, Duree GC, Bollinger BM. Ultrashort laser pulse heating of nanoparticles: comparison of theoretical approaches. *Adv Opt Technol* 2008;**2008**:251718.
73. Merabia S, Shenogin S, Joly L, Keblinski P, Barrat J-L. Heat transfer from nanoparticles: a corresponding state analysis. *Proc Natl Acad Sci USA* 2009;**106**:15113–8.
74. Baffou G, Quidant R, García de Abajo FJ. Nanoscale control of optical heating in complex plasmonic systems. *ACS Nano* 2010;**4**:709–16.
75. Bartczak D, Muskens OL, Nitti S, Sanchez-Elsner T, Millar TM, Kanaras AG. Interactions of human endothelial cells with gold nanoparticles of different morphologies. *Small* 2011; **7**:388–94.
76. Bartczak D, Muskens OL, Millar TM, Sanchez-Elsner T, Kanaras AG. Laser-induced damage and recovery of plasmonically targeted human endothelial cells. *Nano Lett* 2011; **11**:1358–63.
77. Cole JR, Mirin NA, Knight MW, Goodrich GP, Halas NJ. Photothermal efficiencies of nanoshells and nanorods for clinical therapeutic applications. *J Phys Chem C* 2009;**113**: 12090–4.
78. Chen H, Shao L, Ming T, Sun Z, Zhao C, Yang B, et al. Understanding the photothermal conversion efficiency of gold nanocrystals. *Small* 2010;**6**:2272–80.
79. Roper DK, Ahn W, Hoepfner M. Microscale heat transfer transduced by surface plasmon resonant gold nanoparticles. *J Phys Chem C* 2007;**111**:3636–41.

80. Urban AS, Fedoruk M, Horton MR, Raüdler JO, Stefani FD, Feldmann J. Controlled nanometric phase transitions of phospholipid membranes by plasmonic heating of single gold nanoparticles. *Nano Lett* 2009;**9**:2903–8.
81. Bendix PM, Nader S, Reihani S, Oddershede LB. Direct measurements of heating by electromagnetically trapped gold nanoparticles on supported lipid bilayers. *ACS Nano* 2010; **4**:2256–62.
82. Kyrsting A, Bendix PM, Stamou DG, Oddershede LB. Heat profiling of three-dimensionally optically trapped gold nanoparticles using vesicle cargo release. *Nano Lett* 2010;**11**:888–92.
83. Stehr J, Hrelescu C, Sperling RA, Raschke G, Wunderlich M, Nichtl A, et al. Gold nanostoves for microsecond DNA melting analysis. *Nano Lett* 2008;**8**:619–23.
84. Bakhtiari ABS, Hsiao D, Jin G, Gates BD, Branda NR. An efficient method based on the photothermal effect for the release of molecules from metal nanoparticle surfaces. *Angew Chem Int Ed* 2009;**48**:4166–9.
85. Jain PK, Qian W, El-Sayed MA. Ultrafast cooling of photoexcited electrons in gold nanoparticle-thiolated DNA conjugates involves the dissociation of the gold–thiol bond. *J Am Chem Soc* 2006;**128**:2426–33.
86. Yan C, Pattani V, Tunnell JW, Ren P. Temperature-induced unfolding of epidermal growth factor (EGF): insight from molecular dynamics simulation. *J Mol Graph Model* 2010; **29**:2–12.
87. Csaki A, Garwe F, Steinbruck A, Maubach G, Festag G, Weise A, et al. A parallel approach for subwavelength molecular surgery using gene-specific positioned metal nanoparticles as laser light antennas. *Nano Lett* 2007;**7**:247–53.
88. Pitsillides CM, Joe EK, Wei XB, Anderson RR, Lin CP. Selective cell targeting with light-absorbing microparticles and nanoparticles. *Biophys J* 2003;**84**:4023–32.
89. Wu G, Mikhailovsky A, Khant HA, Fu C, Chiu W, Zasadzinski JA. Remotely triggered liposome release by near-infrared light absorption via hollow gold nanoshells. *J Am Chem Soc* 2008;**130**:8175–7.
90. Tong L, Zhao Y, Huff TB, Hansen MN, Wei A, Cheng JX. Gold nanorods mediate tumor cell death by compromising membrane integrity. *Adv Mater* 2007;**19**:3136–41.
91. Zharov VP, Letfullin RR, Galitovskaya EN. Microbubbles-overlapping mode for laser killing of cancer cells with absorbing nanoparticle clusters. *J Phys D Appl Phys* 2005; **38**:2571–81.
92. O'Neal DP, Hirsch LR, Halas NJ, Payne JD, West JL. Photo-thermal tumor ablation in mice using near infrared-absorbing nanoparticles. *Cancer Lett* 2004;**209**:171–6.
93. Park J-H, von Maltzahn G, Xu MJ, Fogal V, Kotamraju VR, Ruoslahti E, et al. Cooperative nanomaterial system to sensitize, target, and treat tumors. *Proc Natl Acad Sci USA* 2010; **107**:981–6.
94. Sershen SR, Westcott SL, Halas NJ, West JL. Temperature-sensitive polymer-nanoshell composites for photothermally modulated drug delivery. *J Biomed Mater Res* 2000; **51**:293–8.
95. You J, Shao R, Wei X, Gupta S, Li C. Near-infrared light triggers release of paclitaxel from biodegradable microspheres: photothermal effect and enhanced antitumor activity. *Small* 2010;**6**:1022–31.
96. Diagaradjane P, Shetty A, Wang JC, Elliott AM, Schwartz J, Shentu S, et al. Modulation of in vivo tumor radiation response via gold nanoshell-mediated vascular-focused hyperthermia: characterizing an integrated antihypoxic and localized vascular disrupting targeting strategy. *Nano Lett* 2008;**8**:1492–500.

97. Baffou G, Quidant R, de Abajo FJG. Nanoscale control of optical heating in complex plasmonic systems. *ACS Nano* 2010;**4**:709–16.
98. Gobin AM, Lee MH, Halas NJ, James WD, Drezek RA, West JL. Near-infrared resonant nanoshells for combined optical imaging and photothermal cancer therapy. *Nano Lett* 2007;**7**:1929–34.
99. Hirsch LR, Stafford RJ, Bankson JA, Sershen SR, Rivera B, Price RE, et al. Nanoshell-mediated near-infrared thermal therapy of tumors under magnetic resonance guidance. *Proc Natl Acad Sci USA* 2003;**100**:13549–54.
100. Bartczak D, Muskens OL, Nitti S, Sanchez-Elsner T, Millar TM, Kanaras AG. Interactions of human endothelial cells with gold nanoparticles of different morphologies. *Small* 2012;**8**:122–30.
101. Tong L, Cheng J-X. Gold nanorod-mediated photothermolysis induces apoptosis of macrophages via damage of mitochondria. *Nanomedicine* 2009;**4**:265–76.
102. Baffou G, Girard C, Quidant R. Mapping heat origin in plasmonic structures. *Phys Rev Lett* 2010;**104**:136805.

Chapter 14

Nanocarriers as Nanomedicines: Design Concepts and Recent Advances

Valeria Grazú* and María Moros*
*Nanotherapy and Nanodiagnostics Group (GN^2), Instituto de Nanociencia de Aragón, Universidad de Zaragoza, Edificio I+D, Mariano Esquillor s/n, Campus Rio Ebro, Zaragoza, Spain

Christian Sánchez-Espinel†
†NanoImmunoTech SL, Pza. Fernando Conde, Montero Ríos 9,36201 Vigo, Spain

1. INTRODUCTION

To date, one of the main limitations of many currently available medicines is their lack of specificity toward the pathological site. Typically, only 1% of the administered dose reaches the target cell while the rest of the drug is distributed throughout the body.[1,2] In addition, nearly half of all new chemically based drugs are insoluble, or poorly soluble, in water. Both factors, therefore, require the application of a large total dose of a drug to achieve high local concentration, thereby causing numerous off-target toxic effects. In this sense, *targeted drug delivery* can improve drug concentration at the target site and maximize therapeutic response. The concept of drug targeting suggested by Paul Erlich in 1906 was the development of a "magic bullet" consisting in one component that should recognize and bind the target while a second component should provide the therapeutic function. Currently, a combination of a targeting moiety and external forces or spontaneous forces—due to physiological features of the target area—is used to achieve a selective and quantitative accumulation of the therapeutic agent in the required area of the body relative to others.[3] The advantages over conventional drug delivery systems are many and wide ranging: from the lower frequency of administration, along with the concomitant simplification of the drug administration protocol, to the significant reduction of the quantity of drug required to achieve a therapeutic effect; not to mention the reduction of the drug side effects as well as the cost of therapy.[3]

From the second half of the twentieth century, a great deal of research has been devoted to the development of nanomaterial drug delivery systems, and

Frontiers of Nanoscience, Vol. 4. DOI: 10.1016/B978-0-12-415769-9.00014-5

the term nanomedicine has been widely used in many publications. The field of nanomedicine has been recently defined as "the comprehensive monitoring, control, construction, repair, defense and improvement of all human biological systems, working from the molecular level using engineered devices and nanostructures, ultimately to achieve medical benefit" by the Medical Standing Committee of the European Science Foundation. Nanocarriers, due to their small size, have the potential to improve the site-specific delivery and therapeutic effect of drugs that suffer from poor solubility, poor stability, and unwanted toxicity by changing their tissue distribution and improving their pharmacokinetics.[4]

A versatile library of nanomedicines including carbon nanotubes (CNTs), organic nanostructures (e.g., liposomes, dendrimers, and polymer-based nanocarriers), and inorganic nanoparticles (e.g., metal, silica, and semiconductor quantum dots, QDs) has already been constructed. These extensive libraries are composed of an assortment of different sizes, shapes, and materials with various chemical and surface properties. But as the field of nanotechnology is under constant and rapid growth, new additions continue to supplement those already known. It is important to clarify that while there is an ongoing global debate as to what really constitutes a nanomaterial, not all nanomedicines are nanoparticles.[5] Further, popular size thresholds (e.g., 1–00 nm) fail to adequately include all nanomaterials used for medical applications; for example, nanomedicines for medical applications are defined as particles with a size between 1 and 1000 nm.

Out of the controversy in this debate, the key advantages of the use of nanosystems for drug delivery can be identified: (i) improved drug bioavailability by enhancing its aqueous solubility, (ii) enhanced interaction with cell membranes and proteins as nanosystems are similar in size, (iii) high surface area to conjugate high loads and multiple therapeutic agents, (iv) high functionality that provides the potential to engineer multifunctional platforms including coatings leading to prolonged circulation in the bloodstream, targeting ligands that introduce the ability of crossing different biological membranes in the body, (v) the possibility of tailoring their physical properties by changing their size which allows the possibility of achieving an active remote triggering of the drug release, and finally, (vi) the possibility of developing multimodal all-in-one systems that combine different properties.

Nanosystems, however, have different drawbacks that have to be overcome. Although drugs are usually of small size and inert to the immune system (with the exception of allergic reactions), nanomedicines are in general larger, and therefore they can be recognized by the immune system or be toxic by themselves. Additionally, nanomedicines have to deal with different barriers once placed *in vivo*. Thus, except when specific uptake of these complexes by immune cells is required (e.g., in the development of vaccines), the ideal drug nanocarrier should (i) ensure that the drug arrives and acts preferentially at the selected target, reducing systemic levels of the drug; (ii) be stable in the complex biological environment with an extended blood circulation lifetime, maximizing drug action; (iii) be nontoxic to blood cellular components; and (iv) be invisible to the immune system.[6] In order to build an ideal nanocarrier, various physicochemical attributes of drug carriers such as composition, size, surface chemistry, shape, and mechanical flexibility can be tuned to optimize stability, extend blood circulation, targeting specificity, cargo release kinetics, facilitate cellular uptake, and intracellular drug delivery into the correct subcellular compartment.[7] However, all these issues increase the difficulty to unveil how these nanomaterials interact with biological systems.[8]

Many articles addressing the design of novel nanomaterials with potential for therapeutic or theragnostic ('therapeutic and diagnostic' function) applications are published every month. Yet to truly generate useful data from biological assays designed to unravel the interaction of these materials with living matter, these assays must be carefully and rationally designed. For example, it is necessary to use cell lines to which the material will most likely be exposed, and analytical techniques in which the nanomaterial does not interfere with the readout or introduce artifacts. For instance, the exclusive use of optical microscopy or transmission electron microscopy to define intracellular location of nanocarriers could lead to misinterpretation. Likewise, fixation and permeabilization processes during fluorescent colocalization studies could affect the intracellular distribution pattern of the nanocarrier.[9] Furthermore, the nanomedicines to be analyzed must be reproducibly manufactured, well characterized, and tested for sterility. For this last purpose, a sterilizing methodology that does not affect the intrinsic physiochemical properties of the nanocarrier is required.[10] In this sense, the historical background clearly shows that the most effective candidates arise from a rational design and an interdisciplinary interface among different scientific disciplines, academia, industry, and regulatory agencies.

This book chapter offers a general overview of the different aspects that should be considered for the rational design of nanosystems for drug delivery, as well as examples of organic and inorganic nanosystems that can be used for this purpose.

2. NANOCARRIERS FATE: BARRIERS TO OVERCOME

Despite many advances in recent years, nanomedicines still present several drawbacks, such as the difficulty to selectively reach their target site, poor stability, and rapid clearance.[11,12] These challenges are mainly posed by the biological barriers that these medicines must cross, the inherent physicochemical properties of the carriers, and the administration route, to name a few.[13] The pharmacokinetic profile of the nanomedicine, that is, the adsorption, distribution, metabolism, and excretion (ADME) processes should be studied in advance as they could provide valuable information for the design of more effective therapeutics (Figure 14.1).

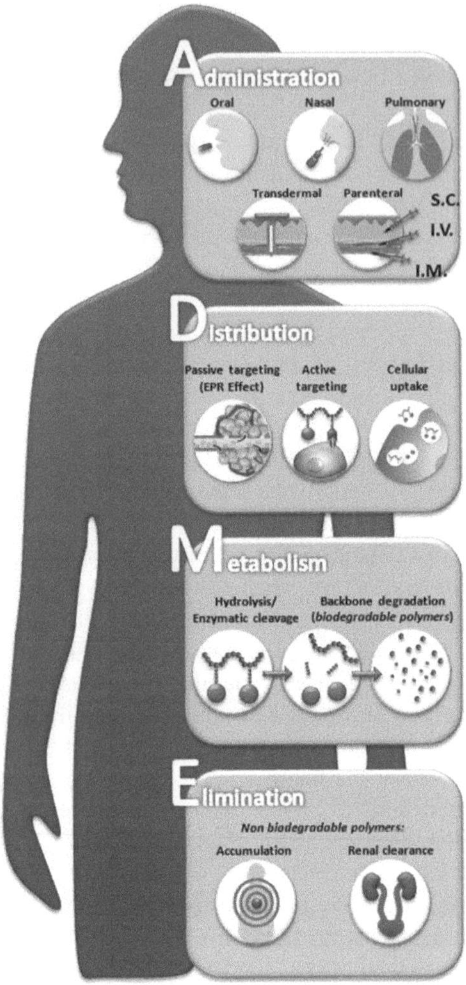

FIGURE 14.1 Nanomedicine ADME route in the body. Administration (different routes)→Di tion→Internalization→Metabolism→Elimination (ent routes). Adapted from Ref. 14. Copyright 201 permission from Elsevier.

2.1. Absorption and Administration Route Implications

Nanomedicines can be formulated for several administration routes, namely, oral, dermal, pulmonary, ocular, and parenteral, among others, and the initial biological barriers to overcome will depend on the route.

The *oral* route is the most convenient for drug administration due to patient compliance, ease of administration, and low cost.[15] However, the oral bioavailability of many drugs is limited, mainly because of their low solubility, low stability, or poor permeability to cross the gut epithelia, which is protected by mucus layers.[16] Nanocarriers such as polymer and lipid nanosystems can protect the drug against degradation and increase the permeability of the drug across this barrier.[17,18] Consequently, nanosystems intended for oral administration should be designed to resist the harsh conditions of the gastrointestinal tract, such as the acidic pH, enzyme activities (presystemic metabolism), and the first-pass metabolism by the liver, all of which could hamper drug delivery.[19,20] Nanocarriers must adhere to the mucus before crossing the epithelia; therefore, formulations that increase residence time at the mucosal surface (mucoadhesive), nanosystems able to diffuse faster through it or to break the mucus lining, are some of the approaches that are gaining relevance in this field.[16]

Pulmonary delivery is ideal for the noninvasive administration of drugs to systemic circulation because of the large surface area of alveoli, the highly permeable epithelium, and the absence of harsh pH conditions.[21] Furthermore, this delivery route bypasses first-pass hepatic metabolism, and consequently, the dose can be reduced. Nanosystems can offer many advantages for the purposes of pulmonary drug delivery, although they may be rapidly cleared once administered. In fact, the upper section of the airway (tracheobronchial) is protected by a mucus layer that removes particles by mucociliary movement. The lower airway (alveolar), in contrast, removes the particles *via* macrophage phagocytosis.[22] Carrier size is crucial for pulmonary delivery, as once these molecules have been inhaled they can be deposited in distinct tissues: particles from 1 to 5 μm can sediment in the tracheobronchial region, whereas larger particles are deposited in the *oropharyngeal* airways. Particles <1 μm are exhaled, while particles <100 nm are effectively deposited in the alveolar regions.[23]

To date, most of the nanomedicines ongoing clinical trials or already marketed have been formulated for *parenteral administration* (intravenous, subcutaneous, intramuscular, intradermal, and intraperitoneal). However, these injection methods hold some disadvantages, such as the poor patient compliance. Nanocarriers administered through nonintravenous injection are transported into regional lymph nodes before distribution to the blood circulation.[24] These routes allow for slow release of the drug from the injection site to the bloodstream, and therefore, less frequent injections are required.[25] Conversely, the intravenous administration is the most straightforward method, as the nanomedicine enters the bloodstream and is distributed throughout the body immediately. For intravenously administered nanomedicines, their immediate contact with a high ionic strength solution such as blood may lead to their aggregation.[26] The formation of larger particles may cause capillary blockade and embolism. Thus given all these consideration, the stability of nanocarriers in biological media is one of the key issues to be addressed when formulating nanomedicines.[27,28]

Nanocarriers can be stabilized by two main mechanisms, either by interparticle electrostatic repulsive force or by steric stabilization. Electrostatic stabilization concerns the surface charge density created on the particle surface immersed in liquid, which induces particles to repel each other when they have the same charge. Electrostatic stabilization has some drawbacks, as it depends on the pH and the ionic strength of the media.[29,30]

In the case of steric stabilization, polymers and surfactants are widely used,[31–33] and the repulsion between particles arises from the interaction between their layers. Steric stabilization results in greater protection against changes to high salt concentrations than electrostatically stabilized particles, so their use is preferred for nanomedicines intended for *in vivo* uses. Both methods can be combined to stabilize particles.

After systemic administration, nanomedicines travel through the blood, where they could be rapidly cleared by the reticuloendothelial system (RES), one of the greatest challenges for drug delivery systems. Long-circulating nanomedicines must then extravasate from the blood through the vascular endothelium and reach the cells across the interstitium, where high pressure (in solid tumors for instance) may also impede their movement.[34] Other barriers include the cellular, endosomal, and nuclear membranes, depending on the target site (Figure 14.2).

2.2. Biodistribution

After administered by distinct routes, nanocarriers can be distributed to several tissues, where nonbiodegradable nanocarriers can be accumulated for prolonged periods of time. Nanocarriers size, shape, or surface among others can be tailored to prolong blood circulation and enhance in some extend specific organ targeting, although not specific targeting is mostly unavoidable.

Once the particles are in the body, their size and surface charges (which could differ from the original ones) are strong determinants of their removal. Anatomical filters such as the kidneys, liver, and spleen may determine the residence time of nanoparticles in the blood. In general, it is accepted that small NPs pass through glomerular filtration, thus being removed through the urine. In fact, QDs with hydrodynamic sizes <5.5 nm are excreted by

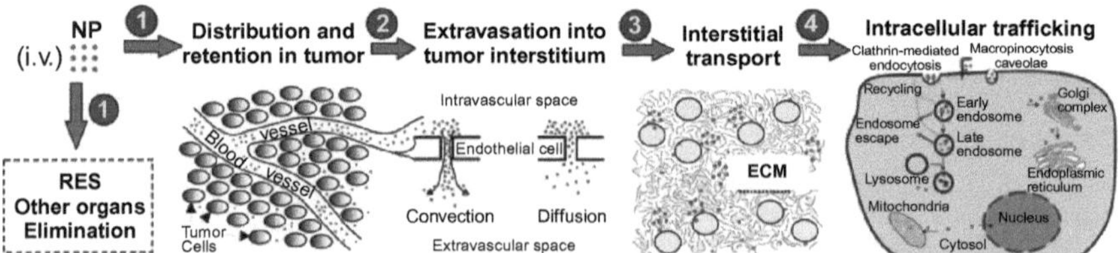

FIGURE 14.2 Distribution of nanomedicines from injection site to their destination site, for instance, a tumor. (1) Transport and distribution to organs via systemic circulation, including elimination by cells of reticuloendothelial system (RES). (2) Extravasation from endothelial vasculature. (3) Interstitial transport to reach individual cells through the extracellular matrix (ECM). (4) Endocytosis and intracellular trafficking to subcellular organelles (early and late endosomes, lysosomes, Golgi complex, endoplasmic reticulum, cytosol, mitochondria, nucleus). Reprinted from Ref. 35. Copyright 2011 with permission from Elsevier.

urine, even when they are functionalized with ligands to target tumor cells, but larger ones are retained for longer periods in the organism.[36,37] Larger particles that are not cleared by the kidneys can be subjected to alternative processes such as metabolism (for biodegradable components), biliary excretion after liver uptake, or biodistribution into tissues.

Although it is difficult to establish general rules for nanocarriers biodistribution due to the many factors implied,[38] it is likely that smaller nanocarriers are distributed in more tissues than larger particles, and that nanocarriers that exceed 200 nm and are not deformable enough will be cleared by the spleen.[39,40] Nanocarriers shape can also influence the biodistribution pattern. In fact, nonspherical shape of particles could be shown to enhance circulation time and influence the accumulation.[41] Likewise, the blood half-life of a nanocarrier can be influenced by the number of attached ligands that it exhibits on its surface.[42] On the other hand, the liver and spleen are the organs which retain most NPs after intravenous injection.[43] For instance, the accumulation of gold NPs (AuNPs) ranging from 1.4 to 200 nm is increased with size, while in other organs there is little size dependency.[44]

At present, there are no standardized protocols for evaluating nanomedicine distribution, and it is difficult to compare results between studies[45]: the initial dose, the administration route, the experimental model, and time range may influence the biodistribution. As it will be discussed, biodistribution of nanocarriers depends on many factors, such as the physicochemical properties of the particles themselves, their interactions with plasma proteins, their capacity to extravasate from the blood vessel, and their removal from circulation.

2.2.1. Protein Adsorption

Once nanomedicines are in the blood, protein binding to their surface occurs, giving rise to an interface organization called "protein corona."[46] Certain components of this corona, called opsonins, induce the nanomedicines uptake by the RES, and consequently their removal from the bloodstream.[33] Therefore, the binding of plasma proteins to nanocarriers can influence their biodistribution and therapeutic efficacy.

The composition of the protein corona is complex and variable, although it seems that apolipoproteins are the main type of proteins that coat liposomes and polymeric particles, whereas inorganic and charged particles preferentially absorb blood contact factors.[47] The composition also depends on nanocarrier size, its surface charge, shape, roughness, and the protein affinity for it[48–52]: For instance, it has been shown that 30-nm AuNPs adsorb more proteins on their surface than 50-nm ones, and the absorbed plasma proteins differ between them.[53] Neutrally charged nanomedicines adsorb fewer proteins on their surface than do their charged counterparts, thus showing a direct correlation between surface charge and protein binding.[54] Finally, the hydrophobicity of the nanocarrier surface gives rise to a larger degree of adsorbed proteins.[46]

Therefore, hydrophilic and neutral carriers will adsorb fewer proteins, and consequently, they are candidates to remain longer in the bloodstream.

2.2.2. Removal by the RES

For drug delivery applications, nanomedicines must reach their target site before being removed. However, recognition of nanomedicines by the RES can lead to rapid removal from the bloodstream, thus limiting their therapeutic action.[45,55] The RES is comprised of organs and cells (monocytes and macrophages) that help defend the body against foreign organisms, and they are responsible for eliminating damaged cells and other degraded products. Mature macrophages are present in the liver (Kupffer cells), spleen, lungs, bone marrow, and lymph nodes and can remove nanomedicines by phagocytosis within minutes. The binding of opsonins such as fibrinogen, immunoglobulins, and complement components to the surface of nanocarriers increases their phagocytosis and uptake via cells of the RES. Therefore, it is important to design stealth nanocarriers, that is, nanocarriers with surfaces that prevent as far as possible the binding of these proteins.

Inorganic NPs synthesized with various materials have been coated with poly(ethylene glycol) (PEG),[56,57] block copolymers such as poloxamers[58] and poloxamine,[59] dextran,[60] amphipathic polymers[61] and chitosan,[62] among others, to create nanomedicines with prolonged half-lives.

The first attempts to obtain stealth liposomes were focused on the manipulation of lipid membrane components in order to modify bilayer fluidity. Conventional liposomes were then formulated with saturated phospholipids and cholesterol in order to prolong their circulation time.[63] However, these liposomes did not overcome opsonin binding. Liposome composition has been modified by the addition of ganglioside GM1 and glucuronide,[64] although the incorporation of PEG on the liposome surface appears to be the most effective way of obtaining stealth characteristics.[63] Similarly, among polymer therapeutics, PEG is the most widely used macromolecule to prevent opsonization.[65]

PEG is a relatively inert hydrophilic polymer that provides satisfactory steric hindrance for preventing protein binding. It is considered nontoxic and nonimmunogenic and has been approved by the U.S. Food and Drug Administration (FDA).[66] In order to prolong nanocarriers half-life, the molecular weight of PEG and its grafting density must be optimized. When the polymer density increases, the space between each polymer is reduced, changing the PEG molecule configuration from a “mushroom” to a “brush” state.[67] The mushroom configuration will lead to a greatly enhanced RES uptake of NPs than the brush state.[68] For instance, magnetic NP uptake by macrophages *in vitro* differs depending on the molecular weight of the PEG: a PEG higher than 3000 Da does not give rise to macrophage uptake, while PEG600-coated NPs are extensively taken up, an observation that could be attributable to their less covered particle surface.[69] Similarly, the biodistribution of gadolinium

NPs (hydrodynamic diameter < 20 nm) and polylysine dendrimers *in vivo* after intravenous injection in mice strongly depends on the surface charge and PEG length.[70–72]

Research conducted with polymer–drug conjugates already on the market shows the same tendency. Two polymer–drug conjugates formulated with PEG-interferon-α (Pegasys® and PEG Intron®) and approved as treatments for hepatitis C include slightly different proteins and conjugation chemistry; however, differences are particularly notable in the molecular weight and structure of their PEGs.[73,74] Consequently, their activity and plasma half-lives differ: interferon PEGylated with a 40-kDa PEG has a higher residence time, while interferon PEGylated with a 12-kDa has a faster but shorter therapeutic response.

In spite of abundant examples of PEGylated nanomedicines in the literature, PEGs are not biodegradable. Although they can be excreted in urine or feces,[75] at high molecular weights, PEGs accumulate, leading to their deposition in lysosomes and renal tubular vacuolation.[76] Furthermore, several research groups have reported that the repeated intravenous injections of PEGylated liposomes can generate immune responses and elicit specific antibodies directed against PEG, resulting in the loss of their long-circulating characteristics and leading to accumulation in liver.[6] Therefore, many other natural and synthetic polymers are being developed as alternatives.[77]

In addition to the size and surface composition of nanocarriers, *shape* is also relevant regarding blood circulation time. It has been shown that non-spherical particles have longer circulation times than their spherical counterparts because they are less susceptible to RES capture.[41,78]

Finally, recent publications have reported that *mechanical flexibility* is also a relevant parameter to consider when designing an efficient drug nanocarrier. Unlike blood cells, which can squeeze through capillaries that are smaller than their own diameter, most nanosystems are often quite rigid.[79] Several articles have described that flexible nanocarriers not only exhibit longer circulation in the vasculature but are also taken up to a lesser extent by macrophages than rigid ones.[80]

In summary, in most nanomedicine applications, RES uptake is a major drawback as extended duration in blood circulation is required for a therapeutic effect. Therefore, nanomedicine design should focus on avoiding rapid clearance. However, shorter blood half-lives may be desirable in other applications, such as the delivery of contrast agents. Likewise, although it is outside the scope of this review, there are some instances where NP targeting to cells of the RES, particularly macrophages, is desirable. Macrophages play a central role in host–defense against infectious agents and in inflammation processes, and therefore, these cells may be suitable targets for drug delivery to treat these diseases (reviewed in Refs. 81,82). Many examples of nanocarriers with targeting ligands to increase macrophage uptake such as peptides, mannose[83] or its derivades,[84] or without[85] these molecules can be found in the literature.[86]

2.2.3. Vascular Endothelium

The vascular endothelium is a cellular monolayer that lines the surface of blood vessels and has vital functions, including vascular smooth muscle tone, angiogenesis, and host–defense reactions.[87] It is a semipermeable barrier that controls the transport of macromolecules and fluids between the blood and the interstitial space, thereby assuring tissue fluid homeostasis.

Molecules can cross the vessel endothelium either between the individual cells (paracellular route) or through the cells by an active mechanism (transcytosis) (Figure 14.3). Alternatively, small molecules can diffuse through the cells, driven by concentration gradients.[35]

Typically, most of the endothelium is continuous (muscle, skin, lung, fat, connective tissue, and nervous system) and presents tight junctions that leave a gap between endothelial cells of only ~3 nm (reviewed in Ref. 88).[89] In contrast, some nonpathological tissues exhibit fenestrated endothelium (glands, kidney, and intestinal mucosa) with gaps of <100 nm.[90] Discontinuous endothelium with pores of >100 nm is found in the liver, spleen, and bone marrow.[90] Consequently, these tissues permit passive extravasation of nanocarriers from the circulation into them. On the other hand, intercellular

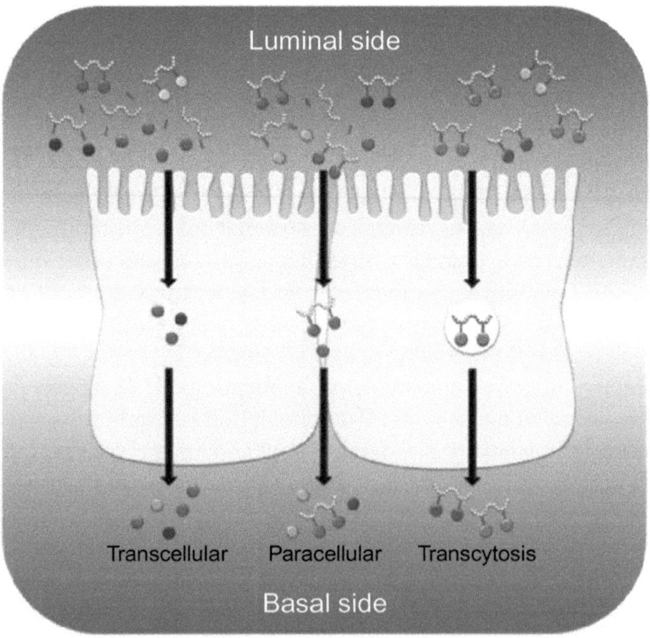

FIGURE 14.3 Transepithelial delivery. Nanomedicines can cross the epithelium through either transcytosis (through the cells) or via paracellular (between the cells) route. Small molecules, such as free drugs, can also diffuse through the cells (transcellular route). Reprinted from Ref. 14. Copyright 2011 with permission from Elsevier.

tight junctions between vascular endothelial cells in the brain are especially restrictive and could even reduce flux of small molecules across the blood–brain barrier. Readers are referred to recent reviews for further details.[91–93]

Paracellular transport of nanocarriers through the tight junctions between the individual cells is difficult to achieve when endothelium is continuous. However, it has been reported that some dendrimers are able to disrupt and open the tight junctions to facilitate paracellular (and also transcellular) transport.[94,95]

However, if the tissue target of a given nanomedicine lies beyond the vascular wall, it must have the capacity to cross the endothelium. Transcellular transport of molecules in vesicles (transcytosis) through the cells is an alternative approach to achieve this goal, as due to the size limit of passive diffusion, most nanocarriers cannot enter by diffusion (reviewed in Ref. 96). In fact, large molecules, such as plasma proteins with sizes >3 nm, can exploit this route. Albumin, for instance, is transported using a caveolae-mediated pathway,[87] as endothelial cells are particularly rich in these lipid rafts. Caveolae are flask-shaped membrane invaginations that can pinch off from the luminal membrane carrying the molecules, cross the cell, and discharge them at the other side of the endothelial cell layer.

Size and charge of the nanocarriers influence the paracellular and transcellular transport, as have been demonstrated for AuNPs *in vitro*.[97] The neutral and negatively charged polymer-coated NPs with a size of 5 nm showed a significantly higher translocation than the positive or bigger nanoparticles. In addition to nanocarrier size, *shape* also influences the circulation time and extravasation: nonspherical carriers remain predominately in the proximity of the endothelium rather than in the center of the blood vessel, and, consequently, extravasate more easily or exit faster to smaller capillaries.[98]

Moreover, the vascular endothelium cells have a major role in several diseases involving acute inflammation (reviewed in Ref. 99). Inflammation mediators, such as vascular endothelial growth factor (VEFG or VEFG-A), are induced, changing the organization of the intercellular tight junction fenestrations and increasing vascular permeability and plasma protein extravasation.[100,101] Deregulated vascular remodeling and vessel growth (aberrant angiogenesis) occur in many pathologies, including tumor growth and metastasis, and in many inflammatory and autoimmune disorders, such as rheumatoid arthritis, atherosclerosis, and psoriasis, among others.[102] Therefore, the vascular endothelium itself is becoming an important target for drug delivery. To enhance the binding of nanocarriers to the endothelium, ligands recognizing endothelial molecules exposed in diseased areas can be used. For instance, P- and E-selectins, vascular cell adhesion molecule-1, and intercellular adhesion molecule-1 (ICAM-1) are targeting molecules expressed in the endothelium.[103,104]

After extravasation, in order to reach cells, nanomedicines must cross the *interstitium*, the space between cells and tissues and surrounded by cell membranes and blood vessel walls. The transport of a carrier in the interstitium is influenced by the physiological and physicochemical properties of the interstitial space (i.e., pressure, size, charge, structure, composition) as well as the physicochemical properties of the carrier.[105] Tumor interstitium differs greatly from that of healthy tissues, as the undesirable accumulation of vascular contents in the tumor caused by leaky vasculature and dysfunctional lymphatics leads to interstitial hypertension (IFP) (reviewed in Ref. 106). Tumor high IFP relative to the surrounding tissues results in a net fluid flow from the high-pressure core to the tumor periphery, which may hamper the interstitial transport of carriers to cells in the interior portions.[107] Several strategies to decrease tumor IFP have been developed, including the normalization of tumor vasculature by inhibiting angiogenic growth factors,[108] the reduction of the swelling pressure of the extracellular matrix using hyaluronan-degrading enzymes, and the lowering of vascular resistance with nicotinamide, among others (reviewed in Refs. 106,109).

2.2.4. Cell Internalization: Endocytic Mechanisms

If cellular uptake is desired, more barriers should be taken into account. Endocytosis is the process by which a cell internalizes membrane and extracellular materials.

Generally, endocytic pathways can be divided in two broad categories, phagocytosis (cell-eating) and pinocytosis (cell-drinking) (Figure 14.4). The former is a critical step that occurs mainly in specialized cells (neutrophils, macrophages, monocytes), and it is intended to clear large pathogens such as bacteria or debris such as dead cells.[110,111] Nonphagocytic pathways occur in virtually all cells and can be divided into clathrin-mediated endocytosis (CME) and non-CME. Non-CME includes caveolar endocytosis, macropinocytosis, and clathrin- and caveolae-independent pathways; the latter are subclassified as Arf6-, flotillin-, Cdc42-, and RhoA dependent,[112] although ongoing research may lead to the characterization of other routes (Figure 14.4).

Phagocytosis and macropinocytosis involve long-range remodeling of membrane and the cytoskeleton lying beneath it. The common organizing structure is a cup-shaped invagination of the plasma membrane that becomes the phagosome or macropinosome, respectively.

Particle phagocytosis is initiated by the interaction of specific receptors on the surface of the phagocyte with certain molecules on the surface of the particle, such as opsonized immunoglobulins G.[114] This leads to the internalization of the particle via an actin-based mechanism.[115] The phagosome then matures by a series of fusion and fission events with components of the endocytic pathway (endosomes, lysosomes), which ultimately leads to the contents digestion.[116]

Macropinocytosis generates vesicles that are heterogeneous in size (up to 5 μm in diameter), and, like phagocytosis, membrane ruffling occurs in response to directed actin polymerization.[117] Macropinocytosis is receiving

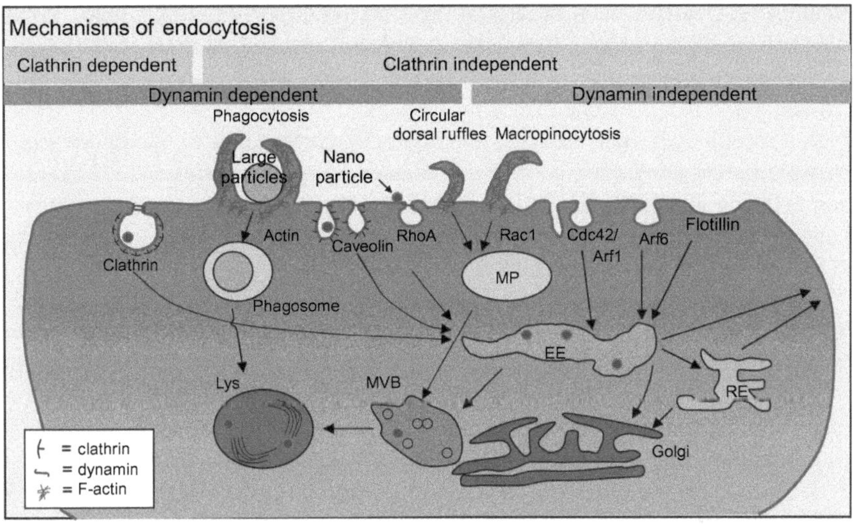

FIGURE 14.4 Model of endocytic mechanisms and intracellular transport. Nanomedicines (gray dots) taken up by endocytosis are enclosed within the early endosomes (EE), phagosomes, or macropinosomes (MP). These vesicles with particles then mature down the degradative pathway and become multivesicular bodies/late endosomes (MVB) which fuse with lysosomes (Lys). Alternatively, the components may be transported back to the cell surface either directly from EE or through the recycling endosomes (RE). The pH drops gradually from the cell surface to lysosomes where pH is 4.0–5.5. The lysosomes contain proteases and other enzymes that degrade most biological substances. Reprinted from Ref. 113. Copyright 2011 with permission from Elsevier.

increasing attention not only due to its involvement in immune defense and clearance of apoptotic bodies but also as a key route for drug delivery, as numerous cell-penetrating peptides enter the cell using this pathway (reviewed in Refs. 118–120).

Clathrin-mediated endocytosis, previously known as receptor-mediated endocytosis, is the best known pathway, and for most cells, it is the main mechanism for the internalization of macromolecules, such as transferrin and low-density lipoproteins.[121] During endocytosis, cargos are recognized by protein complexes, which lead to clathrin polymerization and the formation of a characteristic coated pit (reviewed in Refs. 122,123). These pits invaginate and pinch off from the plasma membrane (mediated by dynamin), to form endocytic vesicles with an average size of ~150 nm.[124] Vesicles fuse with early endosomes, where receptors are sorted for either recycling or transit to late endosomes and lysosomes.[125]

Caveolae are characteristic flask-shaped membrane invaginations (50–80 nm) and a subset of lipid rafts, a heterogeneous domain of the membrane that is highly dynamic and enriched in sterol and sphingolipids.[126] The key feature that distinguishes caveolae from other lipid rafts is the presence of caveolin. Ligands known to be internalized through caveolae-mediated endocytosis include folic acid, albumin, and viruses, such as Simian Virus 40 (SV40).[127] For many years, it has been believed that ligands entering the cell *via* this mechanism ended in an organelle called the caveosome before being transported to the smooth endoplasmic reticulum and that, in some instances, they could bypass lysosomes.[128,129] However, a recent study indicates that this organelle corresponds to late endosomes/lysosomes, and consequently, the term caveosome is not appropriate.[130] In addition to their role in intracellular transport, caveolae are involved in other physiological processes,[131–134] being one of the most prominent functions the regulation of transcytosis in endothelial cells.[135,136]

Other clathrin-independent endocytic routes include Arf6-, flotillin-, Cdc42- (CLIC/GEEG), and RhoA-dependent pathways, which have been extensively reviewed.[137–140]

For drug delivery, many nanomedicines must enter the cell and reach a specific intracellular location, such as the nucleus or the cytosol. Nanomedicines can gain access to the cell following several pathways or multiple endocytic pathways simultaneously. Given this feature, a greater understanding of these pathways and the subsequently intracellular localization is highly relevant from the clinical point of view.

When the particle is a liposome, apart from the internalization by endocytic pathways, other processes have been proposed: fusion of the liposome with the cell membrane and exchange of lipids between the liposome and the plasma membrane.[141,142] In the case of other nanocarriers, cellular internalization by endocytosis generally occurs. The rate and mechanism of nanocarrier uptake are cell-type dependent[143] and vary with carrier size, shape, charge, aggregation state, and other surface characteristics, such as the ligand arrangement or length.[144–149]

The *size* of nanocarriers is crucial for endocytic uptake, although it is not possible to establish a general trend. For instance, cells from the conjunctival epithelium display a greater uptake for 100 nm poly(D,L-lactide-*co*-glycolide) (PLGA) NPs compared to ones measuring 800 nm or 10 μm.[150] Carrier size may also directly affect the mode of endocytosis, as reported when comparing the uptake mechanism of fluorescent latex bed by mouse melanoma B16 cells *in vitro*. The uptake of smaller particles (up to 200 nm) was clathrin mediated, while 500-nm particles appeared to follow a nondegradative, caveolae-mediated pathway.[151] In fact, clathrin-mediated endocytosis has been traditionally proposed as the main route for NPs smaller than 200 nm.[146,152] However, recent results suggest that caveolae can accommodate nanocarriers of up to 100 nm in diameter.[153]

Particle *shape* also plays a key role in their cellular uptake, as recently been demonstrated by Yoo and Mitragotri.[154] They designed PLGA particles whose shape can be switched in real time in response to a stimulus. The external stimuli (pH, temperature, or chemical) can induce a change in shape, from an elliptical disk to a sphere. This change is achieved by appropriate selection of the polymer molecular weight, particle size, and the strength of the stimulus. Interestingly, elliptical disk-shaped particles are not phagocytosed by macrophages; once the particles become spherical shaped, the macrophages quickly internalize them. Other results concerning the shape of the particles are contradictory. For instance, while DeSimone's group found that the cellular uptake of rod-like NPs occurs faster and to a greater extent than their spherical counterparts (even when they share the same volume),[155] Chan's group reported the opposite.[156]

The *surface charge* of particles is a crucial parameter that explains the rates and mechanism of uptake. In general, due to the negatively charged cell plasma membrane, cationic particles show better internalization rates than their negative or neutral counterparts.[157–159] This finding could be attributable to the fact that neutral and negatively charged NPs are adsorbed to a lesser extent on the negatively charged cell membrane and, consequently, show lower levels of internalization, as demonstrated for AuNPs.[160] The charge may also affect the route of endocytosis, as reported for polylactide NPs incubated with HeLa cells.[161] While positively charged NPs internalize rapidly via the clathrin-mediated pathway, negatively charged ones show an inferior rate of endocytosis and use an alternative pathway, unrelated to the clathrin or caveolae routes.

Interestingly, not only the surface charge of the particle influences endocytosis, but also the *ligand arrangement*. Verma *et al.* studied the impact of diverse arrangements of capping ligands onto AuNPs on cell uptake.[148] While NPs with molecules in an ordered ribbon-like arrangement (striped) penetrated the cell at 4 °C and under conditions where active internalization processes are blocked, equivalent NPs coated with the same moieties but in a random distribution were not internalized in these conditions. In addition, the *striped* particles effectively reached the cytosol without generating holes in the plasma membrane, while *random* particles remained trapped in vesicular bodies.

In addition to the physicochemical properties of the carriers, the endocytosis rate can be modified by adding *ligands* on their surface, such as transferrin,[162] glucose,[163] folic acid,[164] or peptides (cell-penetrating peptides, CPPs), among others. Many CPPs have been synthesized based on the sequences of protein transduction domains because of their ability to deliver a range of molecules into cells (reviewed in Refs. 165–167). Initial reports excluded endocytic pathways as the main mechanism responsible for the uptake of these peptides,[168] proposing a direct interaction of the positively charged CPP with negatively charged membrane components.[169] Unfortunately, subsequent studies suggested that this apparent membrane translocation of CPPs via an energy-independent mechanism was caused by artifacts related to cell fixation.[170,171] It is now widely accepted that CPP internalization is largely mediated by various endocytic processes, such as macropinocytosis, and clathrin- and caveolae- or other lipid raft-mediated pathways,[169,172] depending on the type of CPP, cargo molecules, and target cell.[173]

Although the results reported in the literature are inconsistent, endosomal entrapment is a significant hurdle for CPP-mediated cytoplasmic delivery of therapeutics.[174,175] For instance, various types of particles functionalized with

Tat peptide have been shown to accumulate in different subcellular regions: while 3-nm AuNPs are localized in the nucleus and in the cytoplasm,[176] QDs remain trapped in vesicles.[177] Interestingly, 14-nm AuNPs (Tat functionalized) can be initially found in the cytosol, nucleus, and mitochondria; however, after being processed to vesicles, these particles escape both by penetration of the vesicle membrane and by membrane rupture (Figure 14.5).[178]

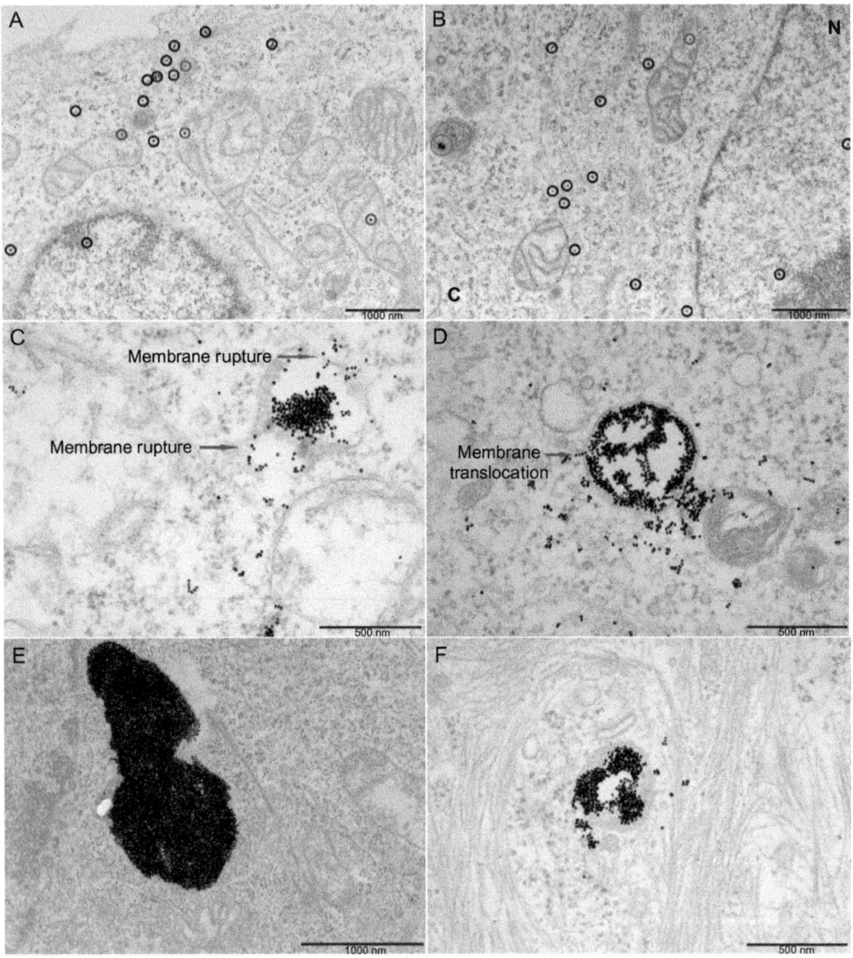

FIGURE 14.5 Transmission electron microscopy (TEM) of HeLa cells incubated for different times with TAT-modified AuNPs: (A) 2-h incubation showing particles in the cytoplasm, mitochondria, the nucleus, and vesicles, (B) a similar scenario but, in addition, vesicles containing a larger number of particles are found after 10 h, (C) densely filled vesicle found after 24 h, releasing particles by membrane rupture and (D) by direct membrane translocation, and (E) massively packed vesicles found after 24 h and (F) after 48 h. Reprinted with permission from Ref. 178. Copyright 2011 American Chemical Society.

The discrepancies found between studies can be explained by variations in the size of the carriers used, as it has been reported that the cellular uptake efficiency and the intracellular location of AuNPs are dependent on both the CPP and particle size.[179] To overcome the endosomal barrier, several strategies have been developed *in vitro*, such as the use of the endosomolytic reagent chloroquine or dimethyl sulphoxide in combination with CPPs.[172,180] However, these kinds of chemical reagents may not be suitable for use *in vivo* because they may be cytotoxic. As an alternative, the combined use of CPPs and peptides that promote the destabilization of the endosome membrane or the chemical modification of CPPs to render peptides that are also able to overcome the endosomal membrane are gaining relevance.[181–183]

Finally, the lack of cellular specificity of CPPs[184,185] can be overcome through combining these peptides with specific targeting ligands, such as antibodies,[186] transferrin,[162] carbohydrates, aptamers,[187] folic acid,[164] and other small molecules,[188] that are currently being used to target cells *in vitro* and *in vivo*, as reviewed in this chapter.

2.2.5. *Intracellular Barriers*

Internalized endocytic vesicles have several potential destinations, mainly the lysosomal compartment for degradation, or recycling back to the cell surface after release of receptor ligands. The entrapment of nanocarriers in endosomal vesicles and lysosomal degradation are major barriers for successful drug delivery. Therefore, the delivery of carriers using pathways that avoid endosomal internalization or promote escape from the endosome to the cytosol could overcome this problem. The cytosol is a complex fluid comprising macromolecules as well as other structures, such as the cytoskeleton.[189] The physicochemical and biological properties of the cytosol provide an innate defense system against foreign nucleic acids, which may be a barrier to successful nucleic acid-based drug delivery, for example.[190] In fact, to achieve successful gene therapy, DNA and small interfering RNA (siRNA) must be localized to the nucleus and cytoplasm, respectively. The nucleus, on the other hand, is separated from the cytoplasm by a double lipid bilayer (nuclear envelope), which is perforated by nuclear pore complexes.[191] Small molecules can pass through these complexes *via* passive diffusion driven by a concentration gradient; however, molecules larger than 40–45 kDa require specific targeting signals to gain access to the nucleus.[192]

2.2.5.1. Endosomal Escape

With the exception of carriers that directly penetrate the cytoplasm without being processed to endosomes and lysosomes, or carriers that are targeted to lysosomes, the remaining nanocarriers must have the capacity to escape from the degradative environment of the lysosomes. Endosomal release can be mediated by pore formation in the endosomal membrane, pH-buffering effect ("proton sponge" effect), a fusogenic mechanism, or a photochemical disruption of the endosome membrane (reviewed in Ref. 193).

Many cationic polymers, such as poly(ethylenimine),[194] poly(amidoamine),[195] propylacrylic acid,[196] poly (L-Lysine),[197] and imidazole-containing molecules,[198] can be used to disrupt the endosome. Endosomal escape occurs through the *proton sponge* mechanism, which is an osmotic effect.[199] Materials containing an excess of uncharged amines and, therefore, a high buffering capacity inhibit endosomal acidification by absorbing the protons that are pumped inside. Accumulation of protons subsequently forces an influx of counter ions into endosomes, thereby causing osmotic swelling and rupture of the endosome membrane.

In the case of cationic liposomes (or lipid-based carriers), endosomal escape occurs by a *fusogenic* mechanism, in which the lipid vector fuses with the endosomal membrane. In this process, the positive charge of the liposomes interacts with the anionic phospholipids of the endosomal membrane, causing some of them to displace (flip-flop mechanism).[200] This lipid charge neutralization causes membrane fusion and the release of the contents into the cytosol. The membrane fusion capacity of liposomes is often enhanced through the inclusion of noncationic helper lipids, such as 1,2-dioleyl-*sn*-glycerol-3-phosphoethanolamine (DOPE), thereby facilitating endosomal escape.[201] Surprisingly, negatively charged lipid nanocapsules have also been reported to bypass endosomes. This behavior could be attributed to the presence of hydroxystearate of PEG, a nonionic surfactant that destabilizes the lysosome membrane and neutralizes the pH.[202]

Another mechanism for endosomal escape is the destabilization of the endosomal membrane by *fusogenic peptides*, which are short sequences of amino acids generally comprising a large number of basic residues.[165] These sequences are inspired mainly by viral peptides, which undergo conformational changes triggered by the change in pH inside the endosomes–lysosomes. This new conformation is then able to insert into and fuse with the endosomal membrane, resulting in the delivery of contents to the cytosol.[203] For instance, the influenza virus contains a fusogenic peptide sequence in the haemagglutinin protein that has been used as a model to design amphipathic synthetic peptides, such as GALA and KALA, to disrupt endosomes.[204–206]

Alternatively, *photochemical internalization* using photosensitizer molecules can be applied to disrupt the endosomal membrane (reviewed in Ref. 207). The photosensitizers localize in the membrane of endosomes, and after exposure to light, these molecules induce the formation of reactive oxygen species (ROS) and subsequent damage of the endosomal membrane.

2.2.5.2. Delivery to Nucleus

Reaching the nucleus is essential for many therapies, including gene therapy, and some drugs used for chemotherapy, for instance, cisplatin. The passive diffusion of carriers to the nucleus for most cytosolic nanomedicines is complex as only molecules smaller than 40–45 kDa can diffuse through nuclear pore complexes. Small carriers, such as polymer–drug conjugates (<10 nm), can cross the cell nucleus without targeting sequences.[208] However, larger molecules should be actively target to the nucleus, carrying specific signals such as the nuclear localization sequence (NLS) to cross the nuclear envelope.[209] By coating AuNPs with nucleoplasmin (protein containing two NLSs), and then with the receptor for NLS (which consists of two subunits, importin α and β), it was possible to determine that macromolecules up to 39 nm can cross the NPC.[210] However, the entry of larger particles modified with NLS has also been reported. In fact, doxorubicin (Dox)-loaded PLGA particles with a hydrodynamic diameter of 234 nm entered the nuclei of human breast adenocarcinoma cells *in vitro*.[211] This way to overcome the membrane may involve a collapse of the nuclear membrane during the mitosis (reviewed in Ref. 212). Readers are referred to excellent reviews for further details.[192,213,214]

2.3. Metabolism and Excretion

The nanocarrier used for drug delivery should be eliminated from the body, either by degradation of the biodegradable nanosystems or by excretion of the nondegradable. In this regard, if the nanomedicine is to be used at a high dose or for long-term treatment, this is a key issue to avoid accumulation within the different organs.

For instance, many polymeric nanocarriers are biodegradable, so their bonds can be cleaved hydrolytically or enzymatically. In addition, natural polymers such as collagen can undergo biodegradation along with degradation products of the body, and some synthetic polymers such as poly (aspartic acid) can be degraded by lysosomal enzymes.[25] As nanomedicines are becoming more complex with the years, the potential safety of each metabolite that might arise as a result of their *in vivo* degradation or metabolic fate must be clearly reflected.

Nonbiodegradable nanomedicines should be removed by renal or hepatobiliary clearance in order to avoid bioaccumulation. The renal clearance is influenced by the nanocarriers' molecular weight, the size, and the shape, being the rate of clearance, in general, inversely correlated with the molecular weight and the size.[36] The liver is also a major excretion route for nanocarriers, taking place across hepatocytes through the biliary production pathway. For instance, 50-nm polystyrene NPs were mainly excreted as intact particles in bile after 24 h of intravenous injection.[215]

Many inorganic nanocarriers are very stable and difficult to metabolize, so they could reside in the body for long time periods.[24] For instance, it has been reported that PEGylated QDs are retained in the body for at least 2 years.[216] In this regard, biodegradable nanomedicines are preferred for drug delivery. Otherwise, an early characterization of the elimination routes should be performed.

3. TARGETING STRATEGIES

Since the concept of drug targeting was first postulated by Paul Ehrlich in 1906 with the "magic bullet," drug-targeted therapy has become a major goal for treating many diseases. Drug targeting is important, since the non-specific distribution of drugs throughout the body generally results in a reduction of therapeutic efficacy and an increase in harmful side effects.[217] In general terms, drug targeting can be understood as an approach to ensure the accumulation of a drug in a target organ, tissue, or cell, selectively and quantitatively, without resulting in significant effects on nontarget tissues and independently of the administration route.[218] However, drug targeting remains a major challenge for clinical applications, since it implies many coordinated steps, such as discovering the appropriate target for a particular disease, finding the drug that effectively treats it and carrying the drug in an effective way to that target.

The concept of drug targeting is extensive. This approach to drug delivery does not necessarily involve the use of specific targeting ligands, as in certain circumstances, the physiological features of the target area could be used for targeting the carriers. The main ways to target drugs to diseased tissue include the direct administration of the nanomedicines to the affected zone, the use of magnetic fields to guide a magnetic carrier to the affected area, the passive targeting and accumulation of nanocarriers through the leaky vasculature of tumor or inflammation areas (enhanced permeability and retention (EPR) effect), the passive targeting of RES, and the active targeting of several areas by attaching specific ligands on the surface of the carriers (ligands with an increased affinity toward the site of interest).[218]

From the aforementioned strategies, the *direct administration* of a drug into a target tissue is technically difficult to achieve. Furthermore, the diseased site may not be localized but spread over various tissues or cells.[219] Therefore, this approach has many limitations.

On the other hand, *magnetically* targeting the drug to a particular area is a noninvasive approach that can be achieved using magnetic carriers and by applying an external magnetic field to that area.[220] Furthermore, the magnetic field can increase the cellular uptake of the magnetic particles.[221] Although magnetic targeting has been demonstrated *in vitro* and *in vivo*, this technique also has limitations, as the effectiveness of this method of drug delivery is determined by parameters such as the physicochemical properties of the particles, the field strength, the depth of the target tissue, the rate of blood flow, and vascular supply.[222] In fact, the accumulation of magnetic NPs in smaller blood vessels with a lower rate of blood flow and located closer to the surface is higher than in blood vessels such as the aorta, with a rapid blood flow.[223,224]

To date, the most widely used strategies for drug targeting are passive and active approaches (Figure 14.6).[225]

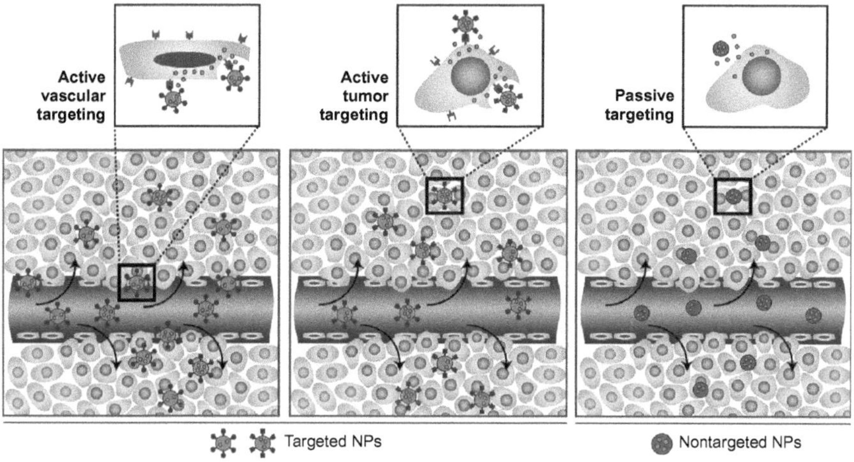

FIGURE 14.6 Passive *versus* active targeting. (Right) Particles tend to passively extravasate through the leaky vasculature, which is characteristic of solid tumors and inflamed tissue, and preferentially accumulate through the EPR effect. (Middle) Once particles have extravasated in the target tissue, the presence of ligands on the particle surface can result in active targeting of particles to receptors that are present on target cell or tissue resulting in enhanced cell uptake through receptor-mediated endocytosis. This process, referred to as "active targeting," can enhance the therapeutic efficacy of drugs, especially those which do not readily permeate the cell membrane and require an intracellular site of action for bioactivity. (Left) The particles can also be engineered for vascular targeting by incorporating ligands that bind to endothelial cell-surface receptors. While the presence of leaky vasculature is not required for vascular targeting, when present, this strategy may potentially work synergistically for drug delivery. Reprinted with permission from Ref. 225. Copyright 2009 American Chemical Society.

3.1. Passive Targeting

Passive targeting of NPs via the EPR effect leads to the accumulation of drug carriers in the tumor region, ischemic tissue, or inflamed area as a result of their extravasation through leaky vasculature.[226,227] In contrast, this passive targeting *via* EPR does not occur in healthy tissues.

Passive targeting has been widely used as an approach to treat cancer; tumor growth is angiogenesis dependent, meaning that a tumor requires the internal growth of a vascular supply from the surrounding tissues to proliferate and metastasize. The new vasculature of tumors is structurally and functionally abnormal and is characterized by its tortuous architecture, irregular blood flow, defective endothelial cells with wide fenestrations, increased vessel leakiness, and heterogeneous hyperpermeability compared to healthy tissues.[228] The increases in vascular permeability and blood vessel fenestration are behind the EPR effect described by Maeda.[229] Nanomedicines have the capacity to passively extravasate in tumor tissues and accumulate in these tissues because of ineffective lymphatic drainage.[227,230,231]

Therefore, long-circulating nanocarriers, larger than 40 kDa (threshold of renal clearance), can accumulate in tumors via the EPR effect.[232] This accumulation leads to increased therapeutic efficacy, combined with a decrease in the side effects associated with systematic chemotherapy.[233] In fact, Dox encapsulated in PEG-liposomes (Doxil®/Caelyx®) demonstrates high efficacy in the treatment of solid tumors (EPR-based) and strongly diminishes the side effects caused by Dox alone.[234]

It is known that the physicochemical features of carriers, such as size and charge, regulate the accumulation of these particles in tumors.[230] Carriers <200 nm with a hydrophilic surface are believed to exhibit an improved EPR effect as a result of their extended blood life, albeit bacteria as large as 2 μm could also be accumulated.[27] In addition, cationic nanocarriers preferentially target tumor vessels and exhibit higher permeability compared with their anionic or neutral counterparts.[235,236]

However, parameters other than carrier size and charge can influence the EPR effect. In fact, heterogenic tumor vascularization, scarce tumor penetration, and high interstitial fluid pressure could restrict nanocarrier accumulation. For example, nontargeted 25- and 60-nm poly(ethylene glycol)-block-poly(ε-caprolactone) (PEG-b-PCL) micelles show low levels of *in vivo* cellular uptake by tumor cells.[237,238] To overcome the weak penetration into the tumor core caused by the dense extracellular matrix of the interstitial space,[239] nanocarriers that can change their size have been proposed.[240] In that study, Wong et al. demonstrated extravasation of 100-nm multistage QD gelatin particles from the leaky blood vessels of tumor vasculature and their capacity to shrink to 10 nm once exposed to the tumor microenvironment. Triggered by proteases expressed in the tumor, the size change leads to a lower diffusional hindrance in the interstitial matrix, and therefore to greater penetration into the tumor parenchyma.

Despite all the advantages described above, passive targeting has several drawbacks that could hamper its use for drug delivery. First of all, the EPR effect is not observed in all types of tumor. In addition, the EPR effect is size dependent, and the great variability from tumor to tumor, from patient to patient, or even in the same tumor could limit the usefulness of this approach.[241,242] In fact, large tumors show great heterogeneity,[230] as the core area shows less accumulation of nanocarriers than other parts, due to high hypovascularity.

Even if the nanocarriers are able to arrive within the tumor, they should face other hurdles for effective drug delivery, such as the high interstitial pressure, the slow blood flow, and the complex extracellular matrix (reviewed in Ref. 243). Different approaches that can be used to increase passive transport of nanocarriers include manipulating the vascular permeability, increasing vascular surface area, and/or decreasing interstitial fluid pressure among others.[230,244]

It is worth mentioning that sometimes passive targeting can be performed to label the RES, as systemically administered particles without an appropriate coating are recognized and accumulated by RES organs, mainly the liver and spleen. In fact, magnetic NPs approved for magnetic resonance imaging (MRI) use this passive targeting strategy to accumulate in the liver.[245]

3.2. Active Targeting

To date, the most universal way to improve the nanocarrier affinity toward a target is the binding of a ligand that selectively recognizes and binds to the target site. For instance, in many pathological states, cells express new surface molecules that are absent in healthy cells or strongly overexpress some of these molecules when compared with the healthy state. These molecules can serve as targets, to which nanocarriers functionalized with specific targeting ligands can be directed. Active targeting, therefore, relies in specific interactions, usually through specific ligand–receptor interactions. Examples of these molecules include antibodies or their fragments, lectins, lipoproteins, hormones, polysaccharides, and some low-molecular-weight ligands such as folate.

Cancer stands out as one of the diseases that is most likely to benefit from targeted drug delivery as, in general, tumor cells express or overexpress many molecules on their surface that distinguish them from healthy cells. It has to be said, however, that active targeting does not necessarily enhance the overall accumulation of nanocarriers in tumors compared with their nontargeted counterparts, but that their cellular uptake into targeted cells is generally increased.[246]

Tumors can be targeted at two levels, namely, the cancer cell and the vascular endothelium. The next section provides a review of several examples of possible targets.

Transferrin (Tf), a glycoprotein that transports iron into the cell, is a popular tumor-targeting ligand because its receptor (TfR) is expressed on many cancer cells at levels severalfold higher than on healthy cells. Overexpression of TfRs is observed in bladder transitional cell carcinomas, breast cancer, gliomas, lung adenocarcinoma, chronic lymphocytic leukemia, and non-Hodgkin's lymphoma, among others, and the expression of these receptors correlates with tumor grade and stage or prognosis.[247,248] To date, many Tf-conjugated nanocarriers have demonstrated enhanced cellular targeting and uptake, and several Tf containing nanocarriers are currently in clinical trials (recently reviewed in Ref. 249). For example, a linear cyclodextrin-based polymer conjugated with Tf achieves intercellular delivery of siRNA to solid tumors in a human phase I clinical trial.[250] Likewise, a oxaliplatin-encapsulated Tf-conjugated liposome (MBP-426) is currently in a phase II of clinical trials (Source: http://www.cancer.gov/clinicaltrials, accessed January 2012).

The folate receptor (FR) is another well-known tumor marker. This receptor binds to folic acid, which is required in one-carbon metabolic reactions, and, consequently, is essential for the synthesis of nucleotide bases. The alpha isoform of the FR is overexpressed in ~40% of human cancers, including breast, bladder, pancreas, colorectal, and ovarian, although its expression is variable in these tumors.[251–253] One advantage of folate is its size (MW ~ 441 kDa). Being this small, it penetrates solid tumors rapidly, whereas larger molecules such as antibodies (MW ~ 160,000 kDa) may fail to reach distant cancer cells.[254] Numerous studies have demonstrated that attachment of folic acid to nanocarriers, such as dendrimers and liposomes, allows faster uptake and higher accumulation in tumor cells when compared to healthy cells, both *in vivo* and *in vitro*.[255,256] For instance, heparin–folate–paclitaxel NPs demonstrated a higher accumulation and a retarded growth tumor in a xenograft model of resistant human squamous cancer compared to nontargeted NPs or free paclitaxel (Figure 14.7).[257]

Despite these promising results, in some cases, targeting with folate has not led to an increased accumulation in tumor compared with nontargeted carriers.[258] In these cases, optimization of the ligand density of targeting molecules and passivating molecules per nanocarrier is required to ensure effective cell targeting; furthermore, the number of receptors expressed on the cell surface should be taken into account.[259,260]

The *asialoglycoprotein receptor* is expressed on the surface of liver hepatocytes and specifically recognizes the terminal β-D-galactose and *N*-acetylgalactosamine residues,[261] although synthetic glycomimetic ligands can also be recognized with even higher affinity.[262] Therefore, this receptor has been studied as a site for drug targeting using nanocarriers containing galactose residues. As an example, paclitaxel-loaded cross-linked micelles

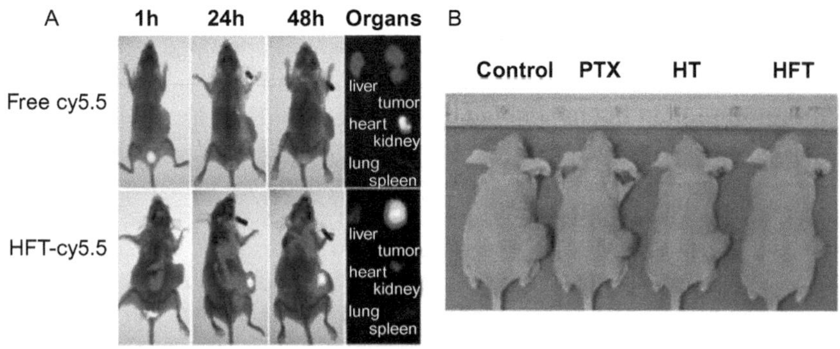

FIGURE 14.7 (A) *In vivo* distribution of heparin–folate–paclitaxel NPs (HFT) in folate receptor overexpressing human epidermal carcinoma (KB-8-5) tumor-bearing mice. NIR dye Cy5.5 or Cy5.5-labeled HFT was injected into and imaged at 1, 24, and 48 h after injection. Imaging of representative major organs including liver, spleen, kidney, heart, lung, and tumors removed from mice 48 h after injection. For HFT-T-treated mice, the greatest fluorescence intensity was observed in the tumor compared with the other tissues. For free Cy5.5-treated mice, the greatest fluorescence intensity was observed in the kidney. (B) The tumor growth of KB-8-5 xenografts was significantly inhibited in the HFT-treated group compared with the control, free paclitaxel, and non-targeted NPs (HT). Adapted with permission from Ref. 257. Copyright 2011 American Chemical Society. (For color version of this figure, the reader is referred to the online version of this chapter.)

coated with galactose specifically accumulate in the liver and inhibit the growth of a hepatoma in mice.[263] In a phase I clinical trial, a Dox copolymer containing galactosamine was shown to deliver a greater amount of Dox to a hepatoma (15- to 20-fold higher) than that achieved by free Dox.[264]

Another possible target for targeted drug delivery is *growth factors* and their receptors, since they play a fundamental role in the regulation of cell growth and replication.[265] The epidermal growth factor receptor (EGFR) belongs to a family of receptor tyrosine kinases that includes three other members (erbB2/HER2, erbB3/HER3, and erbB4/HER4).[265] The EGFR is overexpressed in a variety of human tumors,[266] and furthermore, approximately 20–25% of breast cancers exhibit amplification and/or overexpression of *HER2*; this overexpression is associated with increased tumor invasiveness and metastasis and a worse outcome when compared with HER2-negative breast cancer.[267] Trastuzumab (Herceptin®), a humanized murine monoclonal antibody targeting the extracellular domain of HER2 receptor, has improved the treatment of patients with HER2-positive breast cancer, in both the adjuvant and metastatic settings.[268,269] Other examples of anti-HER2 targeting for drug delivery include topotecan-encapsulated liposomes functionalized with an anti-HER2 single-chain variant fragment (scFv F5) antibody.[270,271] In these studies, HER2-targeted liposomes showed enhanced antitumor activity in a prostate tumor xenograft model compared with nontargeted liposomes and with free topotecan.

While the aforementioned targets direct the nanocarriers to the cancer cell, the design of nanomedicines that actively target *tumor endothelial* cells is another promising strategy. Targeting in tumor vessels has some advantages over tumor cellular targets. For instance, no tissue penetration is required as the binding to receptors can be achieved directly after intravenous injection, the emergence of drug resistance is diminished as endothelial cells are more stable than tumor cells, and most types of endothelial cell markers are expressed or overexpressed by many types of tumor.[272]

VEFG-A (or *VEFG*) is involved in the formation of new blood vessels in tumors (angiogenesis) and hypoxia, and oncogenes upregulate its expression in many human tumors.[273] VEGF can bind to two receptors: VEGFR-1 and VEGFR-2, although VEGFR-2 mainly mediates VEGF signaling in endothelial cells and is highly expressed on endothelial cells in tumor neovasculature. Therefore, VEGF and VEGFR-2 are seen as promising targets to prevent the development of tumor vasculature. The first antiangiogenic molecule introduced into the clinic was bevacizumab (Avastin®), a humanized monoclonal antibody that inhibits VEGF and has proven efficacy in glioblastoma, kidney, colorectal, and non-small cell lung cancer. Other experimental approaches include a PEGylated di-Fab antibody (CDP791) that binds the VEGFR-2 and is currently in phase II clinical trials for non-small cell lung cancer,[274] or Dox loaded liposomes functionalized with antibodies anti-VEGFR-2.[275]

Many other molecules appear as possible targets for targeting the vascular endothelium, such as the ICAM-1, the platelet endothelial cell adhesion molecule-1, matrix metalloproteinase (MMP) receptors, or E-selectin, among others.[103,104,276–278] To date, many nanocarriers are being developed to target tumor vasculature (reviewed in Ref. 279).

The overexpression of *integrins*, such as αvβ3 and αvβ5, in angiogenic vessels makes these molecules potential candidates for active targeting.[280] Peptides based on the amino acid sequence arginine–glycine–aspartate (RGD peptides) have been extensively used in tumor diagnostics and therapeutics.[241,281]

However, if crossing the vascular wall is needed, these convectional peptides could not be used. To solve this drawback, the laboratory of Ruoslahti has described a cyclic peptide iRGD that is able to home to a specific target and, in addition, penetrates into that tissue.[282] They combined the tumor-homing RGD sequence with a tissue penetration motif, which has to be exposed at the C-terminus of the peptide to be active (CendR motif). The homing sequence (RGD) directs the peptide to the tumor vascular endothelium (binding to αv integrins), where the peptide is cleaved by a protease, so that the CendR motif becomes active. The activated CendR motif binds to neuropilin-1 (a trans-membrane receptor) that mediates penetration into tissue and cells. While compounds coupled with iRGD bind to the tumor vessels and spread into the tissue, conventional RGD peptides only deliver the cargo to the blood vessels. Not only drugs conjugated with iRGD can be carried deep into extravascular tumor tissue, but also systemic injection of iRGD improves the therapeutic index of different drugs, such as Dox and trastuzumab.[283]

Targeting of nanocarriers to specific endothelial *caveolar markers* is also emerging as a possible targeting strategy, as it could provide a rapid way to bypass the endothelial barrier, improving tissue-specific delivery of nano-carriers through increased accessibility and tissue penetration (reviewed in Refs. [244,284]).

For instance, Sahay *et al.* reported that polymeric micelles with cross-linked ionic cores of poly(methacrylic acid) and nonionic shell of poly(ethylene oxide) could be internalized *in vitro* in epithelial cancer cells but not in normal epithelial cells that formed tight junctions (Figure 14.8).[285] The micelles were able to enter in cancer cells with absent or dysfunctional tight junctions through caveolae-mediated endocytosis and bypass early endosomes.

In addition, albumin is internalized by interacting with endothelium gp60 receptors through a caveolae-mediated mechanism.[286] Abraxane®, a nanoparticle albumin-bound (nab) form of paclitaxel (130 nm) (approved by the FDA in 1995 to treat breast cancer), takes advantage of this transcytosis for efficient delivery to the tumor sites.[287] Paclitaxel and albumin are then released into the tumor interstitium, where nanoparticles are captured by SPARC (secreted protein, acidic and rich in cysteine), a 43-kDa secreted glycoprotein with high binding affinity to albumin.[288] As SPARC is secreted in tumors, nanoparticles are taken up in tumor cells resulting in selective tumor cytotoxicity.

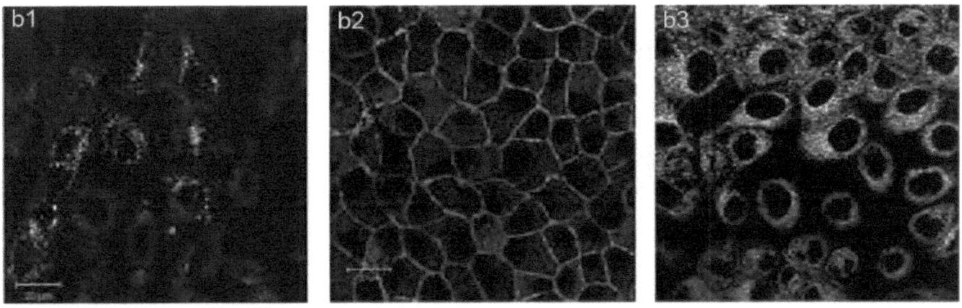

FIGURE 14.8 Cellular uptake of polymeric micelles is regulated by formation of tight junctions. Live cell confocal imaging of epithelial MDCK cells exposed to micelles, performed using non-confluent cells (b1) or confluent cells without (b2) or with calcium-depleted medium (b3), which dysfunctions the tight junctions. The micelles were able to enter in cancer with absent or dysfunctional tight junctions, but not in confluent cells with tight junctions. Reprinted from Ref. 285. Copyright 2010 with permission from Elsevier.

As discussed above, active targeting could have advantages in drug delivery in many diseases, including cancer. In fact, active targeting can improve cellular uptake to cells, which is really useful for carrying agents such as nucleic acids, which are unable to enter the cells by themselves. To date, many therapeutic targets have been described, and targeted products consisting in antibodies and antibodies containing radioisotopes (Zevalin®) have been approved. However, despite many years of effort, no actively targeted drug conjugates have thus far been approved for clinical use. In fact, the only existing targeted drug product (gemtuzumab ozogamicin, Mylotarg®), consisting in a CD33 antibody conjugated to calicheamicin and approved for treating acute myeloid leukemia, has been voluntarily withdrawn by the company (June 2010).

Therefore, active targeting is a promising strategy, but translating it from *in vitro* to *in vivo* is a really difficult task to achieve. In fact, active targeting to cancer shares the same hurdles as passive targeting that could include the presence of smooth muscle cells or pericytes covering tumor blood vessels, a high density of cells, a dense extracellular matrix, high intercellular fluid pressure, or multidrug resistance.[243] But on top of that, new challenges related with the interaction with the receptors appear. Actually, efficient interaction of nanomedicines with the receptors for active targeting depends on a variety of factors that include the degree of receptor's expression on target cells relative to nontarget cells, receptor availability on the target cell surface, heterogeneity expression of the receptors in tumor cells or in different disease stages, number of available ligands on the nanocarrier, possible receptor saturation, etc.[289]

Furthermore, active targeting may be limited by spatial and temporal heterogeneity in the expression of targets, as it is impossible to know what fraction of tumor cells expresses a specific receptor at any time point. Thus, translating the *in vitro* results based on cancer cells which express a unique surface marker to an *in vivo* model may not be representative. Moreover, the animal models that are routinely used in preclinical trials are not representative for a real clinical situation (striking differences in environmental aspects and disease status) and thus, many agents which work well in animal models might not have any success in clinical trials in patients.[289]

4. TRIGGERED RELEASE OF THE THERAPEUTIC AGENT

Once the nanomedicines have reached their target site, drugs must be released from nanocarriers in order to become bioavailable. In fact, it would be most desirable if drugs could be released in a manner that precisely matches physiological needs at the proper times (temporal modulation) and/or at the proper site (site-specific targeting).[290] Release can be governed by passive processes including diffusion, particle erosion, particle degradation, and polymer swelling.[291] Most interestingly, control over the timing and degree of release can be achieved by external stimuli.[292] In this regard and depending on the nanosystem, a wide variety of strategies have been developed to accomplish drug release in response to a change in the nanocarrier environment, such as a change in temperature,[293] pH,[294,295] redox potential,[296,297] presence of certain enzymes,[298] or triggered exogenously by irradiation with light[299] or exposure to an electric or magnetic field.[300,301] However, several challenges have to be faced when designing triggered release technologies, including achieving enough stability to prevent uncontrolled discharge of cargo, and specificity and high drug release efficiency when triggering is promoted.[302]

4.1. Temperature

Among all the approaches available, temperature is one of the most common triggering strategies used. First, several disease states are associated with an increase in temperature.[303] For instance, in the case of cancer, certain tumor microenvironments are characterized by mild hyperthermia (HT, 1–2 °C above healthy tissues).[304] But in addition, different clinical procedures could be used for achieving localized heating (up to 40–43 °C), by application of microwaves, ionizing radiation, continuous-wave ultrasound, magnetic-, or light-triggered hyperthermia, among others.[305] Therefore, a wide variety of thermoresponsive nanodrug carriers have been employed to release their payload within environments with temperatures above the physiological one.[306]

The incorporation of thermoresponsive materials on nanocarriers is one of the most used strategies for achieving this goal. These materials have a sharp transition temperature at which they become either soluble or insoluble. When the transition is from a more soluble to a less soluble state, this temperature is known as the lower critical solution temperature (LCST). Conversely, if the transition is from a less soluble to a more soluble state, this temperature is known as the upper critical solution temperature (UCST). Materials with

LCST, such as *N*-substituted polyacrylamides with the predominant one being poly(*N*-isopropylacrylamide) (PNIPAM), are more widely used than materials with UCST. Cross-linked polymers containing thermoresponsive segments undergo a swollen-to-shrinken transition when temperature increases above the LCST. This has been used to show how thermosensitive micelles assembled from amphiphilic block copolymers could selectively increase their size due to a raise in the temperature above LCST. This temperature difference triggers that the outer shells of these micelles become hydrophobic which, in turn, causes an increase in the size due to their aggregation.[307] In addition, the dehydrated shell of the micelles above the LCST can be uptaken by cells more easily than micelles below the LCST (Figure 14.9).[308] In another example, a significantly higher antitumor efficacy was observed in a human gastric cancer xenograft model treated with thermosensitive docetaxel-loaded micelles accompanied by hyperthermia compared with the conventional docetaxel formulation.[309] HT greatly enhanced the antitumor effect of docetaxel-loaded micelles showing a marked tumor volume and weight growth percentage inhibition of 81.5% in comparison to 32.2% for docetaxel-loaded micelles without HT. In the case of conventional docetaxel formulation, no effect with HT was observed.

In addition to polymeric micelles, thermosensitive liposomes can be used for specific drug release in solid tumors in combination with local HT.[311,312] In this case, one alternative for introducing thermosensitive responsiveness is by tuning the lipid composition. Using this strategy, it was possible to finely tune the temperature at which Dox was released. Different obtained thermoresponsive formulations with release temperatures at 39–40 and 41–43 °C were compared *in vivo* versus a nonthermally active formulation using a human xenograft model in mice.[313,314] In this case, the combined use of the lipid formulations with HT also increased their antitumor activity. However, while

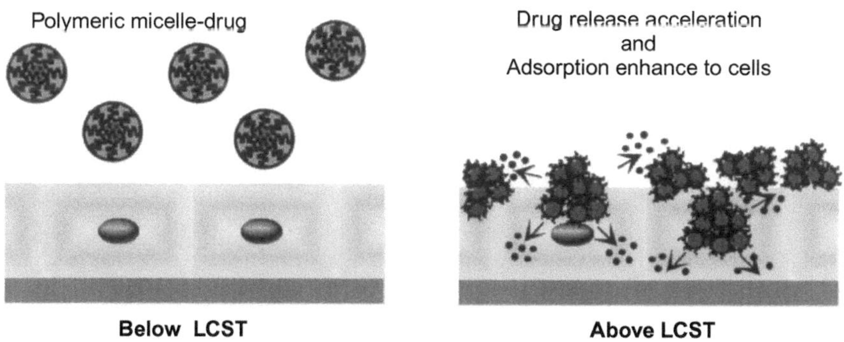

FIGURE 14.9 Interactions between thermoresponsive polymeric micelles constructed using block copolymers of poly(*N*-isopropylacrylamide) and poly(butylmethacrylate) and cells can be modulated by temperature control. Reprinted from Ref. 310. Copyright 1999 with permission from Elsevier.

all lipid formulations extended the rate of survival, the low-temperature release formulations was significantly better. All animals that received this low-temperature formulation survived to the end of the 60 days' observation period along with a complete loss of tumor mass. These formulations are currently being investigated in Phase-III clinical trials for hepatocellular carcinoma in combination with radio frequency ablation to provide localized heating and in Phase-II trials for recurrent chest wall breast cancer in combination with mild heat energy (www.celsion.com). The limitation of these thermoresponsive liposomes of requiring a high thermal dose (that would cause necrosis to the healthy tissue surrounding the tumor) for triggering drug release had been solved by fixing thermosensitive polymers to the liposomal membranes.[315,316]

In addition to micelles and liposomes, other thermoresponsive organic nanostructures such as nanogels, dendrimers, and polymeric NPs, among others, have been developed.[306]

This swelling–shrinking transition of temperature-responsive nanocarriers could also be used to deliver drugs efficiently to the cytosol by physically breaking endosomal vesicles within the cells.[317] In addition, an abrupt volume expansion triggered by externally modulate temperature (cold shock) could also be used to induce ultimate necrotic death of cells by destabilizing whole cell cytoskeleton and membrane structural integrities. For this purpose, temperature-responsive nanogels named "thermally detonatable nanobombs" were developed (Figure 14.10).[318]

Another interesting strategy is to control the generation of heat at the site of interest by site-specific delivery of heat-generating inorganic NPs (e.g., magnetic or gold) followed by remote application of an external alternating magnetic field or light irradiation. The literature cites many examples where combining the magnetism or light-induced heating with the incorporation of thermoresponsive materials (e.g., PNIPAAM) to several nanosystems, including gold, and magnetic NPs or hybrid nanocarriers allowed the release of the drug cargo upon a localized increase in temperature.[319,320] Iron oxide cores

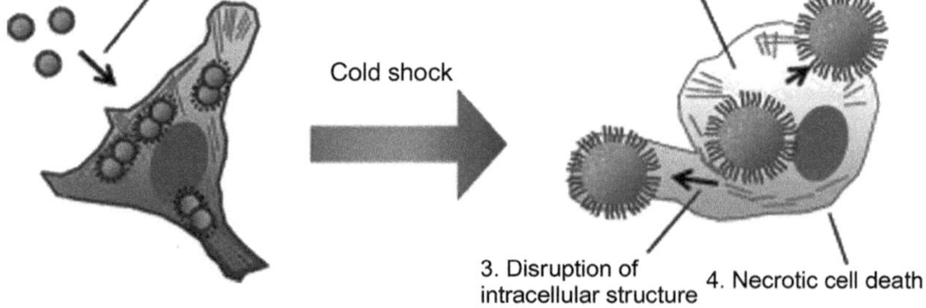

FIGURE 14.10 Schematic illustration of intracellular explosion of volume transition nanogels for necrotic cell death. Reprinted from Ref. 318. Copyright 2009 with permission from Elsevier.

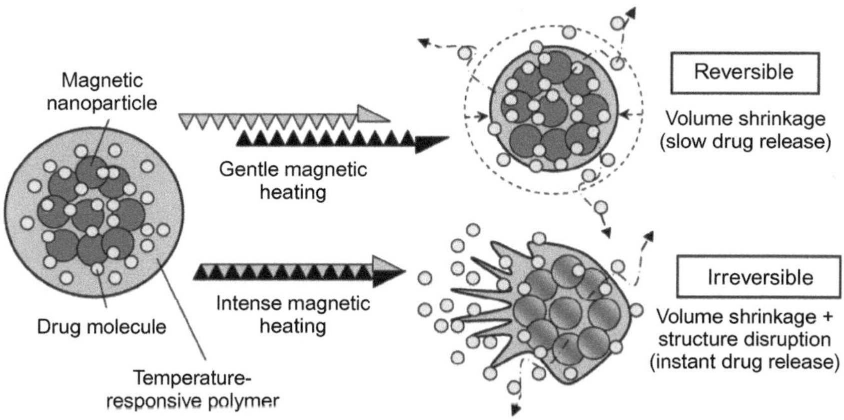

FIGURE 14.11 Two drug release mechanisms under magnetic heating. Gentle magnetic heating causes temperature-responsive polymer to shrink, squeezing drug out from the nanoparticle. Intense magnetic heating additionally ruptures the nanoparticle, triggering a burst-like drug release. Reprinted from Ref. 322. Copyright 2009 with permission from Elsevier.

encapsulated with thermoresponsive polymers have shown either a reversible shrinkage or an irreversible rupture of the thermoresponsive shell under gentle or intense magnetic heating, respectively.[321–323] This allows selecting among two different drug release mechanisms: a slow one or a burst-like drug release (Figure 14.11).

It is important to underline that when the temperature polymer chains are not covalently cross-linked the nanocarrier may undergo liquid (sol) state to solid (gel) state transitions, instead of swelling–shrinking transitions. This was recently used to demonstrate the feasibility of achieving an on–off control of the drug release. For this purpose, a nanocomposite was developed for which Pluronic® F-127 was selected as the sol-gel polymer system and iron oxide nanoparticles were chosen as the nanoscale heating source to heat the polymer to temperatures near or above the gel to solution transition.[301]

Another approach for triggered drug delivery using heat-generating NPs consists on attaching the drug molecule to the nanosystem through a heat labile linker, instead of entrapping the drug within a thermosensitive polymer.[324] Combining this strategy with magnetic NPs, it was possible to achieve complex release profiles of multiple drugs by controlling the power of the electromagnetic field pulses used.[325] (Figure 14.12) In this context, it is worth mentioning the "active drug-elution technology" being developed by Biophan Technologies, Inc. (http://www.biophan.com/).

In the case of AuNPs, a large number of different strategies had been reported to make them thermosensitives. However, in comparison with magnetic ones, relatively little work has been done yet to test them for *in vitro* cellular uptake or *in vivo* thermal targeting against tumors.[306] In this sense, AuNPs (18 ± 11 nm diameter) functionalized with poly(NIPAM) showed a

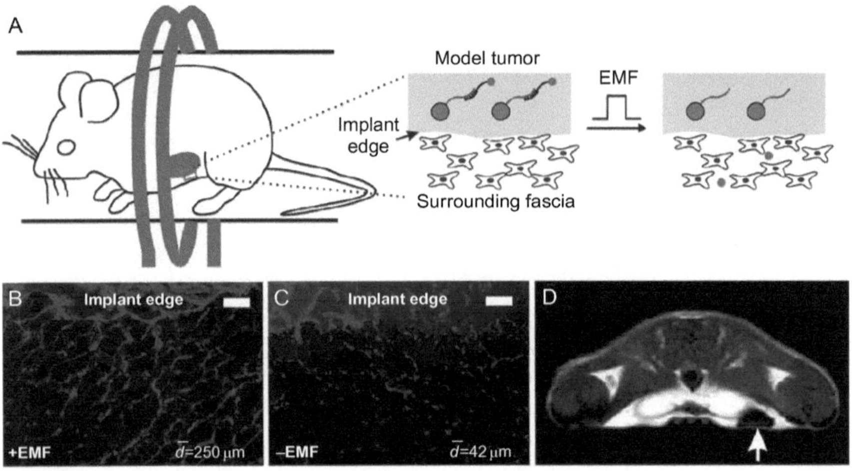

FIGURE 14.12 Remotely triggered release from magnetic NPs *in vivo*. Nanoparticles were mixed with matrigel and injected subcutaneously near the posterior mammary fat pad of mice, forming model tumors (A). Application of alternating electromagnetic field (EMF) to implants resulted in release of model drugs and penetration far into surrounding tissue (B) when compared to unexposed controls (C, scale bar = 100 µm). These mice were imaged with a 7T MRI scanner, and transverse section shown in (D) depicts image contrast due to presence of nanoparticles (arrow). Reprinted from Ref. 325. Copyright 2007 with permission from Wiley.

cellular uptake that was dependent of the temperature of the incubation media. As previously observed with micelles based on poly(NIPAM), this behavior was attributed to an increase in the surface hydrophobicity of the NPs triggered by the temperature-induced shrinking of the polymeric shell.[326] Several works with polyelectrolyte capsules with AuNPs in their walls have also shown the feasibility of a photoactivated release of the polymeric cargo to the whole cytosol of living cells (Figure 14.13).[327,328]

In addition to heat-generating inorganic NPs, light-sensitive drug nanocarriers emerge as an alternative for efficient controlled release. These nanocarriers, generally synthesized with photosensitizers containing polymers (e.g., azobenzene, stilbene, triphenylmethane), can be used to increase the local temperature. When light is applied to these nanocarriers, the chromophore absorbs light, and the resulting temperature change alters their swelling behavior. Using this strategy, gold-capped mesoporous silica nanospheres for photoinduced intracellular release of drugs in human cells were recently developed.[329] The 100-nm silica nanospheres were capped with 5-nm gold nanospheres and functionalized with a cationic photoreactive linker. Photoirradiation using ultraviolet (UV) light cleaved the photolabile linker, changing the surface charge property of the AuNPs, from positive to negative. The charge repulsion between the AuNPs and silica nanospheres uncapped the silica mesopores and allowed drug release (Figure 14.14).

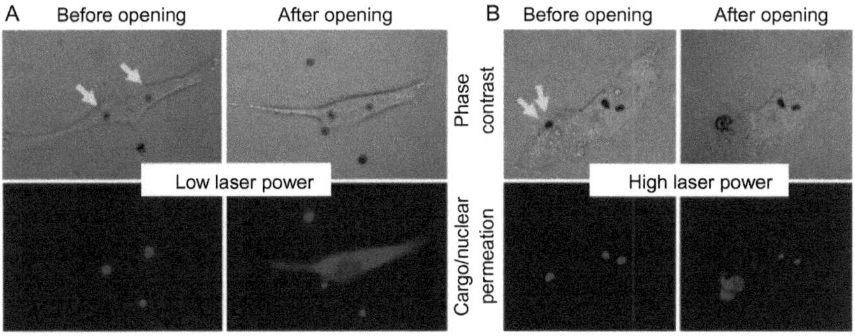

FIGURE 14.13 Cargo release and cytotoxicity experiments with capsules filled with red Alexa Fluor 594 dextran as cargo and gold particles embedded in their walls. Capsules were illuminated with (A) low laser power and (B) high laser power. Phase contrast images show cells that have incorporated capsules (yellow arrows) before and after laser illumination. Red fluorescence images show the cargo release and the nuclear permeation (Ethd-1) in cases where capsules trapped in cells were excited with low laser power and high laser power, respectively. In the case of high power illumination, permeation of the cell membrane leads to loss of fluorescent cargo by diffusion out of the cell. Reprinted with permission from Ref. 327. Copyright 2008 American Chemical Society. (For interpretation of the references to color in this figure legend, the reader is referred to the online version of this chapter.)

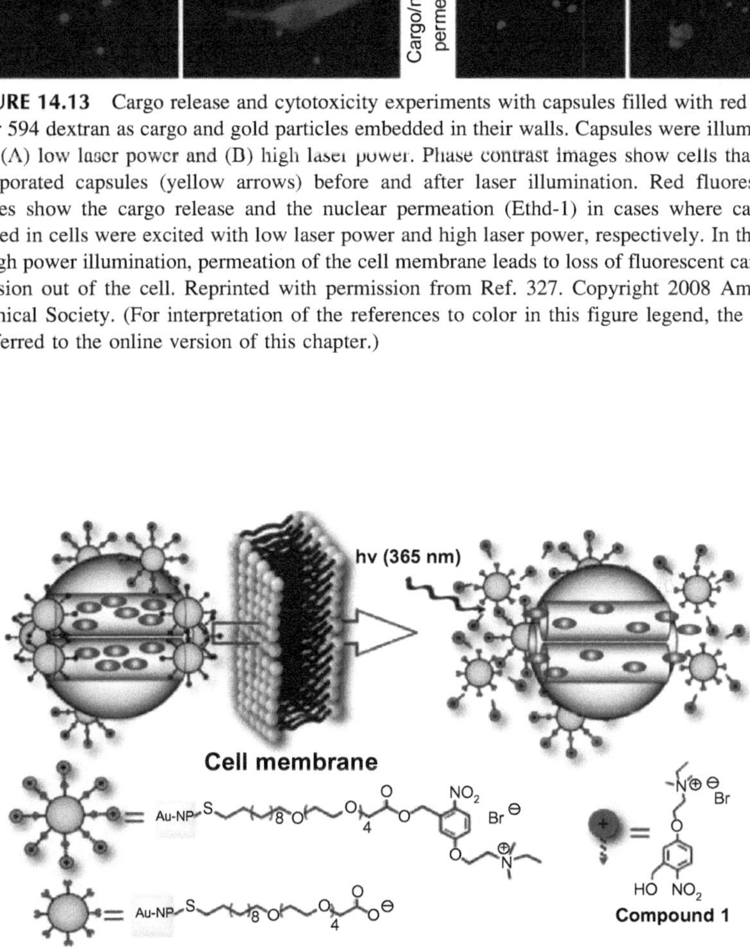

FIGURE 14.14 Schematic illustration of the photoinduced intracellular controlled release of silica nanospheres-gold NPs (AuNPs). Upon UV irradiation, the photolabile linker on the AuNPs was cleaved, changing the surface charge property of these AuNPs from positive to negative. The charge repulsion between the AuNPs and silica nanospheres would then uncap the mesopores and allowed the release of guest molecules. Reprinted with permission from Ref. 329. Copyright 2009 American Chemical Society. (For color version of this figure, the reader is referred to the online version of this chapter.)

4.2. pH

Many pathological processes in various tissues and organs, such as cancer and cystic fibrosis, are accompanied by a local pH decrease.[330] Tumors tend to have lower pH values (as low as 5.7) than normal tissue environment (pH 7.4). Nanocarriers could also face changes in pH if they are internalized via endocytosis. If this occurs, they are sequestered into an early endosome and exposed to a progressive acidification while the vesicle matures into late endosomes and degradative lysosomes (pH 5.0–5.5). These *pH changes* can be exploited as a trigger to achieve site-specific drug cargo release or to achieve nanosystem escape from endosomes.[331] The literature reports on a wide variety of strategies developed to achieve pH-triggered drug release. In this regard, polyelectrolyte polymers that suffer conformation changes, and swell when they are exposed to ionic strength and pH changes, have been introduced into several nanocarriers in order to release the therapeutic drug site specifically, burst the endosome, or promote the association of the nanocarrier with cellular membranes.[332] For instance, in the case of polyelectrolyte micelles, electrostatic forces are responsible for keeping this nanocarrier assembled together, as the balance between positive and negative charges is affected by pH changes. Therefore, upon pH changes, the micelles, which can be loaded with drugs and enzymes, become unstable and release their contents.[333] However, if polymeric polyelectrolytes are covalently cross-linked, the nanocarrier may undergo swelling–shrinking transition upon changes of pH, instead of losing its integrity. This is the case of covalently cross-linked layer-by-layer (LbL) microcapsules made of composite polyelectrolytes (poly(methacrylic acid) coupled with poly (*N*-vinylpyrrolidone) or PMAA/PVPON), which showed a dramatic highly reversible swelling at acidic and basic pH due to excess of positive and negative charges in the capsule wall (Figure 14.15).[334]

Another interesting example of pH-sensitive nanocarrier is a virus-mimetic nanogel developed by Bae and colleagues, which is able to migrate from cell to cell releasing part of their cargo in each one.[335] These nanogels consist of

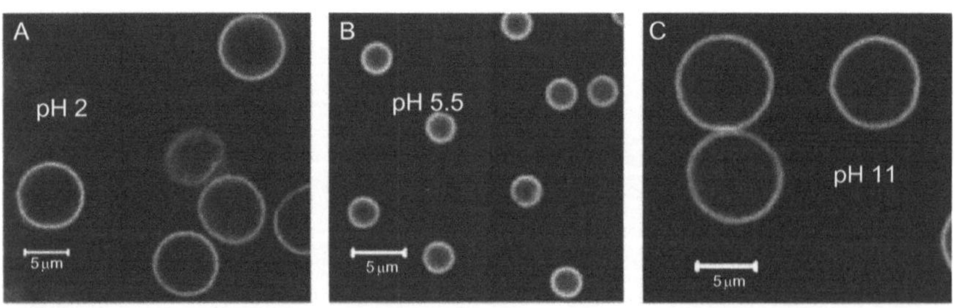

FIGURE 14.15 pH dependence of the diameter of cross-linked PMAA capsules for 22 h. Confocal laser scanning microscopy images show the PMAA capsules at pH 2 (A), pH 5.5 (B), and pH 11 (C). The scale bar is 5 μm. Reprinted with permission from Ref. 334. Copyright 2006 American Chemical Society.

a hydrophobic polymer core made of poly(L-histidine-*co*-phenylalanine) which is loaded with Dox (a model anticancer drug). In addition, the core is surrounded by a PEG–bovine serum albumin (PEG–BSA) double hydrophilic shell that helps to avoid potential immune responses, and subsequently conjugated with folate ligands to target overexpressing tumor cells. Once the virus-mimetic nanogels are internalized, they enter the endosomes, where they experience a pH-triggered volume expansion and, consequently, the drug release. The swollen nanogels can then escape from endosomes by disrupting the endosomal membranes. However, the flexible tethering of the BSA corona to the pH-sensitive core by the PEG units accommodated a large amount of expansion without particle disassembly. Therefore, the nanogels could shrink back to their initial size in the cytosol as they encountered a less acidic pH and migrate then to another cell just as a virus, where the cycle can be repeated.

Another strategy widely used is the development of nanocarriers that degrade in response to a reduction in pH. Thus, acid-labile cross-linkers are usually used for the development of pH-degradable nanogels or micelles made of amphiphilic block copolymers among other nanosystems. This scheme is frequently used to improve the intracellular delivery of the therapeutic agents, as the rapid acidification that occurs along the endolysosomal pathway causes the disruption of the carrier and trigger the payload release.[336–338]

In addition to the development of nanocarriers that hydrolyze or change their conformation due to changes of pH, another strategy to introduce pH sensitivity is the coupling of a PEG shedding to the nanosystem through bonds that are hydrolyzed in acidic conditions, such as diorto esters, vinyl esters, double esters, and hydrazones. The nanocarriers are stable during systemic circulation and at neutral pH, but when the PEG shedding is removed at acidic environments, the destabilization and fusion of the nanocarrier with the endosomal/cellular membranes was promoted (Figure 14.16).[339,340]

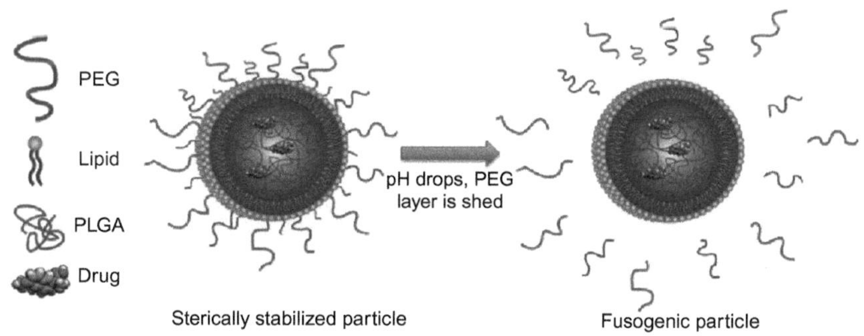

FIGURE 14.16 Schematic of a lipid-polymer hybrid NP with pH-triggered PEG shedding. The hybrid nanoparticle consists of a drug-loaded PLGA polymeric core, a fusogenic DOPE/oleic acid monolayer shell, and a sterically stabilizing PEG corona. The PEG layer comes off in response to environmental acidity, making the particle fusogenic toward lipid membranes. Reprinted with permission from Ref. 339. Copyright 2011 American Chemical Society.

Another methodology takes advantage of a similar concept, namely, direct conjugation of the drug to the nanocarrier by acid-labile linkages, such as the ones mentioned above.[341] The drug is then referred to as a prodrug and it remains inactive until it is cleaved from the carrier.[342–344]

In addition, lipids, polymers, peptides, and proteins that become fusogenic at acidic pH can also be introduced into the nanocarrier.[345] This is a biomimetic approach, simulating viruses that mediate endosomal escape. Highly carboxylated polymers or peptides are hydrophilic at pH 7.4 but become hydrophobic in acidic conditions as a result of their extensive protonation. This promotes interaction with the endosomal membrane and its disruption.[203] In the case of liposomes, there are several examples using fusogenic lipids (e.g., 1,2-dioleoyl-*sn*-glycero-3-phosphoethanolamine or DOPE, diaclglycerol or DAG, phosphatidic acid, etc.), which, after protonation at acidic pH, switch their molecular structure to one that drives the fusion of liposomal membranes with endosomal membranes.[201]

4.3. Ultrasounds

Focused ultrasound allows the deposition of thermal and mechanical energies deep inside the human body. The bioeffects of ultrasound can lead to local tissue heating, cavitation, and radiation force, which can increase the therapeutic effect of drug nanocarriers. In this regard, ultrasounds can be used for (i) local drug release from nanocarriers, (ii) increased extravasation of drugs and/or carriers, and (iii) enhanced diffusivity of drugs.[346,347] Other advantages of this approach are its ability to penetrate deep into the body (better than light), its noninvasiveness, the fact that it can be focused on targeted areas and applied to spatially controlled areas, with a spatial control of about 1 mm, and that it does not damage the surrounding tissues. In addition, it increases the permeability of blood tissue barriers and cell membranes which render target cells more prone to drug uptake, or more susceptible to therapeutic processes.[348]

When using nanocarriers that are sensitive to mechanical forces (the oscillating ultrasound pressure waves) and/or sensitive to temperature, the content of the nanocarriers can be released locally by ultrasounds. In this regard, several acoustically active nanocarriers have been developed including liposomes, nanomicelles, nanoemulsions, and solid lipid nanoparticles.[346] For instance, entrapped bubbles within the shell of lipids, surfactant layers, or polymers that constitute these nanocarriers have been used to release coencapsulated drugs by their ultrasound rupture.[349]

Currently, the most commonly used nanoparticles with ultrasound are polymeric micelles and liposomes. For instance, several *in vivo* studies have reported a significant reduction in the tumor volume, or even tumor regression, when using polymeric micelles loaded with chemotherapeutic drugs in conjunction with acoustic activation, when compared with noninsonated animals.[350–352] As an example, Figure 14.17B shows the dramatic regression of

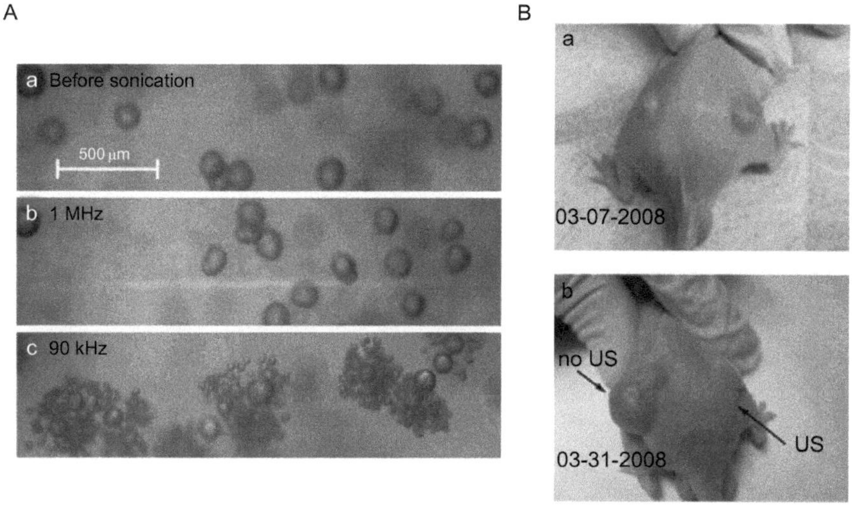

FIGURE 14.17 (A) PFP/PEG-PLLA microbubbles in a plasma clot (a) before and after sonication for 1 min by (b) 1 MHz, 3.4 W/cm^2, and (c) 90 kHz, 2.8 W/cm^2 ultrasound at room temperature. (B) Photographs of a mouse bearing two ovarian carcinoma tumors (a) immediately before and (b) 3 weeks after the treatment; a mouse was treated by four systemic injections of nanodroplet-encapsulated paclitaxel given twice weekly; the right tumor was sonicated by 1-MHz ultrasound delivered 4 h after the injection of the drug formulation. Reprinted from Ref. 353. Copyright 2009 with permission from Elsevier.

ovarian xenografts model tumors through systemic injections of drug-loaded nanoemulsions combined with therapeutic ultrasound.[353] For achieving this goal, nanoemulsions of perfluoropentane (PFP), as ultrasound triggered microbubbles generating agent (Figure 14.17A), were developed from paclitaxel-loaded PEG-*co*-poly(L-lactide) (PEG-PLLA) or PEG-*co*-polycaprolactone (PEG-PCL) micelles. No therapeutic effect from the nanodroplet/ultrasound combination was observed without the drug, indicating that therapeutic effect was caused by the ultrasound-enhanced chemotherapeutic action of the tumor-targeted drug, rather than the mechanical or thermal action of ultrasound itself. The same effect was observed with breast and pancreatic xenograft model tumors. In addition of releasing the drug from the nanocarriers, recent studies have reported that ultrasounds can enhance the uptake of chemotherapeutic drugs by the cell, both *in vitro* and *in vivo*.[354] In this regard, not only the release of drug, but also the cell membrane perturbation may be important parameters to reduce the tumor growth *in vivo*.

Sonosensitivity could also be introduced to liposomes, either by encapsulation of air bubbles or by altering the lipid composition of the liposome membrane.[355] Several acoustically active liposome formulations have showed an enhanced antitumor activity *in vivo* when applied in combination with ultrasound.[356,357]

In addition, AuNPs have been recently reported as efficient photoacoustic contrast agents, and therefore, they appear as promising candidates as acoustically active nanosystems for drug delivery.[358–360]

4.4. Redox

It is well known that the cellular cytosol has a more reducing environment than the extracellular environment. Glutathione (GSH), the most abundant reducing agent in most cells, has a typical intracellular concentration of about 10 mM, whereas its concentration is only about 0.002 mM in the cellular exterior. It should also be noted that endosomal compartment is also redox-active in which the redox potential is modulated by a specific reducing enzyme gamma-interferon-inducible lysosomal thiol reductase (GILT) in the copresence of a reducing agent such as cysteine (but not GSH) (Figure 14.18).[361] These features have been exploited to design *redox responsiveness drug delivery nanosystems* for specific intracellular release of the therapeutic cargo.

Several nanocarriers, such as micelles, NPs, nanocapsules, polymersomes, and nanogels, have been efficiently used to deliver different therapeutic agents in intracellular reducing conditions.[361] As a general strategy,

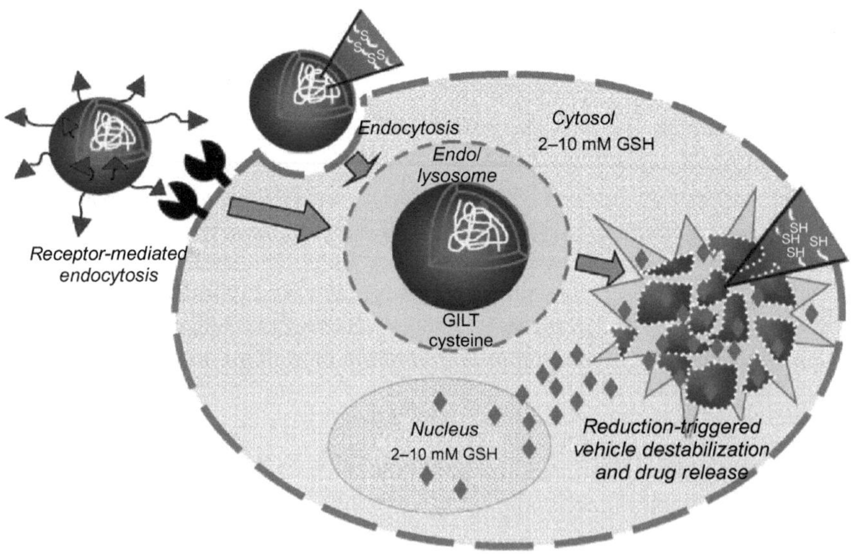

FIGURE 14.18 Schematic illustration of the intracellular trafficking pathway of redox-responsive nanovehicles including steps of cellular internalization, endosomal escape, reduction triggered vehicle degradation, and drug release. The redox potential of the cytosol is primarily determined by GSH/GSSG, while that of the endo/lysosome is modulated by a specific reducing enzyme GILT and cofactor cysteine. Redox-responsive nanovehicles may also be partially degraded in the endo/lysosomal compartments. Reprinted from Ref. 361. Copyright 2011 with permission from Elsevier.

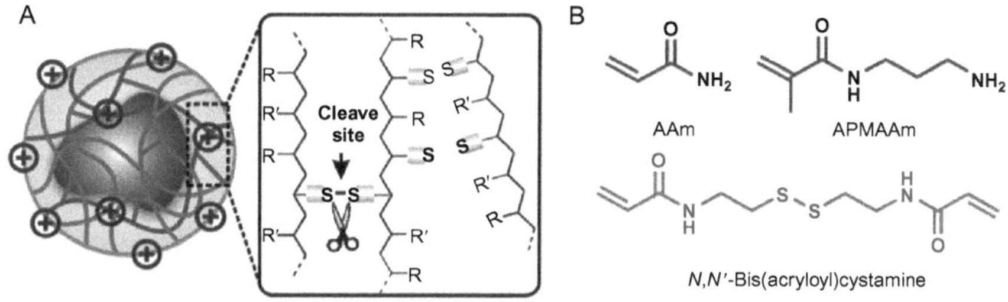

FIGURE 14.19 Example of redox-responsive protein nanocapsules. (A) Schematic of protein nanocapsules with redox-responsive polymeric matrix (R and R′ represent different monomers' moieties) and (B) chemical structures of acrylamide (AA) and *N*-(3-aminopropyl) methacrylamide (APMAAm) monomers and cross-linker for nanocapsules. Reprinted from Ref. 365. Copyright 2011 with permission from Elsevier.

thiol-reducible disulfide bonds have been used to introduce redox responsiveness (Figure 14.19).[362–365]

As an example, disulfide cross-linked (methacrylic acid) (PMASH) hydrogel capsules developed by LbL assembly were applied for *in vitro* and *in vivo* delivery of proteins and peptides for vaccine applications.[366] These capsules were stable in oxidizing conditions, but rapidly disassembled in reducing environments similar to those inside specialized antigen presenting cells such as dendritic cells. Using robust transgenic mice models and ovalbumin (OVA) as a model vaccine, it was reported that OVA encapsulated within the PMASH capsules resulted in at least sixfold greater proliferation of OVA-specific CD8 T cells and 70-fold greater proliferation of OVA-specific CD4 T cells *in vivo* compared to the equivalent amount of OVA protein administered alone.[367]

As oxidation is an almost ubiquitous feature of inflammatory processes, the presence of a wide variety of highly oxidizing compounds, such as superoxide anion, hydrogen peroxide, hypochlorite (ROS), nitric oxide, or peroxynitrite (reactive nitrogen species), is a common hallmark of many diseases. These include neurodegenerative (e.g., Parkinson's Alzheimer), inflammatory bowel diseases (e.g., Crohn's or ulcerative colitis), atherosclerosis, or inflammatory lung diseases (asthma, fibrosis), among others.[368–370] Based on this principle, the development of nanocarriers that respond to the presence of oxidants has been reported. For instance, polysulfide-based nanocarriers exhibit sensitivity to oxidation, so they could be used for localized delivery of anti-inflammatory drugs.[371]

4.5. Enzymes

The presence of enzymes which are specific to an organ or disease state may be useful to regulate the release of drugs. Therefore, compared to redox- and pH-sensitive nanocarriers systems, *enzyme sensitive* ones are considered advantageous because drug release depends on the presence of a particular enzyme, which if only present in the diseased cells will limit drug release to the target cells. The most used enzymes for achieving an enzymatic triggered drug delivery are proteases, as in many diseases their expression are found deregulated when compared with healthy states. Cancerous cells are known to overexpress some proteases such as the cell-surface protease hepsin (prostate tumors),[372] cathepsin B (lung, ovarian, and colorectal tumors),[373] cathepsin D (breast cancer),[374] or MMPs (associated with tumor metastasis, inflammatory, and vascular diseases),[375,376] among others. Although enzyme-responsive drug delivery platforms have been widely explored in the macroscale, few enzyme-responsive nanocarriers have been reported.[7]

Enzyme-cleavable peptides can be used for cross-linking the delivery nanosystem or for coupling the therapeutic agent; the drug release would be then activated by a particular enzyme that is expressed or overexpressed in a specific diseased cell or environment. For instance, a novel enzyme-responsive silica–iron oxide magnetic nanoparticle complex has been developed to selectively induce intracellular drug delivery in tumor cells with high cathepsin B expression.[377] In addition, tumor-environment-responsive ferritin-based nanocarriers that could evolve its surface properties upon sensing the tumor-associated MMP-2 have been reported.[378] In the presence of MMP-2 activity, a NP's segment is cleaved by the enzyme, resulting in the initiation of aggregation and, consequently, in an enhancement of T2 relaxivity for MRI. In another example, PEG-conjugated hyaluronic acid NPs were investigated as carriers for anticancer drugs including Dox and camptothecin (CPT). *In vitro* drug release tests showed that these NPs rapidly release both drugs in the presence of hyaluronidase-1 enzyme, which is abundant in the intracellular compartments of cancer cells.[379] Furthermore, an *in vivo* biodistribution study using these NPs loaded with CPT indicated that they selectively accumulated into tumor sites after systemic administration into tumor-bearing mice, and no significant increase in tumor size was observed for at least 35 days (Figure 14.20).

In addition to enzyme-cleavable peptides, different "saccharide" derivatives have been used to introduce enzyme responsiveness to nanomaterials.[380] This is the case of novel silica mesoporous NPs whose pore outlets were functionalized with a series of hydrolyzed starch products.[381] The release of Dox due to the presence of different lysosomal glycosidases enzymes was shown. Furthermore, a decrease in the cell viability of HeLa cells was observed due to the endocytic internalization of the NP, which triggered the enzyme-dependent opening of the saccharide molecular gate and the consequent release of the cytotoxic agent.

It is important to underline that nanocarriers can also be designed to be *responsive to several different stimuli*, for example, temperature- and redox-sensitive NPs,[382] or temperature, pH, and redox-sensitive block copolymer micelles.[383]

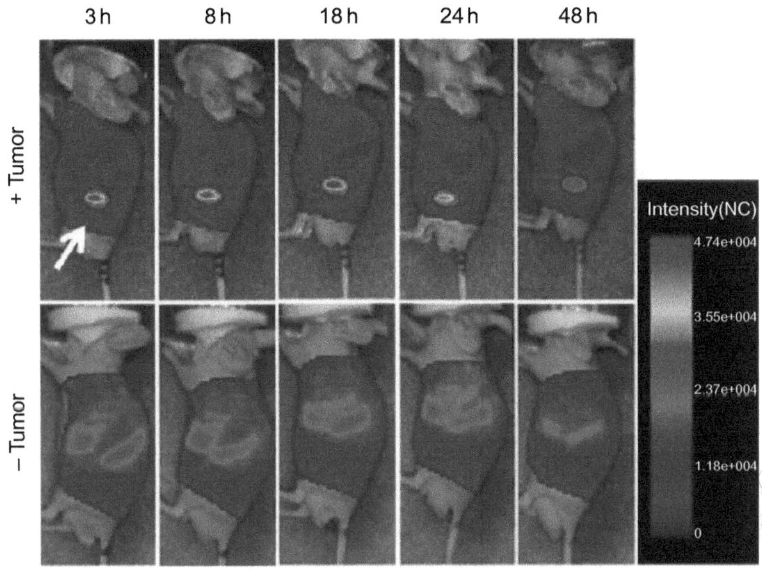

FIGURE 14.20 *In vivo* noninvasive fluorescence images of CPT-PEG-hyaluronic acid NPs in tumor-bearing mice and normal mice. (A) Whole body images of athymic nude mice with and without squamous cell carcinoma tumors after intravenous injection of CPT-P-HA-NPs. The arrow indicates the tumor site. Reprinted with permission from Ref. 379. Copyright 2011 American Chemical Society. (For color version of this figure, the reader is referred to the online version of this chapter.)

4.6. Temporal Control

As exposed above, the development of "smart" nanomaterials that are able to deliver a therapeutic agent based on environmental causes or remote stimulus has introduced new methods to trigger the release of drugs and localizes it within a particular site. However, a smart drug delivery nanosystem should face several hurdles to survive *in vivo*, deliver the cargo, release the drug into the target cells, and match the desired kinetics of the release. Therefore, a step forward in the development of responsive *nanocarriers is to introduce the capability to temporarily adapt* their size, shape, surface chemistry, wettability, adhesive properties, etc., to surrounding environments. This would not only allow triggering the release of the therapeutic cargo or their escape from the endocytic pathway but may also help to overcome several other biological barriers that affect their therapeutic effectiveness.[8,384]

As previously mentioned, nanocarriers that exhibit fast and reversible swelling behavior can be used to break up endosomal membranes. Other possible application of temporal control over the size of nanocarriers is to prevent nanocarrier reentry into the circulation after extravasation to the tumor tissue via EPR. This could be achieved by trapping them by mean of a site-specific

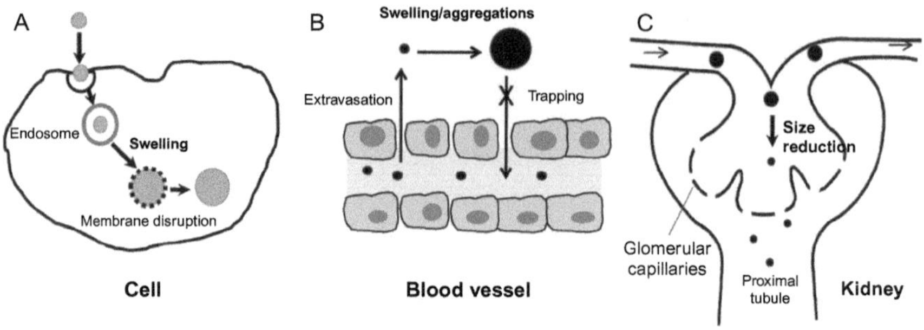

FIGURE 14.21 Possible applications of temporal control over size in drug delivery. (A) Endosomal escape by size expansion; increase in particle size within the endosomes may physically disrupt the endosomal compartment, leading to cytosolic delivery. (B) Caging of particles by size expansion; once particles enter the target area by extravasation, increase in size over the gap junction between adjacent endothelial cells may prevent the particles from reentering the circulation, thus effectively trapping the particles in the target area. (C) Renal excretion by size reduction; timely elimination of particles may be possible by decreasing particle size to <10 nm, which can be filtered through fenestrations in glomerular capillaries of the kidney. Reprinted from Ref. 8. Copyright 2011 with permission from Elsevier.

size expansion by swelling of the nanocarriers or by inducing their aggregation. Besides, timely renal excretion of nanocarriers may be possible by decreasing particle size to <10 nm (Figure 14.21).

Although the validity of these concepts has not been demonstrated *in vivo*, few promising results have been reported indicating the feasibility of triggering changes in the size of the nanocarrier. As an example, nanoparticles composed of bio-based photosensitive poly(4-cinnamic acid)-*co*-poly (3,4-cinnamic acid) (PCA) have been found to exhibit reversible size changes upon irradiation of different wavelengths of UV. The particle size decreased due to photo-cross-linking of PCA with UV irradiation at 280 nm and reversibly increased when using 254 nm by photocleavage (Figure 14.22). Moreover, these nanoparticles showed interesting size change behaviors during their hydrolytic degradation depending on the degree of cross-linking (Figure 14.22).[385]

Furthermore, it was also recently established *in vivo* that temporal control over the surface chemistry of nanocarriers allows the masking of ligands for targeting or cell internalization during circulation to prevent nonspecific binding and RES clearance and unmasking only when they reach target sites (Figure 14.23).[386–388]

Temporal shape switching may also have interesting applications in drug delivery as it would allow the exploitation of the advantages of using both spherical (e.g., better cell internalization ratios) and nonspherical nanocarriers (e.g., lower macrophage uptake, longer circulation in the blood, higher accumulation in target tissues).[389,390] In this respect, an excellent review has

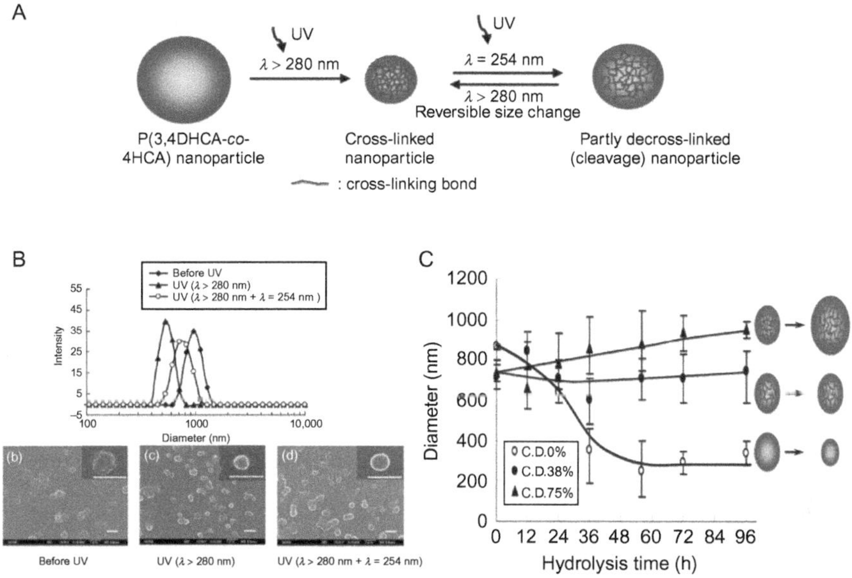

FIGURE 14.22 (A) Schematic representation of size change behavior of poly(4-cinnamic acid)-*co*-poly(3,4-cinnamic acid) NPs with UV irradiation; chemical structure of UV-induced [2+2] cycloaddition formation (cross-linking) and deformation (cleavage). (B) Size distributions of the nanoparticles in THF before and after UV irradiation at $\lambda > 280$ nm for 30 min and subsequent irradiation at $\lambda = 254$ nm for 30 s by dynamic light scattering (DLS) measurements. Scanning electron microscopy images of the nanoparticles before UV irradiation, after UV irradiation at $\lambda > 280$ nm for 30 min, and after subsequent irradiation at $\lambda = 254$ nm for 30 s are shown. The scale bars are 1 μm. (C) Diameter changes of the remaining nanoparticles with different cross-linking degrees during hydrolysis in alkaline buffer solution (NaOH/NaHCO3) at pH 12 at 37 °C for predetermined times by DLS measurements. CD means the cross-linking degree. Reprinted with permission from Ref. 385. Copyright 2008 American Chemical Society. (For color version of this figure, the reader is referred to the online version of this chapter.)

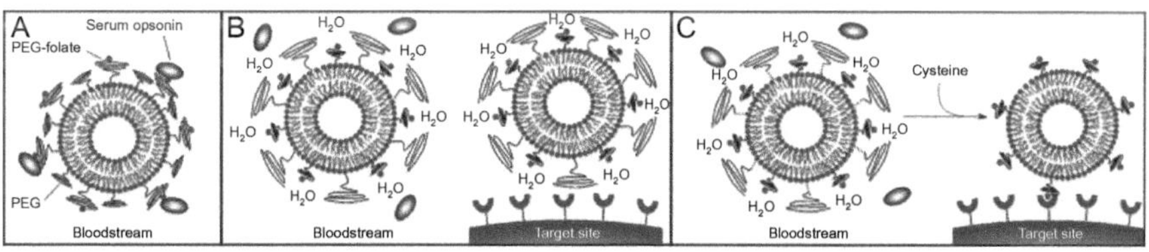

FIGURE 14.23 Schematic depicting folate receptor-targeted nanocarrier options. (A) When attached to PEG chains longer than those incorporated for RES evasion, folate-targeting ligands are readily recognized by the RES resulting in accelerated clearance and reduced extravasation to tumor. (B) Longer PEG chains conceal folate from the RES, but hinder receptor-mediated uptake by target cells overexpressing the folate receptor. (C) Long cleavable PEG–phospholipid conjugates mask folate during circulation to enable passive targeting to tumor but may be detached at the target site. Resultant exposure of folate enables targeting to cells overexpressing the folate receptor. Reprinted from Ref. 386. Copyright 2009 with permission from Elsevier.

recently discussed why the design of more complex "adaptive" nanosystems, whose properties can be changed in real time, either by self-evolution over time or in response to a stimulus externally controlled or not, is a promising strategy to address the complex biological hurdles that hamper the therapeutic efficiency of the simplest "nonadaptive" nanocarriers.[8]

5. EXAMPLES OF NANOSYSTEMS FOR DRUG DELIVERY

Over the past 10 years, a library of nanotechnology-based therapeutic agents, also known as nanodrugs or nanomedicines, has been extensively explored to improve the pharmacokinetic, biodistribution, and therapeutic properties of drugs.[291] The nanomedical approaches developed cover a wide variety of materials and sizes and also diverse shapes, such as CNTs, organic nanostructures (e.g., liposomes, dendrimers, and polymer-based NPs), and inorganic NPs (e.g., metal, silica, and semiconductor QDs) (Figure 14.24). Each of these nanocarriers has distinct size ranges in the nanoscale and different physicochemical features.

A review of the literature highlights that compared with the rapid progression in the development of drug delivery systems using organic nanostructures relatively fewer advances have been made for inorganic-based ones.[391] Along

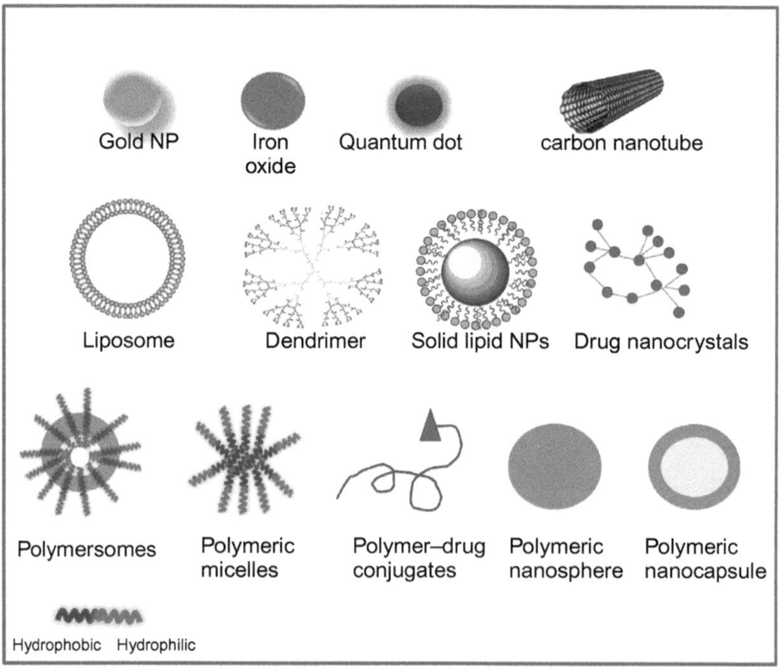

FIGURE 14.24 Scheme showing the different nanocarriers revised in this chapter. (For color version of this figure, the reader is referred to the online version of this chapter.)

these lines, the FDA approved biodegradable polymeric NPs, such as poly (lactic acid) (PLA) and PLGA for human use. Moreover, some organic nanocarriers are already on the market. This is the case of Abraxane®, the first polymeric NP-based product, which consists of albumin-bound paclitaxel NPs. However, despite the slower development of drug nanocarriers based on inorganic nanostructures, recent progress in the field of nanotechnology and nanofabrication has led to the production of a wide variety of inorganic nanosystems with great potential for drug delivery purposes. In this regard, the advantages and disadvantages of several organic and inorganic nanostructures for drug delivery are highlighted below.

5.1. Inorganic Nanoparticles

These NPs are generally composed of an inorganic core, and an organic shell located in the most external part of the particle. Depending on the inorganic core composition, shape, size, and crystallinity, these molecules have a number of tunable physical properties, such as optical absorption (e.g., metallic NPs), fluorescence (semiconductor QDs), and magnetic moment (e.g., iron oxides). Conversely, the organic shell not only confers stability to these nanomaterials in biological media but can also be engineered to provide useful reactive groups for its further functionalization with different biomolecules in order to achieve a particular biological functionality, such as active targeting of tissues or cells. Depending on the type of organic shell, a variety of chemical transformations can be used to introduce functional groups onto the surface of these particles, such as carboxylic acids, thiols, amines, and alcohols.[392] In addition, depending on the chemical nature of the therapeutic agent, it can be stored by physical adsorption, incorporated during the synthesis of the NP, or covalently attached to a reactive surface group. Accumulation of high drug concentrations can be achieved using these NPs because of the surface-area-to-volume ratio (Rsv) of NPs, which can be as high as 3.3 nm^{-1} for a 5-nm perfectly spherical shape NP.[305] For example, ~100 ligands can be covalently conjugated to a AuNP with a core diameter of 2 nm.[393]

Inorganic NPs have several advantages over organic ones for the purpose of drug delivery. First, they are easier to prepare with a defined size and a very narrow size distribution. More interestingly, they often exhibit multiple functions that are useful in medicine, for example, as heat generation and contrast agents. In contrast, organic NPs serve only as drug reservoirs.[391,394,395]

For example, as a result of their intrinsic superparamagnetic character, *iron oxide NPs* are routinely valuable in medicine as efficient transversal relaxation (T2) MRI contrast agents.[222,396] However, the next generation of active targeting MNPs, currently being investigated, has the potential to offer significantly improved tumor detection and localization. For example, it has been recently demonstrated that specific accumulation of chlorotoxin-targeted iron oxide nanoparticles in 9L glioma flank xenografts resulted in significantly contrast enhancement of tumors in comparison to nontargeted control nanoparticles.[397]

Furthermore, under the influence of an alternating external magnetic field, magnetic NPs can release localized heat. This heat can be used to trigger drug delivery (as reviewed in this chapter) or it can be used to destroy target cells.[305] The latter is a minimally invasive procedure called magnetic hyperthermia and is already available to patients at the NanoTherm® therapy center at the Charité-Universitätsmedizin (Berlin, Germany), where several clinical trials are being conducted by MagForce Nanotechnologies AG (http://www.magforce.de/en/unternehmen/ueber-uns.html).[398,399] The treatment consists in injecting the MNPs directly into the tumor tissue with the aim of its thermal destruction or sensitization to radiotherapy after being exposed to an alternating magnetic field.[398,400,401] Depending on the degree of temperature increase, hyperthermia treatment can be classified into different types. When temperature reaches >46 °C (up to 55 °C), the process is named thermo-ablation as cells die by necrosis, coagulation, or carbonization. When the temperature generated is in the range of 41–46 °C (moderate hyperthermia), cells undergo heat stress, which finally results in the activation and/or initiation of many intra- and extracellular degradation mechanisms known as apoptosis, which causes cell death. The generation of lower temperatures (<41 °C) is known as diathermia and is used for the treatment of rheumatic diseases in physiotherapy.[402]

However, the use of heat for thermal ablation has many limitations for *in vivo* applications compared to its use for drug delivery, as very large amounts of heat-generating nanomaterials must be delivered to tumors.[4] Moreover, thermal convection, caused by strong blood perfusion, reduces the temperature of the tumor and hence results in a decrease in hyperthermia efficiency in those tumors located in regions with high perfusion (e.g., within liver, lung, and kidney).[402] In addition, while this localized therapy can be used for treatment of primary tumors and/or debulking prior to surgical removal of a large tumor, it could not be used to treat disseminated metastatic disease (e.g., breast, prostate, colon cancers) responsible for most cancer-related mortality.[5]

Like magnetic ones, AuNPs work as efficient agents for targeted cell ablation and controlled drug release in response to light irradiation.[403] The potential of both, magnetic and AuNPs, to achieve active controlled drug release triggered by exogenous stimuli is a very interesting advantage over organic drug nanocarriers. In fact, in most organic nanosystems, the release of the stored drug is regulated passively by diffusion, particle erosion, particle degradation, or polymer swelling, among others.[391]

In the case of AuNPs, the electric field of light causes the collective oscillation of the conductive-band electrons on the surface of the particles, which is known as surface plasmon resonance (SPR). The intrinsic frequency at which the delocalized electrons in a NP resonate, giving rise to its SPR peak, can be easily tuned from the visible (390–750 nm) to near-infrared (NIR) wavelengths (800–1000 nm) by modifying the NP aspect ratio.[404] When using light radiation with a frequency strongly overlapping the SPR absorption band of the particle, the light absorbed is efficiently converted into heat by rapid electron–phonon and phonon–phonon processes. This highly proficient and localized light-to-heat conversion by AuNPs can be used for triggering drug release or for photothermal therapy. In addition, these unique optical properties of these molecules are also valuable for optical imaging. Optical imaging is a highly sensitive and inexpensive technique, which can yield high-resolution images and is used endoscopically for minimally invasive approaches.[405,406]

Clinical applications for optical imaging and photothermal therapeutics, however, are limited depending on whether the SPR wavelength band of the NP is within the visible or the infrared region of the light spectrum. When the SPR band is in the former, the clinical application of AuNPs is restricted by the small depth of penetration, which makes the imaging or treatment of deep target tissues much more difficult when compared to MRI and magnetic hyperthermia. However, tissue penetration can be improved (up to few centimeters) by minimizing photon absorption by tissue components, which is achieved by tuning the SPR band from the visible to NIR wavelength.[2,407] In this regard, several reports show the viability of AuNPs with diverse shapes (e.g., nanoshells, hollow gold nanospheres, nanorods, nanocages) to achieve targeted cell ablation in response to NIR irradiation.[408,409] For instance, intravenously administrated PEG-coated gold nanoshells have been successfully applied for photothermal ablation therapy in a mouse model of colon carcinoma, without observing toxicity in the surrounding healthy tissue.[403]

Furthermore, through smart design, it has been shown that is possible to combine targeting, therapeutic, and diagnostic functions within a single multimodal nanoshell complex.[410] Similarly, PLGA-Au half-shell NPs have been used to simultaneously deliver Dox (chemotherapy drug) and heat to tumor sites.[411] The synergistic effect of this combined treatment resulted in higher therapeutic efficacy and shorter treatment times. Animal studies have also shown that hyperthermia has a synergistic effect with other therapeutic approaches for cancer.[2] In this sense, PEGylated AuNPs (27 nm) functionalized with recombinant human tumor necrosis factor (rhTNF-α) are currently ongoing early clinical trials as anticancer agent (CYT-6091). TNF-α is a potent antivascular cytokine at high doses and can be used to destroy tumor vasculature. However, its clinical use is limited by its toxicity.[412] It was found that although soluble TNF- and CYT-6091 both reduced tumor perfusion, CYT-6091 could be administered at three-folder higher dose. Besides, combination of CYT-6091 with heat resulted in significant tumor regression when compared to either heating or using CYT-6091 alone.[413] Exploiting the photothermal properties of AuNPs is therefore promising. A reflection of this prospect is a human clinical trial on gold nanoshells for photothermal ablation of recurrent (or refractory) head and neck cancer, which is currently in progress (Nanospectra Biosciences, Inc.).

In addition, due to the efficient conversion of light to heat that is transferred to the environment, AuNPs are able to create higher levels of photoacoustic signals in the NIR region and are therefore being investigated as contrast enhancement agents for photoacoustic tomography (PAT).[414] PAT is a hybrid imaging technique that combines the good soft-tissue contrast of an optical imaging method with the high spatial resolution of ultrasound imaging modalities. In this sense, it is capable of providing optical contrast of an object 5 cm deep in biological tissue with a spatial resolution of approximately 0.2 mm.[415] As an example, it was recently demonstrated the feasibility of using gold nanocages as lymph node tracers for photoacoustic (PA) imaging and mapping of a sentinel lymph node up to 3.3 cm below skin surface (Figure 14.25).[416]

Other applications that could also be combined to achieve active controlled drug release include their use as contrast enhancers for computer tomography and X-ray absorbers for cancer therapy.[417] Further support for the use of AuNPs as carriers for drug delivery comes from evidence that their

FIGURE 14.25 Depth capability of sentinel lymph nodes mapping with gold nanocages. Photoacoustic (PA) images taken (A) before and (B–E) after the injection of AuNCs for (B) 28 min; (C) 126 min with a layer of chicken breast tissue placed on axillary region (imaging depth 10 mm); (D) 165 min with the second layer of chicken breast tissue (imaging depth 21 mm); (E) 226 min with the third layer of chicken breast tissue (imaging depth 33 mm). (F) PA B-scan with 20 times signal averaging, showing the SLN located 33 mm deep. (G) The amplitude variations of PA signals over imaging depths. BV, blood vessel. Adapted with permission from Ref. 416. Copyright 2009 American Chemical Society.

functionalization with multiple ligands can be achieved in a one-pot synthesis as the gold surface reacts with thiol groups to form strong bonds.[418,419] As an example, AuNPs multifunctionalized with glucose and folic acid (dual-ligand AuNPs) have been synthesized to explore the benefits of multivalent interactions to enhance cancer targeting specificity.[420]

Semiconductor nanocrystals, referred to as *QDs*, are also being studied as drug carriers as a result of their fluorescent properties, which also allow their combined use as imaging agents. QDs absorb and reemit light in function of their sizes and chemical composition, and therefore, their fluorescence is tunable from the ultraviolet to the NIR regions of the electromagnetic spectrum.[421] Furthermore, they display superior fluorescent properties (10–50 times larger molar extinction coefficients) with less photobleaching compared with conventional organic dyes and fluorescent proteins. In addition, a broad excitation coupled to a narrow and symmetric emission spectrum renders multicolored QDs. Therefore, QDs can be used as sensitive fluorescent probes for screening biological markers of disease in fluids, as specific labels for classifying tissue biopsies, and as high-resolution agents for multiplexed, quantitative, long-term and real-time medical imaging, and detection.[422–426] In this regard, it has been experimentally demonstrated that QDs can be used, both *in vivo* and *in vitro*, for real-time imaging of cancer cells.[216,427,428]

Besides the function of fluorescence imaging, QDs can be used for drug transporting, as both small molecule hydrophobic drugs and hydrophilic therapeutic agents, including siRNA and antisense oligodeoxynucleotide, have been successfully conjugated to QDs via either noncovalent or covalent bonds.[429] For instance, QDs coated with a polymer, where siRNA was adsorbed by electrostatic attraction, allowed real-time tracking of siRNA delivery *in vitro* and an improvement in gene silence efficiency, when compared with other transfection agents.[430]

However, QD-based drug delivery systems are not as robust and well established as other drug nanosystems. Therefore, a more common approach is the utilization of QDs as fluorescent markers for tagging conventional drug carriers. With this goal, QDs can be used to elucidate the pharmacokinetics, pharmacodynamics, and even the drug release properties of drug candidates, which could provide valuable information for the design principles for drug carriers. In fact, other techniques that are used to analyze the drug delivery profiles, such as chromatography, are difficult to implement in living subjects; hence, it is hard to acquire real-time drug release information in real conditions.[431] Fortunately, QDs can be used to solve these drawbacks, using the fluorescence resonance energy transfer (FRET) mechanism.[432] In this process, a donor QD, initially in its electronic excited state, may transfer energy to an acceptor chromophore when they are in closed vicinity, resulting in the fluorescence quenching of QDs and enhancement of the chromophore. The efficiency of FRET is dependent on the intermolecular separation, making it useful over distances comparable to the dimensions of biological macromolecules. In this regard, Bagalkot *et al.* reported a bi-FRET system, where QDs were functionalized with RNA aptamers which specifically bind to prostate-specific membrane antigen (PSMA).[433] In addition, Dox, an antineoplasic drug with fluorescent properties, was intercalated within the RNA aptamers. The resulting QDs were nonfluorescent as the fluorescence of QD was quenched by the Dox absorbance, and in addition, the fluorescence of Dox was quenched by double-stranded RNA aptamer (Figure 14.26). When the complexes were taken up by the PSMA positive cells *in vitro*, the slow release of Dox resulted in recovery of both QDs and Dox fluorescence, and therefore, cells could be imaged and the delivery of Dox sensed at the same time.

QDs are also being explored as potential delivery systems for photodynamic therapy (PDT).[434] In PDT, cells are treated by applying a photosensitizing (PS) drug, which is activated by light, leading to an energy or electron transfer to oxygen or other molecules. During PDT, reactive oxygen intermediates (ROI) are generated, which can react with and damage molecules in cell organelles, leading to cell damage, mutation, death, and photooxidation

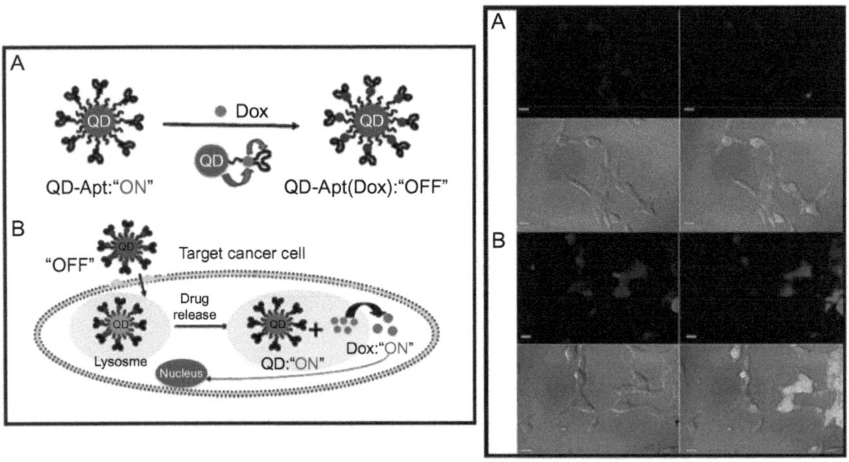

FIGURE 14.26 (Left) (A) Schematic illustration of QD–aptamer-Dox (QD–Apt(Dox)) Bi-FRET system. In the first step, the CdSe/ZnS core–shell QD is surface functionalized with the A10 PSMA aptamer, and the intercalation of Dox within the aptamer results in the formation of the QD–Apt(Dox). This complex is completely quenched, resulting in the "OFF" state. (B) Schematic illustration of specific uptake of QD–Apt(Dox) conjugates into target cancer cell through PSMA mediate endocytosis. The release of Dox from the QD–Apt(Dox) conjugates induces the recovery of fluorescence from both QD and Dox ("ON" state), thereby sensing the intracellular delivery of Dox and enabling the synchronous fluorescent localization and killing of cancer cells. (Right) Confocal laser scanning microscopy images of PSMA expressing cells after incubation with QD–Apt (Dox) conjugates for (A) 0 h and (B) 1.5 h. Dox and QD are shown in red and green, respectively, and the lower right images of each panel represents the overlay of Dox and QD fluorescent. The scale bar is 20 μm. Adapted with permission from Ref. 433. Copyright 2007 American Chemical Society. (For interpretation of the references to color in this figure legend, the reader is referred to the online version of this chapter.)

of cell constituents. The quality of a PS drug for PDT depends on its efficiency for energy and or electron transfer to molecular oxygen and the subsequent production of ROI. QDs that are directly photoactivated are not effective for ROI production. However, if QDs are conjugated to PS drugs, the energy transfer processes from QDs to PS drugs and from PS drugs to oxygen provides a higher efficiency for ROI production.[435]

It is important to note that most examples of inorganic NPs designed for drug delivery applications are filled nanoparticles. As their dimensional characteristics are usually below 100 nm, they present a large Rsv and therefore high drug-storing capacity. However, *hollow inorganic nanostructures with porous cores* raise the possibility of loading a greater amount of therapeutic agent than filled particles. When the inorganic core and/or the organic shell consist of materials such as silica, alumina, or calcium phosphate, porosity can be engineered.[436,437] In addition, porous hollow AuNPs, referred to as nanocages, have recently been synthesized.[438] While porous particles not only increase the drug loading capacity, they also provide a physical encasement to entrap and protect the drug cargo from degradation or denaturation.

Despite all the advantages of inorganic NPs described above, their slow rate of biodegradation or lack of biodegradability raises *safety issues*, especially for applications that require long-term administration.[395,439] For example, in the case of *QDs*, toxicity depends on the chemical composition of their inorganic core. These nanocrystals are usually made of elements from groups III to V (e.g., InP, In from group III, P from group V), or from groups II to VI of the periodic table (e.g., cadmium selenide [CdSe], Cd from group II, Se from group VI).[440] Given their higher quantum yield, the latter are used preferentially over the former. However, they are potentially toxic as a result of the release of ionized heavy metals.[441] These elements could be potent toxins, neurotoxins, and/or teratogens depending on the dosage, complexation, and accumulation in the liver and nervous system.[442] In this case, the selection of an adequate organic shell is critical in order to provide a strong physical barrier between the core and the external environment, thereby reducing their potential toxicity.[421] However, the degradation of coatings as a result of exposure to acidic or oxidative environments, such as those present in endosomes, can result in severe toxicity.[443–447]

It has been shown that QDs biodistribution is highly dependent of their surface functionalization, although to date, not many studies have been conducted to determine the QDs biodistribution after systemic administration. For instance, BSA-conjugated QDs were predominantly cleared by RES in rats while QDs with a zwitterionic cysteine coating showed a predominantly kidney elimination.[36,448] Successful use of QDs for drug delivery will depend on identifying a therapeutic niche where their unique properties can be exploited in an appropriate risk–benefit setting.[5]

In the case of *silica NPs*, critical questions such as their bioaccumulation after administration and their potential toxicity remain unanswered. A number of studies have addressed the fate of these NPs in the body and reported them as biodegradable. However, further research is required in this direction.[449] Also, it has recently been described that the pore architecture of silica NPs greatly influences their biocompatibility and should therefore be carefully designed.[450]

Conversely, *gold and superparamagnetic iron oxide NPs* are generally recognized to have low *in vivo* toxicity.[447] Magnetic NPs can naturally be broken down resulting in the release of ferric iron which can then participate in the normal iron metabolism, and for this reason, they were initially considered to be noncytotoxic. In fact, dextran-coated formulations of iron oxide NPs have been clinically approved as contrast agents for MRI.[451] This advance is encouraging for researchers working in the design of novel and smart drug delivery systems based on inorganic nanomaterials. However, intravenous administration of magnetic NPs may lead to accumulation in a targeted organ and may potentially generate iron overload. Such an excess could result toxic as high free iron levels can cause an imbalance in homeostasis, thereby leading to DNA damage, oxidative stress, and inflammation.[452,453]

In the case of AuNPs, as Au is a noble and inert metal with some therapeutic and even medicinal value, AuNPs have long been considered to be nontoxic. However, the induction of oxidative stress upon their cell internalization has not been clearly established. In fact, a recent study reported that AuNPs caused oxidative stress and upregulation of stress response genes in human lung fibroblasts *in vitro* after their internalization.[454]

In addition, inorganic NPs can interfere with the delicate balance of cellular homeostasis and therefore alter complex intracellular signaling pathways, resulting in a cascade of possible effects. However, to date, this has only scarcely been investigated and contradictory results have been reported. In order to provide more comparative and reliable data on the cytotoxic profile of these nanocarriers, a thorough characterization of the NPs is required. Standardization is also necessary in terms of the assays that are used to study their toxicity.[455] On the basis of these reports, further studies about the fate, accumulation, and corresponding toxicity of inorganic NPs are essential in order to assure their safe application in drug delivery.[447]

5.2. Organic Nanosystems

A wide range of nanostructures based on proteins (e.g., albumin), poly(amino acids), polysaccharides (e.g., dextran, chitosan), surfactants, dendrimers, solid drug crystals (nanocrystals), synthetic polymers, and other organic material have been proposed for use as particulate or macromolecular drug delivery systems. The organic material preferentially chosen to prepare organic nanostructures is biodegradable to nontoxic products and biocompatible. Thanks to these features, these nanostructures are to date those most commonly used for drug delivery, particularly in the case of formulations that require chronic dosing. Like inorganic NPs, the surface functional groups of organic nanosystems can be easily engineered via chemical modification. Moreover, they are suitable for the entrapment and delivery of a wide range of therapeutic agents. Depending on their chemical composition, not only can they encapsulate hydrophilic and/or hydrophobic low-molecular-weight drugs but can also be used for more complex therapeutic agents, such as proteins, peptides, genes, oligonucleotides, and siRNAs.

Undoubtedly, one of the organic nanostructures that has received most attention are *liposomes*. These molecules consist of aqueous compartments surrounded by one or more lipid bilayers comprising amphipathic phospholipids (mainly phosphatidylcholines). Thus, water-soluble drugs can be entrapped in the aqueous phase, and lipophilic or amphiphilic ones can be inserted into the lipid bilayers. In addition, the liposome diameter can be adjusted from a few microns down to about 25 nm depending on the preparation technique selected.[456] Liposomes are widely reported to enhance the half-life of various therapeutic agents. For example, the half-life of soluble Dox in blood is approximately 5 min, whereas its half-life is increased up to 20–30 h when incorporated in liposomes.[447] Another advantage of liposomes is that the phospholipids used for their preparation are also present in natural cell membranes. This feature allows liposomes to fuse with the plasmatic cell membrane of target cells, thereby facilitating drug targeting at a subcellular level.[457] Moreover, their composition can be engineered to achieve a temperature- or pH-sensitive drug carrier. Many therapeutic agents are already on the market in liposomal formulations, these presenting an improved dose/effect ratio and less adverse reactions compared to the free substances at the same concentration.[2,458] Liposomal Dox, an FDA-approved drug, has been shown to inhibit both taxane- and platinum-sensitive and resistant/recurrent ovarian cancers.[459]

However, while liposomes have been extensively used in cosmetics, they have not yet made a significant medical impact. This observation could be attributed to their limited biological stability as they undergo destabilization by interaction with serum proteins as well as clearance by the RES system. Various strategies have been or are currently being developed to extend the stability and circulatory residency times of these molecules and thus allow a better control of drug delivery. The head groups of phospholipids can be modified to introduce not only functional groups that facilitate conjugation to antibodies or other ligands but also polymerizable moieties to improve their stability and *in vivo* circulation times.[447] The glycerol backbone of phospholipids can be modified with the same purpose by converting carbonyl ester bonds into either ether or carbamyl esters.[302] Moreover, phospholipid composition can be altered to provide adequate stability. In this regard, recent liposome preparations have been synthetized using PEGylated phospholipids. PEG is a biologically inert polymeric coating that reduces immunogenicity and also decreases the phagocytosis of NPs by the RES system.[460] Therefore, PEGylation helps to bypass RES, thereby promoting drug accumulation in the target organ.

Other inherent problems of liposomes that hamper their clinical applications are as follows: (i) unsuitability for oral administration routes, (ii) low encapsulation efficiency, (iii) leakage of water-soluble drugs in the presence of blood components, (iv) aggregation and poor storage stability, (v) issues related to batch-to-batch reproducibility, (vi) difficulties concerning scaling up for clinical evaluation, and (vii) toxicity.[2] Furthermore, toxicity can occur mainly due to liposome composition, particle size, or charge.[461,462] For example, cationic liposomes interact with serum proteins, lipoproteins, and the extracellular matrix of tissues, leading to the aggregation or release of the therapeutic agents before reaching the target cells and thus potentially causing systemic toxicity.[463] Moreover, given their net positive charge, cationic liposomes can cause liver damage, which thus limits their repeated dosing.[464] Therefore, the lipid composition and net charge of liposomes should be carefully selected to minimize adverse side effects. In addition, it has been reported that some therapeutic liposomes and lipid excipient-based anticancer drugs are recognized by the immune system leading to a variety of adverse immune response such as complement activation. Complement activation is the major contributing factor to a hypersensitivity syndrome called C activation-related pseudoallergy.[465]

Solid lipid NPs are a pharmaceutical alternative to liposomes due to their higher stability in biological systems.[466,467] This property results from their relatively rigid core, which consists of hydrophobic lipids that are solid at room and body temperatures, surrounded by a monolayer of phospholipids. This increased stability allows their use for drug delivery intravenously, orally, topically, and via inhalation.[468,469] Their mean diameters range from 50 to 1000 nm for colloid drug delivery applications. Another advantage is their ease of biodegradation, which makes them less toxic than polymeric and inorganic NPs. Moreover, depending on the pharmacokinetic parameters required, their synthesis can be engineered to obtain three types of hydrophobic core designs: a homogenous matrix, a drug-enriched shell, or a drug-enriched core. Additional features are the avoidance of organic solvents during their synthesis, feasibility of carrying both lipophilic and hydrophilic drugs, and the lack of problems concerning large-scale production and sterilization as compared with other organic nanosystems such as liposomes.[470] Solid lipid NPs have the potential to overcome chemoresistance in many chemotherapeutic-resistance cancers.[296] However, they also present some disadvantages, such as the presence of simultaneous alternative colloidal structures (micelles, liposomes, micelles, and drug nanocrystals), as well as physical instability during storage and administration.[471]

Nanoparticles prepared from nonsoluble polymers (polymeric nanoparticles) are also a widely used system for the delivery of nanomaterial and can provide an excellent vehicle for a number of biomolecules, drugs, and genes to the site of interest. These NPs are subnanosized colloidal structures composed of synthetic or semisynthetic polymers that vary in size from 10 to 1000 nm.

Significant biodegradable polymeric nanoparticles include those made from natural polymers, such as proteins (e.g., gelatins, albumin) and polysaccharides (e.g., chitosan, dextrans, sodium alginate) and biodegradable synthetic and semisynthetic polymers typically consisting of PLA, poly(glycolic acid), PLGA, or poly(methyl methacrylate), among others. The selection of the base polymer depends on many factors such as: (i) size of the desired NPs; (ii) properties of the drug (aqueous solubility, stability, etc.) to be conjugated or encapsulated; (iii) surface characteristics and functionality; (iv) degree of biodegradability and biocompatibility; and (v) drug release profile of the final product.[472,473] Depending upon the method of preparation, nanospheres, or nanocapsules can be obtained. While the former comprise a solid matrix with different levels of porosity where the molecules are homogeneously distributed across the bulk or attached to the surface, the latter are vesicular systems where the solid or solubilized drug is surrounded by a polymeric membrane.

The copolymer PLGA is the most commonly used biodegradable polymer. It is FDA approved as a compatible biomaterial in humans and has been in use for drug delivery purposes since 1970. Administered for the treatment of advanced prostate cancer, Lupron Depot® (Abbot Laboratories International) is an example of a commercially available drug delivery device composed by PLGA microspheres.[474] In addition to the degradability of PLGA, other characteristics of interest include its biocompatibility, good mechanical properties, and ease of processing. It undergoes hydrolysis of ester linkages in the presence of water, resulting in original monomers, lactic acid, and glycolic acid, which are easily metabolized in the body as they are byproducts of various metabolic pathways.[447] Therefore, there is minimal systemic toxicity associated with the use of nanomaterials made with this copolymer. Moreover, the time of biodegradation of PLGA-based NPs and therefore the release kinetics of the therapeutic agent can be modulated by varying the molecular weight of the copolymer and the constituent monomer ratio used in production. Given all these advantages, PLGA-based NPs have been addressed as drug delivery nanosystems for a wide variety of diseases, such as arthritis, diabetes, pain, and cancer.[447] However, recent studies on the biodistribution and retention of PLGA NPs following oral administration showed that 40% accumulated in liver.[475] Another FDA-approved polymeric NP that is being commercialized is Abraxane®.[476] This product consists of albumin NPs for the active and targeted delivery of chemotherapeutics to treat metastatic breast cancer. In 2009, sales for this drug exceeded $350 million. In addition, Celgene paid $2.9 billion for the company, Abraxis; this has been the largest acquisition deal to date in the field of nanomedicine.[477] Many years of research devoted to biodegradable polymeric NPs indicate that they are promising drug delivery carrier systems because of their versatile formulation, sustained release properties, subcellular size, and biocompatibility with various cells and tissues in the body.[473] However, despite their numerous advantages, polymer NPs also have some drawbacks. The toxicity of preparation solvents, acidity of degradation byproducts, drug release that is frequently biphasic, large-scale production, manufacturing, and batch-to-batch reproducibility remain an issue for many polymeric NPs.[2] In addition, there are still many unknowns regarding their toxicity profile and biological effects. However, there is still much room for overcoming these limitations.

Polymeric micelles are nanostructures in a size range within 10–200 nm in pure water, with narrow size distribution. They are formed by the self-assembly of copolymers with amphiphilic character in water, above a minimal concentration called the critical micelle concentration (CMC). Due to the large differences in solubility between their hydrophilic and hydrophobic segments, these amphiphilic polymers are known to assemble in an aqueous milieu into a core–shell architecture, where hydrophobic segments are segregated from the aqueous exterior to form an inner core surrounded by segments.[478] Micelles are known to have an anisotropic water distribution within their structure, which means that water concentration decreases from the surface toward the core of the micelle, with a completely hydrophobic (water-excluded) core. The spatial distribution of a solubilized drug in a micelle will depend on its polarity; nonpolar molecules will be encapsulated/solubilized in the micellar core, and substances with intermediate polarity within the micellar shell. Based on the core–drug interactions through chemical and physical forces, drugs would display different release profiles.

One advantage of polymeric micelles is that the isolating effect from the environment of the polymeric entity can protect molecules sensitive to chemical or biological triggers from degradation and metabolism. In addition, due to the high molecular weight of the amphiphiles that are used to synthesize them, they display remarkably lower CMC and, therefore, show slower dissociation rates upon dilution than micelles formed by regular surfactants (e.g., polyethoxylated castor oil or Cremophor® EL, polysorbate 80 or Tween® 80). This higher stability allows extended circulation time in blood. Besides, when PEG is the hydrophilic component, significant reduction of the uptake by the macrophages has been reported.[479] In addition, versatility in monomer species, block length ratio, and surface modification provide polymer micelles with multifunctionality.

In spite of these advantages, only few polymer micelles are currently marketed (e.g., Estrasorb® for estrogen therapy commercialized by Novavax). This is consequence of two major drawbacks of these nanocarriers: low drug loading capacity and low stability in aqueous media. Regardless of the higher stability upon dilution compared with simple and small surfactant-based micelles, as a polymer micelle is a physically assembled structure in water, a gradual disassembly of the micelles takes place at concentration below the CMC.[480] However, strong investigative efforts are focused on developing strategies to improve both negative issues and transform them in a useful tool for clinical applications.[481,482]

Polymer–drug and polymer–protein conjugates are one of those nanocarriers, along with other systems such as polymeric micelles, that have been designed as biologically active polymeric drugs long before the term "nanomedicine" was introduced. In 1994, the first polymer anticancer drug conjugate was transferred into clinical trials. From then until now, polymer–drug conjugates are one of the most investigated types of nanocarriers and they are currently in clinical trials as advanced as phase III.[65]

It should be noted that although nano-sized (typically 2–25 nm), such polymer conjugates are quite distinct from polymeric nanoparticles. In this case, the therapeutic agent is covalently linked to the macromolecular backbone of the polymer through a physiologically labile bond. Among what it is considered, polymer conjugates, PEGylated proteins, antibodies, and most recently aptamers are included. The techniques used for drug or protein conjugation are getting more sophisticated with the years. For instance, in the case of polymer–protein and polymer–peptide conjugation, besides chemical conjugation, enzyme-mediated conjugations and recombinant techniques are currently being used.[483] Although initially developed as parenterally administered nanocarriers, there are currently a large number of examples which are on the market or have been transferred into clinical development formulated for oral and topical applications.[484] In this sense, promising results are emerging from a number of ongoing Phase I–III studies involving anticancer conjugates. Furthermore, a significant number of studies are exploring new clinical indications for conjugates already approved in another clinical setting and/or studies on new drug combinations (Tables 14.1–14.4).

Early polymer conjugates were developed as treatments for life-threatening diseases (cancer and infectious diseases). Under these circumstances, the risk–benefit for these complex technologies was considered to be acceptable. Recently, it has been approved their use for the treatment of chronic and debilitating diseases, such as rheumatoid arthritis or accelerated macular degeneration. Results from clinical trials have undoubtedly demonstrated several advantages of polymer conjugates over the corresponding parent drugs, including fewer side effects, enhanced therapeutic efficacy, ease of drug administration, and improved patient compliance. However, future generation of polymer conjugates will have to meet a number of challenges, including the development of novel polymers with modulated rates of degradation, versatile conjugation chemistry for site-specific attachment of targeting moieties, and development of polymerization methods that allow accurate control of polymer molecular weights and molecular weight distributions.[485] Moreover, given the increasing complexity in the design of new generations of polymer conjugates, there is a strongest need for the development of an appropriate regulatory framework designed exclusively for nanomedicines.

The use of *polymersomes* for drug delivery has also been described in recent years. These nanostructures are defined as polymeric vesicles made of amphiphilic block copolymers that self-assemble in aqueous medium forming a bilayer that surrounds a liquid aqueous core.[486] Thus, lipophilic drugs can accommodate across the membrane and water-soluble ones within the

TABLE 14.1 Organic Nanocarrier Drug Delivery Systems on the Market

Carrier	Product Name	Main Component	Targeting Moiety or Indication	Company	Route of Administration
Polymer–protein conjugates	Adagen®	PEG-adenosine deaminase	Enzyme replacement. Severe combined immunodeficiency syndrome	Enzon	Intramuscular
	Cimzia®	PEG-anti-TNFα Fab	Crohn's disease, rheumatoid arthritis	UCB	Subcutaneous
	Neulasta®	PEG-hrGCSF (pegfilgrastim)	Chemotherapy-induced neutropenia	Amgen	Subcutaneous
	Mircera®	PEG-EPO (polyethylene glycol-epoetinbeta)	Anemia associated with chronic kidney disease	Roche	Intravenous/ subcutaneous
	Pegasys®	PEG-interferon alpha 2a	Hepatitis C	Roche	Subcutaneous
	Pegintron®	PEG-interferon alpha 2b	Hepatitis C, VIH	Schering-Plough	Subcutaneous
	Oncaspar®	L-Asparaginase	Acute lymphocytic leukemia (ALL)	Enzon	Intravenous/ intramuscular
	Somavert®	Pegvisomant, antagonist of human growth hormone (GH)	Acromegaly	Pfizer	Subcutaneous
	Zinostatin®	Styrene maleic anhydride-neocarzinostatin (SMANCS)	Hepatocellular carcinoma (HCC)	Astellas Pharma	Local admin. via hepatic artery infusion
Polymer–drug conjugates	Copaxone®	Glatiramer acetate, copolymer L-Glu, L-Ala, L-Lys, L-Tir	Multiple esclerosis	Teva	Subcutaneous
	Macugen®	Pegaptanib sodium, selective anti-VEGF inhibitor	Age-related macular degeneration (AMD)	Eyetech	Intravitrous

	Renagel®	Sevelamer hydrochloride, phosphate-binding polymer	Chronic kidney disease (CKD) patients on hemodialysis	Genzyme	Oral
	Welchol®	Colesevelam (bile acid sequestrant), cholesterol binding polymer	Hyperlipidemia, Type II diabetes mellitus	Genzyme	Oral
Liposomes and lipid NPs	Ambisome®	Liposomal amphotericin B	Fungal and protozoal infections	Astellas Pharma	Intravenous
	Abelcet®	Lipidic amphotericin B	Fungal infections	Sigma-Tau Pharmaceutical	Intravenous
	Amphocil®	Lipidic amphotericin B	Fungal infections, invasive aspergillosis	Beacon Pharmaceuticals	Intravenous
	Daunoxome®	Liposomal daunorubicin	HIV-associated Kaposi's sarcoma	Gilead Science	Intravenous
	Depocyt®	Liposomal (nonconcentric vesicles) Cytarabine	Lymphomatous meningitis	Pacira Pharmaceuticals	Intrathecal
	Depodur®	Morphine sulfate extended-release liposome	Postsurgical analgesia, pain relief	EKR Therapeutics	Epidural
	Doxil®/ Caelyx®	PEGylated liposomal doxorubicin	Breast and ovarian cancer, Kaposi's sarcoma	Johnson & Johnson/ Schering-Plough	Intravenous
	Epaxal®	Liposomal aluminum-free vaccine	Hepatitis A	Berna Biotech	Intramuscular
	Estrasorb®	Micellar estradiol	Menopausal therapy	Novavax	Topical
	Inflexal®	Virosomal adjuvanted vaccine	Influenza	Berna Biotech	Intramuscular

Continued

TABLE 14.1 Organic Nanocarrier Drug Delivery Systems on the Market—Cont'd

Carrier	Product Name	Main Component	Targeting Moiety or Indication	Company	Route of Administration
	Myocet®	Doxorubicin HCL liposome	Patients with HER2-positive breast cancer	Cephalon	Intravenous
	Mepact®	Mifamurtide—Muramyl tripeptide phosphatidylethanolamine (MTP-PE)	Nonmetastatic osteosarcoma	Takeda Pharmaceutical Co.	Intravenous
	Visudyne®	Liposome/lipidic verteporfin	Age-related macular degeneration (AMD) and phtodynamic therapy	Novartis	Intravenous
Nanoparticles (NPs) (Withdrawn 2010)	Abraxane®	Paclitaxel protein-bound NPs. Albumin bound	Breast cancer	Abraxis (Celgene)	Intravenous
	Bexxar®	^{131}I-tositumomab	Patients with CD20 antigen-expressing relapsed or refractory, low-grade, follicular, or transformed NHL, including patients with Rituximab-refractory NHL	Bexxar	Intravenous
	Mylotarg®	Gemtuzumab ozogamicin—anti-CD33 antibody-calicheamicin	Acute myeloid leukemia (AML)	Wyeth/Pfizer	Intravenous
	Zevalin®	^{90}Y-ibritumomab tiuxetan	Indolent form of non-Hodgkin lymphoma (NHL)	Spectrum Pharmaceuticals	Intravenous

TABLE 14.2 Inorganic Nanocarrier Drug Delivery Systems on the Market

Carrier	Product Name	Main Component	Targeting Moiety or Indication	Company	Route of Administration
Magnetic NPs	Cosmofer®	Dextran iron oxide	Anemia	GRY Pharma	Intravenous
	Ferrlecit®	Sodium ferric gluconate complex in sucrose	Anemia	Sanofi-Aventis	Intravenous
	Venofer®	Polynuclear iron (III)—hydroxide core superficially surrounded by noncovalently bound sucrose molecules	Iron deficiency anemia	Fresenius	Intravenous
Nanocrystals	Emend®	Aprepitant. P/neurokinin 1 (NK1) receptor antagonist	Antiemetic agent. prevention of postoperative nausea and vomiting (PONV)	Merck	Oral
	Megace® ES	Megestrol acetate	Appetite loss, severe malnutrition, and significant weight loss in AIDS patients	PAR Phar. Co.	Oral
	Rapamune®	Sirolimus	Immunosuppressive. Prevention of organ rejection in kidney transplant patients aged 13 years or older	Pfizer/Wyetha	Oral
	Tricor®	Fenofibrate	Hypercholesterolemia	Abbott	Oral
	Triglide®	Fenofibrate	Hypercholesterolemia	Sciele Pharma, Inc.	Oral

TABLE 14.3 Organic Nanocarrier Drug Delivery Systems in Clinical Development

Carrier	Product Name	Main Component	Targeting Moiety or Indication	Company	Route of Admin.	Clinical Phase
Liposomes and lipid NPs	2B3101	Glutathione pegylated liposomal doxorubicin	Brain metastases of breast cancer	TO-BBB	Intravenous	Phase 1–2
	Atragen	t-Retinoicacid	Acute pro-AML and advanced renal cancer	Aronex	Intravenous	Phase 1–2
	Brakiva^{TM}	Topotecan HCl, a potent semisynthetic derivative of camptothecin, encapsulated in the aqueous and acidic core of proprietary sphingomyelin-based liposomes called Optisome^{TM}	Small cell lung cancer (SCLC) and ovarian cancer	Talon Therapeutics	Intravenous	Phase 1–2
	CPX-1	(Irinotecan HCl:Floxuridine) Liposome	Colorectal cancer	Celator Pharmaceuticals	Intravenous	Phase 2
	CPX-351	(Cytarabine:Daunorubicin) Liposome	Hematologic malignancies, particularly acute myeloid leukemia (AML)	Celator Pharmaceuticals	Intravenous	Phase 2b
	NLCPT11	Nanoliposomal irinotecan hydrochloride	Recurrent high-grade gliomas	University of California, San Francisco (USA)	Intravenous	Phase 1
	LE-DT	Liposome Encapsulated Docetaxel	Various kinds of cancers. Pancreatic and prostate cancer patients in Phase 2	Neopharm, Inc.	Intravenous	Phase 1

Liposomes and lipid NPs	LEP-ETU	NeoLipid® liposomal with paclitaxel (Taxol)	Various kind of cancers	Neopharm, Inc.	Intravenous	Phase 1
	LN-SN38	NeoLipid® liposomal with SN38, the active metabolite of irinotecan	Advanced colorectal cancer	Neopharm, Inc.	Intravenous	Phase 2
	L-Annamycin	Liposomal annamycin	ALL, AML	Callisto Pharmaceuticals, Inc.	Intravenous	Phase 1–2
	Lipoplatin™	Liposomal cisplatin	Various kinds of cancers, including pivotal EMA Phase III trials for non-small cell lung cancer (NSCLC) and pancreatic cancer	Regulon	Intravenous	Phase 2–3
	Oncotcs	Vincristine sulfate liposomes	Non-Hodgkin's lymphoma (NHL)	Inex/Enzon	Intravenous	Phase 2–3
	Sarcodoxome®	Doxorubicin	Sarcoma	GP Pharm	Intravenous	Phase 2
	SLIT cisplatin	Sustained-Release lipid inhalation targeting (SLIT) cisplatin	Carcinoma of the lung	Transave	Aerosol	Phase 2
	Thermodox®	Doxorubicin enhanced with lysolipid thermally sensitive liposomes (LTSL)	Hepatocellular carcinoma (HCC)	Celsion	Intravenous	Phase 3

Continued

TABLE 14.3 Organic Nanocarrier Drug Delivery Systems in Clinical Development—Cont'd

Carrier	Product Name	Main Component	Targeting Moiety or Indication	Company	Route of Admin.	Clinical Phase
Polymer Based	ADI-PEG20	Arginine deiminase (ADI) with polyethylene glycol 20 kDa	Melanoma, hepatocellular carcinoma, pancreatic cancer, prostate cancer, mesothelioma	Phoenix Pharmalogics-Polaris Group	Intramuscular	Phase 3
	ANG1005	Engineered peptide compounds (EPiC) Angiopep-2 peptide—Paclitaxel	Brain tumor, crossing the blood–brain barrier (BBB)	Angiochem	Intravenous	Phase 2
	AMG223	Polymer which binds phosphate	Hyperphosphatemia in chronic kidney disease (CKD) patients on hemodialysis	Amgen	Oral	Phase 2
	ARC1779	Pegylated aptamer consisting of 40 nucleotides. Anti-platelet agent	Thrombotic thrombocytopenic purpura (TTP)	Archemix	Intravenous	Phase 2a
	CDP791	PEG VEGFR-2/KDR inhibitor, with carboplatin–paclitaxel	Non-small cell lung cancer (NSCLC)	UCB Pharma	Intravenous	Phase 2
	E10030	PEG-anti-PDGF aptamer combination with Lucentis®	Age-related macular degeneration (AMD)	Opthotech	Intravitreal	Phase 2
	Hemospan™ MP4	PEG hemoglobin	Blood transfusions during episodes of temporary blood loss, such as surgery or trauma	Sangart, Inc.	Intravenous	Phase 1–2
	NKTR-102	PEG irinotecan	Metastatic breast cancer	Nektar	Intravenous	Phase 2
	NKTR-105	PEG docetaxel	Various kinds of cancers	Nektar	Intravenous	Phase 1

Polymer Based	NKTR-118	PEG naloxone	Opioid-induced constipation (OIC)	Nektar	Oral	Phase 3
	NKTR-119	PEG naloxone	Analgesic for pain	Nektar	Oral	Phase 3
	Opaxio™ CT2103 (formerly known as Xyotax)	Polyglutamate paclitaxel drug	Various solid tumors, including non-small cell lung, ovarian, breast, and prostate cancers	Cell Therapeutics	Intravenous	Phase 3
	PEG-SN38	PEG camptothecin derivative	Various kinds of cancers	Enzon, Inc.	Intravenous	Phase 2
	Prolindac™	Polymer prodrug of DACH platinum	Various kinds of cancers	Access Pharmaceuticals	Intravenous	Phase 2
	Trascend™	Melanotransferrin (LRP)	Brain tumor, crossing the blood–brain barrier (BBB)	Bioasis Technologies	Intravenous	Preclinical
	Trascend™	VH0455Peptide (LDLR)	Brain tumor, crossing the blood–brain barrier (BBB)	Vect-Horus	Intravenous	Preclinical
	Trascend™	Peptide RAP (LRP-2)	Brain tumor, crossing the blood–brain barrier (BBB)	Raptor Pharmaceutical	Intravenous	Preclinical
	Uricase-PEG20	Pegylated form of the enzyme uricase (urate oxidase)	Refractory gout and tumor lysis syndrome	Enzymerx—Polaris Group	Intravenous	Phase 1
	XMT-1001	Fleximer technology—Polyacetal-camptothecin conjugate	Various kinds of cancers	Mersana	Intravenous	Phase 1–2
	Vivagel®	Lysine-based dendrimer SPL7013	Bacterial vaginosis (BV)	Starpharma	Topical	Phase 1-2

TABLE 14.4 Inorganic Nanocarrier Drug Delivery Systems in Clinical Development

Carrier	Product Name	Main Component	Targeting Moiety or Indication	Company	Route of Admin.	Clinical Phase
Magnetic NPs	Nano-Cancer®	Iron oxide NPs	Glioblastoma (brain tumor) highly malignant	Magforce Nanotechnologies AG	Local	Phase 2
Nanocrystals	CPSI-2364	Synthetic guanylhydrazone semapimod	Inhibits signal transduction pathways that produce proinflammatory cytokines (e.g., TNF-alpha, IL-1, IL-6) and nitric oxide. Psoriasis, Crohn's disease, ERCP-induced pancreatitis	Cytokine Pharmasciences, Inc.	Intravenous	Phase 1–2
	NPI32101	Nanocrystalline silver	Antimicrobial and anti-inflammatory	Nucryst Pharmaceuticals Corp.	Topical	Phase 2
	Paxceed	Micellar paclitaxel	Secondary-progressive multiple sclerosis (MS) and rheumatoid arthritis	Angiotech Pharmaceuticals	Intravenous	Phase 3
	Theralux	Nanocrystal Elan	PDT (photodynamic therapy)—anticancer	Kiadis Pharma	Intravenous	Phase 2

Companies Web site:	
Abbott	http://www.abbott.com
Accelrys	http://accelrys.com/
Access Pharmaceuticals	http://www.accesspharma.com/
AlphaRx, Inc.	http://www.alpharx.com/
Amgen, Inc.	http://www.amgen.com/
Angiotech Pharmaceuticals, Inc.	http://www.angiotech.com/
Aptalis Pharmaceutical Technologies	http://www.aptalispharmaceuticaltechnologies.com/

AQUANOVA AG	http://www.aquanova.de/
Astellas Pharma	http://www.astellas.com/
Beacon Pharmaceuticals	http://www.beacon-pharma.com/
Berna Biotech	http://www.bernabiotech.es/
Bexxar	http://www.bexxar.com/
Biophan Technologies, Inc.	http://www.biophan.com/
BioSante Pharmaceuticals	http://www.biosantepharma.com/
Calando Pharmaceuticals, Inc.	http://www.calandopharma.com/
Camurus AB	http://www.camurus.com/
Capsulution Pharma AG	http://www.capsulution.com/
Celgene Corporation	http://www.celgene.com/
Centre for Research on Adaptive Nanostructures and Nanodevices (CRANN)	http://www.crann.tcd.ie/
Cephalon, Inc.	http://www.cephalon.com/
Ceramisphere Pty Ltd.	http://www.ceramisphere.com/
Cerulean Pharma, Inc.	http://www.ceruleanrx.com/
Copernicus Therapeutics, Inc.	http://168.144.36.118/index/index.asp
CritiTech, Inc.	http://www.crititech.com/
CytImmune Sciences, Inc.	http://www.cytimmune.com/
Cytokine PharmaSciences, Inc.	http://www.cytokinepharmasciences.com/
Debiotech SA	http://www.debiotech.com/
EKR Therapeutics	http://www.ekrtx.com/
Elan Corporation, plc	http://www.elan.com/

Continued

TABLE 14.4 Inorganic Nanocarrier Drug Delivery Systems in Clinical Development—Cont'd

Companies Web site:	
Enzon Pharmaceuticals, Inc.	http://enzon.com/
Eyetech	http://www.eyetech.com/
F. Hoffmann-La Roche Ltd.	http://www.roche.com/
Flamel Technologies SA	http://www.flamel.com/
Flexible Medical Systems	http://www.flexmedsys.com/
Fresenius	http://www.fresenius.com/
Genencor	http://www.genencor.com/
Genzyme	http://www.genzyme.com/
Gilead	http://www.gilead.com/
GlaxoSmithKline	http://www.gsk.com/
IOTA NanoSolutions Ltd.	http://www.iotanano.com/
Izon Science Ltd.	http://www.izon.com/
Kiadis Pharma	http://www.kiadis.com/
Kuecept Ltd.	http://www.kuecept.com/
Lena Nanoceutics Ltd.	http://www.lenanano.com/
Life Technologies	http://www.lifetechnologies.com/
Lypro Biosciences	http://www.lyprobio.com/
Makefield Therapeutics, Inc.	http://www.makefieldtherapeutics.com/
Medical Nanotechnologies, Inc.	http://www.nanomedinc.com/

Merck & Co., Inc.	http://www.merck.com/
Merck Serono	http://www.merckserono.com/
Merck Sharp & Dohme Corp.	http://www.merck.com/
Nano Interface Technology, Inc.	http://www.nanointerfacetech.com/
Nano S Biotechnology	http://www.nano-s.com/
NanoBio® Corporation	http://www.nanobio.com/
NanoBioMagnetics, Inc.	http://www.nanobmi.com/
Nanobiotix	http://www.nanobiotix.com/
NanoCarrier Co. Ltd.	http://www.nanocarrier.co.jp/
NanoMaterials Technology Pte Ltd.	http://www.nanomt.com/
NanoSight Ltd.	http://www.nanosight.com/
Nanosphere, Inc.	http://www.nanosphere.us/
Nanotherapeutics, Inc.	http://www.nanotherapeutics.com/
Nemucore Medical Innovations	http://www.nemucore.com/
nLife Therapeutics, S.L.	http://www.n-life.es/
Novartis Institutes for Biomedical Research (NIBR)	http://www.novartis.com/research.shtml
Novartis Pharmaceuticals	http://www.novartis.com/
Novavax, Inc.	http://www.novavax.com/
NUCRYST Pharmaceuticals Corp.	http://www.nucryst.com/
Pacira Pharmaceutics	http://www.pacira.com/
PAR Pharmaceutical, Inc.	http://www.strativapharma.com/
Particle Sciences, Inc.	http://www.particlesciences.com/

Continued

TABLE 14.4 Inorganic Nanocarrier Drug Delivery Systems in Clinical Development—Cont'd

Companies Web site:	
Pfizer	http://www.pfizer.com/
PharmaNova, Inc.	http://www.pharmanovaco.com/
Pharmidex	http://pharmidex.co.uk/
pSivida	http://www.psivida.com/
Sigma-Tau Pharmaceuticals, Inc.	http://www.sigmatau.com/
SoluBest Ltd.	http://www.solubest.com/
Spectrum Pharmaceuticals, Inc.	http://www.spectrumpharm.com/
Spherics, Inc.	http://www.spherics.com/
Talon Therapeutics	http://www.talontx.com/
Takeda Pharmaceuticals	http://www.takeda.com/
Teva	http://www.tevapharm.com/
The Dow Chemical Company ("Dow")	http://www.dow.com/
UCB Pharma	http://www.ucbpharma.es/
VYOME Biosciences	http://www.vyome.in/

core. In this sense, the efficiently combined loading of two anticancer drugs with different solubility (Dox and placitaxol) in order to achieve improved activity against solid tumors has been reported (Figure 14.27).[487]

The formation of these structures is constrained to a very narrow hydrophilic/hydrophobic ratio; therefore, not all amphiphilic block copolymers

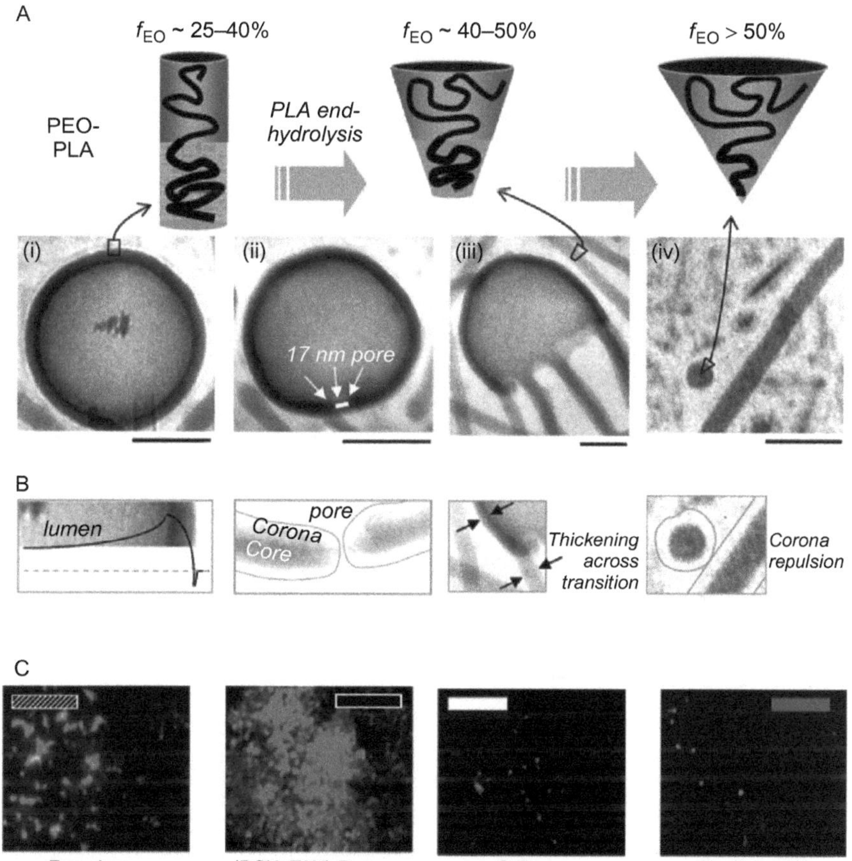

FIGURE 14.27 Nanopolymersome self-assembly, degradation, and accumulation within the tumor. (A) Cryo-TEM images of empty PEG-PLA-based self-assembled aggregates. Image contrast is provided by the hydrophobic core of the polymersome bilayer and not by the hydrated corona. Hydrolysis of PLA in the core of vesicles triggers the growth of pores and "unravels" vesicles into worm-like micelles and spheres. Scale bars are 100 nm. (B) Magnified views show key features of degradable polymersomes, including evidence of the vesicular lumen, the corona-lined pore, the thickening in membrane–micelle transition, and the exclusion zone that indicates an invisible layer of PEG around the micelles. (C) Typical fluorescent microscopy images of TUNEL-labeled tumor tissue, with large increases in TUNEL positive cells observed 48 h post-polymersome-drug treatments ((DOX+TAX)-Psomes). Scale bar 20 μm. Reprinted from Ref. 487. Copyright 2006, with permission from Elsevier.

display polymersome-forming tendency. For example, polystyrene–polyacrylic acid block copolymers are among the ones that are used for this purpose. However, since these derivatives are not biodegradable, molecules combining PEG and polyesters like PLA have been designed.[488,489] Compared to liposomes, polymersomes have superior properties, such as greater toughness, better stability, and tailorable surface properties, while retain their low immunogenicity. In addition, they are relatively easy to prepare and functionalize.[490] Due to the relatively short way these entities have gone, a limited number of patents is available and any polymersome-based product is currently marketed.[491] However, the combination of these nanostructures with biological entities is showing great potential for the devolvement of novel tools to improve theragnosis.[490]

It is estimated that 40% or more of active substances being identified through combinatorial screening programs are poorly soluble in water.[492] The use of *drug nanocrystals* is a very interesting approach that allows high dosages of poorly soluble drugs and improves their biological distribution and bioavailability.[493] The term drug nanocrystals implies a crystalline state of the discrete drug particles, but depending on the production method they can also be partially or completely amorphous.[494] They consist only of drug without any matrix material, and they are stabilized by a thin (2–5 nm) layer consisting of an adsorbed surfactant or polymer.[495,496] This thin layer of coating material can protect from degradation chemically labile drugs.[497] Moreover, nanocrystals have a drug loading capacity of 100% as there is not carrier material as in polymeric nanoparticles for instance. It is well known that if only a limited fraction of the administered carrier reaches the target, only a small fraction of the total amount of drug administered reaches the organ; thus, the therapeutic dose required to cure the disease may not be achieved. Therefore, the higher amount of drug stored by these nanocarriers highly increases the amount of drug that could reach the target organ compared to other parenteral drug delivery systems. In addition, their *in vivo* behavior could be tuned by modifying or changing the stabilizing layer.[493]

Since the beginning of the 1990s, the use of nanocrystals for oral bioavailability enhancement, and for intravenous or pulmonary drug delivery, had been established. Their versatility and ease of scale-up and commercial production had enabled the development of several drug nanocrystals formulations that are currently on the market. Examples of trade names include NanoCrystal® (Élan, prev. Nanosystems), DissoCubes® (SkyePharma, prev. PharmaSol), and smartCrystal® (Soliqs/Abbott, prev. PharmaSol). In spite of all the above mentioned advantages, drug nanocrystals also present some drawbacks, such as long production times and the requirement of crystallization, which are not always possible with some therapeutic compounds.

Other organic nanostructures studied for drug delivery purposes are *dendrimers*. These are hyperbranched polymers, typically symmetric around a central core, with a defined number of generations and terminal functional groups, and they often adopt a spherical three-dimensional morphology.[498] Dendrimers are synthesized stepwise by a repetitive reaction sequence that allows almost full control of their size, shape, surface/interior chemistry, number of branches, flexibility, and topology. They have three major portions: a core, an inner shell, or interior layers (generations) composed of repeating units that are radically attached to the interior core and an outer shell. The mechanisms used for their syntheses allow the introduction of functionalities in each of these portions to control properties such as solubility, thermal stability, and functionalization. Polyamidoamines (PAMAMs) were the first dendrimers to be synthesized and therefore are one of the most popular.

Because of their molecular architecture, dendrimers show some significantly improved physical and chemical properties when compared to traditional linear polymers.[499] These unique physical properties, including monodispersity, water solubility, encapsulation capacity, high degree of branching, and large number of functionalizable peripheral groups, make these nanostructures appropriate candidates for drug delivery applications.[500] Therapeutic agents can be loaded onto these nanostructures in three ways: (i) through covalent attachment to the functional groups present on their periphery to form what is called dendrimer prodrugs; (ii) through coordination to the outer functional groups via reversible interactions, such as ionic ones; and (iii) through encapsulation by means of the formation of a dendrimer–drug supramolecular assembly.

There are several *in vivo* studies reporting the great potential of dendrimers for drug delivery applications.[498,501] In this regard, an asymmetric bis-HMPA PEGylated dendrimer functionalized with Dox via a pH-sensitive acyl hydrazone bond demonstrated outstanding antitumor activity in mice bearing murine C26 colon carcinoma.[502]

Despite their great potential as drug carriers, dendrimers also have several drawbacks. It has recently been reported that those with positively charged surface groups, like amino-terminated PAMAM dendrimers, can cause toxicity. It was observed that they destabilize cell membranes and cause cell lysis. However, this problem can be overcome by careful surface engineering such as functionalization of the terminal groups with PEG or fatty acids.[503,504] In contrast, hydroxy- and methoxy-terminated dendrimers do not show acute or long-term toxicity.[447] Another disadvantage of these nanostructures is that their multistep synthesis still requires great effort and produces low yields for large structures. However, recent success in simplifying and optimizing the synthesis of these molecules could allow a large variety of structures with reduced large-scale production costs.[343]

5.3. Carbon Nanotubes

CNTs are graphene sheets rolled into a seamless cylinder that can be open-ended or capped. The simplest CNTs are composed by a single graphene sheet and are known as single-walled nanotubes (SWNTs). They can also comprise several concentric and nested graphene sheets, and in this case, they are known as multi-walled nanotubes (MWNTs). While they can reach several micrometers in length, their diameter can be as small as 0.7–1.5 nm for SWNTs and 2–50 nm for MWNTs. They can be dispersed single nanotubes or aggregates of nanotube bundles.

The long and cylindrical shape of CNTs presents a large inner volume, in contrast to spherical NPs. Therefore, their total surface area is very high, and not only can the therapeutic agent be conjugated to their external wall but the internal cylindrical area can also be filled with a wide variety of biomolecules, ranging from small derivatives to proteins. To enhance dispersion in physiologically relevant solutions, the surface is often modified by amphiphilic molecules (PEG-lipids) or polymers by noncovalent adsorption (e.g., π–π stacking, hydrophobic interactions, or electrostatic attractions) or by covalent functionalization (e.g., carboxylation, amidation, and cycloaddition).[505,506] Covalently conjugated drug molecules are usually linked via cleavable bonds to functional groups on the CNT surface obtained after oxidation processes or on their polymer coating. In the case of flat aromatic molecules, these are adsorbed on the CNT surface via noncovalent π–π stacking with remarkably high loading capacity. This is the case of Dox, a commonly used chemotherapy drug, for which a loading efficiency of up to 4 g per gram of PEGylated SWNTs was recently reported.[507]

As CNTs show virtually no degradation in biological systems, they may not have clear advantages over biodegradable organic nanostructures if they are used merely as drug carriers. It is important to note, however, that a few studies have reported that SWNTs are biodegradable through natural enzymatic catalysis.[508–510] However, due to their nanometric size, their unique physical properties imply the presence of various functionalities in the same nanostructure and this feature allows the development of interesting remote controlled theragnostic agents.[4,505,506,511] In this regard, high mechanical strength, thermal as well as electrical conductivity, and suitable stiffness and flexibility are some of the physical properties of interest. Furthermore, CNTs display strong optical absorbance in the NIR range, thus resulting in their heating. As biological systems are highly transparent to electromagnetic wavelengths in the spectral window between 800 and 1400 nm, this unique property can be used to trigger the temperature-responsive release of drugs conjugated via covalent bonds that are sensitive to temperature increase or entrapped within temperature-responsive polymers wrapped around CNTs.[512] However, the heat generated by NIR laser pulses can also be used to trigger the endosomal escape of CNTs (via disruption of the endosomal membrane) or induce thermal destruction of target cells containing significant concentrations of CNTs.[513] These molecules also enhance the dielectric contrast between cancerous and healthy cells, thereby facilitating the selective thermal ablation of the former.[514] Furthermore, photothermal ablation can be combined with the release of therapeutic agents (drugs, genes, etc.) codelivered by CNTs, thereby allowing multidrug resistance of several illnesses—as it is the case of current cancer chemotherapies—to be overcome. Alternatively, radiofrequency waves that can intercalate more deeply in the body than NIR light can also be used to produce intense heat energy from CNTs.[515] Other intrinsic optical properties of CNTs, such as Raman scattering, [516] NIR photoluminescence,[517] or photoacoustic signal,[518] are also useful in biomedical imaging. Moreover, these nanotubes are spontaneously internalized by a wide variety of cell types, although the exact uptake mechanism may depend on their size, aggregation state, and surface chemistry.[4,519,520] The combination of these properties confers CNTs high potential as a scaffold for the fabrication of smart nanodrugs.

Despite various encouraging preclinical results reported by several research groups, a number of obstacles must be overcome before the clinical application of CNTs.[521] Their fiber-like shape and size can cause cytotoxicity, platelet aggregation, lung inflammation, mitochondrial dysfunction, generation of ROS, and DNA damage.[505,522–526] Furthermore, the effects of CNTs on fertility, nerve systems, and immune responses after administration to animals have not been studied systematically and require future attention.[505] Conversely, there are also many reports showing how the *in vivo* toxicity and immunogenicity of CNTs are remarkably reduced by altering their degree of functionalization. As an example, highly PEGylated CNTs exhibited a reasonably long blood circulation half-life and remarkably reduced *in vivo* toxicity when compared with nonfunctionalized raw CNTs.[527] These studies highlight the importance of correct selection and control of the surface coatings, chemistry, diameters, and lengths of CNTs in order to achieve safer CNT-based biomedical applications.

Contrasting results from studies on the effects of the physicochemical properties of CNTs on cell internalization, biodistribution, excretion, and toxicity could be attributed to the fact that these molecules are currently synthesized and used in a wide distribution of diameters, lengths, chiralities, and electronic types. Moreover, most industrially produced CNTs are manufactured by chemical vapor deposition, and therefore, they often contain a considerable amount of impurities, such as metal catalysts (e.g., Co, Fe, Ni, Mo), organic materials (e.g., amorphous or microstructured residual organic carbon), and supporting materials (e.g., alumina, magnesium oxide, or silica). Such impurities also contribute to CNT potential toxicity.[506,527] Recent success in separating CNTs into highly purified fractions of metal impurities and with well-defined length, chiralities, and electronic type is allowing researchers to explore how these features influence CNT interactions with living matter.[528–530] This line of investigation is expected to provide insights into the true potentiality of chemically modified CNTs as efficient and safe delivery systems.[527]

Although it is clear that further research is required, the great number of conflicting toxicological studies to date suggests that under certain conditions, especially those involving chronic exposure, CNTs pose a serious risk to human health. Given the potentially harmful effects reported in some of these studies, it has been suggested that, until further research has been performed,

CNTs should not be recommended for any application involving inhalation or applications where they might be filtered through the kidneys.[505]

5.4. Hybrid Nanocarriers

Hybrid nanocarriers are a new class of nanomaterials that is being developed for drug delivery applications and that is composed of at least two different nanomaterials. This kind of nanocarriers not only allows combining the advantages from existing nanocarriers but also possesses unique advantages for imaging and therapeutic applications. As an example of the great potential of hybrid nanomaterials, the limited effect of gold nanorods (GNRs) on *in vivo* photothermal cancer therapy was overcome recently by the development of a novel hybrid nanocarrier (Figure 14.28).[531] Although ligand-conjugated GNRs have been effective for photothermal killing of cancer cells *in vitro*, desirable photothermal therapeutic effects in an *in vivo* animal model have been limited due to a high liver uptake during circulation. However, when GNRs were encapsulated on chitosan-modified Pluronic-based temperature-sensitive nanocarriers, it resulted in a very impressive therapeutic effect *in vivo* in nude mice bearing tumors, as an apparently complete resorption of the tumor was achieved. This higher efficiency of GNRs as photothermal agents was due to the flexible, hydrophilic, and soft characteristics of Pluronic encapsulation which provided a longer circulation time, a better tumor accumulation, and a lower liver uptake in comparison with nonencapsulated GNRs (Figure 14.28).

In spite of all the advantages that are being proposed, it is important to underline that they also have other challenges, including reproducible manufacture, characterization, and safety. Readers that are interested in these novel nanomaterials are encouraged to consult several articles published recently in the literature.[442,532–534]

6. TRANSLATION FROM BASIC RESEARCH TO THE HEALTHCARE MARKET

The pharmaceutical industry is a sector seeking radical innovation, as more than $120 billion pharmaceutical products are losing their patent protection between 2009 and 2014.[477] Experts have predicted the nanotechnology market will reach close to a trillion dollars by 2015 and the healthcare will be one of highest growth. In this context, it is predicted that nanocarriers will account for 40% of a $136 billion of NDD (nanotechnology-enabled drug delivery) market by 2021, with a 60/40 split between nanocrystals and nanocarriers, respectively.[535] Nowadays, the enormous effort and investment in basic research in this area (in the past 11 years, governments around the world have invested more than US$67 billion) translates into new targeted delivery mechanisms and nanocarriers that allow to achieve know-how and more value for

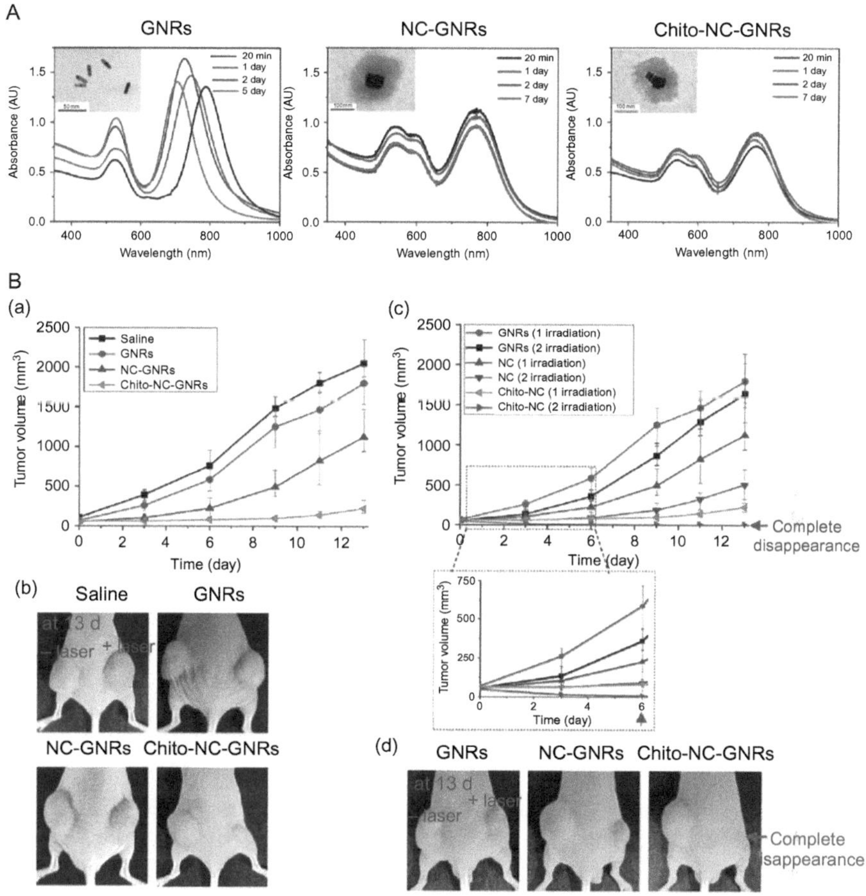

FIGURE 14.28 (A) Absorption spectra and transmission electron microscope (TEM) images (inset scale bar = 0 nm in the case of gold nanorods (GNRs) and 100 nm in the case of other groups) of GNRs themselves, Pluronic (NC-GNRs) and chitosan-conjugated form of Pluronic-based nanocarriers (Chito-NC-GNRs). (B) (a) Changes in tumor volumes and (b) the tumor images after onetime NIR laser irradiation (808 nm, 4 W/cm^2) for 4 min at 24 h after the i.v. injection of the nanomaterials. (c) Change in tumor volumes (an enlarged graph at initial time) and (d) the tumor images after NIR laser irradiations at 24 and 48 h after single i.v. injection of the nanomaterials. Adapted with permission from Ref. 531. Copyright 2011 American Chemical Society. (For color version of this figure, the reader is referred to the online version of this chapter.)

companies. However, the healthcare market is complex and very long, and cost clinical development is required.

Failure in clinical development is the major drawback for new agent approval. In fact, the number of new agents approved annually as new medicines is similar to that seen 60 years ago.[536] Compounds fail for many reasons, such as poor oral bioavailability, pharmacokinetic properties, or toxicity issues, that are not predicted by animal pharmacology models or by preclinical ADME studies. From the aforementioned, poor pharmacokinetic properties or bioavailability had become only a minor cause of overall attrition ($\leq$ 10–20%) by 2000, whereas lack of efficacy and low margins of safety were the major causes of Phase II (66%) and III (30%) attrition.[537]

Therefore, it is crucial that during the basic research phase, the design of the nanomedicine is made considering its proposed clinical use, route of administration, likely dose, and frequency of dosing. In this regard, if the nanomedicine is to be used at a high dose or for long-term treatment, verification of its biodegradability and biocompatibility calls for robust *in vitro* and *in vivo* scientific experimentation, in which doses and time frames are chosen to match likely clinical pharmacokinetics. Furthermore, nanomedicines are becoming more complex (e.g., combination of distinct nanomaterials, need for complex chemical functionalization strategies, etc.), and therefore, the potential safety of each component and/or metabolite that might arise as a result of their *in vivo* degradation or metabolic fate must be clearly reflected. However, if these medicines are nondegradable, their renal and/or hepatobiliary removal should be confirmed at an early stage after their administration. Moreover, if the nanocarrier requires cell internalization and trafficking to a specific intracellular compartment before the therapeutic cargo is released, it is crucial to understand and optimize its internalization rate and intracellular fate, as these are key factors that determine clinical outcome.

It is true that most of the nanomaterials that continuously appear with great potential for drug delivery will never enter patient clinical trials, but their rational design taking into account all the above mentioned issues will greatly improve their chances. Figure 14.29 summarizes the several empirical stop–go checkpoints that have to be taken into account to design, optimize, and select a nanomedicine candidate for preclinical development. An excellent review recently written by Duncan and Gaspar critically reviewed past and modern nanomedicines highlighting their current status and clinical relevance.[5]

During 2000–2010, the scientific output and creation of intellectual property about nanomaterials for drug delivery have shown a steady growth. This has not been matched, however, with a similar pattern in investment, product development, and commercialization. This discrepancy between knowledge production and market introduction is a common hallmark of innovative and emerging technologies. In spite of this, several successful nanocarrier drug delivery systems are already on the market, showing a market size of more than $2.6 billion in 2009.[477] The total number found is 38:30 organic nanocarriers (Table 14.1) and 8 inorganic nanocarriers (Table 14.2). On the other way, there are 40 nanocarriers in clinical development: 35 organic nanocarriers (Table 14.3) and 5 inorganic nanocarriers (Table 14.4) (please refer Companies website section for the websites).[5,538,539]

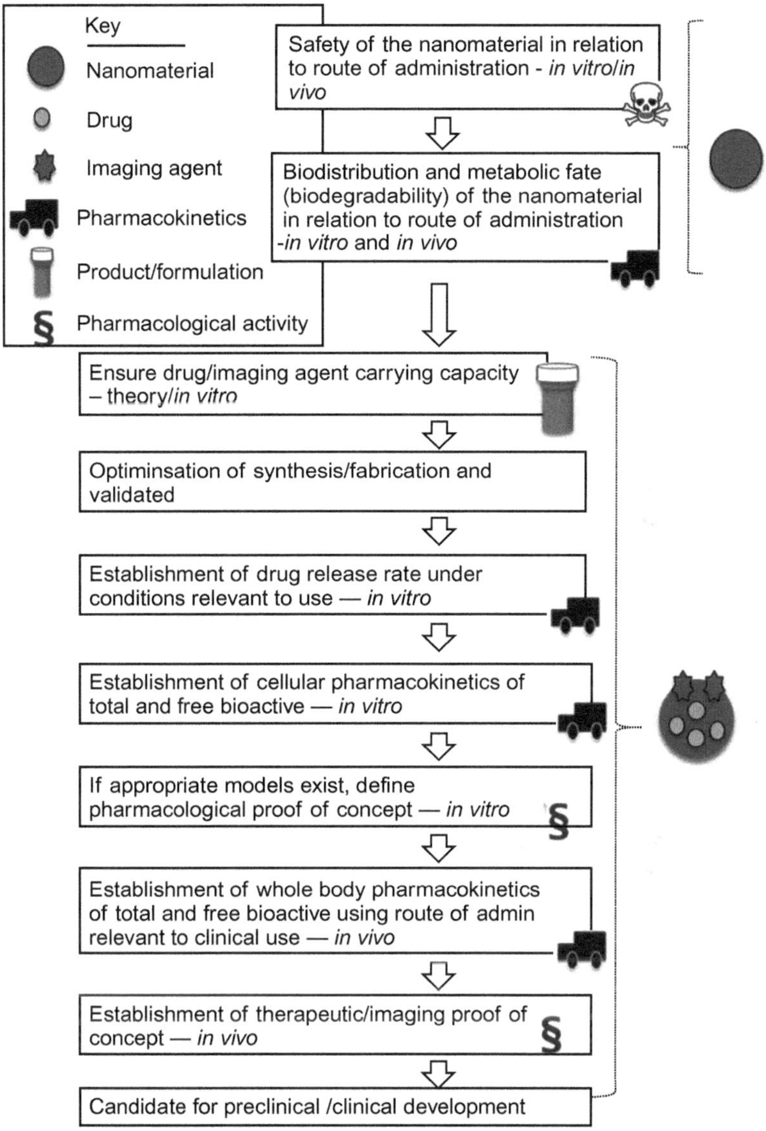

FIGURE 14.29 Summary of stop–go checkpoints for nanomedicine design, optimization, and candidate selection for preclinical development. Reprinted with permission from Ref. 5. Copyright 2011 American Chemical Society.

7. SCOPE

Nanomedicine is one of the fastest growing areas in modern nanotechnology research. However, despite nanomedicines having been used routinely in clinics since 1994, there is still the need for the development of a new generation of multifunctional nanotherapeutics with varied attributes: from efficient tissue penetration and site-specific drug delivery that can be triggered remotely to those that facilitate the imaging of their location in the body. As a consequence, one of the primary goals of nanomedicine over the next 5–10 years must be the development of a new generation of valuable nanotherapeutics that offer significant advances in patient screening, diagnosis, staging, treatment, and monitoring in all stages of cancer and other diseases.[477]

Further, the development of novel specialized treatments for cancer will be a high-impact area of Nanomedicine as no significant changes in the life expectancy has been reported for several cancers (e.g., brain, lung, pancreas, liver, ovarian, etc.) over the past 50 years. In this sense, several reports forecast that at least 50% of all drugs used in 2020 will be nanotechnology based. However, it is important to underline that translation of nanomedicines from the basic research phase to clinical use in patients not only is expensive and time consuming but also requires appropriate funding. To this end, several barriers must be overcome before a potential nanomedicine could enter phase III clinical trials with regard to demonstrating favorable pharmacokinetics, biodistribution profiles, stability, and drug release rates; factors that all contribute to a better understanding of the bio-nano interface (toxicity, endocytosis/exocytosis, and intracellular trafficking where relevant) among other issues.

There are of course numerous other key issues which must be considered, for example, the choice of the optimal carrier may differ for each particular application and will depend on several factors, such as the clinical target, the optimal route of administration, likely dose and frequency of dosing, the hydrophilic/hydrophobic ratio of the type of drug required, the target population and patient ease of use, treatment duration, and the long-term operational stability required. Duncan and Gaspar recently argued that the key to success is not multidisciplinary but a true interdisciplinary collaboration between the core scientific disciplines, academia, industry, and regulatory agencies under the goal of developing "high-quality nanomedicines by rational design."[5]

REFERENCES

1. Kaplun AP, Bezrukov DA, Shvets VI. Rational design of nano- and micro-size medicinal forms of biologically active substances. *Appl Biochem Microbiol* 2011;**47**:711–7.
2. Fernandez-Fernandez A, Manchanda R, McGoron AJ. Theranostic applications of nanomaterials in cancer: drug delivery, image-guided therapy, and multifunctional platforms. *Appl Biochem Biotechnol* 2011;**165**:1628–51.
3. Torchilin VP. Passive and active drug targeting: drug delivery to tumors as an example. In: Schäfer-Korting M, editor. *Drug delivery. Handbook of experimental pharmacology*. Berlin: Springer; 2010.
4. Mahmood M, Casciano D, Xu Y, Biris AS. Engineered nanostructural materials for application in cancer biology and medicine. *J Appl Toxicol* 2012;**32**:10–9.
5. Duncan R, Gaspar R. Nanomedicine(s) under the microscope. *Mol Pharm* 2011;**8**:2101–41.

6. Cavadas M, González-Fernández A, Franco R. Pathogen-mimetic stealth nanocarriers for drug delivery: a future possibility. *Nanomedicine* 2011;**7**:730–43.
7. Caldorera-Moore M, Guimard N, Shi L, Roy K. Designer nanoparticles: incorporating size, shape and triggered release into nanoscale drug carriers. *Expert Opin Drug Deliv* 2010; **7**:479–95.
8. Yoo J-W, Doshi N, Mitragotri S. Adaptive micro and nanoparticles: temporal control over carrier properties to facilitate drug delivery. *Adv Drug Deliv Rev* 2011;**63**:1247–56. doi:10.1016/j.addr.2011.05.004.
9. Williams Y, et al. Comparison of three cell fixation methods for high content analysis assays utilizing quantum dots. *J Microsc* 2008;**232**:91–8.
10. França A, et al. Sterilization matters: consequences of different sterilization techniques on gold nanoparticles. *Small* 2010;**6**:89–95.
11. Cabral H, Nishiyama N, Kataoka K. Supramolecular nanodevices: from design validation to theranostic nanomedicine. *Acc Chem Res* 2011;**44**:999–1008.
12. Peer D, et al. Nanocarriers as an emerging platform for cancer therapy. *Nat Nanotechnol* 2007;**2**:751–60.
13. Petros RA, DeSimone JM. Strategies in the design of nanoparticles for therapeutic applications. *Nat Rev Drug Discov* 2010;**9**:615–27.
14. Markovsky E, et al. Administration, distribution, metabolism and elimination of polymer-drug conjugates. *J Control Release* 2011. doi:10.1016/j.jconrel.2011.12.021.
15. Calleja P, et al. Molecular buckets: cyclodextrins for oral cancer therapy. *Ther Deliv* 2012;**3**:43–57.
16. Ensign LM, Cone R, Hanes J. Oral drug delivery with polymeric nanoparticles: the gastrointestinal mucus barriers. *Adv Drug Deliv Rev* 2012;**64**:557–70.
17. Wong TW. Design of oral insulin delivery systems. *J Drug Target* 2010;**18**:79–92.
18. Plapied L, Duhem N, des Rieux A, Préat V. Fate of polymeric nanocarriers for oral drug delivery. *Curr Opin Colloid Interface Sci* 2011;**16**:228–37.
19. Arruebo M, Fernández-Pacheco R, Ibarra MR, Santamaría J. Magnetic nanoparticles for drug delivery. *Nano Today* 2007;**2**:22–32.
20. Roger E, Lagarce F, Garcion E, Benoit J-P. Biopharmaceutical parameters to consider in order to alter the fate of nanocarriers after oral delivery. *Nanomedicine (Lond)* 2010;**5**:287–306.
21. Fu J, Fiegel J, Krauland E, Hanes J. New polymeric carriers for controlled drug delivery following inhalation or injection. *Biomaterials* 2002;**23**:4425–33.
22. Zhang J, Wu L, Chan H-K, Watanabe W. Formation, characterization, and fate of inhaled drug nanoparticles. *Adv Drug Deliv Rev* 2011;**63**:441–55.
23. Sung JC, Pulliam BL, Edwards DA. Nanoparticles for drug delivery to the lungs. *Trends Biotechnol* 2007;**25**:563–70.
24. Li M, Al-Jamal KT, Kostarelos K, Reineke J. Physiologically based pharmacokinetic modeling of nanoparticles. *ACS Nano* 2010;**4**:6303–17.
25. Markovsky E, et al. Administration, distribution, metabolism and elimination of polymer-drug conjugates. *J Control Release* 2011. doi:10.1016/j.jconrel.2011.12.021.
26. Veiseh O, Gunn JW, Zhang M. Design and fabrication of magnetic nanoparticles for targeted drug delivery and imaging. *Adv Drug Deliv Rev* 2010;**62**:284–304.
27. Acharya S, Sahoo SK. PLGA nanoparticles containing various anticancer agents and tumour delivery by EPR effect. *Adv Drug Deliv Rev* 2011;**63**:170–83.
28. Wu L, Zhang J, Watanabe W. Physical and chemical stability of drug nanoparticles. *Adv Drug Deliv Rev* 2011;**63**:456–69.

29. Brewer SH, Glomm WR, Johnson MC, Knag MK, Franzen S. Probing BSA binding to citrate-coated gold nanoparticles and surfaces. *Langmuir* 2005;**21**:9303–7.
30. Sperling RA, Parak WJ. Surface modification, functionalization and bioconjugation of colloidal inorganic nanoparticles. *Philos Transact A Math Phys Eng Sci* 2010;**368**:1333–83.
31. Lourenco C, Teixeira M, Simões S, Gaspar R. Steric stabilization of nanoparticles: size and surface properties. *Int J Pharm* 1996;**138**:1–12.
32. Gref R, et al. The controlled intravenous delivery of drugs using PEG-coated sterically stabilized nanospheres. *Adv Drug Deliv Rev* 1995;**16**:215–33.
33. Owens III DE, Peppas NA. Opsonization, biodistribution, and pharmacokinetics of polymeric nanoparticles. *Int J Pharm* 2006;**307**:93–102.
34. Ferrari M. Frontiers in cancer nanomedicine: directing mass transport through biological barriers. *Trends Biotechnol* 2010;**28**:181–8.
35. Li Y, Wang J, Wientjes MG, Au JL. Delivery of nanomedicines to extracellular and intracellular compartments of a solid tumor. *Adv Drug Deliv Rev* 2012;**64**:29–39. doi:10.1016/j.addr.2011.04.006.
36. Choi HS, et al. Renal clearance of quantum dots. *Nat Biotechnol* 2007;**25**:1165–70.
37. Choi HS, et al. Design considerations for tumour-targeted nanoparticles. *Nat Nanotechnol* 2010;**5**:42–7.
38. Li Y-F, Chen C. Fate and toxicity of metallic and metal-containing nanoparticles for biomedical applications. *Small* 2011;**7**:2965–80.
39. De Jong WH, et al. Particle size-dependent organ distribution of gold nanoparticles after intravenous administration. *Biomaterials* 2008;**29**:1912–9.
40. Moghimi SM, Hunter A, Murray JC. Long-circulating and target-specific nanoparticles: theory to practice. *Pharmacol Rev* 2001;**53**:283–318.
41. Daum N, Tscheka C, Neumeyer A, Schneider M. Novel approaches for drug delivery systems in nanomedicine: effects of particle design and shape. *Wiley Interdiscip Rev Nanomed Nanobiotechnol* 2012;**4**:52–65.
42. Park J, et al. Systematic surface engineering of magnetic nanoworms for in vivo tumor targeting. *Small* 2009;**5**:694–700.
43. Lasagna-Reeves C, et al. Bioaccumulation and toxicity of gold nanoparticles after repeated administration in mice. *Biochem Biophys Res Commun* 2010;**393**:649–55.
44. Hirn S, et al. Particle size-dependent and surface charge-dependent biodistribution of gold nanoparticles after intravenous administration. *Eur J Pharm Biopharm* 2011;**77**:407–16.
45. Phillips MA, Gran ML, Peppas NA. Targeted nanodelivery of drugs and diagnostics. *Nano Today* 2010;**5**:143–59.
46. Aggarwal P, Hall JB, McLeland CB, Dobrovolskaia MA, McNeil SE. Nanoparticle interaction with plasma proteins as it relates to particle biodistribution, biocompatibility and therapeutic efficacy. *Adv Drug Deliv Rev* 2009;**61**:428–37.
47. Karmali PP, Simberg D. Interactions of nanoparticles with plasma proteins: implication on clearance and toxicity of drug delivery systems. *Expert Opin Drug Deliv* 2011;**8**:343–57.
48. Monopoli MP, et al. Physical-chemical aspects of protein corona: relevance to in vitro and in vivo biological impacts of nanoparticles. *J Am Chem Soc* 2011;**133**:2525–34.
49. Dell'Orco D, Lundqvist M, Oslakovic C, Cedervall T, Linse S. Modeling the time evolution of the nanoparticle-protein corona in a body fluid. *PLoS One* 2010;**5**:e10949.
50. Asuri P, Bale SS, Karajanagi SS, Kane RS. The protein–nanomaterial interface. *Curr Opin Biotechnol* 2006;**17**:562–8.
51. Walczyk D, Bombelli FB, Monopoli MP, Lynch I, Dawson KA. What the cell "sees" in bionanoscience. *J Am Chem Soc* 2010;**132**:5761–8.

52. Mahmoudi M, Serpooshan V. Large protein absorptions from small changes on the surface of nanoparticles. *J Phys Chem C* 2011;**115**:18275–83.
53. Dobrovolskaia MA, et al. Interaction of colloidal gold nanoparticles with human blood: effects on particle size and analysis of plasma protein binding profiles. *Nanomedicine* 2009;**5**:106–17.
54. Lundqvist M, et al. Nanoparticle size and surface properties determine the protein corona with possible implications for biological impacts. *Proc Natl Acad Sci USA* 2008;**105**: 14265–70.
55. Storm G, Belliot SO, Daemen T, Lasic DD. Surface modification of nanoparticles to oppose uptake by the mononuclear phagocyte system. *Adv Drug Deliv Rev* 1995;**17**:31–48.
56. Lipka J, et al. Biodistribution of PEG-modified gold nanoparticles following intratracheal instillation and intravenous injection. *Biomaterials* 2010;**31**:1–8.
57. Karakoti AS, Das S, Thevuthasan S, Seal S. PEGylated inorganic nanoparticles. *Angew Chem Int Ed Engl* 2011;**50**:1980–94.
58. Choi SW, Kim WS, Kim JH. Surface modification of functional nanoparticles for controlled drug delivery. *J Dispers Sci Technol* 2003;**24**:475–87.
59. Moghimi S, Pavey K, Hunter A. Real-time evidence of surface modification at polystyrene lattices by poloxamine 908 in the presence of serum: in vivo conversion of macrophage-prone nanoparticles to stealth entities by poloxamine 908. *FEBS Lett* 2003;**547**:177–82.
60. Laurent S, et al. Magnetic iron oxide nanoparticles: synthesis, stabilization, vectorization, physicochemical characterizations, and biological applications. *Chem Rev* 2008;**108**: 2064–110.
61. Torchilin VP, Trubetskoy VS. Which polymers can make nanoparticulate drug carriers long-circulating? *Adv Drug Deliv Rev* 1995;**16**:141–55.
62. Arias JL, Reddy LH, Couvreur P. Superior preclinical efficacy of gemcitabine developed as chitosan nanoparticulate system. *Biomacromolecules* 2011;**12**:97–104.
63. Immordino ML, Dosio F, Cattel L. Stealth liposomes: review of the basic science, rationale, and clinical applications, existing and potential. *Int J Nanomedicine* 2006;**1**:297–315.
64. Moghimi SM, Szebeni J. Stealth liposomes and long circulating nanoparticles: critical issues in pharmacokinetics, opsonization and protein-binding properties. *Prog Lipid Res* 2003;**42**:463–78.
65. Duncan R. Polymer therapeutics as nanomedicines: new perspectives. *Curr Opin Biotechnol* 2011;**22**:492–501.
66. Veronese FM, Pasut G. PEGylation, successful approach to drug delivery. *Drug Discov Today* 2005;**10**:1451–8.
67. Dufort S, Sancey L, Coll J-L. Physico-chemical parameters that govern nanoparticles fate also dictate rules for their molecular evolution. *Adv Drug Deliv Rev* 2012;**64**:179–89. doi10.1016/j.addr.2011.09.009.
68. Li SD, Huang L. Stealth nanoparticles: high density but sheddable PEG is a key for tumor targeting. *J Control Release* 2010;**145**:178–81.
69. Xie J, Xu C, Kohler N, Hou Y, Sun S. Controlled PEGylation of monodisperse Fe3O4 nanoparticles for reduced non-specific uptake by macrophage cells. *Adv Mater* 2007;**19**:3163–6.
70. Faure A-C, et al. Control of the in vivo biodistribution of hybrid nanoparticles with different poly(ethylene glycol) coatings. *Small* 2009;**5**:2565–75.
71. Kaminskas LM, et al. The impact of molecular weight and PEG chain length on the systemic pharmacokinetics of PEGylated poly L-lysine dendrimers. *Mol Pharm* 2008;**5**:449–63.
72. Kaminskas LM, et al. Pharmacokinetics and tumor disposition of PEGylated, methotrexate conjugated poly-L-lysine dendrimers. *Mol Pharm* 2009;**6**:1190–204.

73. Wang YS, et al. Structural and biological characterization of pegylated recombinant interferon alpha-2b and its therapeutic implications. *Adv Drug Deliv Rev* 2002;**54**:547–70.
74. Reddy LH, Murthy RSR. Pharmacokinetics and biodistribution studies of doxorubicin loaded poly(butyl cyanoacrylate) nanoparticles synthesized by two different techniques. *Biomed Pap Med Fac Univ Palacky Olomouc Czech Repub* 2004;**148**:161–6.
75. Webster R, et al. PEGylated proteins: evaluation of their safety in the absence of definitive metabolism studies. *Drug Metab Dispos* 2007;**35**:9–16.
76. Bendele A, Seely J, Richey C, Sennello G, Shopp G. Renal tubular vacuolation in animals treated with polyethylene-glycol-conjugated proteins. *Toxicol Sci* 1998;**42**:152–7.
77. Barz M, Luxenhofer R, Zentel R, Vicent MJ. Overcoming the PEG-addiction: well-defined alternatives to PEG, from structure–property relationships to better defined therapeutics. *Polym Chem* 2011;**2**:1900.
78. Huang X, Teng X, Chen D, Tang F, He J. The effect of the shape of mesoporous silica nanoparticles on cellular uptake and cell function. *Biomaterials* 2010;**31**:438–48.
79. Nielsen LE, Landel RF. *Mechanical properties of polymers and composites.* New York: Marcel Dekker, Inc.; 1994.
80. Doshi N, Zahr AS, Bhaskar S, Lahann J, Mitragotri S. Red blood cell-mimicking synthetic biomaterial particles. *Proc Natl Acad Sci USA* 2009;**106**:21495–9.
81. Kelly C, Jefferies C, Cryan S-A. Targeted liposomal drug delivery to monocytes and macrophages. *J Drug Deliv* 2011;**2011**:727241.
82. Ahsan F, Rivas IP, Khan MA, Torres Suarez AI. Targeting to macrophages: role of physicochemical properties of particulate carriers-liposomes and microspheres-on the phagocytosis by macrophages. *J Control Release* 2002;**79**:29–40.
83. Jiang H-L, et al. The potential of mannosylated chitosan microspheres to target macrophage mannose receptors in an adjuvant-delivery system for intranasal immunization. *Biomaterials* 2008;**29**:1931–9.
84. Singodia D, Verma A, Verma RK, Mishra PR. Investigations into an alternate approach to target mannose receptors on macrophages using 4-sulfated N-acetyl galactosamine more efficiently in comparison with mannose-decorated liposomes: an application in drug delivery. *Nanomedicine* 2012;**8**:468–77. doi10.1016/j.nano.2011.07.002.
85. Aouadi M, et al. Orally delivered siRNA targeting macrophage Map4k4 suppresses systemic inflammation. *Nature* 2009;**458**:1180–4.
86. Kim S-S, et al. Targeted delivery of siRNA to macrophages for anti-inflammatory treatment. *Mol Ther* 2010;**18**:993–1001.
87. Mehta D, Malik AB. Signaling mechanisms regulating endothelial permeability. *Physiol Rev* 2006;**86**:279–367.
88. Bazzoni G, Dejana E. Endothelial cell-to-cell junctions: molecular organization and role in vascular homeostasis. *Physiol Rev* 2004;**84**:869–901.
89. Pappenheimer J, Renkin E, Borrero LM. Filtration, diffusion and molecular sieving through peripheral capillary membranes. A contribution to the pore theory of capillary permeability. *Am J Physiol* 1951;**167**:13–46.
90. Gentile F, Ferrari M, Decuzzi P. The transport of nanoparticles in blood vessels: the effect of vessel permeability and blood rheology. *Ann Biomed Eng* 2008;**36**:254–61.
91. Agarwal A, et al. Nanoparticles as novel carrier for brain delivery: a review. *Curr Pharm Des* 2009;**15**:917–25.
92. Chen Y, Liu L. Modern methods for delivery of drugs across the blood-brain barrier. *Adv Drug Deliv Rev* 2012;**64**:640–65.
93. Wong HL, Wu XY, Bendayan R. Nanotechnological advances for the delivery of CNS therapeutics. *Adv Drug Deliv Rev* 2012;**64**:686–700.

94. Sweet DM, Kolhatkar RB, Ray A, Swaan P, Ghandehari H. Transepithelial transport of PEGylated anionic poly(amidoamine) dendrimers: implications for oral drug delivery. *J Control Release* 2009;**138**:78–85.
95. Sadekar S, Ghandehari H. Transepithelial transport and toxicity of PAMAM dendrimers: implications for oral drug delivery. *Adv Drug Deliv Rev* 2012;**64**:571–88.
96. Predescu SA, Predescu DN, Malik AB. Molecular determinants of endothelial transcytosis and their role in endothelial permeability. *Am J Physiol Lung Cell Mol Physiol* 2007;**293**:L823–L842.
97. Lin I-C, et al. Interaction of densely polymer-coated gold nanoparticles with epithelial Caco-2 monolayers. *Biomacromolecules* 2011;**12**:1339–48.
98. Decuzzi P, Pasqualini R, Arap W, Ferrari M. Intravascular delivery of particulate systems: does geometry really matter? *Pharm Res* 2009;**26**:235–43.
99. Pober JS, Sessa WC. Evolving functions of endothelial cells in inflammation. *Nat Rev Immunol* 2007;**7**:803–15.
100. Roberts WG, Palade GE. Increased microvascular permeability and endothelial fenestration induced by vascular endothelial growth factor. *J Cell Sci* 1995;**108**:2369–79.
101. Esser S, et al. Vascular endothelial growth factor induces endothelial fenestrations in vitro. *J Cell Biol* 1998;**140**:947–59.
102. Schonthaler HB, Huggenberger R, Wculek SK, Detmar M, Wagner EF. Systemic anti-VEGF treatment strongly reduces skin inflammation in a mouse model of psoriasis. *Proc Natl Acad Sci USA* 2009;**106**:21264–9.
103. Chacko A-M, Hood ED, Zern BJ, Muzykantov VR. Targeted nanocarriers for imaging and therapy of vascular inflammation. *Curr Opin Colloid Interface Sci* 2011;**16**:215–27.
104. Simone E, Ding B-S, Muzykantov V. Targeted delivery of therapeutics to endothelium. *Cell Tissue Res* 2009;**335**:283–300.
105. Jain RK. Transport of molecules in the tumor interstitium: a review. *Cancer Res* 1987;**47**:3039–51.
106. Heldin C-H, Rubin K, Pietras K, Östman A. High interstitial fluid pressure—an obstacle in cancer therapy. *Nat Rev Cancer* 2004;**4**:806–13.
107. Holback H, Yeo Y. Intratumoral drug delivery with nanoparticulate carriers. *Pharm Res* 2011;**28**:1819–30.
108. Jain RK, Tong RT, Munn LL. Effect of vascular normalization by antiangiogenic therapy on interstitial hypertension, peritumor edema, and lymphatic metastasis: insights from a mathematical model. *Cancer Res* 2007;**67**:2729–35.
109. Danquah MK, Zhang XA, Mahato RI. Extravasation of polymeric nanomedicines across tumor vasculature. *Adv Drug Deliv Rev* 2011;**63**:623–39.
110. Rabinovitch M. Professional and non-professional phagocytes: an introduction. *Trends Cell Biol* 1995;**5**:85–7.
111. Conradt B. With a little help from your friends: cells don't die alone. *Nat Cell Biol* 2002;**4**: E139–E143.
112. Mayor S, Pagano RE. Pathways of clathrin-independent endocytosis. *Nat Rev Mol Cell Biol* 2007;**8**:603–12.
113. Iversen T-G, Skotland T, Sandvig K. Endocytosis and intracellular transport of nanoparticles: present knowledge and need for future studies. *Nano Today* 2011;**6**:176–85.
114. Swanson JA. Shaping cups into phagosomes and macropinosomes. *Nat Rev Mol Cell Biol* 2008;**9**:639–49.
115. Chimini G, Chavrier P. Function of Rho family proteins in actin dynamics during phagocytosis and engulfment. *Nat Cell Biol* 2000;**2**:E191–E196.
116. Aderem A, Underhill DM. Mechanisms of phagocytosis in macrophages. *Annu Rev Immunol* 1999;**17**:593–623.

117. Swanson JA, Watts C. Macropinocytosis. *Trends Cell Biol* 1995;**5**:424.
118. Lim JP, Gleeson PA. Macropinocytosis: an endocytic pathway for internalising large gulps. *Immunol Cell Biol* 2011;**89**:836–43.
119. Kerr MC, Teasdale RD. Defining macropinocytosis. *Traffic* 2009;**10**:364–71.
120. Jones AT. Macropinocytosis: searching for an endocytic identity and role in the uptake of cell penetrating peptides. *J Cell Mol Med* 2007;**11**:670–84.
121. Conner SD, Schmid SL. Regulated portals of entry into the cell. *Nature* 2003;**422**:37–44.
122. McMahon HT, Boucrot E. Molecular mechanism and physiological functions of clathrin-mediated endocytosis. *Nat Rev Mol Cell Biol* 2011;**12**:517–33.
123. Kirchhausen T. Clathrin. *Annu Rev Biochem* 2000;**69**:699–727.
124. Damke H. Dynamin and receptor-mediated endocytosis. *FEBS Lett* 1996;**389**:48–51.
125. Rappoport JZ. Focusing on clathrin-mediated endocytosis. *Biochem J* 2008;**412**:415–23.
126. Michel V, Bakovic M. Lipid rafts in health and disease. *Biol Cell* 2007;**99**:129–40.
127. Bareford LM, Swaan PW. Endocytic mechanisms for targeted drug delivery. *Adv Drug Deliv Rev* 2007;**59**:748–58.
128. Pelkmans L, Kartenbeck J, Helenius A. Caveolar endocytosis of simian virus 40 reveals a new two-step vesicular-transport pathway to the ER. *Nat Cell Biol* 2001;**3**:473–83.
129. Parton RG, Richards AA. Lipid rafts and caveolae as portals for endocytosis: new insights and common mechanisms. *Traffic* 2003;**4**:724–38.
130. Hayer A, et al. Caveolin-1 is ubiquitinated and targeted to intralumenal vesicles in endolysosomes for degradation. *J Cell Biol* 2010;**191**:615–29.
131. Ortegren U, Aboulaich N, Ost A, Strålfors P. A new role for caveolae as metabolic platforms. *Trends Endocrinol Metab* 2007;**18**:344–9.
132. Harvey RD, Calaghan SC. Caveolae create local signalling domains through their distinct protein content, lipid profile and morphology. *J Mol Cell Cardiol* 2012;**52**:366–75.
133. Parton RG, Simons K. The multiple faces of caveolae. *Nat Rev Mol Cell Biol* 2007;**8**:185–94.
134. Bastiani M, Parton RG. Caveolae at a glance. *J Cell Sci* 2010;**123**:3831–6.
135. Frank PG, Pavlides S, Lisanti MP. Caveolae and transcytosis in endothelial cells: role in atherosclerosis. *Cell Tissue Res* 2009;**335**:41–7.
136. Oh P, et al. Live dynamic imaging of caveolae pumping targeted antibody rapidly and specifically across endothelium in the lung. *Nat Biotechnol* 2007;**25**:327–37.
137. Sandvig K, Torgersen ML, Raa HA, van Deurs B. Clathrin-independent endocytosis: from nonexisting to an extreme degree of complexity. *Histochem Cell Biol* 2008;**129**:267–76.
138. Hansen CG, Nichols BJ. Molecular mechanisms of clathrin-independent endocytosis. *J Cell Sci* 2009;**122**:1713–21.
139. Howes MT, Mayor S, Parton RG. Molecules, mechanisms, and cellular roles of clathrin-independent endocytosis. *Curr Opin Cell Biol* 2010;**22**:519–27.
140. Sandvig K, Pust S, Skotland T, van Deurs B. Clathrin-independent endocytosis: mechanisms and function. *Curr Opin Cell Biol* 2011;**23**:413–20.
141. Torchilin VP. Recent advances with liposomes as pharmaceutical carriers. *Nat Rev Drug Discov* 2005;**4**:145–60.
142. Dutta D, Pulsipher A, Luo W, Mak H, Yousaf MN. Engineering cell surfaces via liposome fusion. *Bioconjug Chem* 2011;**22**:2423–33.
143. Xia T, Kovochich M, Liong M, Zink JI, Nel AE. Cationic polystyrene nanosphere toxicity depends on cell-specific endocytic and mitochondrial injury pathways. *ACS Nano* 2008;**2**:85–96.
144. Albanese A, Chan WCW. Effect of gold nanoparticle aggregation on cell uptake and toxicity. *ACS Nano* 2011;**5**:5478–89.

145. Dos Santos T, Varela J, Lynch I, Salvati A, Dawson KA. Quantitative assessment of the comparative nanoparticle-uptake efficiency of a range of cell lines. *Small* 2011;**7**:3341–9.
146. Hillaireau H, Couvreur P. Nanocarriers' entry into the cell: relevance to drug delivery. *Cell Mol Life Sci* 2009;**66**:2873–96.
147. Wang J, Byrne JD, Napier ME, DeSimone JM. More effective nanomedicines through particle design. *Small* 2011;**7**:1919–31. doi:10.1002/smll.201100442.
148. Verma A, et al. Surface-structure-regulated cell-membrane penetration by monolayer-protected nanoparticles. *Nat Mater* 2008;**7**:588–95.
149. Cai J, et al. Effect of chain length on cytotoxicity and endocytosis of cationic polymers. *Macromolecules* 2011;**44**:2050–7.
150. Qaddoumi MG, et al. The characteristics and mechanisms of uptake of PLGA nanoparticles in rabbit conjunctival epithelial cell layers. *Pharm Res* 2004;**21**:641–8.
151. Rejman J, Oberle V, Zuhorn IS, Hoekstra D. Size-dependent internalization of particles via the pathways of clathrin and caveolae-mediated endocytosis. *Biochem J* 2004;**377**.159–69.
152. Chithrani BD, Chan WCW. Elucidating the mechanism of cellular uptake and removal of protein-coated gold nanoparticles of different sizes and shapes. *Nano Lett* 2007;**7**:1542–50.
153. Wang Z, Tiruppathi C, Minshall RD, Malik AB. Size and dynamics of caveolae studied using nanoparticles in living endothelial cells. *ACS Nano* 2009;**3**:4110–6.
154. Yoo J-W, Mitragotri S. Polymer particles that switch shape in response to a stimulus. *Proc Natl Acad Sci USA* 2010;**107**:11205–10.
155. Gratton SEA, et al. The effect of particle design on cellular internalization pathways. *Proc Natl Acad Sci USA* 2008;**105**:11613–8.
156. Chithrani BD, Ghazani AA, Chan WCW. Determining the size and shape dependence of gold nanoparticle uptake into mammalian cells. *Nano Lett* 2006;**6**:662–8.
157. Arvizo RR, et al. Effect of nanoparticle surface charge at the plasma membrane and beyond. *Nano Lett* 2010;**10**:2543–8.
158. Harush-Frenkel O, Rozentur E, Benita S, Altschuler Y. Surface charge of nanoparticles determines their endocytic and transcytotic pathway in polarized MDCK cells. *Biomacromolecules* 2008;**9**:435–43.
159. Dausend J, et al. Uptake mechanism of oppositely charged fluorescent nanoparticles in HeLa cells. *Macromol Biosci* 2008;**8**:1135–43.
160. Cho EC, Xie J, Wurm PA, Xia Y. Understanding the role of surface charges in cellular adsorption versus internalization by selectively removing gold nanoparticles on the cell surface with a I2/KI etchant. *Nano Lett* 2009;**9**:1080–4.
161. Harush-Frenkel O, Debotton N, Benita S, Altschuler Y. Targeting of nanoparticles to the clathrin-mediated endocytic pathway. *Biochem Biophys Res Commun* 2007;**353**:26–32.
162. Bartlett DW, Su H, Hildebrandt IJ, Weber WA, Davis ME. Impact of tumor-specific targeting on the biodistribution and efficacy of siRNA nanoparticles measured by multimodality in vivo imaging. *Proc Natl Acad Sci USA* 2007;**104**:15549–54.
163. Moros M, et al. Monosaccharides versus PEG-functionalized NPs: influence in the cellular uptake. *ACS Nano* 2012;**6**:1565–77. doi:10.1021/nn204543c.
164. Low PS, Henne WA, Doorneweerd DD. Discovery and development of folic-acid-based receptor targeting for imaging and therapy of cancer and inflammatory diseases. *Acc Chem Res* 2008;**41**:120–9.
165. Bolhassani A. Potential efficacy of cell-penetrating peptides for nucleic acid and drug delivery in cancer. *Biochim Biophys Acta* 2011;**1816**:232–46.
166. Deshayes S, Morris MC, Divita G, Heitz F. Cell-penetrating peptides: tools for intracellular delivery of therapeutics. *Cell Mol Life Sci* 2005;**62**:1839–49.

167. Said Hassane F, Saleh aF, Abes R, Gait MJ, Lebleu B. Cell penetrating peptides: overview and applications to the delivery of oligonucleotides. *Cell Mol Life Sci* 2010;**67**:715–26.
168. Järver P, Langel U. Cell-penetrating peptides-a brief introduction. *Biochim Biophys Acta* 2006;**1758**:260–3.
169. Madani F, Lindberg S, Langel U, Futaki S, Gräslund A. Mechanisms of cellular uptake of cell-penetrating peptides. *J Biophys* 2011;**2011**:414729.
170. Richard JP, et al. Cell-penetrating peptides. A reevaluation of the mechanism of cellular uptake. *J Biol Chem* 2003;**278**:585–90.
171. Ziegler A. Thermodynamic studies and binding mechanisms of cell-penetrating peptides with lipids and glycosaminoglycans. *Adv Drug Deliv Rev* 2008;**60**:580–97.
172. Trabulo S, Cardoso AL, Mano M, de Lima MCP. Cell-penetrating peptides—mechanisms of cellular uptake and generation of delivery systems. *Pharmaceuticals* 2010;**3**:961–93.
173. Endoh T, Ohtsuki T. Cellular siRNA delivery using cell-penetrating peptides modified for endosomal escape. *Adv Drug Deliv Rev* 2009;**61**:704–9.
174. Fonseca SB, Pereira MP, Kelley SO. Recent advances in the use of cell-penetrating peptides for medical and biological applications. *Adv Drug Deliv Rev* 2009;**61**:953–64.
175. Caron NJ, Quenneville SP, Tremblay JP. Endosome disruption enhances the functional nuclear delivery of Tat-fusion proteins. *Biochem Biophys Res Commun* 2004;**319**:12–20.
176. de la Fuente JM, Berry CC. Tat peptide as an efficient molecule to translocate gold nanoparticles into the cell nucleus. *Bioconjug Chem* 2005;**16**:1176–80.
177. Ruan G, Agrawal A, Marcus AI, Nie S. Imaging and tracking of tat peptide-conjugated quantum dots in living cells: new insights into nanoparticle uptake, intracellular transport, and vesicle shedding. *J Am Chem Soc* 2007;**129**:14759–66.
178. Krpetić Z, et al. Negotiation of intracellular membrane barriers by TAT-modified gold nanoparticles. *ACS Nano* 2011;**5**:5195–201.
179. Oh E, et al. Cellular uptake and fate of PEGylated gold nanoparticles is dependent on both cell-penetration peptides and particle size. *ACS Nano* 2011;**5**:6434–48.
180. Wang H, Zhong C-Y, Wu J-F, Huang Y-B, Liu C-B. Enhancement of TAT cell membrane penetration efficiency by dimethyl sulphoxide. *J Control Release* 2010;**143**:64–70.
181. Ye S-F, et al. Synergistic effects of cell-penetrating peptide Tat and fusogenic peptide HA2-enhanced cellular internalization and gene transduction of organosilica nanoparticles. *Nanomedicine* 2011;**2**:1–9.
182. Abes S, et al. Vectorization of morpholino oligomers by the (R-Ahx-R)4 peptide allows efficient splicing correction in the absence of endosomolytic agents. *J Control Release* 2006;**116**:304–13.
183. Martin ME, Rice KG. Peptide-guided gene delivery. *AAPS J* 2007;**9**:E18–E29.
184. Vivès E, Schmidt J, Pèlegrin A. Cell-penetrating and cell-targeting peptides in drug delivery. *Biochim Biophys Acta* 2008;**1786**:126–38.
185. Schwarze SR, Ho A, Vocero-Akbani A, Dowdy SF. In vivo protein transduction: delivery of a biologically active protein into the mouse. *Science* 1999;**285**:1569–72.
186. Song E, et al. Antibody mediated in vivo delivery of small interfering RNAs via cell-surface receptors. *Nat Biotechnol* 2005;**23**:709–17.
187. McNamara JO, et al. Cell type-specific delivery of siRNAs with aptamer-siRNA chimeras. *Nat Biotechnol* 2006;**24**:1005–15.
188. Kim KS, et al. Bioimaging for targeted delivery of hyaluronic acid derivatives to the livers in cirrhotic mice using quantum dots. *ACS Nano* 2010;**4**:3005–14.
189. Kao HP, Abney JR, Verkman A. Determinants of the translational mobility of a small solute in cell cytoplasm. *J Cell Biol* 1993;**120**:175–84.

190. Lechardeur D, Verkman AS, Lukacs GL. Intracellular routing of plasmid DNA during non-viral gene transfer. *Adv Drug Deliv Rev* 2005;**57**:755–67.
191. Wente SR. Gatekeepers of the nucleus. *Science* 2000;**288**:1374–7.
192. Wagstaff KM, Jans DA. Nuclear drug delivery to target tumour cells. *Eur J Pharmacol* 2009;**625**:174–80.
193. Varkouhi AK, Scholte M, Storm G, Haisma HJ. Endosomal escape pathways for delivery of biologicals. *J Control Release* 2011;**151**:220–8.
194. Boussif O, et al. A versatile vector for gene and oligonucleotide transfer into cells in culture and in vivo: polyethylenimine. *Proc Natl Acad Sci USA* 1995;**92**:7297–301.
195. Richardson SC, Pattrick NG, Man YK, Ferruti P, Duncan R. Poly(amidoamine)s as potential nonviral vectors: ability to form interpolyelectrolyte complexes and to mediate transfection in vitro. *Biomacromolecules* 2001;**2**:1023–8.
196. Cheung CY, Murthy N, Stayton PS, Hoffman AS. A pH-sensitive polymer that enhances cationic lipid-mediated gene transfer. *Bioconjug Chem* 2006;**12**:906 10.
197. Benns JM, Choi JS, Mahato RI, Park JS, Kim SW. pH-sensitive cationic polymer gene delivery vehicle: N-Ac-poly(L-histidine)-graft-poly(L-lysine) comb shaped polymer. *Bioconjug Chem* 2000;**11**:637–45.
198. Chang K-L, Higuchi Y, Kawakami S, Yamashita F, Hashida M. Efficient gene transfection by histidine-modified chitosan through enhancement of endosomal escape. *Bioconjug Chem* 2010;**21**:1087–95.
199. Huang JG, Leshuk T, Gu FX. Emerging nanomaterials for targeting subcellular organelles. *Nano Today* 2011;**6**:478–92.
200. Zelphati O, Szoka FC. Mechanism of oligonucleotide release from cationic liposomes. *Proc Natl Acad Sci USA* 1996;**93**:11493–8.
201. Morille M, Passirani C, Vonarbourg A, Clavreul A, Benoit JP. Progress in developing cationic vectors for non-viral systemic gene therapy against cancer. *Biomaterials* 2008;**29**:3477–96.
202. Paillard A, Hindré F, Vignes-Colombeix C, Benoit J-P, Garcion E. The importance of endo-lysosomal escape with lipid nanocapsules for drug subcellular bioavailability. *Biomaterials* 2010;**31**:7542–54.
203. Mastrobattista E, et al. Functional characterization of an endosome-disruptive peptide and its application in cytosolic delivery of immunoliposome-entrapped proteins. *J Biol Chem* 2002;**277**:27135–43.
204. Li W, Nicol F, Szoka FC. GALA: a designed synthetic pH-responsive amphipathic peptide with applications in drug and gene delivery. *Adv Drug Deliv Rev* 2004;**56**:967–85.
205. Wyman TB, et al. Design, synthesis, and characterization of a cationic peptide that binds to nucleic acids and permeabilizes bilayers. *Biochemistry* 1997;**36**:3008–17.
206. Nakase I, Kobayashi S, Futaki S. Endosome-disruptive peptides for improving cytosolic delivery of bioactive macromolecules. *Biopolymers* 2010;**94**:763–70.
207. Selbo PK, et al. Photochemical internalization provides time- and space-controlled endolysosomal escape of therapeutic molecules. *J Control Release* 2010;**148**:2–12.
208. Zhou Z, et al. Charge-reversal drug conjugate for targeted cancer cell nuclear drug delivery. *Adv Funct Mater* 2009;**19**:3580–9.
209. Escoffre J-M, Teissié J, Rols M-P. Gene transfer: how can the biological barriers be overcome? *J Membr Biol* 2010;**236**:61–74.
210. Panté N, Kann M. Nuclear pore complex is able to transport macromolecules with diameters of 39 nm. *Mol Biol Cell* 2002;**13**:425–34.
211. Misra R, Sahoo SK. Intracellular trafficking of nuclear localization signal conjugated nanoparticles for cancer therapy. *Eur J Pharm Sci* 2010;**39**:152–63.

212. Symens N, et al. Intracellular partitioning of cell organelles and extraneous nanoparticles during mitosis. *Adv Drug Deliv Rev* 2012;**64**:78–94. doi:10.1016/j.addr.2011.11.012.
213. Sui M, Liu W, Shen Y. Nuclear drug delivery for cancer chemotherapy. *J Control Release* 2011;**155**:227–36.
214. Pouton CW, Wagstaff KM, Roth DM, Moseley GW, Jans DA. Targeted delivery to the nucleus. *Adv Drug Deliv Rev* 2007;**59**:698–717.
215. Furumoto K, et al. Biliary excretion of polystyrene microspheres depends on the type of receptor-mediated uptake in rat liver. *Biochim Biophys Acta* 2001;**1526**:221–6.
216. Ballou B, et al. Sentinel lymph node imaging using quantum dots in mouse tumor models. *Bioconjug Chem* 2007;**18**:389–96.
217. Davis ME, Chen ZG, Shin DM. Nanoparticle therapeutics: an emerging treatment modality for cancer. *Nat Rev Drug Discov* 2008;**7**:771–82.
218. Torchilin VP. Drug targeting. *Eur J Pharm Sci* 2000;**11**(Suppl. 2):S81–S91.
219. Torchilin VP. *Drug Delivery (Handbook of Experimental Pharmacology 197).* Berlin Heidelberg: Springer-Verlag; 2010.
220. McBain SC, Yiu HHP, Dobson J. Magnetic nanoparticles for gene and drug delivery. *Int J Nanomedicine* 2008;**3**:169–80.
221. Prijic S, Sersa G. Magnetic nanoparticles as targeted delivery systems in oncology. *Radiol Oncol* 2011;**45**:1–16.
222. Sun C, Lee JSH, Zhang M. Magnetic nanoparticles in MR imaging and drug delivery. *Adv Drug Deliv Rev* 2008;**60**:1252–65.
223. Barakat NS. Magnetically modulated nanosystems: a unique drug-delivery platform. *Nanomedicine* 2009;**4**:799–812.
224. Nacev A, Beni C, Bruno O, Shapiro B. Magnetic nanoparticle transport within flowing blood and into surrounding tissue. *Nanomedicine* 2010;**5**:1459–66.
225. Farokhzad OC, Langer R. Impact of nanotechnology on drug delivery. *ACS Nano* 2009;**3**:16–20.
226. Hartner WC, Verma DD, Levchenko TS, Bernstein EA, Torchilin VP. ATP-loaded liposomes for treatment of myocardial ischemia. *Wiley Interdiscip Rev Nanomed Nanobiotechnol* 2009;**1**:530–9.
227. Maeda H. The enhanced permeability and retention (EPR) effect in tumor vasculature: the key role of tumor-selective macromolecular drug targeting. *Adv Enzyme Regul* 2001;**41**:189–207.
228. Jain RK. Normalization of tumor vasculature: an emerging concept in antiangiogenic therapy. *Science* 2005;**307**:58–62.
229. Maeda H, Sawa T, Konno T. Mechanism of tumor-targeted delivery of macromolecular drugs, including the EPR effect in solid tumor and clinical overview of the prototype polymeric drug SMANCS. *J Control Release* 2001;**74**:47–61.
230. Fang J, Nakamura H, Maeda H. The EPR effect: unique features of tumor blood vessels for drug delivery, factors involved, and limitations and augmentation of the effect. *Adv Drug Deliv Rev* 2011;**63**:136–51.
231. Yuan F, Leunig M, Huang SK. Mirovascular permeability and interstitial penetration of sterically stabilized (stealth) liposomes in a human tumor xenograft. *Cancer Res* 1994;**54**:3352–6.
232. Torchilin V. Tumor delivery of macromolecular drugs based on the EPR effect. *Adv Drug Deliv Rev* 2011;**63**:131–5.
233. Jain KK. Advances in the field of nanooncology. *BMC Med* 2010;**8**:83.

234. Maruyama K. Intracellular targeting delivery of liposomal drugs to solid tumors based on EPR effects. *Adv Drug Deliv Rev* 2011;**63**:161–9.
235. Krasnici S, et al. Effect of the surface charge of liposomes on their uptake by angiogenic tumor vessels. *Int J Cancer* 2003;**105**:561–7.
236. Dass CR, Choong PFM. Targeting of small molecule anticancer drugs to the tumour and its vasculature using cationic liposomes: lessons from gene therapy. *Cancer Cell Int* 2006;**6**:17.
237. Lee H, Hoang B, Fonge H, Reilly RM, Allen C. In vivo distribution of polymeric nanoparticles at the whole-body, tumor, and cellular levels. *Pharm Res* 2010;**27**:2343–55.
238. Lee H, Fonge H, Hoang B, Reilly RM, Allen C. The effects of particle size and molecular targeting on the intratumoral and subcellular distribution of polymeric nanoparticles. *Mol Pharm* 2010;**7**:1195–208.
239. Minchinton AI, Tannock IF. Drug penetration in solid tumours. *Nat Rev Cancer* 2006;**6**: 583–92.
240. Wong C, et al. Multistage nanoparticle delivery system for deep penetration into tumor tissue. *Proc Natl Acad Sci USA* 2011;**108**:2426–31.
241. Ruoslahti E, Bhatia SN, Sailor MJ. Targeting of drugs and nanoparticles to tumors. *J Cell Biol* 2010;**188**:759–68.
242. Hobbs SK, et al. Regulation of transport pathways in tumor vessels: role of tumor type and microenvironment. *Proc Natl Acad Sci USA* 1998;**95**:4607–12.
243. Jain RK, Stylianopoulos T. Delivering nanomedicine to solid tumors. *Nat Rev Clin Oncol* 2010;**7**:653–64.
244. Chrastina A, Massey Ka, Schnitzer JE. Overcoming in vivo barriers to targeted nanodelivery. *Wiley Interdiscip Rev Nanomed Nanobiotechnol* 2011;**3**:421–37.
245. Okuhata Y. Delivery of diagnostic agents for magnetic resonance imaging. *Adv Drug Deliv Rev* 1999;**37**:121–37.
246. Kirpotin DB, et al. Antibody targeting of long-circulating lipidic nanoparticles does not increase tumor localization but does increase internalization in animal models. *Cancer Res* 2006;**66**:6732–40.
247. Kondo K, et al. Transferrin receptor expression in adenocarcinoma of the lung as a histopathologic indicator of prognosis. *Chest* 1990;**97**:1367–71.
248. Daniels TR, Delgado T, Rodriguez JA, Helguera G, Penichet ML. The transferrin receptor part I: biology and targeting with cytotoxic antibodies for the treatment of cancer. *Clin Immunol* 2006;**121**:144–58.
249. Daniels TR, et al. Transferrin receptors and the targeted delivery of therapeutic agents against cancer. *Biochim Biophys Acta* 2012;**1820**:291–317.
250. Davis ME, et al. Evidence of RNAi in humans from systemically administered siRNA via targeted nanoparticles. *Nature* 2010;**464**:1067–70.
251. Low PS, Kularatne SA. Folate-targeted therapeutic and imaging agents for cancer. *Curr Opin Chem Biol* 2009;**13**:256–62.
252. Shia J, et al. Immunohistochemical expression of folate receptor alpha in colorectal carcinoma: patterns and biological significance. *Hum Pathol* 2008;**39**:498–505.
253. Parker N, et al. Folate receptor expression in carcinomas and normal tissues determined by a quantitative radioligand binding assay. *Anal Biochem* 2005;**338**:284–93.
254. Sega EI, Low PS. Tumor detection using folate receptor-targeted imaging agents. *Cancer Metastasis Rev* 2008;**27**:655–64.
255. Zhang W, Tong L, Yang C. Cellular binding and internalization of functionalized silicon nanowires. *Nano Lett* 2012;**12**:1002–6. dx.doi.org/10.1021/nl204131n.

256. Kukowska-Latallo JF, et al. Nanoparticle targeting of anticancer drug improves therapeutic response in animal model of human epithelial cancer. *Cancer Res* 2005;**65**:5317–24.
257. Wang X, et al. A folate receptor-targeting nanoparticle minimizes drug resistance in a human cancer model. *ACS Nano* 2011;**5**:6184–94.
258. McNeeley KM, Annapragada A, Bellamkonda RV. Decreased circulation time offsets increased efficacy of PEGylated nanocarriers targeting folate receptors of glioma. *Nanotechnology* 2007;**18**:385101.
259. Kawano K, Maitani Y. Effects of polyethylene glycol spacer length and ligand density on folate receptor targeting of liposomal Doxorubicin in vitro. *J Drug Deliv* 2011;**2011**:160967.
260. Saul JM, Annapragada A, Natarajan JV, Bellamkonda RV. Controlled targeting of liposomal doxorubicin via the folate receptor in vitro. *J Control Release* 2003;**92**:49–67.
261. Ruiz NI, Drickamer K. Differential ligand binding by two subunits of the rat liver asialoglycoprotein receptor. *Glycobiology* 1996;**6**:551–9.
262. Mamidyala SK, et al. Glycomimetic ligands for the human asialoglycoprotein receptor. *J Am Chem Soc* 2012;**134**:1978–81.
263. Yang R, et al. Galactose-decorated cross-linked biodegradable poly(ethylene glycol)-b-poly (ε-caprolactone) block copolymer micelles for enhanced hepatoma-targeting delivery of paclitaxel. *Biomacromolecules* 2011;**12**:3047–55.
264. Seymour LW, et al. Hepatic drug targeting: phase I evaluation of polymer-bound doxorubicin. *J Clin Oncol* 2002;**20**:1668–76.
265. Scaltriti M, Baselga J. The epidermal growth factor receptor pathway: a model for targeted therapy. *Clin Cancer Res* 2006;**12**:5268–72.
266. Arteaga CL. Epidermal growth factor receptor dependence in human tumors: more than just expression? *Oncologist* 2002;**7**(Suppl. 4):31–9.
267. Sachdev JC, Jahanzeb M. Blockade of the HER family of receptors in the treatment of HER2-positive metastatic breast cancer. *Clin Breast Cancer* 2011;**12**:19–29.
268. Martini M, Vecchione L, Siena S, Tejpar S, Bardelli A. Targeted therapies: how personal should we go? *Nat Rev Clin Oncol* 2011;**9**:87–97.
269. Mukohara T. Role of HER2-targeted agents in adjuvant treatment for breast cancer. *Chemother Res Pract* 2011;**2011**:730360.
270. Nielsen UB, et al. Therapeutic efficacy of anti-ErbB2 immunoliposomes targeted by a phage antibody selected for cellular endocytosis. *Biochim Biophys Acta* 2002;**1591**:109–18.
271. Drummond DC, et al. Development of a highly stable and targetable nanoliposomal formulation of topotecan. *J Control Release* 2010;**141**:13–21.
272. Gosk S, Moos T, Gottstein C, Bendas G. VCAM-1 directed immunoliposomes selectively target tumor vasculature in vivo. *Biochim Biophys Acta* 2008;**1778**:854–63.
273. Ferrara N, Kerbel RS. Angiogenesis as a therapeutic target. *Nature* 2005;**438**:967–74.
274. Pasut G, Veronese FM. PEG conjugates in clinical development or use as anticancer agents: an overview. *Adv Drug Deliv Rev* 2009;**61**:1177–88.
275. Wicki A, et al. Targeting_tumor-associated endothelial cells: anti-VEGFR2-immunoliposomes mediate tumor-vessel disruption and inhibit tumor growth. *Clin Cancer Res* 2012;**18**:454–64.
276. Linkous AG, Yazlovitskaya EM. Novel therapeutic approaches for targeting tumor angiogenesis. *Anticancer Res* 2012;**32**:1–12.
277. Jubeli E, Moine L, Vergnaud J, Barratt G. E-selectin as a target for drug delivery and molecular imaging. *J Control Release* 2011;**158**:194–206.
278. Byrne JD, Betancourt T, Brannon-Peppas L. Active targeting schemes for nanoparticle systems in cancer therapeutics. *Adv Drug Deliv Rev* 2008;**60**:1615–26.

279. Banerjee D, Harfouche R, Sengupta S. Nanotechnology-mediated targeting of tumor angiogenesis. *Vasc Cell* 2011;**3**:3.
280. Desgrosellier JS, Cheresh DA. Integrins in cancer: biological implications and therapeutic opportunities. *Nat Rev Cancer* 2010;**10**:9–22.
281. Kim Y-H, et al. Tumor targeting and imaging using cyclic RGD-PEGylated gold nanoparticle probes with directly conjugated iodine-125. *Small* 2011;**7**:2052–60.
282. Sugahara KN, et al. Tissue-penetrating delivery of compounds and nanoparticles into tumors. *Cancer Cell* 2009;**16**:510–20.
283. Sugahara KN, et al. Coadministration of a tumor-penetrating peptide enhances the efficacy of cancer drugs. *Science* 2010;**328**:1031–5.
284. Wang Z, Tiruppathi C, Cho J, Minshall RD, Malik AB. Delivery of nanoparticle: complexed drugs across the vascular endothelial barrier via caveolae. *IUBMB Life* 2011;**63**: 659–67.
285. Sahay G, Kim JO, Kabanov AV, Bronich TK. The exploitation of differential endocytic pathways in normal and tumor cells in the selective targeting of nanoparticulate chemotherapeutic agents. *Biomaterials* 2010;**31**:923–33.
286. Schnitzer JE. gp60 is an albumin-binding glycoprotein expressed by continuous endothelium involved in albumin transcytosis. *Am J Physiol Heart Circ Physiol* 1992;**262**:H246–H254.
287. Hawkins MJ, Soon-Shiong P, Desai N. Protein nanoparticles as drug carriers in clinical medicine. *Adv Drug Deliv Rev* 2008;**60**:876–85.
288. Kratz F, Elsadek B. Clinical impact of serum proteins on drug delivery. *J Control Release* 2012. doi:10.1016/ j.jconrel.2011.11.028.
289. Bae YH, Park K. Targeted drug delivery to tumors: myths, reality and possibility. *J Control Release* 2011;**153**:198–205.
290. Qiu Y, Park K. Environment-sensitive hydrogels for drug delivery. *Adv Drug Deliv Rev* 2001;**53**:321–39.
291. Mudshinge SR, Deore AB, Patil S, Bhalgat CM. Nanoparticles: emerging carriers for drug delivery. *Saudi Pharm J* 2011;**19**:129–41.
292. Ganta S, Devalapally H, Shahiwala A, Amiji M. A review of stimuli-responsive nanocarriers for drug and gene delivery. *J Control Release* 2008;**126**:187–204.
293. Kim J-H, Lee TR. Thermo- and pH-responsive hydrogel-coated gold nanoparticles. *Chem Mater* 2004;**16**:3647–51.
294. Sato K, Yoshida K, Takahashi S, Anzai J-I. pH- and sugar-sensitive layer-by-layer films and microcapsules for drug delivery. *Adv Drug Deliv Rev* 2011;**63**:809–21.
295. Bigall NC, et al. Magnetic nanocarriers with tunable pH dependence for controlled loading and release of cationic and anionic payloads. *Adv Mater* 2011;**23**:5645–50.
296. Kang KW, et al. Doxorubicin-loaded solid lipid nanoparticles to overcome multidrug resistance in cancer therapy. *Nanomedicine* 2010;**6**:210–3.
297. Luo Z, et al. Mesoporous silica nanoparticles end-capped with collagen: redox-responsive nanoreservoirs for targeted drug delivery. *Angew Chem Int Ed Engl* 2011;**50**:640–3.
298. de la Rica R, Aili D, Stevens MM. Enzyme-responsive nanoparticles for drug release and diagnostics. *Adv Drug Deliv Rev* 2012. doi:10.1016/j.addr.2012.01.002.
299. Bae B-C, Na K. Development of polymeric cargo for delivery of photosensitizer in photodynamic therapy. *Int J Photoenergy* 2012;**2012**:431975.
300. Ge J, Neofytou E, Cahill TJ, Beygui RE, Zare RN. Drug release from electric-field-responsive nanoparticles. *ACS Nano* 2012;**6**:227–33.
301. Hawkins AM, Bottom CE, Liang Z, Puleo DA, Hilt JZ. Magnetic nanocomposite sol-gel systems for remote controlled drug release. *Adv Healthc Mater* 2012;**1**:96–100.

302. Loomis K, McNeeley K, Bellamkonda RV. Nanoparticles with targeting, triggered release, and imaging functionality for cancer applications. *Soft Matter* 2011;**7**:839.
303. Sutton D, Nasongkla N, Blanco E, Gao J. Functionalized micellar systems for cancer targeted drug delivery. *Pharm Res* 2007;**24**:1029–46.
304. Vaupel P, Kallinowski F, Okunieff P. Blood flow, oxygen and nutrient supply, and metabolic microenvironment of human tumors: a review. *Cancer Res* 1989;**49**:6449–65.
305. Goya GF, Grazu V, Ibarra MR. Magnetic nanoparticles for cancer therapy. *Curr Nanosci* 2008;**4**:16.
306. Abulateefeh SR, et al. Thermoresponsive polymer colloids for drug delivery and cancer therapy. *Macromol Biosci* 2011;**11**:1722–34.
307. Wei H, Cheng S-X, Zhang X-Z, Zhuo R-X. Thermo-sensitive polymeric micelles based on poly(N-isopropylacrylamide) as drug carriers. *Prog Polym Sci* 2009;**34**:893–910.
308. Chung JE, et al. Thermo-responsive drug delivery from polymeric micelles constructed using block copolymers of poly(N-isopropylacrylamide) and poly(butylmethacrylate). *J Control Release* 1999;**62**:115–27.
309. Liu B, et al. The antitumor effect of novel docetaxel-loaded thermosensitive micelles. *Eur J Pharm Biopharm* 2008;**69**:527–34.
310. Chung J, et al. Thermo-responsive drug delivery from polymeric micelles constructed using block copolymers of poly(N-isopropylacrylamide) and poly(butylmethacrylate). *J Control Release* 1999;**62**:115–27.
311. Gong W, et al. Improving efficiency of adriamycin crossing blood brain barrier by combination of thermosensitive liposomes and hyperthermia. *Biol Pharm Bull* 2011;**34**:1058–64.
312. Agarwal A, Mackey MA, El-Sayed MA, Bellamkonda RV. Remote triggered release of doxorubicin in tumors by synergistic application of thermosensitive liposomes and gold nanorods. *ACS Nano* 2011;**5**:4919–26.
313. Needham D, Anyarambhatla G, Kong G, Dewhirst MW. A new temperature-sensitive liposome for use with mild hyperthermia: characterization and testing in a human tumor xenograft model. *Cancer Res* 2000;**60**:1197–201.
314. Kong G, et al. Efficacy of liposomes and hyperthermia in a human tumor xenograft model: importance of triggered drug release. *Cancer Res* 2000;**60**:6950.
315. Kono K, et al. Temperature sensitization of liposomes by use of thermosensitive block copolymers synthesized by living cationic polymerization: effect of copolymer chain length. *Bioconjug Chem* 2005;**16**:1367–74.
316. Chen W-H, Regen SL. Thermally gated liposomes. *J Am Chem Soc* 2005;**127**:6538–9.
317. Choi SH, Lee SH, Park TG. Temperature-sensitive pluronic/poly(ethylenimine) nanocapsules for thermally triggered disruption of intracellular endosomal compartment. *Biomacromolecules* 2006;**7**:1864–70.
318. Lee Y, Park SY, Kim C, Park TG. Thermally triggered intracellular explosion of volume transition nanogels for necrotic cell death. *J Control Release* 2009;**135**:89–95.
319. Tai L-A, et al. Thermosensitive liposomes entrapping iron oxide nanoparticles for controllable drug release. *Nanotechnology* 2009;**20**:135101.
320. Brazel CS. Magnetothermally-responsive nanomaterials: combining magnetic nanostructures and thermally-sensitive polymers for triggered drug release. *Pharm Res* 2009;**26**:644–56.
321. Liu T-Y, et al. Instantaneous drug delivery of magnetic/thermally sensitive nanospheres by a high-frequency magnetic field. *Langmuir* 2008;**24**:13306–11.
322. Liu T-Y, Hu S-H, Liu D-M, Chen S-Y, Chen I-W. Biomedical nanoparticle carriers with combined thermal and magnetic responses. *Nano Today* 2009;**4**:52–65.

323. Liu T-Y, Liu K-H, Liu D-M, Chen S-Y, Chen I-W. Temperature-sensitive nanocapsules for controlled drug release caused by magnetically triggered structural disruption. *Adv Funct Mater* 2009;**19**:616–23.
324. McGill SL, Cuylear CL, Adolphi NL, Osiński M, Smyth HDC. Magnetically responsive nanoparticles for drug delivery applications using low magnetic field strengths. *IEEE Trans Nanobioscience* 2009;**8**:33–42.
325. Derfus AM, et al. Remotely triggered release from magnetic nanoparticles. *Adv Mater* 2007;**19**:3932–6.
326. Salmaso S, et al. Cell up-take control of gold nanoparticles functionalized with a thermoresponsive polymer. *J Mater Chem* 2009;**19**:1608.
327. Muñoz Javier A, et al. Photoactivated release of cargo from the cavity of polyelectrolyte capsules to the cytosol of cells. *Langmuir* 2008;**24**:12517–20.
328. Skirtach AG, et al. Laser-induced release of encapsulated materials inside living cells. *Angew Chem Int Ed Engl* 2006;**45**:4612–7.
329. Vivero-Escoto JL, Slowing II, Wu C-W, Lin VS-Y. Photoinduced intracellular controlled release drug delivery in human cells by gold-capped mesoporous silica nanosphere. *J Am Chem Soc* 2009;**131**:3462–3.
330. Tannock IF, Rotin D. Acid pH in tumors and its potential for therapeutic exploitation. *Cancer Res* 1989;**49**:4373–84.
331. Wu W, et al. Multi-functional core-shell hybrid nanogels for pH-dependent magnetic manipulation, fluorescent pH-sensing, and drug delivery. *Biomaterials* 2011;**32**:9876–87.
332. Murthy N, Campbell J, Fausto N, Hoffman AS, Stayton PS. Bioinspired pH-responsive polymers for the intracellular delivery of biomolecular drugs. *Bioconjug Chem* 2003;**14**: 412–9.
333. Kakizawa Y, Kataoka K. Block copolymer micelles for delivery of gene and related compounds. *Adv Drug Deliv Rev* 2002;**54**:203–22.
334. Kozlovskaya V, Sukhishvili SA. Amphoteric hydrogel capsules: multiple encapsulation and release routes. *Macromolecules* 2006;**39**:6191–9.
335. Lee ES, Kim D, Youn YS, Oh KT, Bae YH. A virus-mimetic nanogel vehicle. *Angew Chem Int Ed Engl* 2008;**47**:2418–21.
336. Chan Y, Wong T, Byrne F, Kavallaris M, Bulmus V. Acid-labile core cross-linked micelles for pH-triggered release of antitumor drugs. *Biomacromolecules* 2008;**9**:1826–36.
337. Qiao Z-Y, Zhang R, Du F-S, Liang D-H, Li Z-C. Multi-responsive nanogels containing motifs of ortho ester, oligo(ethylene glycol) and disulfide linkage as carriers of hydrophobic anti-cancer drugs. *J Control Release* 2011;**152**:57–66.
338. Ding C, Gu J, Qu X, Yang Z. Preparation of multifunctional drug carrier for tumor-specific uptake and enhanced intracellular delivery through the conjugation of weak acid labile linker. *Bioconjug Chem* 2009;**20**:1163–70.
339. Clawson C, et al. Synthesis and characterization of lipid-polymer hybrid nanoparticles with pH-triggered poly(ethylene glycol) shedding. *Langmuir* 2011;**27**:10556–61.
340. Yuan Z, et al. PEG-detachable and acid-labile cross-linked micelles based on orthoester linked graft copolymer for paclitaxel release. *Nanotechnology* 2011;**22**:335601.
341. Lavignac N, Nicholls JL, Ferruti P, Duncan R. Poly(amidoamine) conjugates containing doxorubicin bound via an acid-sensitive linker. *Macromol Biosci* 2009;**9**:480–7.
342. Kaminskas LM, et al. Characterisation and tumour targeting of PEGylated polylysine dendrimers bearing doxorubicin via a pH labile linker. *J Control Release* 2011;**152**:241–8.
343. van der Poll DG, et al. Design, synthesis, and biological evaluation of a robust, biodegradable dendrimer. *Bioconjug Chem* 2010;**21**:764–73.

344. Lu C, Xing MMQ, Zhong W. Shell cross-linked and hepatocyte-targeting nanoparticles containing doxorubicin via acid-cleavable linkage. *Nanomedicine* 2011;**7**:80–7.
345. Kakudo T, et al. Transferrin-modified liposomes equipped with a pH-sensitive fusogenic peptide: an artificial viral-like delivery system. *Biochemistry* 2004;**43**:5618–28.
346. Husseini GA, Pitt WG. Micelles and nanoparticles for ultrasonic drug and gene delivery. *Adv Drug Deliv Rev* 2008;**60**:1137–52.
347. Deckers R, Moonen CTW. Ultrasound triggered, image guided, local drug delivery. *J Control Release* 2010;**148**:25–33.
348. Price RJ, Skyba DM, Kaul S, Skalak TC. Delivery of colloidal particles and red blood cells to tissue through microvessel ruptures created by targeted microbubble destruction with ultrasound. *Circulation* 1998;**98**:1264–7.
349. Liu Y, Miyoshi H, Nakamura M. Encapsulated ultrasound microbubbles: therapeutic application in drug/gene delivery. *J Control Release* 2006;**114**:89–99.
350. Nelson JL, Roeder BL, Carmen JC, Roloff F, Pitt WG. Ultrasonically activated chemotherapeutic drug delivery in a rat model. *Cancer Res* 2002;**62**:7280–3.
351. Myhr G, Moan J. Synergistic and tumour selective effects of chemotherapy and ultrasound treatment. *Cancer Lett* 2006;**232**:206–13.
352. Howard B, Gao Z, Lee S-W, Seo M-H, Rapoport N. Ultrasound-enhanced chemotherapy of drug-resistant breast cancer tumors by micellar-encapsulated paclitaxel. *Am J Drug Deliv* 2006;**4**:97–104.
353. Rapoport NY, Kennedy AM, Shea JE, Scaife CL, Nam K-H. Controlled and targeted tumor chemotherapy by ultrasound-activated nanoemulsions/microbubbles. *J Control Release* 2009;**138**:268–76.
354. Husseini GA, et al. Release of doxorubicin from unstabilized and stabilized micelles under the action of ultrasound. *J Nanosci Nanotechnol* 2007;**7**:1028.
355. Evjen TJ, Hagtvet E, Nilssen EA, Brandl M, Fossheim SL. Sonosensitive dioleoylphosphatidylethanolamine-containing liposomes with prolonged blood circulation time of doxorubicin. *Eur J Pharm Sci* 2011;**43**:318–24.
356. Hagtvet E, Evjen TJ, Olsen DR, Fossheim SL, Nilssen EA. Ultrasound enhanced antitumor activity of liposomal doxorubicin in mice. *J Drug Target* 2011;**19**:701–8.
357. Ueno Y, et al. Combination of ultrasound and bubble liposome enhance the effect of doxorubicin and inhibit murine osteosarcoma growth. *Cancer Biol Ther* 2011;**12**:270–7.
358. Moon GD, et al. A new theranostic system based on gold nanocages and phase-change materials with unique features for photoacoustic imaging and controlled release. *J Am Chem Soc* 2011;**133**:4762–5.
359. Li W, et al. Gold nanocages covered with thermally-responsive polymers for controlled release by high-intensity focused ultrasound. *Nanoscale* 2011;**3**:1724–30.
360. Zhang Q, et al. Gold nanoparticles as a contrast agent for in vivo tumor imaging with photoacoustic tomography. *Nanotechnology* 2009;**20**:395102.
361. Cheng R, et al. Glutathione-responsive nano-vehicles as a promising platform for targeted intracellular drug and gene delivery. *J Control Release* 2011;**152**:2–12.
362. Liu J, et al. Redox-responsive polyphosphate nanosized assemblies: a smart drug delivery platform for cancer therapy. *Biomacromolecules* 2011;**12**:2407–15.
363. Wang Y-C, Wang F, Sun T-M, Wang J. Redox-responsive nanoparticles from the single disulfide bond-bridged block copolymer as drug carriers for overcoming multidrug resistance in cancer cells. *Bioconjug Chem* 2011;**22**:1939–45.

364. Ghosh S, Irvin K, Thayumanavan S. Tunable disassembly of micelles using a redox trigger. *Langmuir* 2007;**23**:7916–9.
365. Zhao M, et al. Redox-responsive nanocapsules for intracellular protein delivery. *Biomaterials* 2011;**32**:5223–30.
366. Chong S-F, et al. A paradigm for peptide vaccine delivery using viral epitopes encapsulated in degradable polymer hydrogel capsules. *Biomaterials* 2009;**30**:5178–86.
367. Sexton A, et al. A protective vaccine delivery system for in vivo T cell stimulation using nanoengineered polymer hydrogel capsules. *ACS Nano* 2009;**3**:3391–400.
368. Emerit J, Edeas M, Bricaire F. Neurodegenerative diseases and oxidative stress. *Biomed Pharmacother* 2004;**58**:39–46.
369. Navab M, et al. The oxidation hypothesis of atherogenesis: the role of oxidized phospholipids and HDL. *J Lipid Res* 2004;**45**:993–1007.
370. Harris ML, et al. Free radicals and other reactive oxygen metabolites in inflammatory bowel disease: cause, consequence or epiphenomenon? *Pharmacol Ther* 1992;**53**:375–408.
371. Khutoryanskiy VV, Tirelli N. Oxidation-responsiveness of nanomaterials for targeting inflammatory reactions. *Pure Appl Chem* 2008;**80**:1703–18.
372. Dhanasekaran SM, et al. Delineation of prognostic biomarkers in prostate cancer. *Nature* 2001;**412**:822–6.
373. Werle B, et al. Cathepsin B in infiltrated lymph nodes is of prognostic significance for patients with nonsmall cell lung carcinoma. *Cancer* 2000;**89**:2282–91.
374. Liaudet-Coopman E, et al. Cathepsin D: newly discovered functions of a long-standing aspartic protease in cancer and apoptosis. *Cancer Lett* 2006;**237**:167–79.
375. Koblinski JE, Ahram M, Sloane BF. Unraveling the role of proteases in cancer. *Clin Chim Acta* 2000;**291**:113–35.
376. Egeblad M, Werb Z. New functions for the matrix metalloproteinases in cancer progression. *Nat Rev Cancer* 2002;**2**:161–74.
377. Yang Y, et al. Enzyme-responsive multifunctional magnetic nanoparticles for tumor intracellular drug delivery and imaging. *Chem Asian J* 2011;**6**:1381–9.
378. Matsumura S, Aoki I, Saga T, Shiba K. A tumor-environment-responsive nanocarrier that evolves its surface properties upon sensing matrix metalloproteinase-2 and initiates agglomeration to enhance T2 relaxivity for magnetic resonance imaging. *Mol Pharm* 2011;**8**:1970–4.
379. Choi KY, et al. Smart nanocarrier based on PEGylated hyaluronic acid for cancer therapy. *ACS Nano* 2011;**5**:8591–9.
380. Thomas M, et al. A galactosidase-responsive "trojan horse" for the selective targeting of folate receptor-positive tumor cells. *ChemMedChem* 2011;**6**:1006–10.
381. Bernardos A, et al. Enzyme-responsive intracellular controlled release using nanometric silica mesoporous supports capped with "saccharides" *ACS Nano* 2010;**4**:6353–68.
382. Morimoto N, Winnik FM, Akiyoshi K. Botryoidal assembly of cholesteryl-pullulan/poly(N-isopropylacrylamide) nanogels. *Langmuir* 2007;**23**:217–23.
383. Klaikherd A, Nagamani C, Thayumanavan S. Multi-stimuli sensitive amphiphilic block copolymer assemblies. *J Am Chem Soc* 2009;**131**:4830–8.
384. Li L, et al. Self-quenchable biofunctional nanoparticles of heparin–folate-photosensitizer conjugates for photodynamic therapy. *Carbohydr Polym* 2011;**86**:708–15.
385. Shi D, Matsusaki M, Kaneko T, Akashi M. Photo-cross-linking and cleavage induced reversible size change of bio-based nanoparticles. *Macromolecules* 2008;**41**:8167–72.

386. McNeeley KM, Karathanasis E, Annapragada AV, Bellamkonda RV. Masking and triggered unmasking of targeting ligands on nanocarriers to improve drug delivery to brain tumors. *Biomaterials* 2009;**30**:3986–95.
387. Terada T, Iwai M, Kawakami S, Yamashita F, Hashida M. Novel PEG-matrix metalloproteinase-2 cleavable peptide-lipid containing galactosylated liposomes for hepatocellular carcinoma-selective targeting. *J Control Release* 2006;**111**:333–42.
388. Mok H, Bae KH, Ahn C-H, Park TG. PEGylated and MMP-2 specifically dePEGylated quantum dots: comparative evaluation of cellular uptake. *Langmuir* 2009;**25**:1645–50.
389. Hwang DK, et al. Stop-flow lithography for the production of shape-evolving degradable microgel particles. *J Am Chem Soc* 2009;**131**:4499–504.
390. Yoo J-W, Mitragotri S. Polymer particles that switch shape in response to a stimulus. *Proc Natl Acad Sci USA* 2010;**107**:11205–10.
391. Murakami T, Tsuchida K. Recent advances in inorganic nanoparticle-based drug delivery systems. *Mini Rev Med Chem* 2008;**8**:175–83.
392. Yiu HHP. Engineering the multifunctional surface on magnetic nanoparticles for targeted biomedical applications: a chemical approach. *Nanomedicine (Lond)* 2011;**6**:1429–46.
393. Hostetler MJ, et al. Alkanethiolate gold cluster molecules with core diameters from 1.5 to 5.2 nm: core and monolayer properties as a function of core size. *Langmuir* 1998;**14**:17–30.
394. Xie J, Lee S, Chen X. Nanoparticle-based theranostic agents. *Adv Drug Deliv Rev* 2010;**62**: 1064–79.
395. Huang H-C, Barua S, Sharma G, Dey SK, Rege K. Inorganic nanoparticles for cancer imaging and therapy. *J Control Release* 2011;**155**:344–57.
396. Xu C, Sun S. Superparamagnetic nanoparticles as targeted probes for diagnostic and therapeutic applications. *Dalton Trans* 2009;(29):5583–91. doi:10.1039/b900272n.
397. Sun C, et al. In vivo MRI detection of gliomas by chlorotoxin-conjugated superparamagnetic nanoprobes. *Small* 2008;**4**:372–9.
398. Thiesen B, Jordan A. Clinical applications of magnetic nanoparticles for hyperthermia. *Int J Hyperthermia* 2008;**24**:467–74.
399. Maier-Hauff K, et al. Intracranial thermotherapy using magnetic nanoparticles combined with external beam radiotherapy: results of a feasibility study on patients with glioblastoma multiforme. *J Neurooncol* 2007;**81**:53–60.
400. Maier-Hauff K, et al. Efficacy and safety of intratumoral thermotherapy using magnetic iron-oxide nanoparticles combined with external beam radiotherapy on patients with recurrent glioblastoma multiforme. *J Neurooncol* 2011;**103**:317–24.
401. Maier-Hauff K, et al. Intracranial thermotherapy using magnetic nanoparticles combined with external beam radiotherapy: results of a feasibility study on patients with glioblastoma multiforme. *J Neurooncol* 2007;**81**:53–60.
402. Kumar CSSR, Mohammad F. Magnetic nanomaterials for hyperthermia-based therapy and controlled drug delivery. *Adv Drug Deliv Rev* 2011;**63**:789–808.
403. O'Neal DP, Hirsch LR, Halas NJ, Payne JD, West JL. Photo-thermal tumor ablation in mice using near infrared-absorbing nanoparticles. *Cancer Lett* 2004;**209**:171–6.
404. Sardar R, Funston AM, Mulvaney P, Murray RW. Gold nanoparticles: past, present, and future. *Langmuir* 2009;**25**:13840–51.
405. Huang X, Jain PK, El-Sayed IH, El-Sayed MA. Gold nanoparticles: interesting optical properties and recent applications in cancer diagnostics and therapy. *Nanomedicine (Lond)* 2007;**2**:681–93.
406. Daniel M-C, Astruc D. Gold nanoparticles: assembly, supramolecular chemistry, quantum-size-related properties, and applications toward biology, catalysis, and nanotechnology. *Chem Rev* 2004;**104**:293–346.

407. Sau TK, Rogach AL, Jäckel F, Klar TA, Feldmann J. Properties and applications of colloidal nonspherical noble metal nanoparticles. *Adv Mater* 2010;**22**:1805–25.
408. Melancon MP, Zhou M, Li C. Cancer theranostics with near-infrared light-activatable multimodal nanoparticles. *Acc Chem Res* 2011;**44**:947–56.
409. Rai P, et al. Development and applications of photo-triggered theranostic agents. *Adv Drug Deliv Rev* 2010;**62**:1094–124.
410. Bardhan R, Lal S, Joshi A, Halas NJ. Theranostic nanoshells: from probe design to imaging and treatment of cancer. *Acc Chem Res* 2011;**44**:936–46.
411. Park H, et al. Multifunctional nanoparticles for combined doxorubicin and photothermal treatments. *ACS Nano* 2009;**3**:2919–26.
412. Burton ER, Libutti SK. Targeting TNF-alpha for cancer therapy. *J Biol* 2009;**8**:85.
413. Libutti SK, et al. Phase I and pharmacokinetic studies of CYT-6091, a novel PEGylated colloidal gold-rhTNF nanomedicine. *Clin Cancer Res* 2010;**16**:6139–49.
414. Yang X, Stein EW, Ashkenazi S, Wang LV. Nanoparticles for photoacoustic imaging. *Wiley Interdiscip Rev Nanomed Nanobiotechnol* 2009;**1**:360–8.
415. Ku G, Wang LV. Deeply penetrating photoacoustic tomography in biological tissues enhanced with an optical contrast agent. *Opt Lett* 2005;**30**:507.
416. Song KH, Kim C, Cobley CM, Xia Y, Wang LV. Near-infrared gold nanocages as a new class of tracers for photoacoustic sentinel lymph node mapping on a rat model. *Nano Lett* 2009;**9**:183–8.
417. Hainfeld JF, Slatkin DN, Smilowitz HM. The use of gold nanoparticles to enhance radiotherapy in mice. *Phys Med Biol* 2004;**49**:N309–N315.
418. Perumal S, Hofmann A, Scholz N, Rühl E, Graf C. Kinetics study of the binding of multivalent ligands on size-selected gold nanoparticles. *Langmuir* 2011;**27**:4456–64.
419. Rana S, Bajaj A, Mout R, Rotello VM. Monolayer coated gold nanoparticles for delivery applications. *Adv Drug Deliv Rev* 2012;**64**:200–16. doi:10.1016/j.addr.
420. Li X, et al. Enhancement of cell recognition in vitro by dual-ligand cancer targeting gold nanoparticles. *Biomaterials* 2011;**32**:2540–5.
421. Tan A, et al. Quantum dots and carbon nanotubes in oncology: a review on emerging theranostic applications in nanomedicine. *Nanomedicine (Lond)* 2011;**6**:1101–14.
422. Alivisatos P. The use of nanocrystals in biological detection. *Nat Biotechnol* 2004;**22**:47–52.
423. Voura EB, Jaiswal JK, Mattoussi H, Simon SM. Tracking metastatic tumor cell extravasation with quantum dot nanocrystals and fluorescence emission-scanning microscopy. *Nat Med* 2004;**10**:993–8.
424. Alivisatos AP, Gu W, Larabell C. Quantum dots as cellular probes. *Annu Rev Biomed Eng* 2005;**7**:55–76.
425. Klostranec JM, Chan WCW. Quantum dots in biological and biomedical research: recent progress and present challenges. *Adv Mater* 2006;**18**:1953–64.
426. Ghaderi S, Ramesh B, Seifalian AM. Fluorescence nanoparticles "quantum dots" as drug delivery system and their toxicity: a review. *J Drug Target* 2011;**19**:475–86.
427. Smith AM, Dave S, Nie S, True L, Gao X. Multicolor quantum dots for molecular diagnostics of cancer. *Expert Rev Mol Diagn* 2006;**6**:231–44.
428. Diagaradjane P, et al. Imaging epidermal growth factor receptor expression in vivo: pharmacokinetic and biodistribution characterization of a bioconjugated quantum dot nanoprobe. *Clin Cancer Res* 2008;**14**:731–41.
429. Shao L, Gao Y, Yan F. Semiconductor quantum dots for biomedicial applications. *Sensors* 2011;**11**:11736–51.
430. Yezhelyev MV, Qi L, O'Reagan RM, Nie S, Gao X. Proton-sponge coated quantum dots for siRNA delivery and intracellular imaging. *J Am Chem Soc* 2008;**130**:9006–12.

431. Wang Y, Chen L. Quantum dots, lighting up the research and development of nanomedicine. *Nanomedicine* 2011;**7**:385–402.
432. Gill R, Zayats M, Willner I. Semiconductor quantum dots for bioanalysis. *Angew Chem Int Ed Engl* 2008;**47**:7602–25.
433. Bagalkot V, et al. Quantum dot-aptamer conjugates for synchronous cancer imaging, therapy, and sensing of drug delivery based on bi-fluorescence resonance energy transfer. *Nano Lett* 2007;**7**:3065–70.
434. Biju V, Mundayoor S, Omkumar RV, Anas A, Ishikawa M. Bioconjugated quantum dots for cancer research: present status, prospects and remaining issues. *Biotechnol Adv* 2010;**28**: 199–213.
435. Shao L, Gao Y, Yan F. Semiconductor quantum dots for biomedicial applications. *Sensors* 2011;**11**:11736–51.
436. Li X, Xie QR, Zhang J, Xia W, Gu H. The packaging of siRNA within the mesoporous structure of silica nanoparticles. *Biomaterials* 2011;**32**:9546–56.
437. Giri S, Trewyn BG, Lin VSY. Mesoporous silica nanomaterial-based biotechnological and biomedical delivery systems. *Nanomedicine (Lond)* 2007;**2**:99–111.
438. Cheng K, Sun S. Recent advances in syntheses and therapeutic applications of multifunctional porous hollow nanoparticles. *Nano Today* 2010;**5**:183–96.
439. Kunzmann A, et al. Toxicology of engineered nanomaterials: focus on biocompatibility, biodistribution and biodegradation. *Biochim Biophys Acta* 2011;**1810**:361–73.
440. Reiss P, Protière M, Li L. Core/Shell semiconductor nanocrystals. *Small* 2009;**5**:154–68.
441. Templeton DM, Liu Y. Multiple roles of cadmium in cell death and survival. *Chem Biol Interact* 2010;**188**:267–75.
442. Vivero-Escoto JL, Huang Y-T. Inorganic-organic hybrid nanomaterials for therapeutic and diagnostic imaging applications. *Int J Mol Sci* 2011;**12**:3888–927.
443. Hardman R. A toxicologic review of quantum dots: toxicity depends on physicochemical and environmental factors. *Environ Health Perspect* 2006;**114**:165–72.
444. Choi AO, Cho SJ, Desbarats J, Lovrić J, Maysinger D. Quantum dot-induced cell death involves Fas upregulation and lipid peroxidation in human neuroblastoma cells. *J Nanobio technol* 2007;**5**:1.
445. Shiohara A, Hoshino A, Hanaki K-I, Suzuki K, Yamamoto K. On the cyto-toxicity caused by quantum dots. *Microbiol Immunol* 2004;**48**:669–75.
446. El-Ansary A, Al-Daihan S. On the toxicity of therapeutically used nanoparticles: an overview. *J Toxicol* 2009;**2009**:754810.
447. Sharma A, Madhunapantula SV, Robertson GP. Toxicological considerations when creating nanoparticle-based drugs and drug delivery systems. *Expert Opin Drug Metab Toxicol* 2012;**8**:47–69.
448. Fischer HC, Liu L, Pang KS, Chan WCW. Pharmacokinetics of nanoscale quantum dots: in vivo distribution, sequestration, and clearance in the rat. *Adv Funct Mater* 2006;**16**: 1299–305.
449. Fadeel B, Garcia-Bennett AE. Better safe than sorry: understanding the toxicological properties of inorganic nanoparticles manufactured for biomedical applications. *Adv Drug Deliv Rev* 2010;**62**:362–74.
450. Lee S, Yun H-S, Kim S-H. The comparative effects of mesoporous silica nanoparticles and colloidal silica on inflammation and apoptosis. *Biomaterials* 2011;**32**:9434–43.
451. Qiao R, Yang C, Gao M. Superparamagnetic iron oxide nanoparticles: from preparations to in vivo MRI applications. *J Mater Chem* 2009;**19**:6274–93.

452. Tang M-F, Lei L, Guo S-R, Huang W-L. Recent progress in nanotechnology for cancer therapy. *Chin J Cancer* 2010;**29**:775–80.
453. Buyukhatipoglu K, Clyne AM. Superparamagnetic iron oxide nanoparticles change endothelial cell morphology and mechanics via reactive oxygen species formation. *J Biomed Mater Res A* 2011;**96**:186–95.
454. Li JJ, Hartono D, Ong C-N, Bay B-H, Yung L-YL. Autophagy and oxidative stress associated with gold nanoparticles. *Biomaterials* 2010;**31**:5996–6003.
455. Soenen SJ, et al. Cellular toxicity of inorganic nanoparticles: common aspects and guidelines for improved nanotoxicity evaluation. *Nano Today* 2011;**6**:446–65. doi:10.1016/j.nantod.2011.08.001.
456. Maherani B, Arab-Tehrany ER, Mozafari M, Gaiani C, Linder M. Liposomes: a review of manufacturing techniques and targeting strategies. *Curr Nanosci* 2011;**7**:436–52.
457. Chou LYT, Ming K, Chan WCW. Strategies for the intracellular delivery of nanoparticles. *Chem Soc Rev* 2011;**40**:233–45.
458. Guterres SS, Alves MP, Pohlmann AR. Polymeric nanoparticles, nanospheres and nanocapsules, for cutaneous applications. *Drug Target Insights* 2007;**2**:147–57.
459. Alberts DS, et al. Randomized trial of pegylated liposomal doxorubicin (PLD) plus carboplatin versus carboplatin in platinum-sensitive (PS) patients with recurrent epithelial ovarian or peritoneal carcinoma after failure of initial platinum-based chemotherapy (Southwest Onc.). *Gynecol Oncol* 2008;**108**:90–4.
460. Maitani Y. PEGylated lipidic systems with prolonged circulation longevity for drug delivery in cancer therapeutics. *J Drug Deliv Sci Technol* 2011;**21**:27–34.
461. De Jong WH, Borm PJA. Drug delivery and nanoparticles: applications and hazards. *Int J Nanomedicine* 2008;**3**:133–49.
462. Zolnik BS, González-Fernández A, Sadrieh N, Dobrovolskaia MA. Nanoparticles and the immune system. *Endocrinology* 2010;**151**:458–65.
463. Lv H, Zhang S, Wang B, Cui S, Yan J. Toxicity of cationic lipids and cationic polymers in gene delivery. *J Control Release* 2006;**114**:100–9.
464. Akhtar S, Benter I. Toxicogenomics of non-viral drug delivery systems for RNAi: potential impact on siRNA-mediated gene silencing activity and specificity. *Adv Drug Deliv Rev* 2007;**59**:164–82.
465. Szebeni J, Muggia F, Gabizon A, Barenholz Y. Activation of complement by therapeutic liposomes and other lipid excipient-based therapeutic products: prediction and prevention. *Adv Drug Deliv Rev* 2011;**63**:1020–30.
466. Bunjes H. Structural properties of solid lipid based colloidal drug delivery systems. *Curr Opin Colloid Interface Sci* 2011;**16**:411–5.
467. Müller RH, Mäder K, Gohla S. Solid lipid nanoparticles (SLN) for controlled drug delivery—a review of the state of the art. *Eur J Pharm Biopharm* 2000;**50**:161–77.
468. Wissing SA, Kayser O, Müller RH. Solid lipid nanoparticles for parenteral drug delivery. *Adv Drug Deliv Rev* 2004;**56**:1257–72.
469. Harde H, Das M, Jain S. Solid lipid nanoparticles: an oral bioavailability enhancer vehicle. *Expert Opin Drug Deliv* 2011;**8**:1407–24.
470. Harms M, Müller-Goymann CC. Solid lipid nanoparticles for drug delivery. *J Drug Deliv Sci Technol* 2011;**21**:89–99.
471. Müller RH, Radtke M, Wissing SA. Solid lipid nanoparticles (SLN) and nanostructured lipid carriers (NLC) in cosmetic and dermatological preparations. *Adv Drug Deliv Rev* 2002;**54**: S131–S155.

472. Matalanis A, Jones OG, McClements DJ. Structured biopolymer-based delivery systems for encapsulation, protection, and release of lipophilic compounds. *Food Hydrocolloids* 2011;**25**:1865–80.
473. Mahapatro A, Singh DK. Biodegradable nanoparticles are excellent vehicle for site directed in-vivo delivery of drugs and vaccines. *J Nanobiotechnol* 2011;**9**:55.
474. Wischke C, Schwendeman SP. Principles of encapsulating hydrophobic drugs in PLA/PLGA microparticles. *Int J Pharm* 2008;**364**:298–327.
475. Semete B, et al. In vivo evaluation of the biodistribution and safety of PLGA nanoparticles as drug delivery systems. *Nanomedicine* 2010;**6**:662–71.
476. Desai N, Trieu V, Damascelli B, Soon-Shiong P. SPARC expression correlates with tumor response to albumin-bound paclitaxel in head and neck cancer patients. *Transl Oncol* 2009;**2**:59–64.
477. Roco MC, Mirkin CA, Hersam MC. *Nanotechnology research directions for societal needs in 2020. Retrospective and outlooks.* Berlin and Boston: Springer; 2011.
478. Xiong X-B, Falamarzian A, Garg SM, Lavasanifar A. Engineering of amphiphilic block copolymers for polymeric micellar drug and gene delivery. *J Control Release* 2011;**155**: 261–8.
479. Uhrich KE, Cannizzaro SM, Langer RS, Shakesheff KM. Polymeric systems for controlled drug release. *Chem Rev* 1999;**99**:3181–98.
480. Kim S, Park K. Polymer micelles for drug delivery. In: Mahato RI, Narang AS, editors. *Targeted delivery of small and macromolecular drugs.* Boca Raton, FL: Taylor & Francis; 2010.
481. Moughton AO, Hillmyer MA, Lodge TP. Multicompartment block polymer micelles. *Macromolecules* 2012;**45**:2–19.
482. Huynh L, Neale C, Pomès R, Allen C. Computational approaches to the rational design of nanoemulsions, polymeric micelles, and dendrimers for drug delivery. *Nanomedicine* 2012;**8**:20–36.
483. Veronese F. *PEGylated protein drugs: basic science and clinical applications (milestones in drug therapy).* Basel, Switzerland: Birkhauser Verlag Press; 2009.
484. Duncan R. Polymer conjugates as anticancer nanomedicines. *Nat Rev Cancer* 2006;**6**: 688–701.
485. Gaikwad D, et al. Polymer-drug conjugates: recent achievements. *Res J Pharm Biol Chem Sci* 2011;**2**:200–8.
486. Discher DE, et al. Emerging applications of polymersomes in delivery: from molecular dynamics to shrinkage of tumors. *Prog Polym Sci* 2007;**32**:838–57.
487. Ahmed F, et al. Biodegradable polymersomes loaded with both paclitaxel and doxorubicin permeate and shrink tumors, inducing apoptosis in proportion to accumulated drug. *J Control Release* 2006;**116**:150–8.
488. Kim S-H, Shum HC, Kim JW, Cho J-C, Weitz DA. Multiple polymersomes for programmed release of multiple components. *J Am Chem Soc* 2011;**133**:15165–71.
489. Prakash Jain J, Yenet Ayen W, Kumar N. Self assembling polymers as polymersomes for drug delivery. *Curr Pharm Des* 2011;**17**:65–79.
490. Tanner P, et al. Polymeric vesicles: from drug carriers to nanoreactors and artificial organelles. *Acc Chem Res* 2011;**44**:1039–49.
491. Sosnik A, Carcaboso ÁM, Chiappetta DA. Polymeric nanocarriers: new endeavors for the optimization of the technological aspects of drugs. *Recent Pat Biomed Eng* 2008;**1**:43–59.
492. Lipinski C. Poor aqueous solubility—an industry wide problem in drug discovery. *Am Pharm Rev* 2002;**5**:82–5.
493. Müller RH, Keck CM. 20 Years drug nanocrystals—where are we, and where to go? *Eur J Pharm Biopharm* 2011;**80**:1–3.

494. Katteboinaa S, Chandrasekhar VSR, Balaji S. Drug nanocrystals: a novel formulation approach for poorly soluble drugs. *Int J PharmTech Res* 2009;**1**:682–94.
495. Müller RH, Shegokar R, Gohla S, Keck CM. Nanocrystals: production, cellular drug delivery, current and future products. In: Prokop A, editor. *Intracellular delivery. Fundamental biomedical technologies*. Netherlands: Springer; 2011;**5**:411–32.
496. Müller RH, Gohla S, Keck CM. State of the art of nanocrystals—special features, production, nanotoxicology aspects and intracellular delivery. *Eur J Pharm Biopharm* 2011;**78**:1–9.
497. Muller RH, Keck CM. Challenges and solutions for the delivery of biotech drugs—a review of drug nanocrystal technology and lipid nanoparticles. *J Biotechnol* 2004;**113**:151–70.
498. Medina SH, El-Sayed MEH. Dendrimers as carriers for delivery of chemotherapeutic agents. *Chem Rev* 2009;**109**:3141–57.
499. Klajnert B, Bryszewska M. Dendrimers: properties and applications. *Acta Biochim Pol* 2001;**48**:199–208.
500. Menjoge AR, Kannan RM, Tomalia DA. Dendrimer-based drug and imaging conjugates: design considerations for nanomedical applications. *Drug Discov Today* 2010;**15**:171–85.
501. Padilla De Jesús OL, Ihre HR, Gagne L, Fréchet JMJ, Szoka FC. Polyester dendritic systems for drug delivery applications: in vitro and in vivo evaluation. *Bioconjug Chem* 2002;**13**:453–61.
502. Lee CC, et al. A single dose of doxorubicin-functionalized bow-tie dendrimer cures mice bearing C-26 colon carcinomas. *Proc Natl Acad Sci USA* 2006;**103**:16649–54.
503. Jevprasesphant R, et al. The influence of surface modification on the cytotoxicity of PAMAM dendrimers. *Int J Pharm* 2003;**252**:263–6.
504. Ziemba B, Matuszko G, Bryszewska M, Klajnert B. Influence of dendrimers on red blood cells. *Cell Mol Biol Lett* 2012;**17**:21–35.
505. Liu Z, Robinson JT, Tabakman SM, Yang K, Dai H. Carbon materials for drug delivery & cancer therapy. *Mater Today* 2011;**14**:316–23.
506. Vashist SK, et al. Delivery of drugs and biomolecules using carbon nanotubes. *Carbon* 2011;**49**:4077–97.
507. Liu Z, Sun X, Nakayama-Ratchford N, Dai H. Supramolecular chemistry on water-soluble carbon nanotubes for drug loading and delivery. *ACS Nano* 2007;**1**:50–6.
508. Allen BL, et al. Mechanistic investigations of horseradish peroxidase-catalyzed degradation of single-walled carbon nanotubes. *J Am Chem Soc* 2009;**131**:17194–205.
509. Allen BL, et al. Biodegradation of single-walled carbon nanotubes through enzymatic catalysis. *Nano Lett* 2008;**8**:3899–903.
510. Kagan VE, et al. Carbon nanotubes degraded by neutrophil myeloperoxidase induce less pulmonary inflammation. *Nat Nanotechnol* 2010;**5**:354–9.
511. Elhissi AMA, Ahmed W, Hassan IU, Dhanak VR, D'Emanuele A. Carbon nanotubes in cancer therapy and drug delivery. *J Drug Deliv* 2012;**2012**:837327.
512. Levi-Polyachenko NH, Merkel EJ, Jones BT, Carroll DL, Stewart IV JH. Rapid photothermal intracellular drug delivery using multiwalled carbon nanotubes. *Mol Pharm* 2009;**6**:1092–9.
513. Liu X, et al. Optimization of surface chemistry on single-walled carbon nanotubes for in vivo photothermal ablation of tumors. *Biomaterials* 2011;**32**:144–51.
514. Mashal A, et al. Toward carbon-nanotube-based theranostic agents for microwave detection and treatment of breast cancer: enhanced dielectric and heating response of tissue-mimicking materials. *IEEE Trans Biomed Eng* 2010;**57**:1831–4.
515. Gannon CJ, et al. Carbon nanotube-enhanced thermal destruction of cancer cells in a noninvasive radiofrequency field. *Cancer* 2007;**110**:2654–65.
516. Zavaleta C, et al. Noninvasive Raman spectroscopy in living mice for evaluation of tumor targeting with carbon nanotubes. *Nano Lett* 2008;**8**:2800–5.

517. Welsher K, et al. A route to brightly fluorescent carbon nanotubes for near-infrared imaging in mice. *Nat Nanotechnol* 2009;**4**:773–80.
518. De la Zerda A, et al. Carbon nanotubes as photoacoustic molecular imaging agents in living mice. *Nat Nanotechnol* 2008;**3**:557–62.
519. Kostarelos K, et al. Cellular uptake of functionalized carbon nanotubes is independent of functional group and cell type. *Nat Nanotechnol* 2007;**2**:108–13.
520. Raffa V, Ciofani G, Vittorio O, Riggio C, Cuschieri A. Physicochemical properties affecting cellular uptake of carbon nanotubes. *Nanomedicine (Lond)* 2010;**5**:89–97.
521. Kostarelos K, Bianco A, Prato M. Promises, facts and challenges for carbon nanotubes in imaging and therapeutics. *Nat Nanotechnol* 2009;**4**:627–33.
522. Rybak-Smith MJ, Sim RB. Complement activation by carbon nanotubes. *Adv Drug Deliv Rev* 2011;**63**:1031–41.
523. Aschberger K, et al. Review of carbon nanotubes toxicity and exposure—appraisal of human health risk assessment based on open literature. *Crit Rev Toxicol* 2010;**40**:759–90.
524. Liu Z, et al. Circulation and long-term fate of functionalized, biocompatible single-walled carbon nanotubes in mice probed by Raman spectroscopy. *Proc Natl Acad Sci USA* 2008;**105**:1410–5.
525. Ding L, et al. Molecular characterization of the cytotoxic mechanism of multiwall carbon nanotubes and nano-onions on human skin fibroblast. *Nano Lett* 2005;**5**:2448–64.
526. Pulskamp K, Diabaté S, Krug HF. Carbon nanotubes show no sign of acute toxicity but induce intracellular reactive oxygen species in dependence on contaminants. *Toxicol Lett* 2007;**168**:58–74.
527. Bottini M, Rosato N, Bottini N. PEG-modified carbon nanotubes in biomedicine: current status and challenges ahead. *Biomacromolecules* 2011;**12**:3381–93.
528. Fagan JA, et al. Carbon nanotubes: measuring dispersion and length. *Adv Mater* 2011;**23**:338–48.
529. Ghosh S, Bachilo SM, Weisman RB. Advanced sorting of single-walled carbon nanotubes by nonlinear density-gradient ultracentrifugation. *Nat Nanotechnol* 2010;**5**:443–50.
530. Hersam MC. Progress towards monodisperse single-walled carbon nanotubes. *Nat Nanotechnol* 2008;**3**:387–94.
531. Choi WI, et al. Tumor regression in vivo by photothermal therapy based on gold-nanorod-loaded, functional nanocarriers. *ACS Nano* 2011;**5**:1995–2003.
532. Vallet-Regí M, Colilla M, González B. Medical applications of organic-inorganic hybrid materials within the field of silica-based bioceramics. *Chem Soc Rev* 2011;**40**:596–607.
533. Sanchez C, Belleville P, Popall M, Nicole L. Applications of advanced hybrid organic-inorganic nanomaterials: from laboratory to market. *Chem Soc Rev* 2011;**40**:696–753.
534. Ali Z, et al. Multifunctional nanoparticles for dual imaging. *Anal Chem* 2011;**83**:2877–82.
535. Cientifica's report, Nanotechnology in Drug Delivery 2011-2021: Global Market for Nanocarriers.
536. Munos B. Lessons from 60 years of pharmaceutical innovation. *Nat Rev Drug Discov* 2009; **8**:959–68.
537. Paul SM, et al. How to improve R&D productivity: the pharmaceutical industry's grand challenge. *Nat Rev Drug Discov* 2010;**9**:203–14.
538. Costantino L, Boraschi D. Is there a clinical future for polymeric nanoparticles as brain-targeting drug delivery agents? *Drug Discov Today* 2012;**17**:367–78. doi:10.1016/j.drudis.2011.10.028.
539. Zhang L, et al. Nanoparticles in medicine: therapeutic applications and developments. *Clin Pharmacol Ther* 2008;**83**:761–9.

Part V

Toxicity and Regulation

Chapter 15

Nanotoxicology

Rosana Simón-Vázquez[*,1], Mercedes Peleteiro[*,1], Tamara Lozano[*,1] and África González-Fernández[*]

[*]*Immunology, Biomedical Research Centre (CINBIO), University of Vigo, Campus Lagoas Marcosende, Vigo, Spain*

Amparo Casal[†]

[†]*SOD Medicina del Lavoro, Azienda Ospedaliera Universitaria Carreggi, Florence, Italy*

1. INTRODUCTION

Nanostructures (Ns) are not new materials.[1] Many Ns have already been used in the past, but in recent years, the emerging field of nanotechnology has rapidly developed into a real boom.[2] To some extent, this quick success is partly due to the development of new techniques able to measure, prepare and see nanoparticles. The design of new and potent microscopes,[3] equipment for preparing nanodevices, nanocantilevers, new biosensors able to detect multianalytes, new methods of synthesis and characterization and many other factors[4] are helping in the rapid development of this discipline. These new developments, together with the different and sometimes unique properties that materials show when at nanometre range, are making the nanofield so attractive to many disciplines.

The potential of Ns is really huge, and many companies are becoming interested in the synthesis, design, development or use of nanostructures, for various applications in numerous sectors, such as cosmetics, food, energy, computing, construction, cars, biomedical products, diagnosis products, etc.[5,6] However, problems related with their biocompatibility (including toxicity at several levels) may condition whether or not to use nanomaterials in the future. Environment, workers, patients, healthy people and any living organisms can be affected in a positive or negative way by the Ns. As a positive example, the drug Abraxane® (Ns containing human albumin and paclitaxel, an antitumoural drug) has been able to decrease the toxicity of Taxol (paclitaxel with cremophor), almost completely avoiding the hypersensitivity reactions that the original drug induced in patients.[7] As a negative example, two young female Chinese workers died and five more were severely affected by chronic

1. These authors are equally contributed.

Frontiers of Nanoscience, Vol. 4. DOI: 10.1016/B978-0-12-415769-9.00015-7

inhalation of polyacrylated nanoparticles at work.[8] Hence, careful toxicological studies should be carried out and the relevant safety issues correctly addressed, before any intention of using Ns for biomedical or food purposes in work places or in the environment.

Sterility is another important aspect that should be considered when Ns are intended to be used for *in vivo* applications, because some sterilization methods could modify their properties due to their surface reactivity.[9]

2. TOXICOLOGY

The new science of nanotechnology is providing a large variety of new nanomaterials, but it is also contributing some problems, especially those related with human and environmental safety. Nanostructures are much more complex than other products, such as small pharmaceutical drugs or materials in micro or macro size range. More importantly, they can be deposited on air, water, land, food or various organisms, including human, and their possible hazardous effects must be thoroughly investigated.

The field of toxicology has a long history and detailed knowledge about how some products effect human health. Some examples include the toxic effects of the chronic exposure of workers to silica dust (inducing silicosis and autoimmune responses), asbestos (inducing asbestosis and pleural cancer)[10] or heavy metals (inducing genotoxicity, neurodegenerative problems, hepatic illness, etc.).[11] The great similarities of these elements with nanostructures (e.g. carbon nanotubes with asbestos, silica nanomaterials with silica dust, quantum dots containing heavy metals, such as cadmium, etc.) have led to the hypothesis that these new nanomaterials could behave in a similar way to those previously described and largely demonstrated toxic elements.[7]

What are the main factors directly involved in the toxicity of nanomaterials? Toxicity does not depend on just one factor, because many are involved, such as size (their small size means that they can interact better with biological systems), form, composition, charge, coating, aggregation or capacity of self-assembly, capacity of conjugation (strong or weak bonds) with other components (proteins, lipids, etc.), catalytic properties or combination with other biological components. In addition, toxicity is strongly dependent on the biodistribution capacity of the nanomaterials (to induce local or systemic problems), on their rate of elimination and tissue localization and, very importantly, on whether the materials used are biodegradable or non-biodegradable.[12]

A nanomaterial can be toxic in one format but not in another. The coating of silica nanoparticles with polyethylene glycol (PEG) decreases their toxicity in human cell lines compared to the naked material.[13] Moreover, a change in the coating can also modify the biodistribution of the nanomaterials. Many naked nanoparticles are usually captured by macrophages in liver and spleen but, when pegylated, they have a better blood biodistribution[14] or, when coated with ApoE, the nanomaterial can sometimes enter into the brain, crossing the blood–brain barrier.[15]

Size is also very important in the biodistribution and elimination of Ns. Those that could cross the renal filter would be eliminated much faster than those of larger sizes, but bigger nanoparticles are better recognized and phagocyted by immune cells.[16] Interaction with the immune system is another important issue, because inflammation, immunotoxicity, allergy reactions or immunodeficiencies could be some of the deleterious effects induced by nanomaterials. In contrast, among some of their beneficial effects, nanostructures could help to decrease allergy reactions, to target tumoural immune cells or to develop new and better vaccines.[17] A correct study of biocompatibility, immunogenicity, biodistribution and possible effects is essential for all nanomaterials produced.

Ns can also be conjugated to various compounds, such as monoclonal antibodies, cytokines, folic acid, peptides and sugars, to direct them to specific targets (such as tumour cells). The combination of antibody–Ns can generate a product with the properties of both systems (versatility of the Ns with the specificity of the antibody) but may also increase immunogenicity or reduce the benefits of therapy (penetration into the tumour area can be decreased using a large conjugate).[18]

Several important questions arise when attempting to determine the potential toxicity of nanomaterials. For example, among many others, the following ones stand out: how to measure toxicity, how to standardize the methods between several laboratories, how to study biodistribution and which type of material should be considered reference material. The lack of reproducibility from batch to batch, the difficulty of scaling up the nanoproducts and the different behaviour of similar nanomaterials (depending on composition, size, charge, stability, temperature, media, presence of serum, endotoxin, coating, level of aggregation, type of functionalization, etc.) make interlaboratory comparisons of toxicity studies difficult. Moreover, many conventional methods for checking cellular toxicity *in vitro* are not suitable for nanomaterials (e.g. some nanomaterials interfere with colorimetric methods). Contradictory results regarding toxicity with similar materials are frequently found in the literature, and much more research is needed to clarify this scenario.

3. REGULATION

The above considerations have emphasized the need for nanotechnology regulations, which are increasingly coming into force worldwide. Reasonable regulatory frameworks are required for a correct innovation in nanotechnology, with safety being the main concern. Harmonization criteria, the standardization of methods and the definition of reference materials are some of the issues that agencies around the world are currently trying to tackle.

Several agencies in Europe, the United States and other countries are working on the regulation, restriction or even banning of some nanotechnology applications. For example, the Committee on the Environment, Public Health and Food Safety of the European Parliament (http://www.europarl.europa.eu/)

has political responsibility for the supervision of the activities of the following agencies: the European Medicines Agency, the European Environment Agency, the European Food Safety Authority, the Food and Veterinary Office and the European Centre for Disease Prevention and Control. One recent European initiative is the prohibition of the use of nanoscale ingredients in food, unless they are adequately assessed for risk and labelled on the products.

In Europe, the European Chemicals Agency (ECHA): (http://echa.europa.eu) is another agency working on the elaboration of guidance documents to provide more information about the applicability of nanomaterials. ECHA was specifically created to oversee implementation of the REACH programme, which regulates the Registration, Evaluation, Authorization and Restriction of Chemicals. The European Committee for Standardization, (http://www.cenorm.be) is working on the standardization and labelling of products containing nanomaterials. In addition, some EU Member states are also taking individual measures regarding the use of nanomaterials. For example, Germany has announced that "nanosilver has no place in food, textiles and cosmetics", while Denmark has stopped the sale of "nanofilm", because research in mice has shown serious lung damage after exposure to the floor sealant spray (information on http://www.sidley.com).

Other European initiatives include several research projects on toxicity receiving funding from the 7th EU programme, and many other activities. Some of these activities are listed below, and we apologize to the many other projects that have not been included in the list:

- *e-ICOON* (www.eicoon.eu): Euro-Indo forum for the coordination of research in nanomaterials and for co-operation of researchers in sustainable energy technologies.
- *FramingNano* (www.framingnano.eu): international multi-stakeholder dialogue platform framing the responsible development of nanosciences and nanotechnologies.
- *Hinamox* (www.hinamox.eu): European project that tests the toxicity of metal and metal oxide nanoparticles.
- *ICPCNanoNet* (www.icpc-nanonet.org): a Web-based repository of nanoscience and nanotechnology publications, database of researchers and online forum, to inform and facilitate networking between EU and International Co-operation Partner Country RTD.
- *NanoCharM* (www.nanocharm.org): characterization of multifunctional nanomaterials using ellipsometry and polarimetry.
- *NanoCode* (www.nanocode.eu): multi-stakeholder dialogue providing inputs to implement the European Code of Conduct for research in nanosciences and nanotechnologies.
- *Nanoforum* (www.nanoforum.org): Europe's Gateway to Nanotechnology.
- *NanoMedNet* (www.nano.org.uk/nanomednet): bringing together key players in nanomedicine.

- *NanoMicroClub* (www.nanomicroclub.com): new nano and micro companies meet customers, investors and advisors.
- *Nano-org* (www.nano.org.uk/members.htm): supporting professional and corporate members to access opportunities in nanotechnology.
- *Nanosafety cluster* (http://www.nanosafetycluster.eu): it includes experts from several fields trying to classify and compile the results from different toxicology groups. It is also working on the standardization of protocols, in the definition of reference material and in the development of reliable tests to be performed with nanomaterials.
- *NanoSustain* (www.nanosustain.eu): development of sustainable solutions for nanotechnology-based products, based on hazard characterization and LCA.
- *Nano-TV* (www.youris.com): video news releases describing EU-funded projects for European television and news broadcasters.
- *NEAT* (www.nano.org.uk/nanomasters/index.htm): Nano, Enabling and Advanced Technologies postgraduate course directory—find the right course for you.
- *observatoryNANO* (www.observatory-nano.eu): European observatory for providing scientific, technical and economic analysis of nanotechnologies, and to consider their ethical, societal, EHS and regulatory issues.

In the United States, 25 federal agencies participate in the National Nanotechnology initiative (www.nano.gov) including, among others, the Food and Drug Administration (FDA) (http://www.fda.gov), the National Institute of Standards and Technology (NIST) (http://www.nist.gov), the National Aeronautics and Space Administration (http://www.nasa.gov) and the National Institute of Environmental Health Sciences (http://www.niehs.nih.gov). As outlined on the FDA web page: "The FDA is working with the White House, the National Nanotechnology Initiative, other U.S. government agencies, and international regulators to focus on generating data and coordinating policy approaches to ensure the safety and effectiveness of products using nanomaterials" (http://www.fda.gov/RegulatoryInformation/Guidances/ucm257698.htm).

Several important questions arise because there is no agreement between agencies. The first one is how to define a nanomaterial? For example, the US Environmental Protection Agency considers nanomaterial to be in a range from 1 to 100 nm, but the US Department of Agriculture defines a range from 1 to 300 nm. Recently, the FDA has taken a first step by releasing in June 2011 draft guidance toward greater regulatory certainty regarding nanotechnology. The FDA considers to be "nano" any engineered material or end product with at least one dimension smaller than 100 nm, or when an engineered material or end product exhibits properties or phenomena, including physical or chemical properties or biological effects, attributable to its dimension(s), even if these dimensions fall outside the nanoscale range, up to one micrometre.

The second question: how to evaluate toxicity? As nanomaterials can induce toxicity at several levels, many tests *in vitro* and *in vivo* should be performed to understand their behaviour, before their intended use. However, as indicated above, this is not an easy task, and new methods and techniques are urgently needed. Efforts by researchers and companies should go in the direction of standardization of techniques. In this sense, the Nanotechnology Characterization Laboratory in the USA was established as an interagency collaborative effort between the National Cancer Institute, the NIST and the FDA (http://ncl.cancer.gov) to help regarding the standardization of nanomaterials for use in cancer treatment. They propose cascade assays starting from a deep physicochemical characterization of the nanomaterials, followed by studies on endotoxin content, prior to performing several tests that could last 1 year. The cascade assay includes studies of platelet aggregation and haemolysis (blood), phagocytosis, complement activation and several *in vitro* cytotoxic assays (analysis of hepatic, heart or renal toxicity), genotoxicity and immunotoxicology studies, before using Ns *in vivo* with animal models.

At the International level, the Organization for Economic Co-operation and Development is conducting environmental, health and safety testing on manufactured nanomaterials, and it has published guidance documents on testing nanomaterials (www.oecd.org). The International Organization for Standardization (ISO) (http://www.iso.org) has recently elaborated 14 published standards and/or projects under the TC 229 Secretariat in collaboration with other agencies. Specific tasks include developing standards for terminology and nomenclature; metrology and instrumentation, including specifications for reference materials, test methodologies, modelling and simulations; as well as science-based health, safety and environmental practices (http://www.iso.org/iso/iso_technical_committee.html?commid=381983).

4. TOXICITY OF BIODEGRADABLE NANOSTRUCTURES

Several biodegradable nanomaterials, because of their unique characteristics, have many proven and potential applications, especially in the biomedical field. They can be used like therapeutic delivery systems, because they can improve solubility, extend the half-life, reduce toxicity and accomplish a controlled release and targeted delivery of the drug. They are also used in diagnosis.[19] Some biodegradable nanostructures are already on the market (Table 15.1) and many others are in different clinical trials, such as VivaGel, which is a dendrimer made from poly-L-lysine for the prevention of genital herpes or NB-001; the liposomal vincristine, for Non-Hodgkin's lymphoma; or the Polyglutamate paclitaxel, for several cancers.[20]

Biodegradable nanoparticles can be classified into four different groups:

- Lipid-based nanostructures: liposomes and solid lipid nanoparticles (SLN)
- Dendrimers
- Polymeric nanoparticles
- Albumin-based nanoparticles

TABLE 15.1 Nanostructures Already on the Market

Composition		Trade Name	Company	Indication	Route
Liposomes	L. amphotericin B	Abelcet®	Enzon	Fungal infection	i.v.
	L. amphotericin B	AmBisome®	Gilead Sciences	Fungal and protozoal infection	i.v.
	L. Cytarabine	DepoCyt®	SkyePharma	Lymphomatous meningitis	i.t.
	L. Daunorubicin	DaunoXome®	Gilead Sciences	Kaposi's sarcoma	i.v.
	L. Doxorubicin	Myocet®	Zeneus	Breast cancer	i.v.
	Liposomal IRIV vaccine	Epaxal®	Berma Biotech	Hepatitis A	i.m.
	Liposomal IRIV vaccine	Inflexal®	Berma Biotech	Influenza	i.m.
	L. morphine	DepoDur™	SkyePharma, Endo	Postsurgical analgesia	Epidural
	L. verteporfin	Visudyne®	QLT, Novartis	Macular degeneration	i.v.
	Liposome PEG-doxorubicin	Doxil®/Caelyx®	Ortho Biotech, Schering-Plough	Kaposi's sarcoma, breast and ovarian cancer	i.m.
	Micellular estradiol	Estrasorb™	Novavax	Menopausal therapy	Topical
Polymers	Copolymer L-Glu, L-Ala, L-Lys, L-Tir	Copaxone®	TEVA Pharmaceuticals	Multiple sclerosis	s.c.
	Methoxy-PEG-poly (D,L-lactide) taxol	Genexol-PM®	Samyang	Breast cancer	i.v.
	PEG-adenosine deaminase	Adagen®	Enzon	Severe combined immunodeficiency	i.m.

continued

TABLE 15.1 Nanostructures Already on the Market—Cont'd

Composition		Trade Name	Company	Indication	Route
	PEG-anti VEGF aptamer	Macugen®	OSI Pharmaceuticals	Macular degeneration	i.r.
	PEG-α-interferon 2[a]	Pegasys®	Nektar, Hoffmann-La Roche	Hepatitis C	s.c.
	PEG-GCSF	Neulasta®	Amgen	Neutropenia	s.c.
	PEG-HGF	Somavert®	Nektar, Pfizer	Acromegaly	s.c.
	PEG-L Asn	Oncaspar®	Enzon	Acute lymphoblastic leukaemia	i.v. i.m.
	PEG-anti TNF-α antibody fragment	Cimzia®	Nektar	Rheumatoid arthritis and Crohn's disease	s.c.
	Poly(allylamine hydrochloride)	Renagel®	Genzyme	Kidney disease	Oral
Other platforms	Albumin-bound Paclitaxel	Abraxane®	Abraxis, Bioscience, AstraZeneca	Breast cancer	i.v.
	Nanocrystalline aprepitant	Emend®	Elan, Merk	Antiemetic	Oral
	Nanocrystalline fenofibrate	Tricor®	Elan, Abbott	Antihyperlipidemic	Oral
	Nanocrystalline sirolimus	Rapamune®	Elan, Wyeth Pharmaceuticals	Immunosupressor	Oral
	Nanocrystalline paliperidone palmitate	Xeplion®	Elan, Jonhson & Jonhson	Schizophrenia	i.m.

Combined with metal Ns	Ferric carboxymaltose (Europe)	Ferrinject®	ViforPharma	Iron deficiency anaemia	i.v.
	Iron isomaltoside (Europe)	Monofer®	Pharmacosmos	Iron deficiency anaemia	i.v.
	Ferumoxytol (USA)	Feraheme®	AMAG Pharmaceuticals	Iron-deficiency anaemia	i.v.
	Mangafodipir trisodium	Teslascan®	GE healthcare	Contrast agent for MRI	i.v.
	Gadopentetate dimeglumine	Magnevist®	Bayer Schering Pharma	Contrast agent for MRI	i.v.
	Silicone-coated SPIO	Lumirem®	AMAG Pharmaceuticals	Contrast agent for gastrointestinal tract	Oral
	Dextran-coated ferumoxide	Endorem™ (Europe), Feridex® (USA)	Guerbet, AMAG Pharmaceuticals	Contrast agent for liver or spleen diseases	i.v.
	Dextran-coated USPIO	Ferumoxtran	AMAG Pharmaceuticals	Contrast agent for lymph node and MRI	i.v.

i.m., intramuscular; i.v., intravenous; i.t., intrathecal; s.c., subcutaneous; i.r., intravitreus; MRI, magnetic resonance imaging.

Since the late 1960s, there has been evidence regarding the potential toxicity of some nanoparticles, especially liposomes. The toxicity is largely dependent on, among other characteristics, the doses, application and the nature of the nanoparticles (NP). Their properties, applications and possible cytotoxicity will be discussed below.

4.1. Lipid-Based Nanostructures: Liposomes and Solid Lipid Nanoparticles

Liposomes were discovered in the 1960s as spherical vesicles, made mainly by natural or synthetic phospholipids, which are organized as the lipid bilayer of cell membranes. They are amphipathic molecules with polar and nonpolar regions.[21] There are other nanostructures, similar to liposomes, which are called micelles and are made from a monolayer. As they are very easy to prepare, these nanostructures have multiple applications in the pharmaceutical, cosmetics and food industries: they can be used in drug and gene delivery, as models for artificial cells, carriers for the delivery of dyes to textiles,[22] pesticides to plants, enzymes and nutritional supplements to foods and contrast agents.

One concern regarding their use in medical applications is their toxicity. It is known that cationic liposomes are more cytotoxic than neutral or anionic liposomes. This can be explained because the positive charge of the liposomes allows non-specific interactions with cell membranes.[23] However, it has been observed that anionic liposomes activate platelets and induce platelet aggregation, but cationic or neutral ones do not.[24] It has been shown by *in vitro* studies that cationic liposomes are less cytotoxic than *in vivo*, suggesting that other factors involved *in vivo* must be responsible for the toxicity. When cationic liposomes are administered by the pulmonary route, a dose-dependent toxicity and pulmonary inflammation is observed. As a simultaneous increase in reactive oxygen intermediates (ROI) was found, it was suggested that ROI may play a key role in cationic lipid-mediated toxicity.[23] Yet, no complement activation was observed with different cationic liposomes.[25]

Nevertheless, cationic liposomes can be more efficient when they are used as nonviral vectors, because they spontaneously interact with negatively charged nucleic acids, and the excess of positive charge allows interactions with cell membranes. This could be useful in therapies for treating some cancers.[26,27] Moreover, there is a higher rate of capture of these liposomes by leaky tumour vasculature, but there is also some evidence that they are eliminated very rapidly from blood, as they are captured by macrophages in the liver and spleen, while they also bind to negatively charged serum proteins. These side effects can be reduced by including PEG[28] on the surface of the nanostructure. Similarly, cationic liposomes can be used for the design of vaccines, even though they are more cytotoxic, because they interact better with antigen-presenting cells.[29]

Another important issue is that liposomes can reduce the toxicity of drugs and simultaneously increase their solubility and activity. Drug loading into liposomes can be achieved through different methods. If the drug is hydrophilic, it can be trapped in the central core or in the outer membrane; if it is lipid soluble, it would be trapped in the hydrophobic part of the liposome.[30] The cytotoxicity of the drug *in vivo* seems to be lower when encapsulated than in free conditions. Several examples are bleomycin, whose administration with a neutral liposome reduces lung toxicity[31]; cisplatin, when encapsulated in neutral liposomes and administered intraperitoneally induces low cytotoxicity[32] or other insoluble drugs like mitoxantrone, ceramides or betulinic acid.[33–35] Regarding the biodistribution of these compounds, after systemic administration, conventional liposomes are normally eliminated from the systemic circulation mostly by macrophages of the reticuloendothelial system and accumulated primarily in the liver. This can be useful for the delivery of antiparasite or antimicrobial drugs for treating infections of the mononuclear phagocytic system. However, when the target is different, liposomes can be modified to reduce this uptake by introducing PEG or a targeting ligand to obtain a longer half-life.[36]

SLN are particles made from solid lipids (i.e. lipids that are solid at room temperature and also at body temperature) and stabilized by surfactants.[37] Lipids that are commonly used are triglycerides, complex glyceride mixtures or even waxes. There are two types of nanocomposites derived from SLN that have been designed to improve drug-loading capacity.[37] They are (i) nanostructured lipid carriers, which are produced from a blend of solid and liquid lipids (they are solid at body temperature) and (ii) lipid-derived drug conjugates. The most frequent application of these nanostructures is drug delivery, but there are some studies related to their adjuvant effect on vaccines.[38] Although they are known to be well tolerated *in vitro* and *in vivo*, it is important to take into account the potentially harmful effect of the lipid, the surfactant or both. When cetyl palmitate SLN with different surfactants was tested, no acute toxicity and no increase in weight were observed in liver and spleen. The interaction with red blood cells is quite dependent on the surfactant used; for example, Tween 80 and Poloxamer bind to erythrocytes to a lesser extent than Span 85.[39] It was also observed that the interaction of SLN with serum albumin, which is the major circulatory protein, is very labile.[40] Another important issue is that depending on their characteristics, they can act as adjuvants: decreasing particle size and increasing hydrophobicity improves the adjuvant capacity of these particles.[38] Their potential allergenic role has been studied by some authors, both *in vivo* and *in vitro*, and they did not find anaphylactic shock in animals or induction of pro-inflammatory cytokines in macrophages.[39]

4.2. Dendrimers

Dendrimers are globular, monodisperse macromolecules with three-dimensional polymeric architectures, made from synthetic polymers consisting of a core and several layers with active terminal groups. These layers are formed by repetition of the polymer, each called a generation, with the zero generation core and each layer generation 1, 2, 3, etc. Dendrimers have defined composition, high molecular mass and highly branched architecture. This molecular structure provides many applications, both pharmaceutical and non-pharmaceutical. Among the pharmaceutical applications, they can work as a carrier for drug delivery (transdermal, oral, ocular, pulmonary) and, more specifically, they can target tumour cells with drugs by incorporating tumour-affinity molecules (folic acid, anti-CD-14, prostate-specific membrane antigen, etc.). Dendrimers can enhance the solubility of molecules that are not soluble, in photodynamic and photothermal therapy, gene transfection (they are more efficient than liposomes, because of their p*K* dendrimers can buffer the pH change in the endosomal compartment), as contrast agents for Magnetic Resonance Imaging (MRI) by carrying small molecules like gadolinium and as scaffolds for tissue repair. Non-pharmaceutical applications include use in the development of sensors or multiple bioconjugates (dendritic scaffold); in light harvesting, as catalysts, artificial enzymes and nanocomposites; in inkjet inks and toners, as molecular weight and size standards[41] and also in the development of organic light-emitting diodes (LEDs).[42] There are different types of dendrimers, such as polyamidoamine (PAMAM), poly(propylene imine) (PPI) and poly-L-lysine, of which PAMAM dendrimers are the most intensively investigated from a technological and toxicological point of view.[43]

Dendrimer chemistry was first introduced in the 1970s and dendrimers are being extensively studied, because of their potential in the biomedical field. There are reports that relate the concentration and generation of dendrimers and the free amine groups in the periphery with their toxicity. It is known that those dendrimers that have a positive membrane charge are more cytotoxic than others with a negative or neutral membrane charge,[44] as they can induce disruption, including the formation of pores in membrane, apoptosis caused by mitochondrial dysfunction, or caspase-dependent apoptosis. In terms of uptake of these compounds, it is important to know that the anionic ones are internalized by caveolae-mediated endocytosis. However, cationic and neutral dendrimers are internalized by a non-specific endocytosis mechanism. In addition, cationic dendrimers interact more strongly with albumin, which is the most abundant protein in plasma.[43] Concerning the complement activation, it was observed that PAMAM dendrimer 5.0 G complexed with DNA-induced strong complement activation.[25]

There are other important issues, such as haemolytic toxicity due to the interaction of free cationic terminal groups of these nanostructures with erythrocytes and the polycationic nature of dendrimers that would lead to haemolysis.

Moreover, dendrimers with higher generation (with a greater overall cationic charge) have greater haemolytic toxicity.[45] In addition to haemolytic toxicity, dendrimers have effects on different blood parameters. It was observed that PPI induces a decrease in red blood corpuscle count, haemoglobin content and the mean corpuscular haemoglobin values.[46]

In vivo toxicity studies have again shown that the behaviour of dendrimers depends on their structural core, the number of generations and surface groups.[47] A relationship was found between the increase in dendrimer generation and toxicity. The 7.0 G PAMAM dendrimers produce biological complications that are not observed for smaller dendrimers (3.0 or 5.0 G), but in no case has there been evidence of immunogenicity.[48] Melamine dendrimers were found to induce animal death 6–12 h post intraperitoneal (i.p.) injection, but only at a very high dose (160 mg/kg). At a lower, but still high dose (40 mg/kg), hepatotoxicity was detected.[49]

Charge is very important in relation to the biodistribution of these nanostructures. It has been shown that cationic PAMAM can be cleared very rapidly from circulation (only 0.1–1% was detected in blood just 1 h after administration, with accumulation of dendrimers in the liver), but longer circulation times have been reported for anionic PAMAM.[50] The toxic effects induced by cationic dendrimers could be due to interactions with negatively charged biological membranes, causing membrane disruption by inducing formation of nanoholes or producing membrane thinning or erosion. This leads to leakage of cytosolic enzymes that, in turn, can lead to cell lysis.[47]

Some strategies are being developed to reduce toxicity. Dendrimers could be less toxic if more biocompatible materials are used in their synthesis, such as polyether, polyester, polyether imine, polyether-copolyester, phosphate citric acid, melamine, peptide or triazine. Another possible strategy is to mask the cationic charge by engineering the surface of the dendrimer. Some examples of this strategy include adding polyethilenglycol or carbohydrates, undergoing acetylation, half generation of anionic dendrimers, conjugation with amino acids or peptides. Moreover, the toxicity of dendrimers can be reduced by functionalizing them to simultaneously obtain other beneficial properties for different biomedical applications, such as conjugation with DNA, drugs, antibodies or folic acid, whose receptor is overexpressed in many types of tumour cells.[47]

4.3. Polymeric Ns

Polymeric nanostructures are those made from natural or synthetic polymers or a combination of the two (synthetic–synthetic, synthetic–natural or natural–natural). Depending on the method of preparation, they can be:

- Nanospheres
- Nanocapsules
- Nanoparticles

Their biocompatibility and biodegradability makes them very suitable for many biomedical applications: as nano-vaccines[51]; in drug and gene delivery (controlling the release, improving the efficiency, reducing the toxicity and addressing the drug); in tissue engineering; in medical supplies, like sutures or wound dressings; as imaging agents; in diagnosis, as part of biosensors. They are also used in cosmetics (encapsulation and transport of active ingredients), the food and beverage industry and packaging. As they are highly biocompatible, when some toxicity is observed, this is usually attributed to the drug, surfactant or ions included in the polymers, but not the polymers themselves.

A distinction can be made between synthetic and natural polymeric Ns. The most common synthetic polymeric Ns are:

- Those made from poly (α-hydroxy ester) like polylactide (PLA), polyglycolide or their copolymer poly-DL-lactide-coglycolide. These biodegradable polymers have been approved by the FDA for human use, and they do not induce cytotoxicity *in vivo* or *in vitro*,[52,53] although a decrease in pH (due to the breakdown products) can lead to some irritation.[54]
- Polyorthoesters and polyanhydrides are both well known to be very biocompatible.[54–56] Indeed, there is a drug delivery system (GLIADEL®) based on a polyanhydride that is approved by FDA for the treatment of malignant glioma.
- PEG, a highly biocompatible polyether, has been frequently used as a surface modifier of other nanoparticles to reduce their cytotoxicity.

The most important natural polymeric Ns are:

- Polyamino acids show good biocompatibility,[57] but their use is limited by their antigenic potential.[58]
- Collagen type I is present in large amounts in bone, tendon or skin. It is the most widely used polymer, normally extracted from bovine or porcine animals. This polymer exhibits biodegradability, weak antigenicity and biocompatibility.[59]
- Hyaluronic acid (HA), which is a minor component of the bone extracellular matrix, is typically obtained from rooster combs. HA is a natural non-toxic mucoadhesive polysaccharide that is negatively charged and biodegradable.[60]

In addition to the high cost of production, all these natural materials from animal origin have the risk of disease transmission. Hence, other natural polymers are also being used, such as:

- Polyglucoses like starch or dextran, whose biocompatibility is well known,[61] and they are sometimes used to reduce the cytotoxicity of other compounds.[62] However, it has been observed that certain configurations of dextran-coated Ns can activate the complement cascade *in vitro*, because some reactive groups bind to the C3 complement protein.[63]

- Chitosan is a derived natural carbohydrate polymer prepared by the deacetylation of crustacean natural biopolymer chitin. In most cases, chitosan Ns have been shown to be biocompatible, although, at high concentrations, they can reduce cell viability in human liver cells.[64] Chitosan has been approved by the FDA only for wound dressing, because it achieves haemostasis and induces normal tissue regeneration.[65] It is also used as a satiating food, because it is not absorbed or digested by the body. As described above for dextran-coated Ns, certain chitosan-coated Ns activate the complement cascade.[63] It is also interesting to note that chitosan and its derivatives seem to be toxic to several bacteria, fungi and parasites.[66]
- Alginate is a natural polysaccharide extracted from brown algae and made from mannuronic acid and guluronic acid. As a nanoparticle, it is normally prepared by blending another polymeric component with alginate, showing null or low toxicity.[67,68]
- Polyhydroxybutyrate is a biopolyester produced as granules by microorganisms. It is normally used combined with hydroxyapatite, and it has been seen to be well tolerated by the body when used in implants.[69,70] However, it has been reported that this compound can cause inflammatory responses.[71,72]

Biodistribution studies of polymeric nanoparticles have shown that they are normally opsonized and removed from the bloodstream by macrophages in the liver and spleen. The clearance is very rapid (usually a few minutes) if they are not coated with PEG (pegylated). When they are pegylated, their half-life is longer and they migrate preferentially to the spleen, while smaller particles concentrate in bone marrow.[14]

There is a special nanostructure called nanogel, which is formed by cross-linked polymeric particles. Nanogels can be considered as hydrogels if they contain water-soluble polymer chains. They possess high water content and desirable mechanical properties,[73] as well as showing biocompatibility. These features make nanogels suitable for various applications. They can be used for the following: drug delivery,[74] the measurement of intracellular temperature (highly hydrophilic fluorescent nanogel stays in the cytoplasm and emits stronger fluorescence at a higher temperature),[75] gene delivery, in nanoreactors (e.g. those nanogels that include AuNs in their composition), as skin-specific nanocatalysts for reactive oxygen species (ROS) (containing PtNs),[76] in regenerative medicine and in bio-imaging and chaperone-like functions.[77] As these nanostructures are based on polymeric compounds, they have inherent properties of biodegradability and biocompatibility.

4.4. Albumin-Based Ns

Albumin is the most abundant plasma protein (35–50 g/L human serum) with a molecular weight of 66.5 kDa. Albumin is a multifunctional protein. One of its most relevant functions is the transport of long-chain fatty acids, which makes it essential for the metabolism of lipids. Albumin is also capable of binding a large number of therapeutic drugs including, among many others, penicillin, sulphonamides, indole compounds and benzodiazepines.

Human serum albumin is one of the smallest proteins in the human plasma and has an average half-life of 19 days. It is very stable over a wide range of pH (4–9) and resilient to thermal stress, as it can be heated to 60 °C for up to 10 h without harmful effects. Albumin is soluble in water and ethanol, two viable solvents for intravenous administration. These properties, together with its preferential uptake in tumours and inflamed tissue, its ready availability, biodegradability and its lack of toxicity and immunogenicity, make it an ideal candidate for drug delivery.[78] Albumin can be either obtained directly from human blood or synthesized as recombinant albumin.

The company American Bioscience Inc. has developed an albumin-based nanoparticle technology that is ideal for the encapsulation of hydrophobic compounds. Albumin nanoparticles loaded with a drug are prepared by mixing both compounds in an aqueous solvent and forcing the mixture to pass under high pressure, through a jet, to form a liposome-like structure of 100–200 nm. This company also used this technology to develop Abraxane™, as an alternative for the transport of an antitumoural agent (paclitaxel) into tumour tissues. The encapsulation of paclitaxel in albumin Ns significantly reduces the toxicity of paclitaxel.[79,80]

The increased permeability of cancer tissues allows the accumulation of plasma proteins in the tumour for anabolic processes, due to their high metabolic demand. Albumin undertakes a process of transcytosis via caveolae, initiated by the albumin receptor gp60.[81,82] There is also speculation that Abraxane transport into tumour cells is facilitated by secreted protein acidic rich in cysteine or osteonectin, two extracellular matrix glycoproteins that are overexpressed in a variety of cancers.[82]

In addition to cancer treatment, albumin nanoparticles also have potential applications in brain diseases, as their combination with PEG–PLA allows them to cross the blood–brain barrier,[83] while the addition of apolipoproteins can facilitate transcytosis.[84] This broad spectrum of applications makes them an interesting nanomaterial with medical applications, particularly in cancer, inflammation and brain-related diseases.

4.5. Excipients and Emulsifiers

The toxicity of some nanomaterial-based therapeutic drugs has been shown to be induced by some excipients or emulsifiers used in the formulations, but not by the nanostructure or the drug itself. A well-known case is Cremophor EL (CrEL) (BASF Corp.), a non-ionic micellar solvent from castor (ricinus) oil and a complex mixture of different molecules, which has been used as a vehicle for the transport of paclitaxel, a drug for anticancer treatment and other water-insoluble components, such as certain vitamins and steroids. Although hypersensitivity reactions have been described for all these compounds dissolved in CrEL,[85,86] the most characterized compound is Taxol. The toxicity of Taxol was initially reported during the first trials,[87] and it has since been the subject of considerable research. The effects caused by Taxol were determined *in vitro* by studying complement activation and making a comparison with an equivalent amount of CrEL.[88] In both cases, there was a significant increase in the levels of complement factors C5b-9 and Bb in the sera of both normal and cancer patients, indicative of activation of the complement cascade by the alternative route. This complement activation is responsible for a pseudoallergy reaction, which still persists in 2–7% of patients, even when they have been pretreated with corticosteroids.[89] Although the mechanism of complement activation by CrEL is not completely understood, factor C3 could be triggered by the interaction with some hydrophilic chains of the molecule. This has been previously observed with some surfactants (L101 and L121), which activate the complement alternative pathway by binding to C3 to its hydrophilic adhesive surface.[90] The toxicity of CrEL encouraged intensive research to find a new and non-toxic vehicle for paclitaxel administration.[91–94] This research resulted in Abraxane® (American Pharmaceutical Partners Inc. and American Bioscience Inc.), the first alternative to be approved by the FDA in 2005 for the treatment of metastatic breast cancer. As previously mentioned above, Abraxane® is an albumin-based nanoparticle paclitaxel formulation that provokes less hypersensitivity reactions than Taxol, even at 50% higher doses, and no pretreatment with corticosteroids is required.[79,80]

Another example of toxicity caused by the solvents is that of poloxamer 188, which is used as an additive in Fluosol DA, a perfluorocarbon-based blood substitute. Poloxamer 188, also known as Pluronic F-68, induces complement activation via the alternative pathway in human plasma. This complement activation is not dependent on the micelles formed by the solvent, but on both the nature of the polymer and its interaction with plasma lipoproteins.[95]

To sum up, biodegradable Ns are generally quite biocompatible, although some exceptions have been shown among cationic liposomes and large positively charged dendrimers. In addition, some solvents, but not the nanostructure or the drug, can induce pseudoallergy reactions after inducing the activation of the complement cascade.

5. TOXICITY OF NON-BIODEGRADABLE NANOSTRUCTURES

Hitherto, most of the studies related with inorganic nanostructure applications have been carried out *in vitro* or in animal models, but there have been very few in humans, because of the broad toxicity of most of these compounds. In fact, the high toxicity of some surface-modified inorganic Ns directed against tumour cells is being explored as a possible anticancer therapy. Silver, platinum and gold Ns are the main compounds that have been tested for this purpose. Although there has been considerable research into the antitumour properties of platinum, this has mainly been with platinum compounds (*cis*-platinum and derivatives) rather than with nanoparticles.[96] However, platinum nanoparticles in core–shell form have been the focus of recent study, as they appear to be more efficient as antitumour agents.[97]

Iron is very common in nature and in the human body. Iron oxide Ns have been used to treat iron deficiency and as a contrast agent in imaging techniques, because they can be biodegradable and metabolized in the body. Ferumoxytol (Feraheme® Injection, AMAG Pharmaceuticals, Inc.) is a formulation of semi-synthetic carbohydrate-coated superparamagnetic iron oxide nanoparticles, which has been approved by the FDA in 2009 for the treatment of iron deficiency anaemia in patients with chronic kidney disease.[98] The side effects observed are similar to those reported for conventional oral iron therapy.[99]

Inorganic and metal-based Ns are being used for the development of contrast and enhancer agents for imaging techniques. Feridex I.V. or Ferumoxides, Ferumoxtran (an injectable solution of dextran-coated superparamagnetic iron oxide Ns) and Teslascan (mangafodipir trisodium, a manganese injectable solution) are some examples. Feridex and Teslascan are only FDA-approved for the evaluation of liver lesions by MRI (Table 15.1). Apart from Feridex, there are other small paramagnetic iron oxide Ns (SPIONs) that are being tested for MRI. SPIONs have a promising application for stem cell monitoring in tissue engineering following internalization of Ns. However, it remains to be proven that they are safe and do not interfere with cell differentiation. Interference of SPIONs has been described in the chondrogenic differentiation of human mesenchymal stem cells (MSCs).[100] A later work showed that the toxic effect was caused by the transfection agent and not by the Ns.[101] Yet, more recently, Farrell and colleagues have reported subtle effects in the phenotype of the differentiated MSC *in vivo* and *in vitro*, without inhibition of the multilineage MSC differentiation.[102] *In vitro* studies with surface-modified SPIONs have shown an increased biocompatibility, as well as the possibility of addressing the Ns to monitor tumoural cells and to combine their role as a contrast agent with that of drug delivery and/or hyperthermia.[103] The integrated use of MSCs, gene therapy and SPIONs in cancer diagnosis and therapeutics is currently being studied. MSCs are able to interact with cancer cells and avoid the immune system, and, for this reason, they are being tested for gene therapy using suicide-, apoptosis-, anti-angiogenesis-, immune-stimulatory genes or oncolytic viral vectors.[104]

For imaging purposes, quantum dots (QDs) are also being tested. QDs are semiconductor Ns with a strong surface plasmon resonance (SPR) band that, as mentioned above for gold Ns, can be tailored by changing their size and composition. However, QDs have a high potential toxicity due precisely to their composition and their possible solubilization. Cadmium, a common element in QDs, is highly toxic and can be released from Ns that contain this element. Moreover, this is not the only source of the toxicity caused by QDs *in vitro* and *in vivo*[105] (see Section 5.4.1).

The cosmetics industry is another field, besides biomedicine, where numerous applications have been found for inorganic Ns. For instance, ZnO and TiO_2 Ns have being widely used as sunscreen agents. Recently, fullerene has been proposed as an anti-wrinkle agent based on its antioxidant properties.[106]

Some of the mechanisms of the toxicity caused by non-biodegradable Ns are now being revealed. Although oxidative stress has been suggested as the main common mechanism of NP-induced toxicity, several other physiological processes and bystander effects implicated in the toxicity caused by Ns are being described.[107] For example, the interaction with the body fluids and cellular microenvironment favours the formation of the protein corona.[108,109] This corona is important in NP uptake by cells and could lead to the activation of specific signalling pathways, inducing cell differentiation. It has recently been shown that negatively charged gold Ns bind to and induce unfolding of fibrinogen. Unfolded fibrinogen promotes interaction with the integrin receptor Mac-1 and activates the nuclear factor-kappa β (NF-κβ) signalling pathway, with the subsequent release of inflammatory cytokines.[110]

Although there is still considerable research devoted to understanding how Ns may interfere with cells and physiological processes in humans and the environment, the toxicological effects of some of their components are largely known. Examples are the toxicity of heavy metals in solution and the well-documented occupational diseases related to exposure to inorganic materials, such as asbestos (asbestosis) and silica (silicosis).

5.1. Toxicity of Heavy Metals and Trace Metals

The term "heavy metals" includes the transition metals, some metalloids, lanthanides and actinides. These metals provoke environmental pollution via sources such as leaded petrol, industrial effluents and leaching of metal ions from the soil into lakes and rivers by acid rain.[111]

Cadmium, lead and mercury are the most prevalent contaminants due to their widespread use in industry. Chromium, cobalt, copper, iron and other metals are toxic, because of their high surface reactivity and redox potential, which induce fenton reactions. Disruption of metal ion homeostasis may lead to oxidative stress, a state where increased formation of ROS overwhelms body antioxidant protection and, subsequently, induces DNA damage, lipid peroxidation, protein modification and other effects. These effects are all symptomatic for numerous diseases, including cancer, cardiovascular disease, diabetes, atherosclerosis, neurological disorders (Alzheimer's disease, Parkinson's disease), chronic inflammation and other disorders.[11] For redox inactive metals like mercury, cadmium, lead and arsenic, the mechanism of toxicity implies bonding to sulphydryl groups of proteins and depletion of glutathione.[112] In general, metals activate signalling pathways, and the carcinogenic effect of metals has been related to activation of mainly redox-sensitive transcription factors, involving NF-κβ, AP-1 and p53.

Absorption of metals may have some effects on the body, and these effects depend on the particular metal. Some metals are inherently toxic, whereas others are essential for the normal functioning of the body (trace elements), even though at high concentrations they may cause adverse effects. In addition, metals are non-biodegradable elemental species and such indestructibility leads to accumulation, which, in turn, explains their toxicity. For some of these metals, a more detailed description follows of their routes of absorption and excretion, their accumulation in the body and the toxicity induced in humans.[113–115]

5.1.1. Zinc

Zinc is a trace element and, among other physiological functions, has a relevant role in the immune system. Zn deficiency is more common than Zn accumulation, as its homeostasis is tightly controlled through its role as inter- and intracellular messenger.[112] In humans, around 20–50% is absorbed by the gastrointestinal tract. This absorption decreases when zinc is consumed with calcium and phosphorus but is enhanced when it is consumed with animal proteins. Zinc also forms complexes with metallothionein. Once absorbed, it is distributed throughout the body and mainly deposited in the prostate, muscle, bone, skin/hair, liver, retina and pancreas. It is excreted in urine and faeces, having a half-life of 300 days. Although very rare, toxic effects induced by ingestion of large amounts of zinc have been reported (diarrheic and gastrointestinal effects). The most common effect after inhalation of Zinc oxide is "metal-fume fever", which is characterized by fatigue, fever, chest pain, chills, cough, dyspnoea, nausea, muscle soreness and leukocytosis.

Chronic (long-term) human exposure to lower doses may lead to neuronal and pancreatic toxicity. Moreover, most zinc salts cause dermatitis and skin ulcerations, as well as upper respiratory and bronchial irritation.

5.1.2. Nickel

Nickel is mainly used in manufacturing for the production of nickel steels, although it can also be used as a catalyst for hydrogenation. In some plants and microorganisms, nickel is required for the activation of some of their enzymes, as it is an essential nutrient. In humans, nickel is absorbed by inhalation, skin and oral routes. Once absorbed, it is transported bound to albumin. Its biologic half-life varies from 1 day (for nickel sulphate) to 100 days (for nickel oxide), and it is excreted through urine.

Nickel is a well-known and common allergen (that induces contact dermatitis), which has recently been described as being recognized throughout the Toll-like receptor-4 on the surface of immune cells.[116] This receptor has two regions that contain histidine. Nickel might bind to the protein histidines triggering the signalling cascade that promotes the recruitment of T cells to the exposure region and induces the subsequent allergic reaction. Moreover, nickel has been recognized as a human carcinogen in the respiratory tract.

5.1.3. Mercury

Mercury is liquid at room temperature, and due to its properties, it has been widely used in thermometers, manometers or barometers, lighting, and dental restoration. However, due to its toxic effects, mercury is being replaced by other components. Its absorption and effects vary depending on the form of mercury: elemental, inorganic or organic.

Elemental mercury is mainly absorbed by inhalation (80%). After absorption, it is distributed to all tissues, as it is able to cross the blood–brain barrier and the placenta, which explains its neurotoxicity and potential embryonic toxicity compared with the effects induced by inorganic mercury. Moreover, it is not absorbed through the gastrointestinal tract. When high concentrations of elemental mercury are inhaled, it may produce corrosive bronchitis, and pneumonitis associated with effects on the central nervous system. Chronic exposure causes mercurialism, which has three major manifestations: tremors, gingivitis and erethism, as well as irritability, haematological changes, dermographism and some central nervous symptoms.[117]

The major toxic effect of the methylmercury is neurotoxicity. It is well absorbed orally and 10% of the absorbed methylmercury is distributed to the brain, and 5% remains in the blood, concentrated in the erythrocytes. Methylmercury also accumulates in hair, has a half-life of 45–70 days and is mainly eliminated through faeces.

Inorganic mercury is poorly absorbed by the oral route, and does not cross the blood–brain barrier or the placenta, but it has direct toxic effects on the kidneys. It has a half-life of 2 months and is excreted through urine and faeces.

5.1.4. Chromium

Chromium occurs in nature in two different states: hexavalent chromium (Cr VI) and trivalent chromium (Cr III), the latter is the more stable form and is an essential metal for body metabolism.

The absorption of chromium through inhalation or through the skin depends on the chemical form, the molecular vehicle and the condition of the skin, but hexavalent compounds have a higher absorption than trivalent chromium. Once in the body, chromium compounds are distributed mainly to the liver, spleen and kidneys and are primarily excreted through urine. High exposure to chromium may cause severe irritation of exposed tissues, while the accidental ingestion of high concentrations of chromium may cause acute renal failure.[118]

Hexavalent chromium is corrosive and may cause chronic ulceration and perforation of the nasal septum, chronic ulceration of other skin surfaces, as well as contact dermatitis and asthma.[118] Moreover, hexavalent chromium compounds are considered to be human carcinogens.[119]

5.1.5. Lead

Lead is a heavy metal widely used in construction, batteries, bullets and many other compounds. At certain exposure levels, it is very toxic, mainly damaging the nervous system.

Human adults absorb 5–15% orally, whereas children absorb 42% of the ingested lead. The absorption is enhanced by low dietary levels of calcium and iron. Lead inhalation is not as important and depends on the form (vapour vs. particle), particle size and concentration. Once absorbed, 99% of the lead in the blood stream binds to haemoglobin in erythrocytes, and only 15% of the circulating lead is available for tissue distribution. The half-life of lead in red blood cells is 20–40 days, and in bone tissue can be found for much longer periods (20–30 years), and it is mainly eliminated through the kidneys.

Lead toxicity may affect a number of body systems, especially the nervous system and blood. The main symptoms after a high exposure to lead are inattention, hallucinations, delusions, poor memory and irritability. Occasionally, patients have seizures or other symptoms and signs of increased intracranial pressure.[120]

Overexposure to lead is associated with myriad chronic effects. The most frequently reported are vague gastrointestinal and central nervous system complaints. However, it may also cause haematological effects, kidney toxicity, reproductive effects, developmental toxicity (affecting the nervous system and the neurobehavioural function), cardiovascular effects, immunotoxicity, bone effects and carcinogenicity.

5.1.6. Arsenic

There are two forms of arsenic, trivalent (As3) and pentavalent (As5) arsenic. These inorganic forms are biologically transformed into organoarsenicals in different organisms, including humans.

Inorganic arsenic is mainly absorbed by ingestion (80–90%), but it may also be absorbed through the skin. It has a body half-life of around 10–30 h, and it is excreted through urine, as well as by desquamation of skin and through sweat. It also concentrates in fingernails (Mess' line) and hair.

Symptoms of acute intoxication include local irritation of the skin, mucous membranes and conjunctivae, gastrointestinal inflammation and irritation with possible haemorrhage. It can also induce a delayed sensory loss in the peripheral nervous system, stupor, convulsions and death.[121]

For chronic toxicity, skin is the major target organ and induces dermatitis, hyperpigmentation and hyperkeratosis. Other diseases include liver injury that may lead to cirrhosis, respiratory diseases, peripheral neuropathy, renal necrosis and cancer. Arsenic is considered to be a human carcinogen, which has been associated with skin, lung and urinary bladder tumours.

5.1.7. Cadmium

Cadmium is a metal widely used as a pigment, in steel (to avoid corrosion), solar panels and batteries. In recent years, the use of cadmium has decreased as it is replaced by other components.

Human oral absorption is quite limited (5–10%), although it is increased by dietary deficiencies in calcium or iron and by low protein diets. The absorption of the exposure by inhalation is higher (between 10% and 50%).

Cadmium binds to albumin and to a low-molecular-weight protein, metallothionein. It is stored in the liver, from where it is slowly released. The metal is then transported to the kidneys, where it accumulates, potentially leading to renal failure. Excretion of cadmium is very slow, both by the renal and faecal routes, and it has a biological half-life of 10–40 years.

Overexposure may cause severe irritation of the upper gastrointestinal epithelium that may lead to pulmonary oedema, bronchopneumonia and death. The major effects of chronic toxicity are renal injury, obstructive pulmonary diseases, skeletal abnormalities (Itai-Itai disease), cardiovascular diseases and lung cancer.[122]

5.2. Asbestosis and Silicosis

Among human diseases associated with occupational exposure to materials, the most characteristic are Asbestosis and Silicosis, which are linked to asbestos and crystalline silica, respectively.

Asbestos is the name given to a group of fibres, naturally occurring silicate minerals, which due to their interesting properties (there are thin and long fibres), have been widely used as construction materials for many years. People with long-term exposure to asbestos can suffer from Asbestosis, a respiratory problem, a form of pneumoconiosis caused by inhalation of asbestos fibres, which elicit potent inflammatory responses in the lung parenchyma. The disease is characterized by lung interstitial fibrosis, varying from scattered sites to extensive scarring of the alveolar interstitium. Moreover, asbestos-exposed patients may suffer malignant complications, such as pleural mesothelioma and lung cancer.[123]

Silicosis is another form of pneumoconiosis resulting from inhalation of dust containing crystalline form of silica, usually in the form of quartz. Amorphous silica is relatively non-toxic, although not completely inert. Silicosis is an irreversible disease with an incessant progression even when the stimulus stops.[124] Silica-exposed patients develop not only respiratory diseases but also lung cancer and certain immunological disorders. Exposed workers are particularly more susceptible to suffer autoimmune diseases, such as systematic scleroderma, rheumatoid arthritis or systemic lupus erythematosus.

Crystalline silica particles interact with macrophages through the scavenger receptors.[125] Several works point to MARCO as the receptor implicated in particle uptake,[126,127] and other works suggest the participation of scavenger receptor class A (SR-A).[128–130] MARCO receptor seems also to be implicated in the clearance of other environmental particles like TiO_2,[127] and SR-A has been recently related with the internalization of negatively charged gold colloids.[131]

Histologically, silicosis is characterized by hyalinized and fibrotic nodules, thickening of the alveolar interstitium and accumulation of inflammatory cells, such as alveolar macrophages (AMs) and lymphocytes. The pathogenesis of silicosis has been related to the accumulation of inflammatory cells that produce fibrogenic and inflammatory cytokines and growth factors, including TNF-α, IL-1, TGF-β, macrophage inflammatory protein MIP-1 and MIP-2, platelet derived-, insulin like- and fibroblast growth factors. AMs are thought to be key inflammatory cells in silicosis, as they produce most of these fibrogenic factors in silicotic lung. The pro-inflammatory cytokine TNF-α plays a pivotal role in silicosis by mediating a widespread inflammatory reaction and late fibrogenic reaction.[132]

Tripathi *et al.*[133] have recently reported that various factors including free radicals, ROS, lipid peroxide, IL-1, IL-6, IL-8 and IFN-γ take part especially in the inflammatory reaction, whereas TGF-β and IL-4 seem to be implicated in the fibrotic processes. TNF-α plays an important role in inflammation by controlling other cytokines, such as IL-1, IL-4, IL-6 and IL-8, while IL-6 regulates the fibrosis by modulating the expression of Th2 cytokines.

5.3. Toxicity of Carbon-Based Ns

5.3.1. Carbon Nanotubes

A carbon nanotube (CNT) can be defined as a graphite sheet rolled up into a nanoscale tube. These nanotubes can be single-walled (SWNT), or multi-walled, if they have additional graphene tubes rolled around.[134] CNTs have many different and interesting properties. Therefore, CNTs have many potential applications in the field of materials, including nanoelectronic devices, sensors, structural materials, gas storage containers, fillers and storage materials, and in different areas, such as aerospace, energy and the automobile and chemical industries.[134–137] However, as they are highly toxic, it is hoped that advances in functionalization of CNTs will soon be able to make them more biocompatible. Although their toxicity still remains a paramount concern, the diversity of available chemistries and cell-penetrating structures makes CNTs viable candidates as carriers for the delivery of drugs, DNA, proteins and other molecular probes into mammalian cells.

SWNTs have been broadly investigated as imaging agents for tumour targeting, localization and elimination *in vitro* and *in vivo*.[138]

Lung and epithelial cells represent the primary routes of exposure to CNTs. *In vitro* studies have shown the probable absorption of the CNTs across the cutaneous barrier, as well as some negative effects on lung and dermal cells.[139] The exposure of cultured human skin cells to SWNTs caused oxidative stress and loss of cell viability by formation of free radicals, accumulation of peroxidative products and antioxidant depletion. Exposure to SWCNT also produces ultrastructural and morphological changes in cultured skin cells, indicating that dermal exposure may lead to skin diseases.[140] Perhaps, this is to be expected because graphite and carbon materials have been associated with increased dermatitis and keratosis.[141] Moreover, the cytotoxic response of human dermal fibroblasts in culture seems to be dependent on the degree of functionalization of the SWNT; the response is less cytotoxic when the degree of sidewall functionalization is increased.[142] Other studies suggest that the treatment of keratinocytes with SWCNTs can activate the NF-κβ and the release of IL-8 in a dose-dependent manner.[143,144] It has recently been found that SWCNTs also induce genetic damage by centrosome fragmentation, mitotic spindle damage and aneuploidy.[145]

Exposure of human skin fibroblast cells to MWCNTs and multi-walled carbon nano-onions at cytotoxic doses induces cell cycle arrest and increases apoptosis/necrosis. In addition, expression array analysis shows that multiple cellular pathways are perturbed after the exposure, showing material-specific toxigenomic profiles.[146]

There have been several studies with instilled CNTs in rodents. In a similar way to asbestos exposure, CNTs have been shown to cause pulmonary inflammation, granuloma formation and fibrosis after exposure by inhalation/instillation/aspiration. Treatment with SWCNTs in mice reveals a severe inflammation, a fibrotic response and changes in abundance of lung tissue proteins.[147] MWCNTs induce a pulmonary inflammation in rats at high doses and may produce pulmonary lesions.[148] Rat AMs are able to internalize MWCNTs, and they are, therefore, accumulated in the alveoli at least until 6-month post-exposure.[149] Results of lung histopathology revealed dose-dependent focal peribronchiolar lymphoid aggregates, foamy AM accumulation, lymphoplasmacytic infiltration, fibrosis and diffuse alveolar damage.[150] To verify if the functionalization of CNTs reduces their toxicological profile, a comparison was made between pristine MWCNTs and functionalized MW-NH_2 or MW-COO in rats. The study showed a general pulmonary toxicity coupled with inflammatory response after the exposure, independently of the type of functionalization.[151]

The interaction of the CNTs with the immune system has also been studied. CNTs induce the activation of the complement cascade.[152] Several recent studies in mice suggest that CNTs can interfere with the innate and adaptive immune system. T-cell reactions seem to be related with the formation of granulomas on lungs; B-cell activation promotes allergic immune responses with the subsequent production of IgE and cytokines released from lungs. As a consequence, cyclooxygenase reaction is triggered in the spleen, inducing immunosuppression.[43] However, highly water-soluble functionalized CNTs did not seem to influence the functional activity of immunoregulatory cells. Yet, CNTs with less solubility did not affect lymphocyte functionality but promoted the secretion of pro-inflammatory cytokines by macrophages.[153]

5.3.2. Fullerenes

Fullerenes are the third most stable form of carbon after diamond and graphite. The most abundant and representative form is the C_{60}, with 60 carbon atoms forming a spherical structure.[154] This form was discovered by Kroto *et al.* in 1985 and was given the name "buckminsterfullerene" or "buckyball".[155] Years later, higher (C_{70}, C_{76}, C_{78}, etc.) and lower order fullerenes (e.g. C_{28} and C_{36}) were isolated. The properties of fullerene and surface-functionalized fullerene classes of nanomaterials make them suitable for use in optical, electronic, cosmetic and biomedical applications.[156]

The main exposure routes to fullerenes are inhalation, dermal contact, injection routes and ingestion.[157] There have been several *in vitro* investigations about their toxicity. These studies have reported dose-dependent cytotoxicity in human cells incubated with water-soluble C_{60} fullerenes, due to the production of ROS and lipid peroxidation.[158,159] Studies about the uptake of C_{60} by human monocyte and macrophages support these results. In monocytes, C_{60} appears along the plasma membrane, while in macrophages it has also been found in intracellular organelles, free cytoplasm and nucleus. This suggests that peroxidation at intracellular sites may play an important role in toxicity.[160]

The manipulation of surface chemistry and molecular changes has created a large diversity of fullerenes, showing different behaviours. The use of solvents, extended stirring and mechanical processes are necessary to solubilize fullerene in water. These processes also require toxicity assessment, especially when they include the use of solvents, which appear to particularly increase fullerene toxicity.[161] Nevertheless, the surface chemistry modifications for water solubility decreased the *in vitro* cytotoxicity. Pristine C_{60} incubated with two different human cell lines was more cytotoxic than water-soluble fullerene.[158] Another comparison of pristine and polyhydroxylated fullerene, incubated with mouse L929 fibrosarcoma, rat C6 glioma and U251 human glioma cell lines, revealed that pure fullerenes are more cytotoxic and cause rapid ROS and necrosis than polyhydroxylated fullerenes.[162]

There have been very few *in vivo* studies. Polyalkysulfonated C_{60}, a highly water-soluble fullerene, was non-toxic after oral administration in rats. However, after i.p. injection, the lethal dose was determined to be about 600 mg/kg. When rats were injected via i.p. or i.v. with high doses, the primary site of fullerene elimination was the kidney, inducing phagolysosomal nephropathy.[163] In another study, fullerenes were administered as a single dose to determine the acute oral median lethal dose on rats. At the highest dose, 2000 mg/kg, no deaths were observed.[164]

The i.p. administration in mice of high quantities of C_{60} did not result in lethal, acute or subacute toxicity.[165] Indeed, pristine C_{60} can be considered as a powerful liver-protective agent in a dose-dependent manner, when it is administered before carbon tetrachloride (CCL_4), a typical free radical initiator.[166] Previous studies of biodistribution using $^{14}C_{60}$ showed that after oral administration, most C_{60} is excreted in faeces, but after i.v. administration, $^{14}C_{60}$ quickly passed from the blood to the liver, and then to other tissues.[167] Another study has recently reported that C_{60}, after vein tail administration, is distributed between different tissues, with the highest concentration in the lungs, followed by the spleen, liver, kidneys and brain. Moreover, C_{60} was not detected in blood, indicating that clearance may occur mainly in lungs.[168] Although a time-dependent decrease was observed among the tissues, suggesting that C_{60} could be excreted into faeces and/or urine,[167,168] further research is still required about the time-dependent variation in the biodistribution.

Nasal inhalation in rats showed that fullerenes were internalized by AMs, although minimal toxicity and inflammation were observed.[169] Studying the gene expression profiles after exposure via whole-body inhalation, it was found that few genes involved in the inflammatory response, oxidative stress, apoptosis and metalloendopeptidase activity were up-regulated.[170] Genotoxicity has also been demonstrated by some authors, although the mechanism is not clear (probably due to oxidative stress).[171] However, in another study, genotoxicity was assessed in a bacterial reverse mutation assay (Ames test) and a chromosomal aberration test, with no significant increase being found in the frequency of chromosomal aberration.[164]

The use of fullerenes in cosmetics has implications for the human and aquatic environment, as fullerene molecules in creams could be released into wastewater.[172] A C_{60} fullerene derivative coupled to a heptapeptide and a fluorescent tag has been able to pass the porcine dermis,[173,174] while pristine fullerenes have shown varied behaviours in skin penetration depending on the solvents used. For example, when dissolved in toluene, cyclohexane or chloroform, they penetrated deeply into the stratum corneum, whereas they did not penetrate when in mineral oil.[175]

The embryonic zebrafish model has also been employed to evaluate the toxicity of fullerenes. During early embryogenesis, embryonic zebrafish were exposed to different concentrations of fullerenes. At doses of 200 μg/L, there was a significant increase in malformations, pericardial oedema and mortality. In this model, pristine fullerene was more toxic than polyhydroxylated fullerene, which corresponds with results obtained in other models.[176]

5.4. Metallic Nanoparticles

5.4.1. Quantum Dots

QDs are semiconductor nanocrystals whose electronic and optical properties are dependent on their size and shape, which can be modified during their synthesis.[177,178] Some of the most interesting properties are the tunable fluorescence emission from visible to infrared wavelengths (QDs of the same material, but with different sizes, can emit light of different colours), large absorption coefficients across a wide spectral range and very high levels of brightness and photostability.[179] In the field of materials, QDs are useful for optoelectronic devices,[180] and for light-emitting devices, such as LEDs.

These properties make QDs interesting tools for different applications in molecular, cellular and *in vivo* imaging.[181] QDs have been tested in most biotechnological applications that use fluorescence, including DNA array technology, immunofluorescence assays and cell and animal biology. Some of the most successful uses of QDs have been immunofluorescence labelling of fixed cells and tissues; immunostaining of membrane proteins, microtubules, actin and nuclear antigens; and fluorescence *in situ* hybridization of chromosomes or combed DNA.[177]

For *in vivo* targeting and imaging, the best fluorescent probes are those that emit in the near-infrared region (Near-IR). The correct design of Near-IR QDs makes it possible to obtain high quantum yields and molar absorption coefficients, which has led to the development of a highly luminescent and useful *in vivo* contrast agent.[182] However, QDs may lose the fluorescence when they are administered *in vivo*, but this can be solved by using different techniques of synthesis, coating and bioconjugation.[182] Indeed, many groups are now developing bioconjugated QDs, which exhibit a good stability and highly specific binding to protein targets, as well as offering the possibility of obtaining different sizes and dyes for multiplexed molecular imaging.[179,183] Another possibility is the use of QDs as drug delivery. For this type of application, the QDs that are unstable in the target tissue could offer an advantage, as the drug is dissolved.[184]

The broadly varying physicochemical properties of individual QDs and the variety of QDs dosage/exposure concentrations generate discrepancies in their toxicity. The possible routes of QDs exposure are environmental, workplace, therapeutic or diagnostic administration. Therefore, the main entry pathways are inhalation, dermal contact or ingestion. The toxicity can be determined by various factors such as size, charge, concentration, outer coating bioactivity (capping material and functional groups), as well as oxidative, photolytic and mechanical stability.[185]

Cytotoxicity studies suggest that certain QDs may be toxic, and it seems that this toxicity is dependent on their surface molecule, the processing conditions and the nanoparticle dose.[186,187] However, the mechanism of how QDs induce toxicity is not clear at the cellular level. A study of the cytotoxicity of CdSe and CdSe/CdS QDs in five cell human lines indicated that CdSe/CdS QDs were more toxic than CdSe QDs. This result suggests that coating with a CdS shell not only reduces the cytotoxicity of QDs but also contributes to toxicity through the release of SO_2^- radicals.[188] However, the addition of a ZnS shell decreases cytotoxicity by the reduction of the oxidative degradation of the CdSe core. This degradation leads to the release of free cadmium ions, whose concentration is correlated with cytotoxic effects. Indeed, a study using CdSe core QDs on a human neuroblastoma cell line showed apoptosis in a dose-dependent manner, with loss of mitochondrial membrane potential, mitochondrial discharge of cytochrome *c* and activation of caspase-9 and caspase-3.[186,189] Another example of this type of behaviour has been described for CdTe QDs incubated with human embryonic cells, which penetrate the cellular membranes and accumulate in perinuclear areas, where the release of Cd^{2+} induces significant cytotoxicity.[190]

CdSe/CdS QDs induced acute toxicity when injected in mice at higher doses. The LD_{50} was found at 14.8–17.0 mg/kg body weight, and the release of Cd^{2+} followed by the generation of ROS and oxidative stress seems to be the mechanism of the *in vivo* toxicity. The distribution among the tissues indicates that lung, heart, liver and kidney are the main target organs and that Ns could be excreted after 14 days, via kidneys or intestinal tract.[188]

The biodistribution of different types of CdSe/ZnS QDs targeted with peptides shows an accumulation in mouse liver and spleen, besides other organs, although coating with PEG prevents this non-selective accumulation. After 24-h exposure, acute toxicity did not occur.[184] This is consistent with the study using CdSe/ZnS QDs in rats, which did not show appreciable toxicity even after a long period of exposure. QDs were also found in liver and spleen, and several days later in kidney.[191] The i.v. administration of amphiphilic polyacrylic acid polymer-coated QDs (amp-QDs) and amp-QDs conjugated to PEG-amine groups (mPEG–QDs) showed accumulation in liver, skin and bone marrow in a surface-coating dependent manner, but without signs of necrosis at necropsy.[192]

5.4.2. Metal and Metal Oxide

Metallic and metal oxide nanomaterials have a wide range of applications. Nevertheless, an in-depth study of the potential toxic effects of these Ns is still needed, in order to fulfil the mandatory requirement of ensuring the safety of workers, patients and the general public. Two main mechanisms may be responsible for an eventual toxicity of this type of Ns. First, the intrinsic catalytic activity of such Ns can disturb several processes and intracellular signalling pathways. Second, ions can also be released from the NP, affecting the finely regulated concentration of metallic ions inside the cell.[193]

Silver and gold have been used since ancient times for a wide range of applications: jewellery, utensils, monetary currency, photography, dental prosthesis and therapy (as gold salts against rheumatic diseases).[194,195] In medicine, silver has been used as a bactericide for centuries, but the irreversible pigmentation of the skin and/or eye (argyria and argyrosis) due to silver deposition in long treatments, together with the development of new antibiotics, such as penicillins and cephalosporins, has for a long time now relegated silver to oblivion. Nevertheless, the possibility of producing nanosilver has reinforced and renewed research on silver applications in medicine, including as a bactericide in food contact materials, dental alloys, implants, bone prostheses, surgical instruments, wound dressings and contraceptive devices.[196]

Bulk silver has shown minimal health risk from a perspective of occupational exposure, but soluble silver has more toxic effects, due to its higher absorption.[197] Ion release seems to be the cause of the toxicity of silver Ns, which can interact with body molecules through different exposure pathways. The principal routes are the respiratory tract, the gastrointestinal tract and skin. Another route is the female genital tract, as nanosilver has been integrated into hygiene products.[194,198] *In vitro* and *in vivo* studies indicate that silver Ns can exhibit a significant level of toxicity. Mitochondria are the probable targets for cytotoxicity. The Ag^+ ion seems to induce mitochondrial oxidative stress through interaction with thiol groups of the inner mitochondrial membrane. Although the mechanism is not well defined, when several cell lines were exposed to nanosilver, the function of mitochondria was reduced. It is possible that oxidative stress is responsible for the toxicity, as nanosilver could interact with proteins and enzymes responsible for the activation of antioxidant defence mechanisms.[194] Therefore, the apoptosis is induced via production of ROS.[198] In rats, i.v. administration of high doses of nanosilver revealed haematological and biochemical parameter changes, especially for liver enzymes.[199] Inhaled subacutely, nanosilver in mice induced minimal pulmonary toxicity or inflammation,[200] and dermal exposure in rabbits has demonstrated that nanosilver particles are able to induce dermal erythema and oedema. Histopathology analysis showed significant damage in liver, spleen and skin.[201]

More recently, gold Ns are being extensively characterized for their use against cancer in photothermal therapy or hyperthermia and as a probe for imaging, because of their capability to absorb radiation and their SPR band. These optoelectronic properties can be tailored by changing their size and shape, and/or combining gold with other compounds like silver.[202] Hyperthermia can be produced by near-IR laser irradiation of gold nanoparticles present in tumours, which can thereby induce the killing of tumour cells, via a bystander effect. Although, as a bulk material, gold is biocompatible, the increased reactivity of gold in the nanoscale range has given rise to intensive research, regarding the potential toxicity *in vitro* and *in vivo* of gold nanoparticles. The biosensor field is another area for the relevant application of gold nanoparticles. For instance, gold-based biosensors have been developed for, among other applications, the detection of tumour cells, specific antibodies and glucose levels.[203–205]

Biodistribution studies of gold Ns *in vivo* using mice showed an impaired accumulation in different organs, depending on the particle size and the route of administration.[206] Following multiple experiments carried out by various groups, using different values for NP biodistribution and toxic effects, it is generally agreed that gold Ns of 20 nm, or smaller, are able to cross the blood–brain barrier and that, irrespective of particle size, the most affected organs are the liver and, to a lesser extent, the spleen, where these gold Ns can remain for up to 3 or 4 months. Furthermore, very small Ns (1–2 nm) are potentially very toxic, due to their possible irreversible binding to key biopolymers, such as DNA.[207]

Titanium dioxide (TiO_2), zinc oxide (ZnO) and cerium oxide (CeO_2) are being produced in high quantities by several companies and research groups. TiO_2 and ZnO are widely used in various products (e.g. in sun creams). Moreover, TiO_2 is also used in other commercial products (sport clothes, surface cleaning agents, computer devices), as well as in solar cells.[139,196] In biomedicine, it could be used as a sterilizing agent because of its photocatalytic function, which could be exploited for antibacterial applications.[208]

ZnO is known as a wide band gap semiconductor; therefore, it is a promising material for use in optoelectronic devices, as a catalyst, etc. In biomedicine, ZnO Ns could be useful as a bio-imaging agent and in protein binding assays.[209]

CeO_2 Ns are suitable for a variety of applications as catalysts, ultraviolet adsorbents, gas sensors and abrasive agents, as well as in electrochemistry and optical equipment. In biomedical applications, these nanoparticles have the proven ability to neutralize free radicals, preventing oxidation and giving these Ns an interesting anti-inflammatory activity.[209]

A comparison of the cytotoxicity of these types of Ns reveals different forms of behaviour in cell lines. TiO_2 Ns have been shown to be confined in late endosomal and caveolar compartments, as they are inert under dark conditions, but no cytotoxicity has been found in several cell lines.[210] Similarly, no cytotoxicity was found for CeO_2 Ns and, thanks to their antioxidant properties, they have a cytoprotective effect.[210] In contrast, ZnO Ns have been shown to be the most cytotoxic,[193] inducing the production of pro-inflammatory cytokines and mitochondrial injury. This effect could be explained by the release of Zn^{2+} to the medium and the uptake of the Ns in specific endosomal compartments.[210] It has been recently suggested that cytotoxicity of ZnO nanoparticles depends on particle size and surface charge, with the smaller and positively charged Ns being more toxic.[211]

Biodistribution studies of TiO_2 Ns after oral administration in mice revealed no acute toxicity, but an accumulation in the liver, spleen, kidneys and lung tissues. Changes in biochemical and histological parameters demonstrated important lesions in the liver and kidneys.[212] Intraperitoneal (i.p.) administration induced signs of acute toxicity, such as passive behaviour, loss of appetite, tremor and lethargy. The TiO_2 Ns accumulated mainly in the spleen but were also present in the liver, kidney and lung. Histologically, the spleen showed lesions and thrombosis in the pulmonary vascular system, probably due to the obstruction of blood vessels.[213] Intravenous (i.v.) administration at low doses indicated no inflammatory response or organ toxicity. The accumulation was mainly in the liver, lasting for at least 28 days, and Ns were also found in the spleen, lung and kidney.[214]

Studies *in vivo* with CeO_2 Ns have been performed in rats, with concentration of Ns in the spleen, followed by the liver, blood and brain after i.v.

injection.[215] A recently exhaustive *in vivo* analysis showed that the tissue deposition (on spleen, liver, lungs and kidneys) was greater after i.v. and i.p. administrations, whereas most orally administered Ns were excreted in 24 h. No toxicity was found, and elimination was through faeces.[216]

There have not been many *in vivo* biodistribution studies about ZnO, but the use of radioactive ZnO Ns in mice after i.v. administration showed preferential migration by the Ns to the lungs, with subsequent migration to the intestinal tract, from where they were excreted through faeces.[217]

6. CONCLUSION

In summary, inorganic Ns are more toxic and more limited for *in vivo* applications than biodegradable Ns. However, functionalization may reduce their toxicity and increase their applicability in medicine and cosmetics, although they need to be deeply characterized to ensure safe use and complete removal from the body, after they have undertaken their therapeutic action. Yet, their cytotoxicity is being explored in antitumour therapy and bactericide applications, as mentioned for silver Ns. Nevertheless, the main worry regarding non-degradable Ns is their potential toxicity on workers (by occupational exposure), and their hazardous effects on the environment (contamination of air, river, land, sea and organisms). Hence, the main concerns are related with the manufacture and final use of the nanomaterials and, very importantly, with their life cycle. The difficulties in the characterization and traceability of these nanomaterials, in the estimation of the time they remain in the environment, together with the unknown effects of possible interactions with other nanostructures or with other materials, require intensive research and regulation.

ACKNOWLEDGEMENTS

We greatly appreciate the support from INBIOMED (2009/063, the *Xunta de Galicia*) and HINAMOX (7th EU programme) project.

Conflict of Interests. The authors declare no conflict of interest.

REFERENCES

1. Kreuter J. Nanoparticles—a historical perspective. *Int J Pharm* 2007;**331**(1):1–10.
2. Adlakha-Hutcheon G, Khaydarov R, Korenstein R, Varma R, Vaseashta A, Stamm H, et al. Nanomaterials, nanotechnology: applications, consumer products, and benefits. In: Linkov I, Steevens J, editors. *Nanomaterials: risks and benefits*. Netherlands: Springer; 2009. p. 195–207.
3. Zhang Z, Biskupek J, Kaiser U, Breus VV, Nienhaus GU, Wiedwald U, et al. Structural studies of nanoparticles with an objective-lens Cs-corrected Titan microscope. *Microsc Microanal* 2007;**13**:258–9.
4. Lechuga LM, Zinoviev K, Carrascosa LG, Moreno M. Nanodevices for biosensing: design, fabrication and applications. *Nanotechnol Life Sci* 2007;317–44.

5. Rosi NL, Mirkin CA. Nanostructures in biodiagnostics. *Chem Rev* 2005;**105**(4):1547–62.
6. Chaudhry Q, Scotter M, Blackburn J, Ross B, Boxall A, Castle L, et al. Applications and implications of nanotechnologies for the food sector. *Food Addit Contam Part A* 2008;**25**(3):241–58.
7. Gradishar WJ. Albumin-bound paclitaxel: a next-generation taxane. *Expert Opin Pharmacother* 2006;**7**(8):1041–53.
8. Song Y, Li X, Du X. Exposure to nanoparticles is related to pleural effusion, pulmonary fibrosis and granuloma. *Eur Respir J* 2009;**34**(3):559–67.
9. Franca A, Pelaz B, Moros M, Sanchez-Espinel C, Hernandez A, Fernandez-Lopez C, et al. Sterilization matters: consequences of different sterilization techniques on gold nanoparticles. *Small* 2010;**6**(1):89–95.
10. Mossman BT, Churg A. Mechanisms in the pathogenesis of asbestosis and silicosis. *Am J Respir Crit Care Med* 1998;**157**(5):1666–80.
11. Jomova K, Valko M. Advances in metal-induced oxidative stress and human disease. *Toxicology* 2011;**283**(2–3):65–87.
12. Medina C, Santos-Martinez MJ, Radomski A, Corrigan OI, Radomski MW. Nanoparticles: pharmacological and toxicological significance. *Br J Pharmacol* 2007;**150**(5):552–8.
13. Díaz B, Sánchez-Espinel C, Arruebo M, Faro J, de Miguel E, Magadán S, et al. Assessing methods for blood cell cytotoxic responses to inorganic nanoparticles and nanoparticle aggregates. *Small* 2008;**4**(11):2025–34.
14. Owens 3rd DE, Peppas NA. Opsonization, biodistribution, and pharmacokinetics of polymeric nanoparticles. *Int J Pharm* 2006;**307**(1):93–102.
15. Aggarwal P, Hall JB, McLeland CB, Dobrovolskaia MA, McNeil SE. Nanoparticle interaction with plasma proteins as it relates to particle biodistribution, biocompatibility and therapeutic efficacy. *Adv Drug Deliv Rev* 2009;**61**(6):428–37.
16. Dobrovolskaia MA, McNeil SE. Immunological properties of engineered nanomaterials. *Nat Nanotechnol* 2007;**2**(8):469–78.
17. Zolnik BS, González-Fernández Á, Sadrieh N, Dobrovolskaia MA. Minireview: nanoparticles and the immune system. *Endocrinology* 2010;**151**(2):458–65.
18. Arruebo M, Valladares M, González-Fernández Á. Antibody-conjugated nanoparticles for biomedical applications. J Nanomater 2009; Article ID 439389, 24 pp.
19. Kumari A, Yadav SK, Yadav SC. Biodegradable polymeric nanoparticles based drug delivery systems. *Colloids Surf B Biointerfaces* 2010;**75**(1):1–18.
20. Zhang L, Gu FX, Chan JM, Wang AZ, Langer RS, Farokhzad OC. Nanoparticles in medicine: therapeutic applications and developments. *Clin Pharmacol Ther* 2008;**83**(5):761–9.
21. Bangham AD, Horne RW. Negative staining of phospholipids and their structural modification by surface-active agents as observed in the electron microscope. *J Mol Biol* 1964;**8**:660–8.
22. Barani H, Montazer M. A review on applications of liposomes in textile processing. *J Liposome Res* 2008;**18**(3):249–62.
23. Dokka S, Toledo D, Shi X, Castranova V, Rojanasakul Y. Oxygen radical-mediated pulmonary toxicity induced by some cationic liposomes. *Pharm Res* 2000;**17**(5):521–5.
24. Zbinden G, Wunderli-Allenspach H, Grimm L. Assessment of thrombogenic potential of liposomes. *Toxicology* 1989;**54**(3):273–80.
25. Plank C, Mechtler K, Szoka Jr. FC, Wagner E. Activation of the complement system by synthetic DNA complexes: a potential barrier for intravenous gene delivery. *Hum Gene Ther* 1996;**7**(12):1437–46.

26. Tran MA, Watts RJ, Robertson GP. Use of liposomes as drug delivery vehicles for treatment of melanoma. *Pigment Cell Melanoma Res* 2009;**22**(4):388–99.
27. Santel A, Aleku M, Keil O, Endruschat J, Esche V, Durieux B, et al. RNA interference in the mouse vascular endothelium by systemic administration of siRNA-lipoplexes for cancer therapy. *Gene Ther* 2006;**13**(18):1360–70.
28. Dadashzadeh S, Mirahmadi N, Babaei MH, Vali AM. Peritoneal retention of liposomes: effects of lipid composition, PEG coating and liposome charge. *J Control Release* 2010; **148**(2):177–86.
29. Christensen D, Korsholm KS, Andersen P, Agger EM. Cationic liposomes as vaccine adjuvants. *Expert Rev Vaccines* 2011;**10**(4):513–21.
30. Malam Y, Loizidou M, Seifalian AM. Liposomes and nanoparticles: nanosized vehicles for drug delivery in cancer. *Trends Pharmacol Sci* 2009;**30**(11):592–9.
31. Gwinn WM, Kapita MC, Wang PM, Cesta MF, Martin WJ. Synthetic liposomes are protective from bleomycin-induced lung toxicity. *Am J Physiol Lung Cell Mol Physiol* 2011;**301** (2):L207–L217.
32. Leite EA, Giuberti Cdos S, Wainstein AJ, Wainstein AP, Coelho LG, Lana AM, et al. Acute toxicity of long-circulating and pH-sensitive liposomes containing cisplatin in mice after intraperitoneal administration. *Life Sci* 2009;**84**(19–20):641–9.
33. Chang RS, Kim J, Lee HY, Han SE, Na J, Kim K, et al. Reduced dose-limiting toxicity of intraperitoneal mitoxantrone chemotherapy using cardiolipin-based anionic liposomes. *Nanomedicine* 2010;**6**(6):769–76.
34. Stover TC, Sharma A, Robertson GP, Kester M. Systemic delivery of liposomal short-chain ceramide limits solid tumor growth in murine models of breast adenocarcinoma. *Clin Cancer Res* 2005;**11**(9):3465–74.
35. Mullauer FB, van Bloois L, Daalhuisen JB, Ten Brink MS, Storm G, Medema JP, et al. Betulinic acid delivered in liposomes reduces growth of human lung and colon cancers in mice without causing systemic toxicity. *Anticancer Drugs* 2011;**22**(3):223–33.
36. Sawant RR, Torchilin VP. Liposomes as 'smart' pharmaceutical nanocarriers. *Soft Matter* 2010;**6**(17):4026–44.
37. Mudshinge SR, Deore AB, Patil S, Bhalgat CM. Nanoparticles: emerging carriers for drug delivery. *Saudi Pharm J* 2011;**19**(3):129–41.
38. Olbrich C, Muller RH, Tabatt K, Kayser O, Schulze C, Schade R. Stable biocompatible adjuvants—a new type of adjuvant based on solid lipid nanoparticles: a study on cytotoxicity, compatibility and efficacy in chicken. *Altern Lab Anim* 2002;**30**(4):443–58.
39. Wissing SA, Kayser O, Müller RH. Solid lipid nanoparticles for parenteral drug delivery. *Adv Drug Deliv Rev* 2004;**56**(9):1257–72.
40. Gualbert J, Shahgaldian P, Coleman AW. Interactions of amphiphilic calix[4]arene-based solid lipid nanoparticles with bovine serum albumin. *Int J Pharm* 2003;**257**:69–73.
41. Patel TD, Parikh BN, Gothi GD, Dave JB, Patel CN. A review on pharmaceutical and non pharmaceutical applications on dendrimer. *J Global Pharm Technol* 2010;**2**(2):5–17.
42. Brauge L, Vériot G, Franc G, Deloncle R, Caminade A-M, Majoral J-P. Synthesis of phosphorus dendrimers bearing chromophoric end groups: toward organic blue light-emitting diodes. *Tetrahedron* 2006;**62**(51):11891–9.
43. Kunzmann A, Andersson B, Thurnherr T, Krug H, Scheynius A, Fadeel B. Toxicology of engineered nanomaterials: focus on biocompatibility, biodistribution and biodegradation. *Biochim Biophys Acta* 2011;**1810**(3):361–73.
44. Chen H-T, Neerman MF, Parrish AR, Simanek EE. Cytotoxicity, hemolysis, and acute *in vivo* toxicity of dendrimers based on melamine, candidate vehicles for drug delivery. *J Am Chem Soc* 2004;**126**(32):10044–8.

45. Bhadra D, Yadav AK, Bhadra S, Jain NK. Glycodendrimeric nanoparticulate carriers of primaquine phosphate for liver targeting. *Int J Pharm* 2005;**295**(1–2):221–33.
46. Agashe HB, Dutta T, Garg M, Jain NK. Investigations on the toxicological profile of functionalized fifth-generation poly (propylene imine) dendrimer. *J Pharm Pharmacol* 2006;**58**(11):1491–8.
47. Jain K, Kesharwani P, Gupta U, Jain NK. Dendrimer toxicity: let's meet the challenge. *Int J Pharm* 2010;**394**(1–2):122–42.
48. Roberts JC, Bhalgat MK, Zera RT. Preliminary biological evaluation of polyamidoamine (PAMAM) Starburst dendrimers. *J Biomed Mater Res* 1996;**30**(1):53–65.
49. Neerman MF, Chen HT, Parrish AR, Simanek EE. Reduction of drug toxicity using dendrimers based on melamine. *Mol Pharm* 2004;**1**(5):390–3.
50. Malik N, Wiwattanapatapee R, Klopsch R, Lorenz K, Frey H, Weener JW, et al. Dendrimers: relationship between structure and biocompatibility in vitro, and preliminary studies on the biodistribution of 125I-labelled polyamidoamine dendrimers in vivo. *J Control Release* 2000;**65**(1–2):133–48.
51. Prego C, Paolicelli P, Diaz B, Vicente S, Sanchez A, Gonzalez-Fernandez A, et al. Chitosan-based nanoparticles for improving immunization against hepatitis B infection. *Vaccine* 2010;**28**(14):2607–14.
52. Hong D, et al. Bioconjugated PLGA-4-arm-PEG branched polymeric nanoparticles as novel tumor targeting carriers. *Nanotechnology* 2011;**22**(16):165101.
53. Kalaria D, Sharma G, Beniwal V, Ravi Kumar M. Design of biodegradable nanoparticles for oral delivery of doxorubicin: in vivo; pharmacokinetics and toxicity studies in rats. *Pharm Res* 2009;**26**(3):492–501.
54. Uhrich KE, Cannizzaro SM, Langer RS, Shakesheff KM. Polymeric systems for controlled drug release. *Chem Rev* 1999;**99**(11):3181–98.
55. Wang C, Ge Q, Ting D, Nguyen D, Shen HR, Chen J, et al. Molecularly engineered poly (ortho ester) microspheres for enhanced delivery of DNA vaccines. *Nat Mater* 2004;**3**(3):190–6.
56. Katti DS, Lakshmi S, Langer R, Laurencin CT. Toxicity, biodegradation and elimination of polyanhydrides. *Adv Drug Deliv Rev* 2002;**54**(7):933–61.
57. Itaka K, Ishii T, Hasegawa Y, Kataoka K. Biodegradable polyamino acid-based polycations as safe and effective gene carrier minimizing cumulative toxicity. *Biomaterials* 2010;**31**(13):3707–14.
58. Silva GA, Ducheyne P, Reis RL. Materials in particulate form for tissue engineering. 1. Basic concepts. *J Tissue Eng Regen Med* 2007;**1**(1):4–24.
59. Maeda M, Tani S, Sano A, Fujioka K. Microstructure and release characteristics of the minipellet, a collagen-based drug delivery system for controlled release of protein drugs. *J Control Release* 1999;**62**(3):313–24.
60. Oyarzun-Ampuero FA, Goycoolea FM, Torres D, Alonso MJ. A new drug nanocarrier consisting of polyarginine and hyaluronic acid. *Eur J Pharm Biopharm* 2011;**79**(1):54–7.
61. Marques AP, Reis RL, Hunt JA. The biocompatibility of novel starch-based polymers and composites: in vitro studies. *Biomaterials* 2002;**23**(6):1471–8.
62. Tiyaboonchai W, Woiszwillo J, Middaugh CR. Formulation and characterization of DNA-polyethylenimine-dextran sulfate nanoparticles. *Eur J Pharm Sci* 2003;**19**(4):191–202.
63. Bertholon I, Vauthier C, Labarre D. Complement activation by core–shell poly(isobutylcyanoacrylate)–polysaccharide nanoparticles: influences of surface morphology, length, and type of polysaccharide. *Pharm Res* 2006;**23**(6):1313–23.
64. Loh JW, Yeoh G, Saunders M, Lim LY. Uptake and cytotoxicity of chitosan nanoparticles in human liver cells. *Toxicol Appl Pharmacol* 2010;**249**(2):148–57.

65. Wedmore I, McManus JG, Pusateri AE, Holcomb JB. A special report on the chitosan-based hemostatic dressing: experience in current combat operations. *J Trauma* 2006;**60**(3):655–8.
66. Kean T, Thanou M. Biodegradation, biodistribution and toxicity of chitosan. *Adv Drug Deliv Rev* 2010;**62**(1):3–11.
67. Grillo R, de Melo NFS, de Araújo DR, de Paula E, Rosa AH, Fraceto LF. Polymeric alginate nanoparticles containing the local anesthetic bupivacaine. *J Drug Target* 2010;**18**(9):688–99.
68. Gazori T, Haririan I, Fouladdel S, Namazi A, Nomani A, Azizi E. Inhibition of EGFR expression with chitosan/alginate nanoparticles encapsulating antisense oligonucleotides in T47D cell line using RT-PCR and immunocytochemistry. *Carbohydr Polym* 2010;**80**(4):1042–7.
69. Reis ECC, Borges APB, Fonseca CC, Martinez MMM, Eleotério RB, Morato GO, et al. Biocompatibility, osteointegration, osteoconduction, and biodegradation of a hydroxyapatite-polyhydroxybutyrate composite. *Braz Arch Biol Technol* 2010;**53**:817–26.
70. Carlo EC, Borges AP, Del Carlo RJ, Martinez MM, Oliveira PM, Morato GO, et al. Comparison of in vivo properties of hydroxyapatite-polyhydroxybutyrate composites assessed for bone substitution. *J Craniofac Surg* 2009;**20**(3):853–9.
71. Unverdorben M, Spielberger A, Schywalsky M, Labahn D, Hartwig S, Schneider M, et al. A polyhydroxybutyrate biodegradable stent: preliminary experience in the rabbit. *Cardiovasc Intervent Radiol* 2002;**25**(2):127–32.
72. Alves EG, Rezende CM, Serakides R, Pereira MD, Rosado IR. Orthopedic implant of a polyhydroxybutyrate (PHB) and hydroxyapatite composite in cats. *J Feline Med Surg* 2011;**13**(8):546–52.
73. Oh JK, Drumright R, Siegwart DJ, Matyjaszewski K. The development of microgels/nanogels for drug delivery applications. *Prog Polym Sci* 2008;**33**(4):448–77.
74. Ganta C, Shi A, Battina SK, Pyle M, Rana S, Hua DH, et al. Combination of nanogel polyethylene glycol-polyethylenimine and 6(hydroxymethyl)-1,4-anthracenedione as an anticancer nanomedicine. *J Nanosci Nanotechnol* 2008;**8**(5):2334–40.
75. Gota C, Okabe K, Funatsu T, Harada Y, Uchiyama S. Hydrophilic fluorescent nanogel thermometer for intracellular thermometry. *J Am Chem Soc* 2009;**131**(8):2766–7.
76. Oishi M, Nagasaki Y. Synthesis, characterization, and biomedical applications of core-shell-type stimuli-responsive nanogels—nanogel composed of poly[2-(N, N-diethylamino)ethyl methacrylate] core and PEG tethered chains. *React Funct Polym* 2007;**67**(11):1311–29.
77. Sasaki Y, Akiyoshi K. Nanogel engineering for new nanobiomaterials: from chaperoning engineering to biomedical applications. *Chem Rec* 2010;**10**(6):366–76.
78. Kratz F. Albumin as a drug carrier: design of prodrugs, drug conjugates and nanoparticles. *J Control Release* 2008;**132**(3):171–83.
79. Garber K. Improved paclitaxel formulation hints at new chemotherapy approach. *J Natl Cancer Inst* 2004;**96**(2):90–1.
80. Sparreboom A, Scripture CD, Trieu V, Williams PJ, De T, Yang A, et al. Comparative preclinical and clinical pharmacokinetics of a cremophor-free, nanoparticle albumin-bound paclitaxel (ABI-007) and paclitaxel formulated in Cremophor (Taxol). *Clin Cancer Res* 2005;**11**(11):4136–43.
81. John TA, Vogel SM, Tiruppathi C, Malik AB, Minshall RD. Quantitative analysis of albumin uptake and transport in the rat microvessel endothelial monolayer. *Am J Physiol Lung Cell Mol Physiol* 2003;**284**(1):L187–L196.
82. Desai N, Trieu V, Yao Z, Louie L, Ci S, Yang A, et al. Increased antitumor activity, intratumor paclitaxel concentrations, and endothelial cell transport of cremophor-free, albumin-bound paclitaxel, ABI-007, compared with cremophor-based paclitaxel. *Clin Cancer Res* 2006;**12**(4):1317–24.

83. Lu W, Wan J, She Z, Jiang X. Brain delivery property and accelerated blood clearance of cationic albumin conjugated pegylated nanoparticle. *J Control Release* 2007;**118**(1):38–53.
84. Zensi A, Begley D, Pontikis C, Legros C, Mihoreanu L, Wagner S, et al. Albumin nanoparticles targeted with Apo E enter the CNS by transcytosis and are delivered to neurones. *J Control Release* 2009;**137**(1):78–86.
85. Lassus M, Scott D, Leyland-Jones B. Allergic reactions associated with cremophor containing antineoplastics. *Proc Annu Meet Am Soc Clin Oncol* 1985;**4**:268.
86. Nolte H, Carstensen H, Hertz H. VM-26 (teniposide)-induced hypersensitivity and degranulation of basophils in children. *Am J Pediatr Hematol Oncol* 1988;**10**:308–12.
87. Kris MG, O'Connell JP, Gralla RJ, Wertheim MS, Parente RM, Schiff PB, et al. Phase I trial of taxol given as a 3-hour infusion every 21 days. *Cancer Treat Rep* 1986;**70**(5):605–7.
88. Szebeni J, Muggia FM, Alving CR. Complement activation by Cremophor EL as a possible contributor to hypersensitivity to paclitaxel: an in vitro study. *J Natl Cancer Inst* 1998;**90**(4):300–6.
89. Szebeni J. Complement activation-related pseudoallergy: a new class of drug-induced acute immune toxicity. *Toxicology* 2005;**216**(2–3):106–21.
90. Hunter RL, McNicholl J, Lal AA. Mechanisms of action of nonionic block copolymer adjuvants. *AIDS Res Hum Retroviruses* 1994;**10**(Suppl. 2):S95–S98.
91. Terwogt JM, Nuijen B, Huinink WW, Beijnen JH. Alternative formulations of paclitaxel. *Cancer Treat Rev* 1997;**23**(2):87–95.
92. Lundberg BB. A submicron lipid emulsion coated with amphipathic polyethylene glycol for parenteral administration of paclitaxel (Taxol). *J Pharm Pharmacol* 1997;**49**(1):16–21.
93. Paradis R, Page M. New active paclitaxel amino acids derivatives with improved water solubility. *Anticancer Res* 1998;**18**(4):2711–6.
94. Scripture CD, Figg WD, Sparreboom A. Paclitaxel chemotherapy: from empiricism to a mechanism-based formulation strategy. *Ther Clin Risk Manag* 2005;**1**(2):107–14.
95. Moghimi SM, Hunter AC, Dadswell CM, Savay S, Alving CR, Szebeni J. Causative factors behind poloxamer 188 (Pluronic F68, Flocor)-induced complement activation in human sera. A protective role against poloxamer-mediated complement activation by elevated serum lipoprotein levels. *Biochim Biophys Acta* 2004;**1689**(2):103–13.
96. Hall MD, Mellor HR, Callaghan R, Hambley TW. Basis for design and development of platinum(IV) anticancer complexes. *J Med Chem* 2007;**50**(15):3403–11.
97. Gao J, Liang G, Zhang B, Kuang Y, Zhang X, Xu B. FePt@CoS(2) yolk-shell nanocrystals as a potent agent to kill HeLa cells. *J Am Chem Soc* 2007;**129**(5):1428–33.
98. Schwenk MH. Ferumoxytol: a new intravenous iron preparation for the treatment of iron deficiency anemia in patients with chronic kidney disease. *Pharmacotherapy* 2010;**30**(1):70–9.
99. Auerbach M, Silberstein PT, Webb RT, Averyanova S, Ciuleanu TE, Shao J, et al. Darbepoetin alfa 300 or 500 mug once every 3 weeks with or without intravenous iron in patients with chemotherapy-induced anemia. *Am J Hematol* 2010;**85**(9):655–63.
100. Bulte JW, Kraitchman DL, Mackay AM, Pittenger MF. Chondrogenic differentiation of mesenchymal stem cells is inhibited after magnetic labeling with ferumoxides. *Blood* 2004;**104**(10):3410–2 author reply 3412–3.
101. Arbab AS, Yocum GT, Rad AM, Khakoo AY, Fellowes V, Read EJ, et al. Labeling of cells with ferumoxides-protamine sulfate complexes does not inhibit function or differentiation capacity of hematopoietic or mesenchymal stem cells. *NMR Biomed* 2005;**18**(8):553–9.
102. Farrell E, Wielopolski P, Pavljasevic P, van Tiel S, Jahr H, Verhaar J, et al. Effects of iron oxide incorporation for long term cell tracking on MSC differentiation in vitro and in vivo. *Biochem Biophys Res Commun* 2008;**369**(4):1076–81.

103. Fan C, Gao W, Chen Z, Fan H, Li M, Deng F. Tumor selectivity of stealth multi-functionalized superparamagnetic iron oxide nanoparticles. *Int J Pharm* 2011;**404**(1–2):180–90.
104. Tang C, Russell PJ, Martiniello-Wilks R, Rasko JE, Khatri A. Concise review: nanoparticles and cellular carriers-allies in cancer imaging and cellular gene therapy? *Stem Cells* 2010;**28**(9):1686–702.
105. Lewinski N, Colvin V, Drezek R. Cytotoxicity of nanoparticles. *Small* 2008;**4**(1):26–49.
106. Kato S, Taira H, Aoshima H, Saitoh Y, Miwa N. Clinical evaluation of fullerene-C60 dissolved in squalane for anti-wrinkle cosmetics. *J Nanosci Nanotechnol* 2010;**10**(10):6769–74.
107. Marano F, Hussain S, Rodrigues-Lima F, Baeza-Squiban A, Boland S. Nanoparticles: molecular targets and cell signalling. *Arch Toxicol* 2011;**85**(7):733–41.
108. Lundqvist M, Stigler J, Elia G, Lynch I, Cedervall T, Dawson KA. Nanoparticle size and surface properties determine the protein corona with possible implications for biological impacts. *Proc Natl Acad Sci USA* 2008;**105**(38):14265–70.
109. Nel AE, Madler L, Velegol D, Xia T, Hoek EM, Somasundaran P, et al. Understanding biophysicochemical interactions at the nano-bio interface. *Nat Mater* 2009;**8**(7):543–57.
110. Deng ZJ, Liang M, Monteiro M, Toth I, Minchin RF. Nanoparticle-induced unfolding of fibrinogen promotes Mac-1 receptor activation and inflammation. *Nat Nanotechnol* 2011;**6**(1):39–44.
111. McIntyre T. Phytoremediation of heavy metals from soils. In: Tsao D, editor. *Phytoremediation.* Berlin/Heidelberg: Springer; 2003. p. 97–123.
112. Valko M, Morris H, Cronin MT. Metals, toxicity and oxidative stress. *Curr Med Chem* 2005;**12**(10):1161–208.
113. Harbison RD. *Hamilton & Hardy's industrial toxicology.* 5th ed. St. Louis, MO: Year Book Medical; 1998.
114. Klaassen CD. *Casarett and Doull's toxicology: the basic science of poisons.* 7th ed. New York, USA: McGraw-Hill; 2008.
115. Stacey NH, Winder C. *Occupational toxicology.* 1st ed. London: Taylor & Francis Group; 1995.
116. Schmidt M, Raghavan B, Muller V, Vogl T, Fejer G, Tchaptchet S, et al. Crucial role for human Toll-like receptor 4 in the development of contact allergy to nickel. *Nat Immunol* 2010;**11**(9):814–9.
117. Zahir F, Rizwi SJ, Haq SK, Khan RH. Low dose mercury toxicity and human health. *Environ Toxicol Pharmacol* 2005;**20**(2):351–60.
118. Baruthio F. Toxic effects of chromium and its compounds. *Biol Trace Elem Res* 1992;**32**(1):145–53.
119. Costa M, Klein CB. Toxicity and carcinogenicity of chromium compounds in humans. *Crit Rev Toxicol* 2006;**36**(2):155–63.
120. Papanikolau Nikolas C, Hatzidaki Eleftheria G, Stamatis B, Tzanakakis George N, Tsatsakis Aristidis M. Lead toxicity update. A brief review. *Med Sci Monit* 2005;**11**:329–36.
121. Michael FH. Arsenic toxicity and potential mechanisms of action. *Toxicol Lett* 2002;**133**(1):1–16.
122. Godt J, Scheidig F, Grosse-Siestrup C, Esche V, Brandenburg P, Reich A, et al. The toxicity of cadmium and resulting hazards for human health. *J Occup Med Toxicol* 2006;**1**(1):22.
123. Otsuki T, Maeda M, Murakami S, Hayashi H, Miura Y, Kusaka M, et al. Immunological effects of silica and asbestos. *Cell Mol Immunol* 2007;**4**(4):261–8.
124. Cagle PT, Allen TC. Asbestosis and silicosis. In: Zander DS, et al. editors. *Molecular pathology of lung diseases.* New York: Springer; 2008. p. 658–62.

125. Iyer R, Hamilton RF, Li L, Holian A. Silica-induced apoptosis mediated via scavenger receptor in human alveolar macrophages. *Toxicol Appl Pharmacol* 1996;**141**(1):84–92.
126. Hamilton RF, Thakur SA, Mayfair JK, Holian A. MARCO mediates silica uptake and toxicity in alveolar macrophages from C57BL/6 mice. *J Biol Chem* 2006;**281**(45):34218–26.
127. Arredouani M, Yang Z, Ning Y, Qin G, Soininen R, Tryggvason K, et al. The scavenger receptor MARCO Is required for lung defense against pneumococcal pneumonia and inhaled particles. *J Exp Med* 2004;**200**(2):267–72.
128. Chao SK, Hamilton RF, Pfau JC, Holian A. Cell surface regulation of silica-induced apoptosis by the SR-A scavenger receptor in a murine Lung macrophage cell line (MH-S). *Toxicol Appl Pharmacol* 2001;**174**(1):10–6.
129. Hamilton RF, de Villiers WJS, Holian A. Class A type II scavenger receptor mediates silica-induced apoptosis in Chinese hamster ovary cell line. *Toxicol Appl Pharmacol* 2000;**162**(2):100–6.
130. Beamer CA, Holian A. Scavenger receptor class A type I/II (CD204) null mice fail to develop fibrosis following silica exposure. *Am J Physiol Lung Cell Mol Physiol* 2005;**289**(2):L186–L195.
131. Franca A, Aggarwal P, Barsov EV, Kozlov SV, Dobrovolskaia MA, González-Fernández Á. Macrophage scavenger receptor A mediates the uptake of gold colloids by macrophages in vitro. *Nanomedicine (Lond)* 2011;**6**(7):1175–88.
132. Tiwari R. Biomarkers of silicosis: potential candidates. *Indian J Occup Environ Med* 2005;**9**(3):103–6.
133. Tripathi S, Pandey H, Paul B. Overview of cytokines and receptors in Silicosis. *J Appl Pharm Sci* 2011;**01**(05):01–05.
134. Trojanowicz M. Analytical applications of carbon nanotubes: a review. *TrAC Trends Anal Chem* 2006;**25**(5):480–9.
135. Makar JM, Beaudoin JJ. Carbon nanotubes and their application in the construction industry. *Proceedings of 1st International Symposium on Nanotechnology in Construction.* 2003;331–41.
136. Popov VN. Carbon nanotubes: properties and application. *Mat Sci Eng R Rep* 2004;**43**(3):61–102.
137. Sanles-Sobrido M, Rodriguez-Lorenzo L, Lorenzo-Abalde S, Gonzalez-Fernandez A, Correa-Duarte MA, Alvarez-Puebla RA, et al. Label-free SERS detection of relevant bioanalytes on silver-coated carbon nanotubes: the case of cocaine. *Nanoscale* 2009;**1**(1):153–8.
138. Chen B, Zhang H, Zhai C, Du N, Sun C, Xue J, et al. Carbon nanotube-based magnetic-fluorescent nanohybrids as highly efficient contrast agents for multimodal cellular imaging. *J Mater Chem* 2010;**20**(44):9895–902.
139. Crosera M, Bovenzi M, Maina G, Adami G, Zanette C, Florio C, et al. Nanoparticle dermal absorption and toxicity: a review of the literature. *Int Arch Occup Environ Health* 2009;**82**(9):1043–55.
140. Shvedova A, Castranova V, Kisin E, Schwegler-Berry D, Murray A, Gandelsman V, et al. Exposure to carbon nanotube material: assessment of nanotube cytotoxicity using human keratinocyte cells. *J Toxicol Environ Health A* 2003;**66**(20):1909–26.
141. Bekyarova E, Ni Y, Malarkey EB, Montana V, McWilliams JL, Haddon RC, et al. Applications of carbon nanotubes in biotechnology and biomedicine. *J Biomed Nanotechnol* 2005;**1**(1):3–17.
142. Sayes CM, Liang F, Hudson JL, Mendez J, Guo W, Beach JM, et al. Functionalization density dependence of single-walled carbon nanotubes cytotoxicity in vitro. *Toxicol Lett* 2006;**161**(2):135–42.

143. Manna SK, Sarkar S, Barr J, Wise K, Barrera EV, Jejelowo O, et al. Single-walled carbon nanotube induces oxidative stress and activates nuclear transcription factor-κB in human keratinocytes. *Nano Lett* 2005;**5**(9):1676–84.
144. Zhang LW, Zeng L, Barron AR, Monteiro-Riviere NA. Biological interactions of functionalized single-wall carbon nanotubes in human epidermal keratinocytes. *Int J Toxicol* 2007;**26**(2):103–13.
145. Sargent LM, Shvedova AA, Hubbs AF, Salisbury JL, Benkovic SA, Kashon ML, et al. Induction of aneuploidy by single-walled carbon nanotubes. *Environ Mol Mutagen* 2009;**50**(8):708–17.
146. Ding L, Stilwell J, Zhang T, Elboudwarej O, Jiang H, Selegue JP, et al. Molecular characterization of the cytotoxic mechanism of multiwall carbon nanotubes and nano-onions on human skin fibroblast. *Nano Lett* 2005;**5**(12):2448–64.
147. Teeguarden JG, Webb-Robertson B-J, Waters KM, Murray AR, Kisin ER, Varnum SM, et al. Comparative proteomics and pulmonary toxicity of instilled single-walled carbon nanotubes, crocidolite asbestos, and ultrafine carbon black in mice. *Toxicol Sci* 2011;**120**(1):123–35.
148. Morimoto Y, Hirohashi M, Ogami A, Oyabu T, Myojo T, Todoroki M. *et al.* Nanotoxicology: Pulmonary toxicity of well-dispersed multi-wall carbon nanotubes following inhalation and intratracheal instillation; 2011 Article ID 594912.
149. Kobayashi N, Naya M, Ema M, Endoh S, Maru J, Mizuno K, et al. Biological response and morphological assessment of individually dispersed multi-wall carbon nanotubes in the lung after intratracheal instillation in rats. *Toxicology* 2010;**276**(3):143–53.
150. Reddy ARN, Reddy YN, Krishna DR, Himabindu V. Pulmonary toxicity assessment of multiwalled carbon nanotubes in rats following intratracheal instillation. *Environ Toxicol* 2012;**27**(4):211–9.
151. Roda E, Coccini T, Acerbi D, Barni S, Vaccarone R, Manzo L. Comparative pulmonary toxicity assessment of pristine and functionalized multi-walled carbon nanotubes intratracheally instilled in rats: morphohistochemical evaluations. *Histol Histopathol* 2011;**26**(3):357–67.
152. Salvador-Morales C, Flahaut E, Sim E, Sloan J, Green MLH, Sim RB. Complement activation and protein adsorption by carbon nanotubes. *Mol Immunol* 2006;**43**(3):193–201.
153. Dumortier H, Lacotte S, Pastorin G, Marega R, Wu W, Bonifazi D, et al. Functionalized carbon nanotubes are non-cytotoxic and preserve the functionality of primary immune cells. *Nano Lett* 2006;**6**(7):1522–8.
154. Markovic Z, Trajkovic V. Biomedical potential of the reactive oxygen species generation and quenching by fullerenes (C60). *Biomaterials* 2008;**29**(26):3561–73.
155. Kroto HW, Heath JR, O'Brien SC, Curl RF, Smalley RE. C60: buckminsterfullerene. *Nature* 1985;**318**(14):162–3.
156. Isaacson CW, Kleber M, Field JA. Quantitative analysis of fullerene nanomaterials in environmental systems: a critical review. *Environ Sci Technol* 2009;**43**(17):6463–74.
157. Aschberger K, Johnston HJ, Stone V, Aitken RJ, Tran CL, Hankin SM, et al. Review of fullerene toxicity and exposure—appraisal of a human health risk assessment, based on open literature. *Regul Toxicol Pharmacol* 2010;**58**(3):455–73.
158. Sayes CM, Fortner JD, Guo W, Lyon D, Boyd AM, Ausman KD, et al. The differential cytotoxicity of water-soluble fullerenes. *Nano Lett* 2004;**4**(10):1881–7.
159. Sayes CM, Gobin AM, Ausman KD, Mendez J, West JL, Colvin VL. Nano-C60 cytotoxicity is due to lipid peroxidation. *Biomaterials* 2005;**26**(36):7587–95.
160. Porter AE, Muller K, Skepper J, Midgley P, Welland M. Uptake of C60 by human monocyte macrophages, its localization and implications for toxicity: studied by high resolution electron microscopy and electron tomography. *Acta Biomater* 2006;**2**(4):409–19.

161. Johnston HJ, Hutchison GR, Christensen FM, Aschberger K, Stone V. The biological mechanisms and physicochemical characteristics responsible for driving fullerene toxicity. *Toxicol Sci* 2010;**114**(2):162–82.
162. Isakovic A, Markovic Z, Todorovic-Markovic B, Nikolic N, Vranjes-Djuric S, Mirkovic M, et al. Distinct cytotoxic mechanisms of pristine versus hydroxylated fullerene. *Toxicol Sci* 2006;**91**(1):173–83.
163. Chen HHC, Yu C, Ueng TH, Chen S, Chen BJ, Huang KJ, et al. Acute and subacute toxicity study of water-soluble polyalkylsulfonated C60 in rats. *Toxicol Pathol* 1998;**26**(1):143–51.
164. Mori T, Takada H, Ito S, Matsubayashi K, Miwa N, Sawaguchi T. Preclinical studies on safety of fullerene upon acute oral administration and evaluation for no mutagenesis. *Toxicology* 2006;**225**(1):48–54.
165. Moussa F, Trivin F, Céolin R, Hadchouel M, Sizaret PY, Greugny V, et al. Early effects of C60 administration in Swiss mice: a preliminary account for in vivo C60 toxicity. *Fullerene Sci. Technol.* 1996;**4**(1):21–9.
166. Gharbi N, Pressac M, Hadchouel M, Szwarc H, Wilson SR, Moussa F. [60]Fullerene is a powerful antioxidant in vivo with no acute or subacute toxicity. *Nano Lett* 2005;**5**(12):2578–85.
167. Yamago S, Tokuyama H, Nakamura E, Kikuchi K, Kananishi S, Sueki K, et al. In vivo biological behavior of a water-miscible fullerene: 14C labeling, absorption, distribution, excretion and acute toxicity. *Chem Biol* 1995;**2**(6):385–9.
168. Kubota R, Tahara M, Shimizu K, Sugimoto N, Hirose A, Nishimura T. Time-dependent variation in the biodistribution of C60 in rats determined by liquid chromatography-tandem mass spectrometry. *Toxicol Lett* 2011;**206**(2):172–7.
169. Baker GL, Gupta A, Clark ML, Valenzuela BR, Staska LM, Harbo SJ, et al. Inhalation toxicity and lung toxicokinetics of C60 fullerene nanoparticles and microparticles. *Toxicol Sci* 2008;**101**(1):122–31.
170. Fujita K, Morimoto Y, Ogami A, Myojyo T, Tanaka I, Shimada M, et al. Gene expression profiles in rat lung after inhalation exposure to C60 fullerene particles. *Toxicology* 2009;**258**(1):47–55.
171. Totsuka Y, Kato T, Masuda S-i, Ishino K, Matsumoto Y, Goto S, et al. In Vitro and in vivo genotoxicity induced by fullerene (C60) and kaolin. *Genes Environ* 2011;**33**(1):14–20.
172. Benn TM, Westerhoff P, Herckes P. Detection of fullerenes (C60 and C70) in commercial cosmetics. *Environ Pollut* 2011;**159**(5):1334–42.
173. Rouse JG, Yang J, Ryman-Rasmussen JP, Barron AR, Monteiro-Riviere NA. Effects of mechanical flexion on the penetration of fullerene amino acid-derivatized peptide nanoparticles through skin. *Nano Lett* 2006;**7**(1):155–60.
174. Nielsen GD, Roursgaard M, Jensen KA, Poulsen SS, Larsen ST. In vivo biology and toxicology of fullerenes and their derivatives. *Basic Clin Pharmacol Toxicol* 2008;**103**(3):197–208.
175. Xia XR, Monteiro-Riviere NA, Riviere JE. Skin penetration and kinetics of pristine fullerenes (C60) topically exposed in industrial organic solvents. *Toxicol Appl Pharmacol* 2010;**242**(1):29–37.
176. Usenko CY, Harper SL, Tanguay RL. In vivo evaluation of carbon fullerene toxicity using embryonic zebrafish. *Carbon* 2007;**45**(9):1891–8.
177. Michalet X, Pinaud FF, Bentolila LA, Tsay JM, Doose S, Li JJ, et al. Quantum dots for live cells, in vivo imaging, and diagnostics. *Science* 2005;**307**(5709):538–44.
178. Reimann SM, Manninen M. Electronic structure of quantum dots. *Rev Mod Phys* 2002;**74**(4):1283.
179. Gao X, Cui Y, Levenson RM, Chung LWK, Nie S. In vivo cancer targeting and imaging with semiconductor quantum dots. *Nat Biotechnol* 2004;**22**(8):969–76.

180. Chang H, Lv X, Zhang H, Li J. Quantum dots sensitized graphene: in situ growth and application in photoelectrochemical cells. *Electrochem Commun* 2010;**12**(3):483–7.
181. Mancini MC, Kairdolf BA, Smith AM, Nie S. Oxidative quenching and degradation of polymer-encapsulated quantum dots: new insights into the long-term fate and toxicity of nanocrystals in vivo. *J Am Chem Soc* 2008;**130**(33):10836–7.
182. Jiang W, Papa E, Fischer H, Mardyani S, Chan WCW. Semiconductor quantum dots as contrast agents for whole animal imaging. *Trends Biotechnol* 2004;**22**(12):607–9.
183. Kim S, Lim YT, Soltesz EG, De Grand AM, Lee J, Nakayama A, et al. Near-infrared fluorescent type II quantum dots for sentinel lymph node mapping. *Nat Biotechnol* 2004;**22**(1):93–7.
184. Åkerman ME, Chan WCW, Laakkonen P, Bhatia SN, Ruoslahti E. Nanocrystal targeting in vivo. *Proc Natl Acad Sci USA* 2002;**99**(20):12617–21.
185. Hardman R. A toxicologic review of quantum dots: toxicity depends on physicochemical and environmental factors. *Environ Health Perspect* 2006;**114**(2):165–72.
186. Derfus AM, Chan WCW, Bhatia SN. Probing the cytotoxicity of semiconductor quantum dots. *Nano Lett* 2003;**4**(1):11–8.
187. Hoshino A, Fujioka K, Oku T, Suga M, Sasaki YF, Ohta T, et al. Physicochemical properties and cellular toxicity of nanocrystal quantum dots depend on their surface modification. *Nano Lett* 2004;**4**(11):2163–9.
188. Bao YJ, Li JJ, Wang YT, Yu L, Lou L, Du WJ, et al. Probing cytotoxicity of CdSe and CdSe/CdS quantum dots. *Chin Chem Lett* 2011;**22**(7):843–6.
189. Chan W-H, Shiao N-H, Lu P-Z. CdSe quantum dots induce apoptosis in human neuroblastoma cells via mitochondrial-dependent pathways and inhibition of survival signals. *Toxicol Lett* 2006;**167**(3):191–200.
190. Su Y, Hu M, Fan C, He Y, Li Q, Li W, et al. The cytotoxicity of CdTe quantum dots and the relative contributions from released cadmium ions and nanoparticle properties. *Biomaterials* 2010;**31**(18):4829–34.
191. Hauck TS, Anderson RE, Fischer HC, Newbigging S, Chan WCW. In vivo quantum-dot toxicity assessment. *Small* 2010;**6**(1):138–44.
192. Ballou B, Lagerholm BC, Ernst LA, Bruchez MP, Waggoner AS. Noninvasive imaging of quantum dots in mice. *Bioconjug Chem* 2003;**15**(1):79–86.
193. Lozano T, et al. Cytotoxicity effects of metal oxide nanoparticles in human tumor cell lines. *J Phys Conf Ser* 2011;**304**(1):012046.
194. Chen X, Schluesener HJ. Nanosilver: a nanoproduct in medical application. *Toxicol Lett* 2008;**176**(1):1–12.
195. Hutchings GJ, Brust M, Schmidbaur H. Gold-an introductory perspective. *Chem Soc Rev* 2008;**37**(9):1759–65.
196. Kuang D, Brillet JRM, Chen P, Takata M, Uchida S, Miura H, et al. Application of highly ordered TiO2 nanotube arrays in flexible dye-sensitized solar cells. *ACS Nano* 2008;**2**(6):1113–6.
197. Drake PL, Hazelwood KJ. Exposure-related health effects of silver and silver compounds: a review. *Ann Occup Hyg* 2005;**49**(7):575–85.
198. Chaloupka K, Malam Y, Seifalian AM. Nanosilver as a new generation of nanoproduct in biomedical applications. *Trends Biotechnol* 2010;**28**(11):580–8.
199. Tiwari DK, Jin T, Behari J. Dose-dependent in-vivo toxicity assessment of silver nanoparticle in Wistar rats. *Toxicol Mech Methods* 2011;**21**(1):13–24.
200. Stebounova L, Adamcakova-Dodd A, Kim J, Park H, O'Shaughnessy P, Grassian V, et al. Nanosilver induces minimal lung toxicity or inflammation in a subacute murine inhalation model. *Part Fibre Toxicol* 2011;**8**(1):5.

201. Mohammadkazem K, et al. Assessment of dermal exposure and histopathologic changes of different sized nano-silver in healthy adult rabbits. *J Phys Conf Ser* 2011;**304**(1):012028.
202. Liz-Marzán LM. Nanometals: formation and color. *Mater Today* 2004;**7**(2):26–31.
203. de la Escosura-Muniz A, Sanchez-Espinel C, Diaz-Freitas B, Gonzalez-Fernandez A, Maltez-da Costa M, Merkoci A. Rapid identification and quantification of tumor cells using an electrocatalytic method based on gold nanoparticles. *Anal Chem* 2009;**81**(24):10268–74.
204. de la Escosura-Muniz A, Maltez-da Costa M, Sanchez-Espinel C, Diaz-Freitas B, Fernandez-Suarez J, Gonzalez-Fernandez A, et al. Gold nanoparticle-based electrochemical magnetoimmunosensor for rapid detection of anti-hepatitis B virus antibodies in human serum. *Biosens Bioelectron* 2010;**26**(4):1710–4.
205. Luo XL, Xu JJ, Du Y, Chen HY. A glucose biosensor based on chitosan-glucose oxidase-gold nanoparticles biocomposite formed by one-step electrodeposition. *Anal Biochem* 2004;**334**(2):284–9.
206. Khlebtsov N, Dykman L. Biodistribution and toxicity of engineered gold nanoparticles: a review of in vitro and in vivo studies. *Chem Soc Rev* 2011;**40**(3):1647–71.
207. Semmler-Behnke M, Kreyling WG, Lipka J, Fertsch S, Wenk A, Takenaka S, et al. Biodistribution of 1.4- and 18-nm gold particles in rats. *Small* 2008;**4**(12):2108–11.
208. Daoud WA, Xin JH, Zhang Y-H. Surface functionalization of cellulose fibers with titanium dioxide nanoparticles and their combined bactericidal activities. *Surf Sci* 2005;**599**(1–3):69–75.
209. Byrappa K, Ohara S, Adschiri T. Nanoparticles synthesis using supercritical fluid technology—towards biomedical applications. *Adv Drug Deliv Rev* 2008;**60**(3):299–327.
210. Xia T, Kovochich M, Liong M, Maüdler L, Gilbert B, Shi H, et al. Comparison of the mechanism of toxicity of zinc oxide and cerium oxide nanoparticles based on dissolution and oxidative stress properties. *ACS Nano* 2008;**2**(10):2121–34.
211. Baek M, et al. Factors influencing the cytotoxicity of zinc oxide nanoparticles: particle size and surface charge. *J Phys Conf Ser* 2011;**304**(1):012044.
212. Wang J, Zhou G, Chen C, Yu H, Wang T, Ma Y, et al. Acute toxicity and biodistribution of different sized titanium dioxide particles in mice after oral administration. *Toxicol Lett* 2007;**168**(2):176–85.
213. Chen J, Dong X, Zhao J, Tang G. In vivo acute toxicity of titanium dioxide nanoparticles to mice after intraperitioneal injection. *J Appl Toxicol* 2009;**29**(4):330–7.
214. Fabian E, Landsiedel R, Ma-Hock L, Wiench K, Wohlleben W, van Ravenzwaay B. Tissue distribution and toxicity of intravenously administered titanium dioxide nanoparticles in rats. *Arch Toxicol* 2008;**82**(3):151–7.
215. Yokel RA, Florence RL, Unrine JM, Tseng MT, Graham UM, Wu P, et al. Biodistribution and oxidative stress effects of a systemically-introduced commercial ceria engineered nanomaterial. *Nanotoxicology* 2009;**3**(3):234–48.
216. Hirst SM, Karakoti A, Singh S, Self W, Tyler R, Seal S, et al. Bio-distribution and in vivo antioxidant effects of cerium oxide nanoparticles in mice. *Environ Toxicol* 2011. doi: 10.1002/tox.20704
217. Chen J-K, Shih M-H, Peir J-J, Liu C-H, Chou F-I, Lai W-H, et al. The use of radioactive zinc oxide nanoparticles in determination of their tissue concentrations following intravenous administration in mice. *Analyst* 2010;**135**(7):1742–6.

Chapter 16

Overview of Nanomedicines Regulation in the European Union

Ignasi Gispert
Researcher at the Institute for Law and Technology (Universitat Autònoma de Barcelona) and Advisor of the Legal Observatory of the Center for NanoBioSafety and Sustainability (CNBSS) at the Catalan Institute of Nanotechnology (ICN). Barcelona, Spain

1. INTRODUCTION

Nanotechnology in medicine is used for many purposes, including drug delivery systems, carriers, diagnostics and structures in regenerative medicine. It is expected to "impact the field of medicine much like anaesthetic and antibiotics. The transformation will be so fundamental that we cannot completely comprehend how this change will impact the future of medical care".[a]

Regulatory challenges arising from the use of nanomaterials in the medical sector share similar characteristics to those posed by other emerging technologies in terms of adequacy of existing regulations and guidelines, acceptability of new testing methods and availability of experts. But, in addition, we face particular challenges due to the wide spectrum of their applicability, the diverse characteristics of resulting applications,[b] and a large range of technological, scientific, normative, conceptual and institutional uncertainties.

In this section, our focus will be the regulation of nanomedicine therapeutics (that include "nanopharmaceuticals" and "nano-enabled devices")[c] in the European Union from a product safety perspective.[d]

---

a. Ref. 1.
b. Ref. 2.
c. Although a clear cut distinction is increasingly difficult and there is no common agreed set of definitions or classifications, see Ref. 3. At 26–30. Boisseau and Loubaton discuss the appropriateness of the terms "nanomedicine" considering to be more accurate to use the concept "Nanotechnology in medicine". These authors also classify "nanoPharmaceuticals" in three areas, being: "nanoparticulate Drug Delivery Systems", "Drug delivery (mechanical) Devices" and "Theranostics, combined products".[4]
d. Bowman and Hodges suggest six regulatory frontiers: product safety, privacy and civil liberties, occupational health and safety, intellectual property, international law and environmental law. For each specific frontier, they adopt the "enforcement pyramid" notion, in which a range of regulatory mechanisms are viewed as possible, from "hard law" at the top, through a variety of "soft law" mechanisms further down towards the base of the pyramid.[5,6] At 12.

Frontiers of Nanoscience, Vol. 4. DOI: 10.1016/B978-0-12-415769-9.00016-9

There are no specific regulations for nanotechnologies or nanomaterials at the EU level. Instead, the manufacture, use and disposal of nanomaterials are covered, at least in principle, by a complex set of existing provisions designed to manage risk associated with conventional, bulk-sized materials.[e] Consequently, there is no specific nanomedical regulatory regime in the European Union and the applicable legislation to nanomedicine[f] is covered by legislation on medical products and devices, tissue engineering and other advanced therapies.

The EU pharmaceutical legislation is compiled in Volume 1 (EU pharmaceutical legislation for medical products for human use) and Volume 5 (EU pharmaceutical legislation for medical products for veterinary use) of the publication "The rules governing medicinal products in the European Union". The base legislation is supported by a series of guidelines that are also published in several volumes of the same publication. In addition, medicinal

---

e. Ref. 7. At 359; Ref. 8; Commission of the European Communities (2008) <<Regulatory Aspects of Nanomaterials>>. Commission Staff Working Document. Summary of legislation in relation to health, safety and environment aspects of nanomaterials, regulatory research needs and related measures. SEC(2008)2036. COM(2008)366 final. 17.6.2008.

f. For the purposes of this section, we understand "nanomedicine" as "the application of nanotechnology in view of making a medical diagnosis or treating or preventing diseases. It exploits the improved and often novel physical, chemical and biological properties of materials at nanometre scale".[9]

It is still too early to assess how the Commission Recommendation on a Definition of "nanomaterials"—Commission Recommendation of 18.10.2011 on the definition of nanomaterial (2011/696/UE) OJ L275/38 of 20.10.2011—will impact the pharmaceutical nanoregulation, but there are certain elements that lead us to believe that its impact will be lower in this area than in other sectors (legislation on chemicals or cosmetics, for example):

- The European Medical Agency has underlined that the evaluation of any medicine is based on established principles of benefit/risk analysis, rather than solely on the basis of the technology "*per se*".[10]
- Although the Recommendation states a general threshold of 100 nm, especially in the area of drug delivery, relatively large nanoparticles may be needed for loading a sufficient amount of drug onto the particle. To establish an arbitrary threshold could be contrary to patient's interest and market realities.[11]
- The Recommendation leaves a broad margin for its adaptation to any specific sector. Recital 16 <<The definition set out in this Recommendation should not prejudge nor reflect the scope of application of any piece of Union legislation or of any provisions potentially establishing additional requirements for those materials. (...) It may likewise be necessary to include additional materials, such as some materials with a size smaller than 1 nm or greater than 100 nm in the scope of application of specific legislation provisions suited for a nanomaterial.>>
- Nevertheless, we agree with Chowdhury when stating that "it is important to underline that a regulatory trigger would have to be identified at some stage and it is a prerequisite for the development and implementation of any regulatory guidance in this sector".[12] At 135.

products for paediatric use, orphan, herbal medicinal products and advanced therapies are governed by specific rules.[g]

In the following paragraphs, we will present the basic characteristics of the European Union pharmaceutical product safety regulation (for medicinal products, medical devices and advanced therapy medicinal products—ATMPs), focus afterward on the regulatory gaps arising from the so-called borderline products and conclude with a preliminary assessment of how the precautionary principle might impact the Commission's incremental approach strategy for nanomedicines regulation within the EU.

2. MEDICINAL PRODUCTS

Medicinal products are regulated by Directive 2001/83/EC[h] on the Community Code relating to medicinal products for human use and Regulation (EC) No. 726/2004[i] setting down the Community procedures for the

g. See http://ec.europa.eu/health/documents/eudralex/index_htm [accessed January 2012]. "Guidelines" are not legally enforceable but have to be included in the broad spectrum of soft law voluntary measures that play an extremely important role in nanopharmaceuticals oversight.
Soft Law can be defined as rules of conduct that are laid down in instruments which have not been attributed legally binding force as such, but nevertheless may have certain (indirect) legal effects, and that are aimed at and may produce practical effects.[13] See also Ref. 15.
Soft Law mechanisms have to be seen as a complement (or/and a prelude) to existing mandatory regulatory approaches (interim measures to fill the current risk management gap before our knowledge of the emerging technologies and the associated risk measures are developed) and have been considered "one of the most promising management strategies for ensuring safety and risk control in the short term". Ref. 14 At 81. See also Ref. 15.
It also has to be underlined that Regulation (EC) No. 1907/2006 on the Registration, Evaluation, Authorization and Restriction of Chemicals (REACH) contains broad exemptions for pharmaceuticals but, as a general rule, manufacturers of medicines and medical devices containing nanomaterials should ensure that the raw materials used have been authorized by REACH. See Ref. 16
h. Directive 2001/83/EC of the European Parliament and of the Council of 6 November 2001 on the Community code relating to medicinal products for human use (OJ L311/67, 28/11/2004).
As amended by Directive 2002/98/EC of the European Parliament and of the Council of 27 January 2003 setting standards of quality and safety for the collection, testing, processing, storage and distribution of human blood and blood components and amending Directive 2001/83/EC (OJ L33/30, 08/02/2003); Directive 2004/24/EC of the European Parliament and the Council of 31 March 2004 amending, as regards traditional herbal medicinal products, Directive 2001/83/EC on the Community code relating to medicinal products for human use (OJ L136/85, 30/04/2004); Directive 2004/27/EC of the European Parliament and the Council of 31 March 2004 amending Directive 2001/83/EC on the Community code relating to medicinal products for human use (OJ L136/34, 30/04/2004).
i. Regulation (EC) No 726/2009 of the European Parliament and the Council of 31 March 2004 laying down Community procedures for the authorization and supervision of medical products for human and veterinary use and establishing a European Medicines Agency (OJ L136/1, 30/04/2004).

authorization and supervision of medicinal products and establishing the European Medicines Agency (EMA).

Regulation (EC) No. 726/2004 lays down a centralized Community procedure[j] for the authorization of medical products, for which there is a single evaluation and a single authorization allowing direct access to the market in all Member States (pre-market control system).

Article 3 of the Regulation (EC) No. 726/2004 defines the scope and eligibility of applications for evaluation under the centralized procedure through which medicinal products must (mandatory scope) or may (optional scope, generic/hybrid scope) be authorized by the Community.[k]

The EMA is a decentralized body of the European Union and is in charge of the centralized authorization process, with applications made directly to the Agency. The authorization procedures are based on the risk/

---

j. Alternative authorization mechanisms to the centralized procedure are (a) a "mutual recognition" procedure with applications submitted simultaneously in different Member States; (b) a "decentralized" procedure based on the recognition by the concerned Member States of a marketing authorization already granted by another Member State; (c) a purely "national" authorization procedure for medicinal products to be marketed in only one Member State.

k. (A) Mandatory scope (Art. 3(1))

1. A medical product developed by certain biotechnological processes, namely: recombinant DNA technology; controlled expression of genes coding for biologically active proteins in prokaryotes and eukaryotes including transformed mammalian cells; hybridoma and monoclonal antibody methods.
2. Medicinal products for human use containing a new active substance not authorized in the Community prior to 20.XI.2005, for the treatment of AIDS, cancer, neurodegenerative disorder, diabetes, autoimmune disease and other autoimmune dysfunctions, viral diseases.
3. Orphan medical products as defined in Regulation 141/2000.

(B) Optional scope (Art. 3(2)):

1. A medical product containing a new active substance which, on the day of entry into force of the Regulation (20.11.2005), was not authorized in the Community.
   Biosimilar medicinal products within this category.
   Advanced therapy medicinal product included in Regulation 1394/2007.
2. A medical product which constitutes a significant therapeutic, scientific or technical innovation or that the granting of the authorization is in the interest of patient at Community level;
3. A generic or hybrid medicinal product of a reference medicinal product authorized via the centralized procedure has "automatic" access to the centralized procedure;
4. Certain medical products for paediatric use as defined in Regulation (EC) No 1901/2006.

- In all cases, the optional evaluation has to be requested by the applicant.
- See Ref. 17; see also Eudralex vol. 2A—Procedure for marketing authorization. Chapter 4—Centralized Procedure.

benefit analysis,[l] requires an assessment of toxicology and ecotoxicology and of methodologies used to evaluate toxicity and extensive post-marketing surveillance. This applies also to nanomaterials and nano-related products, even if they are not explicitly mentioned in current provisions.[m]

Medicinal products containing nanoparticles have already been granted Marketing Authorization by the Commission or at national level within the Community based (allegedly) on the fact that "(...) for their manufacture standard processes have often been employed, which are well described and understood" and "should not be considered necessarily as arising from innovative nanotechnology methods".[n]

In an opinion given in 2006, the Committee for Medical Products for Human Use (CHMP)[o] considered that "the evaluation and prevention of potential hazards related to the use of any given nanomedicinal' product is already foreseen under the existing pharmaceutical legislation" and that it may be the case that for borderline products (medicine/device) the development of guidance specific to nanomedicinal products or the update of existing ones may be needed. In 2009, the *Ad hoc* CHMP Expert Group Meeting on "Nanomedicines" already requested to complement existing guidelines.[p] In fact, these statements follow the Commission's positioning of considering that the regulatory framework has to be considered "in principle" as adequate and

l. The European Medicines Agency's opinions are based on balancing the desired effects or "benefits" of a medicine against its undesired effects or "risks". The Agency can recommend the authorization of a medicine whose benefits are judged to be greater than its risks. It is interesting to know that the above mentioned balance is made without any specific methodology, as stated by the EMA "Weighing up the benefits and risks of a medicine is a complex process, since it involves the evaluation of a large amount of data. In addition, there is always some uncertainty around the actual benefits and risks of a medicine, because they can only be determined by looking at the information that is available at a given point in time. (...) The Agency strives towards making its opinions on the balance of benefits and risks as consistent and transparent as possible. To date, however, there is no standard methodology that is used to aid regulatory decisions on the benefits and risks of medicines". Available at: http://www.ema.europa.eu/ema/index.jsp?curl=pages/special_topics/document_listing/document_listing_000314.jsp&mid=WC0b01ac0580223ed6.

It is a matter of further legal research the interplay between the Precautionary principle (a general principle of Community Law) and the risk/benefit methodology: specifically, if and how a proper integration of the precautionary principle is achieved in the risk management stage and, consequently, in the risk-benefit analysis conducted by EMA thought the Market Authorization Application (MAA).

m. Ref. 9. At 3–4.

n. Ref. 9. At 3–4; Ref. 10.

o. The Committee for Medical Products for Human Use (CHMP) forms apart of the European Medical Agency (EMA) and was established by Art 5 of Regulation (EC) No 726/2004.

p. Ref. 2. At 3.

that "current legislation may have to be modified in the light of new information becoming available"[q] and the SCENIHR (Scientific Committee on Emerging and Newly Identified Health Risks) conclusions pointing out several shortcomings of the methods for assessing the potential risks associated with engineered nanoparticles.[r] Faced with these limitations,[24] KELLY and

q. Commission of the European Communities (2008) "Regulatory aspects of nanomaterials". Commission Staff Working Document. Summary of legislation in relation to health, safety and environment aspects of nanomaterials, regulatory research needs and related measures. SEC (2008)2036. COM(2008) 366 final. 17.6.2008. At 3.

r. The European Court of Justice has defined risk assessment as a "scientific process consisting in the identification and characterization of a hazard, the assessment of exposure to the hazard and the characterization of the risk" Case T-13/99 Pfizer Animal Health v Council. At 156.

- In the EU, risk assessment methodology is described in the Technical Guidance Document (TGD) on Risk Assessment. Institute for Health and Consumer Protection. European Chemicals Bureau. Joint Research Centre. European Commission. EUR20418 EN/1. 2003. The guidance given on the TGD is not legally binding. It can be used other methods or approaches if they are considered to be more appropriated, provided that they are scientifically justified. Those methods, including any assumptions, uncertainties and calculations, should be clearly described and justified. See TGD2003. At 1.3.
- The TGD characterizes risk assessment as consisting in four parts, namely, hazard identification, hazard characterization, exposure assessment and risk characterization.
- It is generally agreed that Risk Assessment as described in the TGD cannot be directly extrapolated to nanomaterials. See European Commission's Scientific Committee on Emerging and Newly Identified Health Risks. Ref. 18 At 26.
- In broader terms, risk assessment for nanomaterials precludes going further than identifying hazards—first step of risk assessment—and providing some elements of hazard characterization (dose/response assessment).[19]
- SCENIHR has listed the shortcoming of nanomaterials risk characterization:
  1. Persistence of nanoparticles in the atmosphere, which will depend on rates of agglomeration and disaglomeration, and on degradation;
  2. Relevance of routes of exposure to individual circumstances;
  3. Metrics used for exposure measurements;
  4. Mechanisms of translocation to different parts of the body and the possibility of degradation after nanoparticles enter the body;
  5. Mechanisms of toxicity of nanoparticles;
  6. Phenomenon of transfer between various environmental media.

  These are not simply uncertainties in the values of some traditional parameters, but rather the uncertainties about the potentially unique or significantly modified causal mechanisms themselves. See Ref. 18; Ref. 20.
- In order to overcome these limitations, SCENIHR has proposed to assess the risk of nanomaterials on a case-by-case basis that has been criticized as unworkable. See Ref. 19. At 71.
- Alternative decision-making tools for risk assessment have been proposed, like Multi Criteria Decision Analysis (MCDA), hazard trigger algorithm or life cycle inventory analysis. See Refs. 21–23.

BOGAERT recommend to treat each nanomaterial individually and that "companies should keep abreast of, and comply with, industrial and regulatory standards".[s]

In order to overcome the risk assessment gaps related with nanomaterials, the EMA created the Innovation Task Force (ITF) to ensure EMA-wide coordination of scientific and regulatory competence in the field of emerging therapies and technologies, including nanotechnologies, and to provide for early dialogue with applicants on regulatory, scientific or other issues that might arise. Basically, and in the absence of specific guidance, applicants are encouraged to contact EMA/ITF from the early stage of the development process of their product and seek both Regulatory and Scientific Advice.[t] These include advice on classification issues related with innovative and borderline products. Although EMA is not a classification body and its advice is, in legal terms, a non-binding opinion, to establish an early dialogue with the Regulatory Authority in a collaborative manner is highly advisable: informal dialogue is a good opportunity for EMA to increase its understanding on emerging science, while may provide sponsors with a better insight on potential regulatory implications when developing their prod-

s. Ref. 24. At 453. In broader terms and given the uncertainties related with this area, companies need to be aware of soft law mechanisms which aim at regulating the field of nanomaterials on the basis of voluntary conduct. Among others, its worth mentioning: (1) Commission Recommendation of 07.02.2008 on a code of conduct for responsible nanosciences and nanotechnologies research C(2008) 424 (2008/345/CE) OJ L116/46 of 30.04.2008; (2) Responsible NanoCode (United Kingdom); (3) German NanoKommission; (4) The Swiss Retailers Association (IG DHS) Code of Conduct; (5) Other Industry codes of conduct/practices in line with the Corporate Social Responsibility (CRS): The ICCA (International Council of Chemical Associations) "Responsible Care Global Charter", the BASF "Code of Conduct on Nanotechnology", the Bayer "Code of Good Practice on the Production and on-site-use of Nanomaterials" and the DuPont "Nano Risk Framework are telling examples of this type of initiatives. (6) Certification systems: Closely related with Codes of Conduct it is interesting to mention the Certification systems, a further voluntary measure with regard to the regulation of nanotechnology: (1) Cenarios (first certificable risk management and monitoring system specifically adapted to nanotechnologies); (2) Hohenstein Quality for Nanotechnology for the textile industry; (3) Quality Seal Nano Inside (certifies that a certain product contains nano and that the applicant obliges itself to adhere to the Responsible Nanocode); (4) Assured Nano (accredited scheme of best practice in HSE aspects and safe handling of nanomaterials). For further information: http://www.responsiblenanocode.org/[25] See Ref. 25.

t. The ITF is organized in specialized groups as follows: cell therapy products, gene therapy products, nanomedicines, genomic, borderline combination products including medical devices and medical products. See http://www.emea.europa.eu/pdfs/human/itf/itfmandate.pdf; http://www.ema.europa.eu/ema/index.jsp?curl=pages/regulation/document_listing/document_listing_000210.jsp&murl=menus/regulations/regulations.jsp&mid=WC0b01ac05800ba1da. The reported "encouragement" for the establishment of an early collaborative work between developers and EMA is another example of soft law mechanisms in the (nano)medicine sector.

ucts. Thus, informal dialogue "should increase possibilities for marketing authorisation of novel ideas".[u]

3. MEDICAL DEVICES

Medical devices are regulated by Directive 93/42/EEC (The Medical Device Directive, MDD),[v] Directive 90/385/EEC (The Active Implantable Medical Device Directive, AIMDD)[w] revised and consolidated by Directive 2007/47/EC and Directive 98/79/EC (The *In Vitro* Diagnostic Medical Device Directive)[x] as well as several directives modifying and/or implementing them.

The three mentioned directives are based on the New Approach[y] to technical harmonization and standardization that create a system of market

u. Ref. 26.

v. Council Directive 93/42/EEC of 14 June 1993 concerning medical devices, OJ L169/1.

w. Council Directive 90/385/EEC of 20.06.1990 on the approximation of the laws of the Member States relating to active implantable medical devices. OJ 1990L0385 of 11/10/2007; Directive 2007/47/EC of the European Parliament and of the Council of 5.09.2007 amending Council Directive 90/385/EEC on the approximation of the laws of the Member States relating to active implantable medical devices, Council Directive 93/42/EEC concerning medical devices and Directive 98/8/EC concerning the placing of biocidal products on the market. OJ L247/21 of 21/09/2007.

x. Directive 98/79/EC of the European Parliament and the Council of 27/10/1998 on *in vitro* diagnostic medical devices. OJ L331/1 of 07/12/1998.

y. The principles of the New Approach to technical harmonization and standardization can be summarized as follows:

- Legislative harmonization is limited to essential requirements that products placed on the Community market must meet, if they are to benefit from the free movement within the Community;
- The technical specifications of products meeting the essential requirements set out in the directives are laid down in harmonized standards;
- Application of harmonized or other standards remains voluntary, and the manufacturer may always apply other technical specifications to meet the requirements;
- Products manufactured in compliance with harmonized standards benefit from a presumption of conformity with the corresponding essential requirements.

The operation of the New Approach requires that the standards offer a guaranteed level of protection with regard to the essential requirements established by the directives, and that the national authorities follow up implementation.

European Commission <<Guide to the implementation of directives based on the New Approach and the Global Approach>>. 2000. Available at: http://ec.europa.eu/enterprise/policies/single-market-goods/documents/blue-guide/. [accessed January 2012].

notification. By this system, producers (just) have to prove compliance with the so-called Essential Requirements (that have to be fulfilled in order to reduce risk). The most common way to prove compliance with those "Essential Requirements" is by complying (and obtaining certificates of compliance) with published harmonized European and/or international standards[z] and to implement a risk management or quality system. If products are manufactured following harmonized standards, then they benefit from a presumption of conformity (as it is understood that they comply with the "Essential Requirements" and meet the safety and risk assessment requirements).

Thus, the authorization of medical devices is guaranteed by a Declaration of Conformity issued by the manufacturer and (in most of the cases)[aa] verified by a Certificate of Conformity issued by a Notified Body granting the "EC" marking. The National Authorities are in charge of supervision and monitoring functions.[ab] As we can see, implementation of the directives related with medical devices has been basically left to private authorization bodies (Notified Bodies) and private standards (CEN/CENELEC/ISO), with the supervision of the Member States.

This market notification system applied to medical devices implies a totally different system to the one applied to medicinal products—based on a pre-market control system—that we have seen on the previous section. These fundamentally different procedures for market access pose additional regulatory challenges when dealing with nanomedicine "borderline products" as we will see further on.

In 2007, the Working Group on New and Emerging Technologies in Medical Devices (N&ET Working Group) issued a report assessing the adequacy of the medical device regulatory regime. The Group concluded that the medical device legislation is "in general (. . .) suitable to deal with medical devices

---

z. For ISO standards on nanotechnology from Technical Committee TC229, see http://www.iso.org/iso/iso-technical-committee?commid=381983; new references of harmonized standards for Medical Devices can be found at OJ C242 of 19.08.2011.

aa. The following risk classes are applied:

Class 1 (low risk): products such as adhesive bandage and plaster (manufacturers themselves are responsible for a dossier which ensures quality and safety and no not need to follow the Notified body procedure);

Class 2a and 2b: for example, surgical instruments, diagnostic equipment;

Class 3 (high risk): for example, products which are broken down in the body, contraception, products which end up in the bloodstream and have influence in vital organs, aids with drug components such as nanosilver and implants.

The most important factors taken into account in this risk classification system are potential invasiveness and contact with vital body structures such as the heart, central blood circulation and brain, including the central nervous system. As a result many nanotechnology applications end up in the highest risk class.[27] The essential requirements are stated in Annex I. The other directives on medical devices are applying other classifications but following the same model.

ab. For a global view of the EU medical technology framework, see Ref. 28.

manufactured utilizing nanotechnology. The medical device legislation is based on risk management, and this risk management approach is in principle suitable to address all kinds of risk, including the risk associated with medical devices manufactured using nanotechnology". The Group recommended that "all devices incorporating or consisting of particles, components or devices at the nanoscale are in the highest Class III unless they are encapsulated or bound in such a manner that they cannot be released to the patient's organs, tissues, cells or molecules" based on the fact that such devices require a careful case-by-case risk assessment, and it is desirable that independent verification of the risk assessment should always take place. The Group also recommended the development of specific guidance documents. Finally, the Group underlined the need for an active system of post-marketing surveillance by manufacturers and considered that standards may need to be adjusted to incorporate nanoaspects.[ac] It is to be underlined that strengthening the follow-up and post-market implementation issues for medicines and medical devices already on the market should be viewed as a proper implementation of the Precautionary principle, as we will see further on.

A global assessment of the medicinal device regulatory regime undertaken by D'SILVA and van CALSTER[30] concluded that although the regulation could be considered comprehensive and demanding in terms of safety[ad] and conformity requirements, it has also to be considered that the current risk assessment methodologies "may be in themselves inadequate for assessing the risk posed by these particles" and that the inadequacy could reach "the composition of the regimes themselves",[ae] so opening what ALTENSTETTER[28] named "the agenda for a new risk policy".[af]

4. BIOLOGICAL PRODUCTS

The Biological Products Legislation covers three areas: Directive 2002/98/EC[ag] on blood and blood products, Directive 2004/23/EC[ah] dealing with the donation, procurement, testing, processing, preservation, as well as to manufactured products derived from human tissues and cells and Regulation

ac. Ref. 29.

ad. For instance, a Class III high risk medical devices must undergo a full quality assurance system, examination of the design dossier, a production quality assurance, risk assessment, demonstrate de effectiveness of the device and met the essential safety requirements of Annex I.

ae. For instance, with the borderline products or safety, pollution prevention and waste management. Ref. 31 At 224 and 231; See also Ref. 32.

af. Ref. 28. At 31.

ag. Directive 2002/98/EC. OJ L33/30; Directive 2005/61/EC. OJ L256/32—implementing Directive 2002/98/EC of the European Parliament and of the Council as regards traceability requirements and notification of serious adverse reactions and events.

ah. Directive 2004/23/EC of the European Parliament and of the Council of 31/10/2004 on setting standards of quality and safety for the donation, procurement, testing, processing, preservation, storage and distribution of human tissues and cells. OJ L102/48 of 07/04/2004.

1394/2007/EC[ai] on ATMPs (Advanced Therapy medicinal Products) dealing with new medical products for human use based on gene therapy, somatic-cell therapy or tissue engineering.

In this section, we will be focusing on the ATMP Regulation, not only because nanomaterials play an important role in Advance Therapies, but also because this Regulation complement and/or amends several of the legislation seen so far (medicinal products and devices: Directive 2001/83/EC, Regulation 726/2004, Directive 93/42/EEC, Directive 90/385/EEC and Directive 2004/23/EC—which lays down quality and safety standards in respect of human tissues and cells) and set up a regime that could provide "options of design, mechanisms and approaches that could be applied in the case of nanomedicines" [aj]

Considering that this is a field in which specific expertise is required, the EU has created a compulsory centralized authorization procedure and a Committee for Advanced Therapies (CAT)[ak] at the EMA. The CAT is a multidisciplinary committee of experts dealing with scientific and regulatory aspects of the ATMP legislation.

The CAT is responsible for

1. Primary evaluation of ATMPs marketing authorization applications for the EMA's CHMP;
2. For the classification procedure: Applicants have access to an optional procedure (art. 17 Regulation 1394/2007) by which the CAT assess whether a given product based on genes, cells or tissues meets the scientific criteria that defines ATMPs, with the objective to address at the earliest possible stage questions of borderline products.
3. Certification procedures for SMEs: Small and medium-sizes enterprises can request a scientific evaluation by the CAT of quality and non-clinical data for ATMPs under development. Further to the evaluation, the EMA issues a certificate. The objective of this Certificate is to provide SMEs with an incentive for developing ATMS by offering

ai. Regulation (EC) No 1394/2007 of the European Parliament and of the Council of 13 November 2007 on advanced therapy medicinal products and amending Directive 2001/83/EC and Regulation (EC) No 726/2004, OJ 10.12.2007 L324/137 (2007). In addition, we have to take into consideration Directive 2009/120/EC, Regulation 668/2009 and Regulation 1394/2007.

aj. Ref. 12. At 138.

ak. The CAT is a multidisciplinary committee of experts dealing with scientific and regulatory aspects of the ATMP legislation. Include representative of all Member States and also from the European Free Trade Association, the European Economic Area but also representatives from patient and medical associations.

The setting up of dedicated committees within the EMA has been a standard response in the case of development of specific product regime in pharmaceuticals but so far, not for nanomaterials. As we have seen, it does exist a nanoexpert group within the Innovation Task Force to assist the applicant and advise the Committee on Medical Products for Human Use (CHMP).

them a tool to enter negotiations with larger pharmaceutical companies or attract investors for further development of their product.[al]

5. BORDERLINE AND COMBINED PRODUCTS

It is commonly agreed that potential regulatory gaps will increasingly arise as novel applications of nanotechnologies in nanomedical products are developed. There new devices that combine diagnostic and therapeutic functions will challenge the current classification criteria between medical devices and medical products (including combined ATMPs products that incorporate, as an integral part, a medical device and viable cells and tissues). Also, very likely, classification into the different categories of medical devices.[am]

In the EU regulatory framework, the application of the regulatory regime depends whether the product falls within the definition of medical product,[an] ATMPs[ao] or medical device[ap] (basically, as seen, pre-market control system for medical products and ATMPs on one side, and market notification for medical devices on the other).

---

al. Ref. 33. The objectives of offering incentives to SMEs can also be seen in the establishment of reduced fees for all scientific advise on ATMPs in the field established by art 14 of Regulation 1394/2007 and also reduced marketing authorization fees provided that they can prove that there is a particular public health interest in the Community in the ATMP concerned. Outside the scope of the ATMP regulation, it also has to be mentioned the EMA scientific advice and protocol assistance given to SMEs.

Procedural advice on the Certification and guidelines on the minimum requested for obtaining the certificate, see

http://www.ema.europa.eu/cma/index.jsp?curl=pages/regulation/general/general_content_000300.jsp&mid=WC0b01ac058007f4bd&jsenabled=true;

am. These products can be physically or chemically combined, co-packaged in a kit or separated in cross-label products where all components work as a system and are critical to achieve desired therapeutics effect. The possible combinations are manifold: drug-device; drug-biologic; device-biologic; drug-device-biologic. The final result could be the creation of multi-component systems (that may consist of carrier delivery system—drug-device-, therapeutic agent—drug or biologic-, imaging agent or targeting agent) implantable microchip-based delivery systems that deliver different drugs under controlled conditions or injectable delivery systems (transdermal microneedles). Ref. 34

an. Art. 1.1 Directive 2007/47/EC. Any instrument, apparatus, appliance, software, material or other article, whether used alone or in combination, together with any accessories, including the software intended by its manufacturer to be used specifically for diagnostic and/or therapeutics purposes and necessary for its proper application, intended by the manufacturer to be used for human beings, for diagnostic or other purposes. It achieves its principal intended action in or on the human body by mechanical means.

ao. Regulation 1394/2007/EC. Any medicinal product for human use that includes gene therapy medicinal products, somatic cells therapy products and tissue-engineered products.

ap. Directive 2004/27/EC. Art. 1 (1b 2) Any substance having properties for treating or preventing diseases in human beings or that may be used in or administered to human beings with a view to retorning, correcting or modifying physiological functions by exerting principally a pharmacological, immunological or metabolic action.

When a classification issue exists between medical products and devices, the primary mode of action is the criteria for determining the applicable regulatory regime: nanomedicines with a primary mechanical action and secondary pharmaceutical action are brought under the medical device regulation regime. An extensive set of guidelines and recommendations exist for helping the applicants in deciding which is the applicable regime.[aq] There is no need to underline the importance to clarify this issue on the earliest possible stage because, as we have seen, there are different regulatory requirements for their development.

For applicants that could be in doubt of classification to be given to their product development, it is advisable to contact EMA for regulatory advise on the eligibility procedure, specifically[ar]:

- when there are uncertainties on whether the concerned therapeutic product would qualify as medicinal product;
- for borderline products, having characteristics belonging to diverse legal framework (medicines and medical devices, food supplements);
- for medical substances incorporated in medical devices for which the ancillary functions are borderline

This general framework related to borderline and combined products suffered a fundamental change with Regulation 1394/2007/EC on ATMPs. For products regulated under it, classification by risk category of medical devices and borderline products was suspended and replaced by a product-by-product risk management approach: "Combined advanced therapy medicinal product" is considered an "advanced medicinal product" and, consequently, the provisions of Regulation 1394/2007/EC should apply.[as] In fact, the medical device component has to comply with the essential requirements of the relevant medical device directive (AIMDD

---

aq. http://ec.europa.eu/health/medical-devices/documents/borderline/index_en.htm. Especially MEDDEV 2.1/3rev3.

ar. Ref. 35.

as. Art 2(d) Regulation 1394/2007/EC. "Combined advanced therapy medicinal product" means an advanced therapy medicinal product that fulfils the following conditions:

it must incorporate, as an integral part of the product, one or more medical devices within the meaning of Article 1(2)(a) of Directive 93/42/EEC or one or more active implantable medical devices within the meaning of Article 1(2)(c) of Directive 90/385/EEC, and

its cellular or tissue part must contain viable cells or tissue, or

its cellular or tissue part containing non-viable cells or tissues must be viable to act upon the human body with action that can be considered as primary to that of the device referred to.

and MDD) (art 6 Regulation 1394/2007/EC) and its compliance is incorporated in the ATMP approval procedure and assessed by the CAT.[at]

6. THE PRECAUTIONARY PRINCIPLE AND THE INCREMENTAL APPROACH STRATEGY

6.1. The Precautionary Principle

Identifying the appropriate response to uncertain risks is a difficult task for policymakers and regulators, as they have to seek a compromise between conflicting interests (innovation, safety, free movement of goods, environmental protection, etc.). Within this framework, the Precautionary Principle may guide decision-makers to balance those goals, so helping them in the shaping of a proper regulatory regime for nanomaterials.

At the EU level, it is the authoritative interpretation work of the Court of Justice that has established the Precautionary principle as a general principle of Community law and shaped its content and scope. From the revision of the basic interpretation and triggering factors of the Precautionary Principle by the Court, it can be concluded that the conditions needed for its application appear to be met for engineered nanoparticles and the products incorporating them: Risk of serious and irreversible damage to health and the environment deemed unacceptable to society, supported by solid and objective scientific reasons, even if uncertain.[au]

Therefore, the European public authorities are not only allowed but also compelled to take provisional measures to anticipate the potential occurrence of these risks, and that those measures must be based on general principles of risk management (and must be proportionate, non-discriminatory, consistent, based on an examination of benefits and costs of action or lack of action, and on an examination of scientific developments).

---

at. Recital 18 Regulation 1394/2007/EC. "Advanced therapy medical products may incorporate medical devices or active implantable medical devices. Those devices should meet the essential requirements laid dawn In Council Directive 93/42/EEC (...) and Council Directive 90/385/EEC(...) respectively, in order to ensure an appropriate level of quality and safety. The results of the assessment(...) by a notified body in accordance with those Directives should be recognized by the Agency in the evaluation of a combined advancer therapy medicinal product carried out under this Regulation".

au. The Commission gave its interpretation on the Precautionary Principle on its Communication on the precautionary principle. COM(2000) 1 of 2.2.2000. In the Communication on Regulatory Aspects of Nanomaterials (COM(2008)366final), the Commission specifically stated the applicability of the Precautionary Principle to nanomaterials: "Where the full extent of a risk is unknown, but concerns are so high that risk management measures are considered necessary, as is currently the case for nanomaterials, measures must be based on the precautionary principle". Ref. 36 at 11.

In the final section and when discussing the incremental approach strategy, we will be reviewing the specific legal consequences of the above conclusion regarding nanomaterials in general and nanomedicines specifically.[av]

6.2. The Incremental Approach Strategy

Based on the Commission and Experts Advisory Committees conclusions that we have seen, it seams that the EU will base nanomedical products and devices regulation under the so-called incremental approach.[aw]

Having in mind that the pharmaceutical sector has the highest level of safety and environmental care standards when compared with other sectors,[ax] it could be said that the incremental approach is, at least, a pragmatic solution[ay] that

---

av. When discussing the character of the measures to be taken, it is important to recall, as done by the Commission that "recourse to the precautionary principle does not necessarily mean adopting final instruments designed to produce legal effects. A wide range of activities or measures can be used, like legally binding measures, initiation of research projects or recommendations". Commission of the European Communities (2008). Regulatory Aspects of Nanomaterials. COM(2008) 366 final. 17.6.2008.[37] At. 96–97.

aw. A comprehensive Incremental Approach strategy can be defined as a regulatory process which uses existing legislative structures to the maximum together with the need of reviewing and amending (when appropriate, on a case by case basis and for specific nanomaterials and their applications) existing legislation (including the development of a specific guidance and standards to support existing regulation) and the introduction of supplementary policy. When a high potential risk is identified, the precautionary principle has to be invoked.

In any case, this approach has to be supplemented by

- Support of research initiatives on HSE issues (especially on exposure, dose–response, toxicology and ecotoxicology);
- Promotion of risk assessment throughout the life cycle of a nanotechnology including conception, R&D, manufacturing, distribution, use and disposal—not only at the macro, ecological level but also within the human body, as in the case of drug delivery devices;
- Setting-up of a dialogue among all stakeholders;
- International coordination: within recognized international bodies, working group on nanotechnology has been set up to coordinate efforts among subjects involved in regulation of nanotechnology at different levels.

  See Communication from the Commission "Towards a European strategy for nanotechnology" COM(2004) 338 final. Ref. 38 At 3.44; Ref. 15;

ax. As Dorbeck-Jung and Chowdhury explain "(. . .) the pharmaceuticals regulation regime has proved to be *robust* in terms of well-established rules that have been kept up to date and relevant by continuous and careful adaptation to new product developments into product safety, quality, and efficacy. The principle of risk-benefit balance that guides the authorization of drugs appears to be capable of accommodate a certain amount of uncertainty. The successful marketing of drugs in the past indicates that EU pharmaceutical approval regulation has been capable of coping with nanomedicines". Ref. 39 At 293.

ay. This pragmatic approach implies that if in the short term the incremental approach is followed, it does not exclude that in the longer term, a new regulatory framework could be set up, after allowing the technology to develop while regulators, industry and society continually re-evaluate regulation in light of the evolving scientific evidence and of citizen concern. In this sense, Ref. 6. At 235.

can help in the recasting and adaptation of the actual regulatory framework. In addition, its implementation can be more efficient (time and resources wise) and focused to address the environmental, health and safety concerns of nanotechnology and its application.

Nevertheless, because the "incremental approach" is multi-layered and flexible in design, a consistent implementation roadmap is vital. Without it, we will be faced with loose measures that could lead to greater legal uncertainty.[az] As per today, "regulatory action is still in the stage of reflection" and we only can note "initial governance steps to reduce technical and scientific uncertainties".[ba] A proper implementation of the Precautionary principle may call for acting in order to avoid or minimize possible risks and not to be satisfied with (just) monitoring the development and to take into account the possible effects of nanotechnology already in the definition of EU policies.[bb]

When assessing regulatory actions to be taken, we would like to highlight the need for reviewing the "primary mode of action" as a classification method between medicines and devices as well as the appropriateness of the dual procedure and requirements. Additionally, we consider that especial attention has to be given to the proposals forwarded by CHOWDHURY[12] of including nanomedicines in the mandatory list of Regulation EC/726/2004 and consequently to subject them to the central authorization procedure and the creation of a Nanomedicine Committee following the model of the CAT under the ATMP Regulation.[bc] To organize the collection of information about manufactured nanoparticles and nanomaterials, their properties, their manufacturers, their uses and the people potentially exposed can be considered a specific legal consequence of the application of the precautionary principle that this proposal could help to create in the nanomedicines sector.[bd]

az. The Royal Commission on Environmental Pollution when defining the key qualities that a governance system must comply states: "4.11 (...) Effective and trustworthy governance arrangements must therefore have al least four key qualities. They must be informed, transparent, prospective and adaptive. To achieve these characteristics, they also need to be supported by skilled regulatory bodies and decision-making processes that deliver proportionate outcomes".[40] For the basic meaning of those key qualities, see Ref. 40.

ba. Ref. 39. At 292.

bb. Ref. 42.

bc. Ref. 12.

bd. For a similar conclusion for the chemical sector, see Ref. 43. At 348.

In any case, the development of a specific regulatory regime for nanomedicines cannot be discarded (especially with the increased number of borderline medicinal products that is being forecast),[be] but as per today, it is not a solution if our objective is to tackle, as soon as possible, the regulatory deficits related with health, safety and environmental concerns caused by nanomaterials.[bf] A proper implementation of the incremental approach strategy calls for strengthening the knowledge base—as a fundamental supplementary measure—by applying adequate funding of the potential adverse effects, based on the available data, and carried out by independent entities. Adequate funding raises the question of financing and who is to bear the burden HASELHAUS.[37] considers that it could be consistent with the precautionary

be. The basic argument for a nano-specific regulation points to the fact that existing regulatory situation is not adequate and the effort needed for adapting it could surpass the benefits. For this main reason, it may be better to set up a mandatory nano-specific regulation. In this sense, Hansen states "at some point, regulatory agencies worldwide will have to address the question of whether it is not more effective to implement a new more authoritative and prescriptive legislative framework compared to having to implement a forever-increasing number of smaller and larger adaptations to existing legislation while simultaneously trying to overcome the limitations of chemical risk assessment". Ref. 23 At 449; also see Ref. 44. At 7. A similar conclusion is reached by GASPAR. See Ref. 45

bf. This opinion is based on several factors that have to be taken into consideration, among them, the complexity of the proposed task, the increased regulatory agenda and the uncertainty regarding projected applications.

To design an overarching regulatory framework that implies multiple sectors and products it is not only complex but also will be time consuming, have to involve all stakeholders, etc. In this sense, it is advisable to analyze in detail the legislative process followed with REACH (from the White Paper on European Union Chemical Policy to the actual REACH Regulation) that can be considered as a model of new regulatory governance.

Van Calster points out that in addition to HSE issues the following topics are being incorporated to the regulatory agenda:

- The requirements imposed by modern environmental and consumer protection law on the regulation of new technologies. This includes the precautionary principle (while contested, this principle at any rate applies generally in European environmental law), liability for environmental damage, etc.;
- The public participation principles of international and European environmental law, as exemplified by the *Aarhus* process: access to information, public participation, access to the courts;
- Other requirements of international environmental governance, such as the plight of developing countries and the need to avoid what the Commission calls a "knowledge apartheid";
- The application of international trade law, in particular, the law of the World Trade Organization.

Ref. 7. At 369.

principle the creation of hybrid funds financed by the public and private sector for sharing the burden of financing basic scientific research in fields like nanotechnology.[bg]

Finally, we conclude by stressing the regulator's need to strengthening a regulatory collaborative process, specially in the pharmaceutical regulatory context where soft law mechanisms already play an important role (i.e. expected to increase with nanomedicines regulation). In highly technological fields (like nanomedicine), there is a tendency to base regulatory decisions exclusively on assessment received from experts committees. As EVERS and D'SILVA[46] pointed out, "there is a general concern regarding the democratization of expert knowledge and the legitimization of lay-expertise in debates over science and technology policy-making (. . .) legitimacy depends on the degree to which those affected by it have been included in the decision-making processes".[bh] The role and composition of the CAT in the ATMP Regulation might be a promising starting development for tackling this democratic deficit.

REFERENCES

1. Ellis-Behnker R. *Way to the future: key ingoing applications in nanosciences and how they apply to pharmaceuticals*, In: *1st international workshop on nanomedicines 2010. Summary report. 21 October 2010*; 2010 EMEA/538503/2010. Available at: http://www.ema.europa.eu/docs/en_GB/document_library/Report/2010/10/WC500098380.pdf [accessed January 2012].
2. *CHMP Committee for Medical Products for Human Use Ad hoc Expert Group Meeting on "Nanomedicine"*; April 2009. Available at: http://www.ema.europa.eu/docs/en_GB/document_library/Report/2010/01/WC500069730.pdf [accessed January 2012].
3. European Technology Platform on Nanomedicine. *Roadmaps in nanomedicine towards 2020*. ETP: Berlin; 2009. Available at: http://www.etp-nanomedicine.eu [accessed January 2012].
4. Boisseau P, Loubaton B. *Nanomedicine, nanotechnology in medicine. Comptes Rendus de l'Académie des Sciences* 2011; Tome 12. Fascicule 7. Septembre 2011. C.R. Physique 12. 620–637. At 625. Available at: http://hal.archives-ouvertes.fr/docs/00/59/89/30/PDF/Boisseau_nanomedicine_CRAS.pdf [accessed January 2012].

---

bg. Ref. 37. At 106.

The possible negative impact on international competitiveness that this mixed financing could lead to might be balanced by additional alternative incentives measures to be added to those already existing: the European Medicines Agency (EMA) has implemented Regulation (EC) No. 2049/2005 by providing incentives to SMEs and setting up a SME office. See http://www.ema.europa.eu/ema/index.jsp?curl=pages/regulation/general/general_content_000060.jsp&murl=menus/regulations/regulations.jsp&mid=WC0b01ac0580024b94 [accessed January 2012].

Regulation (EC) No. 2049/2005 of 15.12.2005 laying down, pursuant to Regulation (EC) No. 726/2004 of the European Parliament and of the Council, rules regarding the payment of fees to and the receipt of administrative assistance from the European Medicines Agency by micro-, small- and medium-sized enterprises. OJ L329/4 of 16.12.2005.

bh. Ref. 46. At 126.

5. Bowman DM, Hodge GA. A small-matter of regulation: an international review of nanotechnology regulation. *Sci Technol Law Rev* 2007;**VIII**:1–40.
6. Bowman D, Hodge GA. Nanotechnology down under": getting on Top of regulatory matters. *Nanotechnol Law Bus* 2007;**4**:225–35.
7. Van Calster G. Regulating nanotechnology in the European Union. *Nanotechnol Law Bus* 2006;**3**:359–74.
8. Mantovani E, Porcari A, Morrison MJ, Geertsma RE. *Developments in Nanotechnologies Regulation and Standards 2010. Report of the Observatory Nano* 2010. www.observatory-nano.eu.
9. *CHMP Committee for Medical Products for Human Use. Reflection paper on nanotechnology-based medicinal products for human use*, 2006 EMEA/CHMP/79769/2006.
10. Papaluca Amati M. Risk assessment of pharmaceuticals, In: *4th Annual Safety for Success Dialogue. Brussels 29–30 March 2011.*
11. De Jong B. Drug delivery and nanoparticles: applications and hazards. *Int J Nanomedicine* 2008;**3**(2):133–49.
12. Chowdhury N. Regulation of nanomedicines in the EU: distilling lessons from the pediatric and the advanced therapy medicinal products approaches. *Nanomedicine* 2010;**5**(1):135–42.
13. Senden L. *Soft law in European community law. Modern studies in European law.*, vol. 1. Hart Publishing: Oxford, UK; 2004.
14. Renn O, Roco M. *White paper on nanotechnology risk governance. White paper no. 2.* IRGC International Risk Governance Council; 2006.
15. Widmer M, Meili C, Mantovani E, Porcari A. *The FramingNano governance platform: a new integrated approach to the responsible development of nanotechnologies. Final Report* 2010; Available at: http://www.framinhnano.eu/ [accessed November 2010].
16. *Covington and Burling. Reach and its impact on pharmaceuticals.* Available at: http://www.cov.com/files7publication/ed03bc97-89de-471d-91634f1a26899ba2/Presentation/Publication Attachment7e60742d7-e808-45a9-bec6-590676c12a9a/REACH%20-And%20Its%20Impact %20on%20Pharmaceuticals%20-%20January%202007.pdf [accessed January 2012].
17. CHMP Committee for Human Medicinal Products . *European Medicines Agency pre-submission procedure advice for users of the centralised procedure.* Available at: http://www.ema.europa.eu/docs/en_GB/document_library/Regulatory_and_procedural_guideline/2009/10/WC500004069.pdf [accessed January 2012].
18. SCENIHR—Scientific Committee on Emerging and Newly-Identified Health Risks. *The appropriateness of risk assessment methodology in accordance with the Technical Guidance Documents for new and existing substances for assessing the risks of nanomaterials.* Adopted on 21–22 June 2007.
19. Hansen SF. *Regulation and risk assessment of nanomaterials—too little, too late? PhD Thesis.* Technical University of Denmark: Department of Environmental Engineering; 2009. Available at: http: www.env.dtu.dk [accessed December 2010].
20. SCENIHR — Scientific Committee on Emerging and Newly Identified Health Risks . *The appropriateness of existing methodologies to assess the potential risks associated with engineered and adventitious products of nanotechnologies. Modified Opinion (after public consultation). 10th plenary meeting of 10 March 2006*; 2006 SCENIHR/002/05.
21. Linkov I, Satterstrom FK, Kiker G, Seager TP, Bridges T, Gardner KH, et al. Multicriteria decision analysis: a comprehensive decision approach for management of contaminated sediments. *Risk Anal* 2006;**26**(1):61–78.
22. Hansen SF. Multicriteria mapping of stakeholder preferences in regulating nanotechnology. *J Nanoparticle Res* 2010;**12**:1959–70.

23. Hansen SF. A global view of regulations affecting nanomaterials. *WIREs Nanomed Nanobiotechnol* 2010;**2**:441–9.
24. Kelly B, Bogaert P. Medical nanotechnology in Europe. *RAJ Pharma* 2008;451–8.
25. Collinson S, Alarcon S, Park B, Dorey R, Rocks S, Friedrichs S, et al. *The responsible NanoCode. Int Labmate* 2010;**35**(6). Available at http://www.labmate-online.com/ [accessed May 2011].
26. Papaluca Amati M, Korakianiti E, Riviere F. *The EMEA innovation task force*. Available at: http://www.ema.europa.eu/docs/en_GB/document_library/Presentation/2009/10/WC500004221.pdf [accessed January 2012].
27. RIVM. Nanotechnology in perspective . *Risk to man and the environment. RIVM 2009*. In: van Zijverden M, Sips AJAM, Leenes RE, editors. *Dimensions of technology regulation*; 2009 Report 601785003/2009. Available at: http://www.rivm.nl/bibliotheek/rapporten/601785003.pdf [accessed January 2012].
28. Altenstetter C. Medical device regulation and nanotechnologies. Determining the role of patient safety concerns in policymaking. *Law Policy* 2011;**33**(2):227–55.
29. Working Group on New and Emerging Technologies in Medical Devices . *Report on nanotechnology to the medical devices expert group. Findings and recommendations.* Available at: http://ec.europa.eu/enterprise/sectors/medical-devices/scientific-technical.assessment/working-group/index_en.htm [accessed January 2012].
30. D'Silva J, van Calster G. Taking temperature—a review of European union regulation in nanomedicine. *Eur J Health Law* 2009;**16**:249–69 At 359.
31. Van Calster G, Bowman D, D'Silva J. Trust me, I'm a regulator. The (in)adequacy of EU legislative instruments for three nanotechnologies categories, In: Goodwin MEA, Koops BJ, Leenes RE, editors. *Dimensions of technology regulation. Proceedings of the TILT Conference 2008, Nimegen,* Netherlands: Wolf Legal Publishing; 2010. p. 207–37.
32. Van Calster G, Bowman DM, D'Silva J. Protecting consumers or failing them? The regulation of nanotechnologies in the EU. Environmental law and consumer protection. In: Cristophe Verdure, editor. *European Journal of Consumer Law/Revue européenne de droit de la consommation (REDC). Larcier Ed.* 2011/1. p. 85–113.
33. CAT Scientific Secretariat . Challenges with advanced therapy medicinal products and how to meet them. *Drug Discov* 2010;**9**:195–201.
34. Nakissa S. *FDA considerations for regulation of nanomaterials containing products*. Available at: http://www.mhra.gov.uk/home/groups/comms-con/documents/websiteresources/com2022823.pdf [accessed January 2012].
35. EMA—European Medical Agency. *Regulatory advice to applicants on the eligibility to EMEA procedures as medicinal products (according to Directive 2001/83/EC). EMA Standard Operating Procedure REF* WC500002986.pdf.
36. Alemanno A. *The shaping of the precautionary principle by European Courts: from scientific uncertainty to legal certainty*. Bocconi Legal Studies Research Paper No. 1007404. In: Cuocolo L, Luparia L, editors. *Valori Costituzionali e nuove politiche del diritto*. Halley: Cahiers Europèens; 2007. Available at: http://ssrn.com/abstract=1007404. At 11. [accessed January 2011].
37. Haselhaus S. Nanomaterials and the precautionary principle in the EU. *J Consum Policy* 2009;**33**:91–108.
38. Commission. *Nanotechnologies: a preliminary risk analysis on the basis of a workshop organized in brussels on 1–2 March 2004 by the health and consumer protection directorate general of the European Commission, European Commission*; 2004—At. 22. Available at: http://europa.eu.int/comm/health/ph_risk/events_risk_en.htm.

39. Dorbeck-Jung B, Chowdhury N. Is the European medical products authorisation regulation equipped to cope with the challenges of nanomedicines? *Law Policy* 2011;**33**(2):276–303.
40. *Royal Commission on Environmental Pollution (RCEP). Report on nanotechnology to the medical devices expert group. Findings and recommendations*, In: *Novel Materials in the Environment: The case of nanotechnology. Twenty-seventh Report of the RCEP presented to the UK Parliament in November 2008*; 2008. Available at: http://www.rcep.org.uk/reports/27-novel%20materials/27-novelmaterials.htm#supp [accessed January 2012].
41. Brown S. The new deficit model. *Nat Nanotechnol* 2009;**4**:609–11.
42. Hansen SF, Maynard A, Baun A, Tickner JH. Late lessons for early warnings for nanotechnology. *Nat Nanotechnol* 2008;**3**(445):447.
43. Desmoulin S. French and European Community Law on the nanometric forms of chemical substances: questions about how the law handles uncertain risks. *Nanotechnol Law Bus* 2008;**5**(3):341–52.
44. EEB . *Small scale, big promises, divisive messages. Position paper on nanotechnologies and nanomaterials.* 2009.
45. Gaspar R. Therapeutic products: regulating drugs and medical devices. In: Hodges GA, Bowman DM, Maynard AD, Hodges GA, Bowman DM, Maynard AD, editors. *International handbook on regulating nanotechnologies.* MA, USA: Edward Elgar Publishing; 2010. p. 291–320.
46. Evers J, D'Silva J. Knowledge transfer from citizens' panel to regulatory bodies en the domain of nano-enabled medical applications. *Innovation: The European Journal of Social Science Research* 2009;**22**(1):125–42.

Index

Note: Page numbers followed by "*f*" indicate figures, and "*t*" indicate tables.

E

F

G

H

I

L

N

O

FSC
www.fsc.org
MIX
Papier | Fördert gute Waldnutzung
FSC® C083411

Kolibri 360 GmbH
Ferdinand-Jühlke-Straße 7
99095 Erfurt, Deutschland
produktsicherheit@kolibri360.de